D1221025

Freshwater Macroinvertebrates of Northeastern North America

Freshwater Macroinvertebrates of Northeastern North America

BARBARA L. PECKARSKY

Department of Entomology, Cornell University

PIERRE R. FRAISSINET

Department of Entomology, Cornell University

MARJORY A. PENTON

Department of Entomology, Cornell University

DON J. CONKLIN, JR.

Chadwick and Associates, Environmental
Consultants, Littleton, Colorado

Comstock Publishing Associates a division of

Cornell University Press | Ithaca and London

First published 1990 by Cornell University Press.

Library of Congress Cataloging-in-Publication Data

Freshwater macroinvertebrates of northeastern North America / Barbara L. Peckarsky . . . [et al.].
p. cm.
Includes bibliographical references.
ISBN 0-8014-2076-8 (alk. paper). — ISBN 0-8014-9688-8 (pbk. : alk. paper)
1. Freshwater invertebrates—North America. I. Peckarsky, Barbara Lynn.
QL151.F75 1990
592.092'97—dc20 89-17468

Printed in the United States of America

♾ The paper used in this publication meets the minimum requirements of the American National Standard for Permanence of Paper for Printed Library Materials Z39.48–1984.

We dedicate this textbook to the memory of two of our colleagues, Karl W. Simpson, who intended to be a coauthor of the chapter on Chironomidae, and Clifford O. Berg, whose influence through his lifetime of teaching pervades every aspect of this book.

He was a teacher.
He was a scholar.
He was a writer.
He was a naturalist.

He was a family man.
He was a storyteller.
He was an outdoorsman.
He was a humanitarian.

He was a wise man.
He was a strong man.
He was a gentle man.
He was our friend.

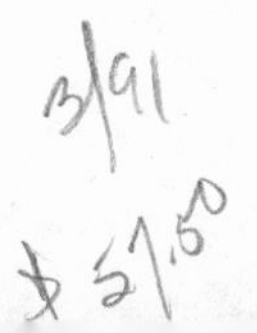

Contents

Acknowledgments

The first version of this textbook was introduced to the aquatic entomology class at Cornell University in spring 1981. Since then, nine classes have used the keys and have provided valuable help with portions that needed improvement. There is no better way to construct effective taxonomic keys than to be present when students are struggling through difficult couplets. We believe that because of our students' participation in the development of the keys, these are among the most "user-friendly" keys to freshwater macroinvertebrates ever published.

I was initially inspired to undertake this project by William Hilsenhoff, whose *Aquatic Insects of Wisconsin* was my introduction to the taxonomy of aquatic insects. Using his text—which was developed in much the same way as this one—as a model, my coauthors and I began the construction of regional keys to the freshwater macroinvertebrates that would be easier to use than texts that covered the taxa of broader geographic ranges. We adapted Hilsenhoff's keys to, first, the fauna of New York State and then to that of northeastern North America. Doing this required the help of many graduate and undergraduate teaching assistants. They were, in chronological order of their tenure, Robert Bukantis, Richard Featherly, James Pakaluk, Laurie Burnham, Carla Delucchi, Bryan Goodwin, David Miller, Steven Nichols, Lance Durfey, Marcia Grimshaw, David Mitchell, LeeAnne Martinez, Joseph McHugh, John Przybyszewski, David Ellsworth, Scott Pitnick, Mark St. Pierre, Andrew Shedlock, Brian Spence, Elizabeth Lynch, Marie Reyes, Sara Tjossem, and Clara Weloth. They endured semesters of continuous alterations in the keys as we incorporated new literature and suggestions from students. Bob and Carla deserve special thanks, since their combined participation spanned eight of the nine years over which this text was developed.

The reference lists that accompany each chapter were originally compiled by my predecessor, Clifford O. Berg, who taught a memorable course on freshwater invertebrates at Cornell for eight years. Cliff gave me these lists in 1979, when I arrived at Cornell. Three of his teaching assistants had added references to the lists as well: Barry OConnor, Sandy Fiance, and Frank Ramberg. I have updated the lists since 1979, but their completeness illustrates Cliff Berg's historical perspective on the literature.

The illustrations have come from many sources. Students and colleagues prepared numerous original drawings.[1] Robert Bode, one of the coauthors of the chironomid key, drew most of the figures in that key. Pierre Fraissinet and Elizabeth Lynch provided original drawings for the keys to the insect orders and Collembola, Ephemeroptera, Plecoptera, Hemiptera, Trichoptera, Lepidoptera, Coleoptera, and Diptera. They also skillfully redrew published figures of many different styles, doing a superb job of standardizing the illustrations for this book. Joseph McHugh prepared original illustrations for the key to the insect orders and Collembola, many of which were redrawn by Elizabeth Lynch to maintain a style consistent with figures in the rest of the book. John Przybyszewski drew illustrations of beetle larvae and adults. Bruce Smith, author of the water mite key, prepared original figures for his key and the key to Hemiptera, and David Strayer, author of the keys to Mollusca and Oligochaeta, drew illustrations for his chapters.

Partial funding for the preparation of the illustrations came from two sources: the Cornell University Women's Development Fund and the College of Agricultural and Life Sciences Book Publication Fund. I am grateful to Warren Johnson for making me aware of the availability of the latter fund.

We are grateful to the copyright holders who graciously granted us permission to redraw or modify their illustrations. A brief credit is given in the figure captions to the source of each illustration, and full citations are presented in the reference list at the end of each chapter. Here we acknowledge these copyright holders. They are Akademie Verlag, Berlin; the American Entomological Society; the American Microscopical Society; *American Midland Naturalist;* the American Museum of Natural History; Ross H. Arnett, Jr.; Yeon J. Bae; E. H. Barman, Jr.; Henri Bertrand; the Biological Society of Washington; R. O. Brinkhurst; Harley P. Brown; John B. Burch; the government of Canada, Department of Fisheries and Oceans; *Canadian Journal of Zoology;* the Canadian Government Publishing Centre; the Carnegie Institute; the Carnegie Museum; the Connecticut State Geological and Natural History Survey; David R. Cook; Cornell University Agricultural Experiment Station; Cornell University Press; *Dansk Naturhistorisk Forening; Entomologica Scandinavica;* the Entomological Society of America; the Entomological Society of Canada; the Entomological Society of New Zealand; the Entomological Society of Ontario; the Entomological Society of Washington; *Florida Entomologist;* Gem Publishing Co.; the Georgia Entomological Society; *Great Basin Naturalist; Great Lakes Entomologist;* Grinnell College; William L. Hilsenhoff; Horton H. Hobbs, Jr.; John R. Holsinger; Illinois State Natural History Survey Division; the Kentucky Academy of Science; Donald J. Klemm; G. W. Krantz; *Malacologia;* Malacological Publications; *Malacological Review; Microentomology;* Midwest Aquatic Enterprises; C. V. Mosby Company; the New York Entomological So-

[1]The original drawings and their artists are as follows. Robert Bode: Ch. 14, Figs. 1–4, 6–13, 15, 19, 20, 24–26, 28–39, 40–51, 55, 56, 58–64, 65, 67, 68, 70, 71, 73, 74, 77–79, 82–89, 90, 91, 93–95, 99, 101–110, 113–115, 116–119, 121–129, 131, 132, 134–137, 140–157, 159–174. Pierre Fraissinet: Ch. 4, Figs. 75, 76; Ch. 8, Fig. 33; Ch. 13, Fig. 40. Elizabeth Lynch: Ch. 2, Fig. 7; Ch. 6, Figs. 29, 30, 37, 40, 52–54, 57, 60; Ch. 7, Figs. 17, 20, 29–32; Ch. 8, Figs. 44, 46, 96, 157; Ch. 9, Figs. 1, 2; Ch. 10 (key to adults), Figs. 48, 49; Ch. 10 (key to larvae), Figs. 40, 41. Joseph McHugh: Ch. 2, Figs. 3, 4, 9, 16. John Przybyszewski, Ch. 10 (key to adults), Figs. 14, 15; Ch. 10 (key to larvae), Fig. 17. Bruce Smith: Ch. 7, Fig. 23; Ch. 16, Fig. 87. David Strayer: Ch. 17, Figs. B–G, 2 (left), 5 (left), 22, 29–31, 66, 67, 75, 76, 120, 125; Ch. 18, Figs. 60, 66, 89, 90.

ciety; *Ohio Journal of Science;* the Pacific Coast Entomological Society; Pennsylvania State University College of Agriculture; Philip D. Perkins; *Psyche;* Purdue University Agricultural Experiment Station; the Ray Society; the Royal Entomological Society of London; the Royal Ontario Museum; the Royal Swedish Academy of Sciences; Guenter A. Schuster; Smithsonian Institution Press; la Società Entomologica Italiana; Societe Entomologique de France; the Society of Systematic Zoology; U.S. Environmental Protection Agency, Environmental Monitoring and Support Laboratory; U.S. National Oceanographic and Atmospheric Administration; University of California Press; University of Minnesota Press; University Presses of Florida; VEB Gustav Fischer Verlag; Virginia Polytechnic Institute and State University; James V. Ward; John Wiley and Sons, Inc.; W. D. Williams; and Zoologiska Institutionen Uppsala Universitet.

I am indebted to Glenn B. Wiggins of the Royal Ontario Museum for permission to use so many of his illustrations of trichopterans as a basis for the modified line drawings in Chapter 8.

Each chapter was rigorously reviewed by a leading expert on that taxon. I am deeply grateful to the following reviewers for their detailed and constructive suggestions: R. W. Merritt (orders), K. A. Christiansen (Collembola), G. F. Edmunds, Jr. (Ephemeroptera), S. W. Dunkle (Odonata), M. J. Westfall, Jr. (Odonata), R. W. Baumann (Plecoptera), A. S. Menke (Hemiptera), O. S. Flint, Jr. (Trichoptera), D. C. Huggins (Lepidoptera), F. N. Young (Coleoptera), E. D. Evans (Megaloptera, Neuroptera), H. J. Teskey (Diptera), L. C. Ferrington (Chironomidae), H. H. Hobbs (Crustacea), D. R. Cooke (Hydrachnidia), W. N. Harman (Mollusca), R. O. Brinkhurst (Oligochaeta), and R. T. Sawyer (Hirudinea). Their contributions have greatly enhanced the value of this book.

A special thanks is owed also to Wayne Gall, who not only helped me locate the New York State records for collections of freshwater invertebrates but also gave me valuable help with the reference lists and wrote part of the crustacean key in the key to the orders.

Finally, I acknowledge six people who put in countless hours preparing the manuscript for publication. They are Kurt Jirka, who edited all aspects of the manuscript before it was submitted to Cornell University Press; Elizabeth Lynch, who in addition to contributing many of the illustrations, provided editorial suggestions; Susan Pohl, who typed the many drafts of the manuscript; and Cathy Cowan, who provided skilled and meticulous library research to meet the high standards of detail and perfection demanded by the two talented copy editors who worked on the manuscript at Cornell University Press, Margo Quinto and Helene Maddux. These six people are to be commended for persevering through this difficult and tedious task.

Barbara L. Peckarsky

Ithaca, New York
April 1989

Freshwater Macroinvertebrates of Northeastern North America

1 Introduction

This textbook is intended for use in northeastern North America by students interested in the natural history and systematics of freshwater macroinvertebrates. Adapted from the laboratory handout materials for Entomology 471, Freshwater Invertebrate Ecology and Systematics, taught at Cornell University, it covers the aquatic insects as well as other macroinvertebrates that are usually collected with them, namely, collembolans, water mites, crustaceans, mollusks, oligochaetes, and leeches.

Each chapter begins with a brief sketch of the natural history of the taxon. Included is information on the classification, life history, habitat, feeding ecology, and respiration of the group. Generalizations are restricted to the level of order or suborder rather than family to avoid duplication of details readily available from other sources. These discussions are not intended as a thorough review of the literature, but as a brief introduction to the taxonomic group that students will learn to identify in the laboratory. We refer users to the references listed in each chapter for more complete descriptions of the natural history of each group. Most of the taxa can be identified by the same techniques: preservation in ethanol and examination under a dissecting microscope without slide-mounting or excessive dissection. We discuss any special collection, preservation, or identification techniques required in the introductory remarks. We also include generalized illustrations of the morphology of each macroinvertebrate group to introduce students to the gross structural features necessary for identification of genera within each taxon.

Each chapter has a checklist of the families and genera of each order (insects) or in some cases higher classification (noninsects) included in the key. Classification is according to Merritt and Cummins (1984) for the insects; specific references used for classifications of noninsects are given in those chapter introductions. Genera included in these checklists are those for which we have found records in northeastern North America as designated in Merritt and Cummins (1984) and elsewhere. We define northeastern North America as inclusive of the northeastern United States and adjacent Canadian provinces. Keys will be effective as far west as Michigan and as far south as Virginia, but are primarily targeted for use in Pennsylvania, New York, New England, Ontario, Quebec, and New Brunswick.

Keys to the genera of each aquatic insect order are limited to the life cycle stages that are aquatic—that is, immatures of most groups except the Hemiptera and Coleoptera, which have semiaquatic or aquatic adults. We do not include keys to pupae or terrestrial adults because they are available in *An Introduction to the Aquatic Insects of North America* by Merritt and Cummins (1984) and because they are beyond the scope of most courses in aquatic entomology or freshwater invertebrate zoology. We include riparian or semiaquatic Coleoptera and Diptera only at the family level, and do not key marine or coastal genera of Collembola, Hemiptera, Diptera, or Coleoptera. Lepidoptera found only in association with emergent vascular hydrophytes are also not included in the keys. Students who need to identify those groups can refer to Merritt and Cummins (1984) or other references given with each chapter.

Couplet halves are denoted by *a* or *b*. For each couplet, the number of the couplet that led to it is given in parentheses. The keys can thus be used in reverse order, to verify identification of a tentatively determined taxon. Family names appear at the first couplet of the series that identifies the genera within each family so that students can easily locate the beginnings of keys to families, a feature that streamlines the keying operation (for those who can identify families by sight).

Each chapter ends with a reference list. Included are systematic studies on the taxa of the region at the generic level or higher. Systematic studies on the taxa of other regions are cited if they contain information on the natural history of taxa found in northeastern North America. The references are intended as a resource for students who wish to initiate more serious research on systematics of particular taxa. Also included in the reference lists are major review papers on the biology of each taxon for those who wish to obtain more detail on the natural history. We have indicated with asterisks those references that were used directly for construction of the keys or from which distribution data were obtained.

The chapters on noninsects follow essentially the same format with a few notable exceptions. Chapter 15 focuses primarily on macrocrustaceans—that is, anostracans, arguloidans, mysids, isopods, amphipods, and decapods. Those groups can be identified to genus without slide-mounting, and they are generally collected by the same methods used for aquatic insects. Microcrustaceans—that is, conchostracans, ostracods, copepods, and cladocerans—can be identified to order without mounting specimens on slides. Therefore, discussion of those groups is limited to the ordinal level.

The key to the genera of water mites (Ch. 16) is comparable to the insect keys except that no level higher than genus is given in the mite key. With his innovative approach to the taxonomy of water mites, the author, Bruce Smith, has eliminated the need for slide-mounting and clearing of specimens by using features of water mites visible on alcohol-preserved specimens. We have tested these keys for several years and have found them effective and usable by students. Detailed studies of aquatic mite taxonomy at the specific level require the classical preparation of specimens, however.

The chapter on mollusks, (Ch. 17) written by David Strayer, gives the geographic location of each species in the key. Dave was able to include this valuable information because of his life-long experience with this group. Since many mollusk species are indigenous, information on location can reliably aid identification. The chapters on mollusks, oligochaetes (Ch. 18, also written by Dave Strayer), and leeches (Ch. 19, written by Donald Klemm) contain keys to species rather than to genera. Because the taxonomy of those groups is better known than is that of the aquatic insects, students can as easily identify them to species as to genera.

One final comment: this textbook is not an original piece of work (with the exception of the keys by Bruce Smith and Dave Strayer). The keys presented here are modified, some slightly, some rather extensively, from other sources. Our intent was not to be original, but to provide an effective and practical regional reference for students and those who assess environmental impact. Although keys intended for larger geographic regions, such as the United States or North America, can be used to identify the northeastern North American fauna, we see several advantages to using regional rather than "global" keys.

Our intent was to provide an alternative to texts that are more inclusive, such as Merritt and Cummins (1984) or Pennak (1978), for students and professionals who are working with regional fauna. Many students find it tedious to work through key couplets that are not applicable to local fauna. Our students and colleagues who work in northeastern North America report that our keys are more efficient because they include only taxa present in this region.

In a text that covers North America, such as Pennak (1978) and Merritt and Cummins (1984), both of which are classics, errors in the construction of taxonomic keys are unavoidable, especially for groups whose taxonomy is poorly known. Characters used to separate genera are rarely reliable at all subregions of the large geographic area covered by the keys. This is especially true if the larvae of many species within a genus remain undescribed. Even in a region such as northeastern North America, we face this problem. But the probability of errors is reduced with keys designed for a smaller region.

One last advantage of regional keys is that they can be tested and new information can be incorporated into a final version. We have worked for nine years to modify these keys in response to problems encountered by the students who have used them. Thus, we are fairly certain that these keys are effective in New York State. We have also tested them, but less extensively, in Pennsylvania, Vermont, Massachusetts, and New Hampshire. We hope that individuals using these keys will bring any problems to our attention, so that improvements can be incorporated into any future editions.

Literature Cited

Merritt, R.W., and K.W. Cummins (eds.). 1984. An introduction to the aquatic insects of North America, 2nd ed. Kendall/Hunt Publ. Co., Dubuque, Iowa. 722 pp.

Pennak, R.W. 1978. Freshwater invertebrates of the United States, 2nd ed. John Wiley and Sons, New York. 803 pp.

2 The Orders of Aquatic Insects and Collembola

Classification

Within the class Insecta, 10 orders of the subclass Pterygota contain species with one or more aquatic life history stages (Table 1), including some of the most primitive insects (infraclass Paleoptera, meaning "old wings" because they cannot be folded over the dorsum) and some more advanced (infraclass Neoptera, meaning "new wings" because they fold over the dorsum). Of the Neoptera, both Exopterygota (wings develop outside the body) and Endopterygota (wings develop inside the body) have aquatic representatives (Daly 1984). The class Collembola, which contains semiaquatic species, was formerly included in the Apterygota, a primitive subclass of Insecta. Collembolans are wingless and can be distinguished from aquatic insects with the key in this chapter.

Table 1. Orders of the Class Insecta that contain aquatic or semiaquatic species, in sequence of ascending evolutionary development

Order	Life stage(s)	Infraclass	Division	Common name
Ephemeroptera[a]	Larvae	Paleoptera		Mayflies
Odonata[a]	Larvae			Dragonflies and damselflies
Plecoptera[a]	Larvae	Neoptera	Exopterygota	Stoneflies
Hemiptera[b,c]	All			True bugs
Trichoptera[a]	Larvae, pupae		Endopterygota	Caddisflies
Lepidoptera[c]	Larvae, pupae			Moths
Coleoptera[c]	Larvae,[c] adults[c]			Beetles
Megaloptera[a]	Larvae			Dobsonflies, fishflies, alderflies
Neuroptera[c]	Larvae			Spongillaflies
Diptera[c]	Larvae, pupae[c]			True flies

[a]All members of the order are aquatic in the life stage given.
[b]Contains semiaquatic species.
[c]Some members of the order are aquatic in the life stage given.

Life History

Four general life history patterns are exhibited by the 10 insect orders and the Collembola that have aquatic or semiaquatic representatives (Table 1). Collembola are *ametabolous*, meaning they do not metamorphose. Ephemeroptera, Odonata, and Plecoptera are *hemimetabolous*; they undergo incomplete metamorphosis from egg to larva (or nymph) to adult with no pupal phase. Hemiptera are *paurometabolous*; their pattern of development is similar to that of hemimetabolous insects except that the nymph and adult of the hemimetabola are very different from each other, whereas the paurometabola undergo very gradual metamorphosis, which results in strong similarity—morphologically, behaviorally, and ecologically—between the nymphs and adults. The only differences are that wings and genitalia are developed in the adults. All the other orders are *holometabolous*, that is, they undergo complete metamorphosis from egg to larva to pupa to adult.

Habitat

Not all life history stages of the orders of aquatic insects occur in (aquatic) or on (semiaquatic) water. Table 1 shows which of the life history stages of each order are aquatic. There are more species of completely terrestrial insects (ca. 100,000) than of either aquatic or semiaquatic insects. Most aquatic orders have aquatic larvae (or nymphs) and a winged terrestrial adult phase (ca. 10,000 species). However, some have species with both aquatic or semiaquatic adults and immatures (Hemiptera, Coleoptera; ca. 1,000 species), and one (Coleoptera) has species with terrestrial larvae and aquatic adults (ca. 100 species). (Species data from Hutchinson 1981.) Because all species of Ephemeroptera, Odonata, Plecoptera, Trichoptera, and Megaloptera have aquatic immatures, those orders are considered *primary invaders*; that is, their larvae are thought to have originally inhabited water (Hynes 1984). The remaining insect orders, which have only small proportions of species with aquatic stages, are *secondary invaders*; that is, subgroups of primarily terrestrial orders invaded the aquatic habitat. However, the fossil record and, hence, the phylogenetic relationship and origins of these groups are poorly known (Wooton 1972, Kristensen 1981, Resh and Solem 1984).

Aquatic insects are found in every conceivable aquatic habitat—from treeholes, mud puddles, and the water trapped inside pitcher plants to large rivers, lakes, and even the open ocean. Cummins and Merritt (1984a) describe two schemes for classifying the microhabitats of aquatic insects. One scheme distinguishes various *habitats* within either still (lentic) or running (lotic) waters (Usinger 1956, Anderson and Wallace 1984). Lotic insects may be associated with riffles (erosional zones), pools (depositional zones), benthic sediments, vascular hydrophytes, or detritus. Lentic insects may be in open water (limnetic), associated with littoral-zone vegetation, or in sediments. Beach-zone insects occur along the shores of large lakes or in the marine intertidal zone. We concentrate here primarily on the freshwater forms.

The other scheme used by Cummins and Merritt (1984a) for classifying aquatic insects distinguishes various "*habits*"; that is, their method of attachment to (if any) or association with the substrate, or their primary mode of locomotion. The various categories of habits for aquatic insects include skating (on water surface), diving, swimming, clinging, climbing (on vegetation), and burrowing.

Feeding

Aquatic insects can be herbivorous, detritivorous (scavengers), carnivorous (predaceous), or parasitic. Cummins and Merritt (1984a) classify the feeding groups of aquatic insects by function because more aquatic insects specialize by method of obtaining food than by food type (the classical system of trophic determination). *Shredders* chew, mine, or gouge coarse particulate detritus or live macrophytes. *Collectors* either gather fine particulate detritus loosely associated with the sediment or from the surface film, or they filter particles suspended in the water column. *Grazers* remove attached periphyton and material closely associated with mineral or organic substrates. *Piercers* pierce and suck the contents of green plants or of animals. *Engulfers* attack live prey and ingest whole or parts of animals. *Parasites* live in or on aquatic animals.

Closer examination of the feeding behavior of aquatic insects reveals that strict categorization into these functional feeding groups is not always valid. Many species change the way they feed as they mature or as the availability of food changes (Cummins and Klug 1979). Facultative feeding habits are probably more common among aquatic insects than is obligate specialization because of the fluctuating nature of the aquatic habitat and food resources.

Respiration

Having evolved from a terrestrial ancestor, all insects originally got oxygen from the air. Air-filled tubes (tracheae) carried oxygen, as O_2 gas, directly to body tissues. Aquatic insects retain these tracheal systems but must either resupply them frequently with fresh (oxygenated) air or reoxygenate the air in them directly from dissolved oxygen in the water. Most lentic insects obtain oxygen from the air; most lotic insects get it from the water. Lentic insects have open tracheal systems that may have many spiracles open (peripneustic or polypneustic), or a few spiracles open (oligopneustic) at the front (propneustic), rear (metapneustic), or at both ends (amphipneustic) of their bodies (Usinger 1956; C. O. Berg, pers. comm., 1980; Eriksen et al. 1984). Some rely totally on gaseous oxygen either by breathing at the water surface or by utilizing the air spaces in plants. Others take dissolved oxygen into air bubbles or gas gills that may or may not be structurally supported. Nonstructurally supported gas gills are compressible; those that are supported by fine, hydrofuge hairs (plastron) are not, a trait that enables organisms to stay submerged for longer periods. Plastron respiration also enables some aquatic insects with open tracheal systems to live in swiftly flowing streams, where they do not have to surface for air.

Most lotic insects, however, have closed tracheal systems (apneustic). They breathe through either their cuticle or their tracheal gills, which are extensions of the cuticle. The tracheal gills are external or internal and, if external, movable or immovable. Moving their tracheal gills is one way aquatic insects behaviorally regulate their oxygen intake (Eriksen et al. 1984). Other behaviors used to regulate oxygen intake are body undulations, "push-ups," "somersaults," or moving to substrate surfaces exposed to well-oxygenated flowing water (Wiley and Kohler 1984).

A few aquatic insects can respire in air and water, a behavior that enables them to tolerate great fluctuations in water levels and to leave the water to metamorphose. They are able to do so either because they have gills and open spiracles (some Diptera, Megaloptera, and Odonata) or because they have spiracular gills (some Coleoptera and Diptera) (C. O. Berg, pers. comm., 1980).

Collection and Preservation

Most aquatic insects can be collected with aquatic D-frame nets of mesh size 1 mm or smaller and can be preserved in 70–95% ethanol. There are, however, numerous other collection methods (see Lattin 1956, Pennak 1978, McCafferty 1981, Merritt et al. 1984) as well as many rearing methods (Merritt et al. 1984) that serious students of aquatic entomology can use. Most groups of freshwater macroinvertebrates can be collected simply by sweeping a D-frame net through vegetation or organic sediments of ponds, or by disturbing mineral or organic sediments in streams and holding the net downstream of the disturbed area. Individual insects can be sorted by hand from debris in the field, or samples containing debris can be preserved in the field and sorted later in the laboratory. We recommend that students learning taxonomy sort their samples in the field because they can observe the behavior and color patterns of live animals before preservation. Such observations are an important supplement to handling preserved material.

Identification

The gross morphological characters used in our key to identify the orders of aquatic insects are segmented thoracic appendages, wings, terminal appendages, gills, specialized mouthparts, numbers of tarsal claws, and structure of antennae. No dissection is required to find these characters, and little special terminology is used. Bentinck (1956), McCafferty (1981), and Cummins and Merritt (1984b) discuss the morphology of aquatic insects in depth and provide detailed illustrations of body parts.

More-Detailed Information

Bentinck 1956, Lattin 1956, Usinger 1956, Pennak 1978, Hilsenhoff 1981, McCafferty 1981, Brigham et al. 1982, Anderson and Wallace 1984, Cummins and Merritt 1984a, 1984b, Daly 1984, Eriksen et al. 1984, Merritt et al. 1984, Resh and Solem 1984.

Key to the Orders of Aquatic Insects and to Collembola

1a. Thorax without segmented legs (except pupae) 2
1b. Thorax with 3 pairs of segmented legs 3

2a (1a). Mummylike, in a case, often silk-cemented and containing vegetable or mineral matter . . . . pupae (not keyed)
2b. Not in a case; mobile larvae, mostly with prolegs or pseudopods on one or more segments (Figs. 1, 2) . . . . **Diptera** (Ch. 13)

3a (1b). With large, functional wings (Figs. 3, 4, 7, 8) 4
3b. Wingless, or with developing wings (wingpads) or brachypterous wings (Figs. 5, 6, 10–12) 6

4a (3a). All wings completely membranous, with numerous veins . . . . nonaquatic adults that may have entered water to oviposit (not keyed)
4b. Mesothoracic wings hardened and shell-like, or leatherlike in basal half (Figs. 3, 4, 7, 8) 5

5a (4b). Mesothoracic wings hard, shell-like meeting along middorsal line (Figs. 3, 7); chewing mouthparts . . . . **Coleoptera** adults (Ch. 10)
5b. Mesothoracic wings hardened in basal half overlapping along middorsal line (except Pleidae) (Figs. 4, 8); sucking mouthparts, formed into a broad or narrow tube (Fig. 9) . . . . **Hemiptera** (in part) (Ch. 7)

6a (3b). Thoracic legs 4-segmented; abdomen never with more than 6 segments; abdominal furcula usually present (Fig. 14) . . . . **Collembola** (Ch. 3)
6b. Thoracic legs 5-segmented; abdomen always with more than 6 segments; abdominal furcula absent 7

7a (6b). With 2 or 3 long, terminal filaments (Figs. 10–12) 8
7b. Terminal filaments absent or consisting of fewer than 10 segments (Figs. 5, 6) 9

8a (7a). Sides of abdomen with platelike, featherlike, or leaflike gills; usually with 3 terminal filaments, occasionally only 2; tarsal claws single (Figs. 11, 12) . . . . **Ephemeroptera** (Ch. 4)
8b. Gills absent from middle abdominal segments; 2 terminal filaments; tarsal claws double (Fig. 10) . . . . **Plecoptera** (Ch. 6)

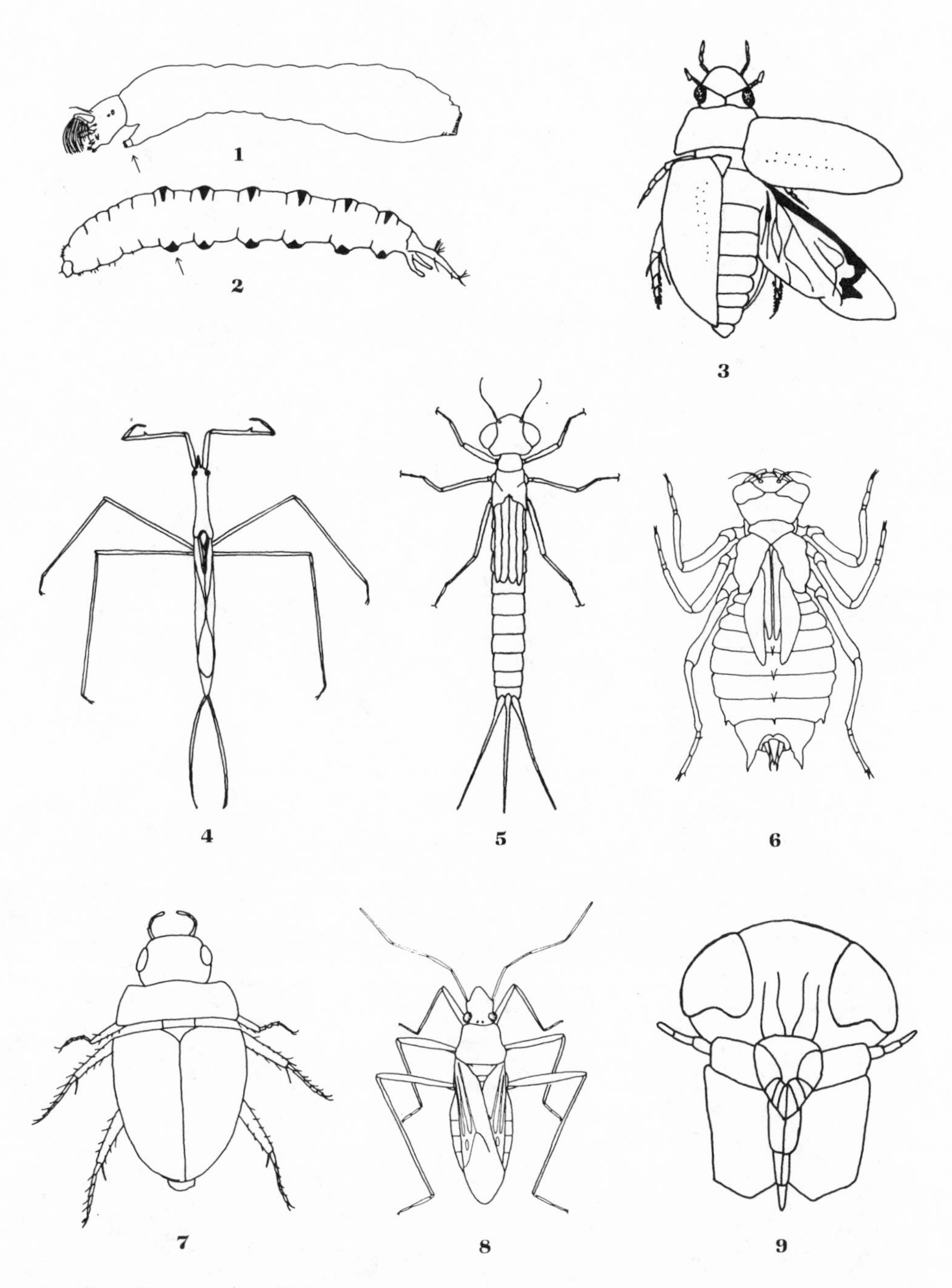

1. *Simulium* (Diptera, Simuliidae) larva, lv. **2.** *Antocha* (Diptera, Tipulidae) larva, lv. **3.** *Tropisternus* (Coleoptera, Hydrophilidae) adult, dv. **4.** *Ranatra* (Hemiptera, Nepidae) adult, dv. **5.** *Coenagrion* (Odonata, Coenagrionidae) nymph, dv. **6.** *Eipcordulia* (Odonata, Corduliidae) nymph, dv. **7.** *Hydrobius* (Coleoptera, Hydrophilidae) adult, dv. **8.** *Mesovelia* (Hemiptera, Mesoveliidae) adult, dv. **9.** Mouthparts of *Notonecta* (Hemiptera, Notonectidae), vv. dv, dorsal view; lv, lateral view; vv, ventral view. (1 modified from Peterson 1981; 2 modified from Alexander and Byers 1981; 5, 6 redrawn from Pritchard and Smith 1956; 8 modified from Pennak 1978.)

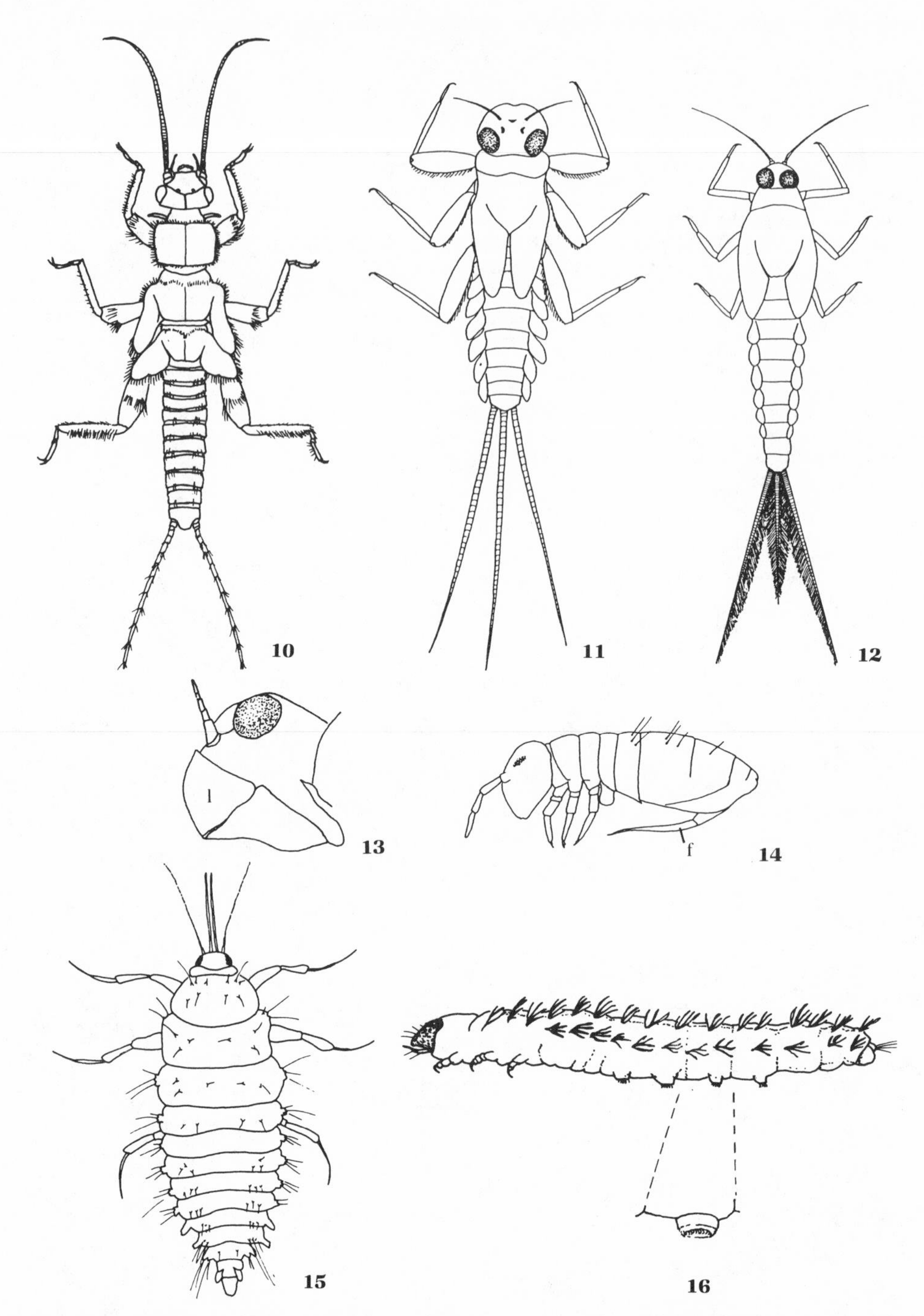

10. *Zapada* (Plecoptera, Nemouridae) nymph, dv. **11.** *Cinygmula* (Ephemeroptera, Heptageniidae) nymph, dv. **12.** *Baetis* (Ephemeroptera, Baetidae) nymph, dv. **13.** Head of Libellulidae (Odonata) nymph, lv: l, labium. **14.** *Isotomurus* (Collembola, Isotomidae), lv: f, furcula. **15.** *Climacia* (Neuroptera, Sisyridae) larva, dv. **16.** *Parapoynx* (Lepidoptera, Pyralidae) larva, lv. dv, dorsal view; lv, lateral view. (10 modified from Baumann et al. 1977; 11 modified from Yoon and Bae 1984; 12 modified from Bae 1985; 13 redrawn from Pennak 1978; 14 redrawn from Waltz and McCafferty 1979; 15 redrawn from Chandler 1956.)

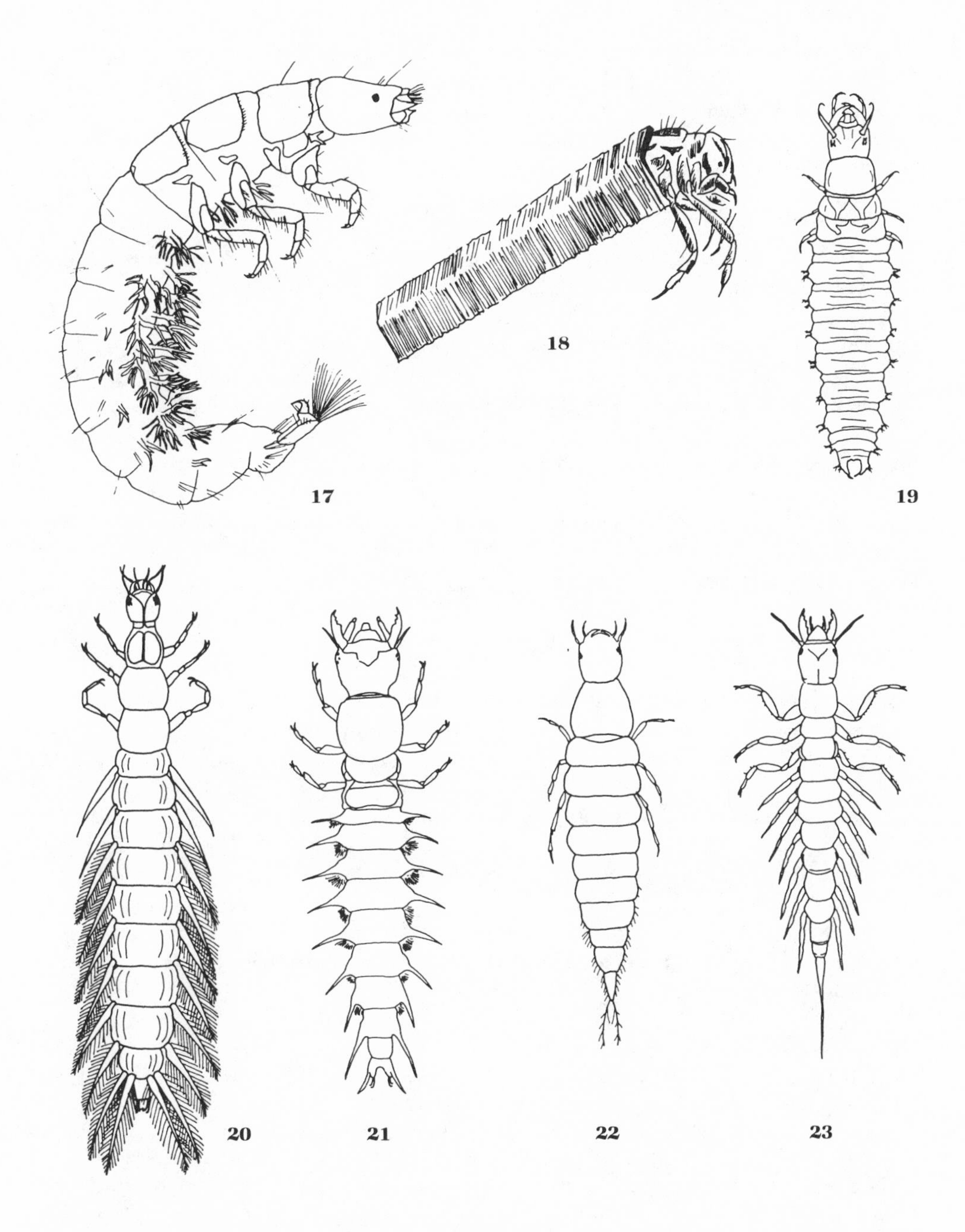

17. *Cheumatopsyche* (Trichoptera, Hydropsychidae) larva, lv. **18.** *Brachycentrus* (Trichoptera, Brachycentridae) larva, lv. **19.** *Tropisternus* (Coleoptera, Hydrophilidae) larva, dv. **20.** *Dineutus* (Coleoptera, Gyrinidae) larva, dv. **21.** *Corydalus* (Megaloptera, Corydalidae) larva, dv. **22.** *Agabus* (Coleoptera, Dytiscidae) larva, dv. **23.** *Sialis* (Megaloptera, Sialidae) larva, dv. dv, dorsal view; lv, lateral view. (17 redrawn from Wiggins 1977; 18 modified from Ross 1944; 19, 20, 22 redrawn from Leech and Chandler 1956; 21, 23 modified from Gurney and Parfin 1959.)

9a (7b). With labium forming an elbowed, extensile grasping organ (Fig. 13) .... **Odonata** (Ch. 5)
9b. With chewing or sucking mouthparts 10

10a (9b). Sucking mouthparts, formed into a narrow tube (as in Fig. 9) 11
10b. Chewing mouthparts, not formed into a narrow tube 12

11a (10a). Parasitic on sponges; all tarsi with 1 claw (Fig. 15) ...... **Neuroptera** (Ch. 12)
11b. Free-living; mesotarsi with 2 claws (Figs. 4, 8) .... **Hemiptera** (in part) (Ch. 7)

12a (10b). Ventral abdominal prolegs each with a ring of fine hooks (crochets) (Fig. 16) .... **Lepidoptera** (Ch. 9)
12b. Abdomen without ventral prolegs, except on terminal segment 13

13a (12b). Antennae extremely small, inconspicuous (Figs. 17, 18), 1-segmented .... **Trichoptera** (Ch. 8)
13b. Antennae elongate, with 3 or more segments (Figs. 19–23) 14

14a (13b). A single claw on each tarsus (Figs. 19, 22) .... **Coleoptera** larvae (in part) (Ch. 10)
14b. Each tarsus with 2 claws (Figs. 20, 21, 23) 15

15a (14b). Without conspicuous lateral filaments (Figs. 19, 22) .... **Coleoptera** larvae (in part) (Ch. 10)
15b. With conspicuous lateral filaments (Figs. 20, 21, 23) 16

16a (15b). Abdomen terminating in 2 slender filaments (Fig. 22) or a median proleg with 4 hooks (Fig. 20) .... **Coleoptera** larvae (in part) (Ch. 10)
16b. Abdomen terminating in a single slender filament (Fig. 23) or in 2 prolegs, each with 2 hooks (Fig. 21) .... **Megaloptera** (Ch. 11)

General References on Freshwater Macroinvertebrate Systematics

(*Used in construction of key.)

*Alexander, C.P., and G.W. Byers. 1981. Tipulidae. *In* J.F. McAlpine, B.V. Peterson, G.E. Shewell, H.J. Teskey, J.R. Vockeroth, and D.M. Wood (coords.). Manual of Nearctic Diptera, vol. 1, pp. 153–190. Res. Branch Agric. Can. Monogr. no. 27. Ottawa.

Anderson, N.H., and J.B. Wallace. 1984. Habitat, life history, and behavioral adaptations of aquatic insects. *In* R.W. Merritt and K.W. Cummins (eds.). An introduction to the aquatic insects of North America, 2nd ed., pp. 38–58. Kendall/Hunt Publ. Co., Dubuque, Iowa.

*Bae, Y.J. 1985. The taxonomic study of Korean Ephemeroptera. Master's thesis. Dept. Biol., Grad. School, Korea Univ. 195 pp.

Ball, G.E. 1982. Current notions about systematics and classification of insects. Manit. Entomol. 13:5–8.

*Baumann, R.W., A.R. Gaufin, and R.F. Surdick. 1977. The stoneflies (Plecoptera) of the Rocky Mountains. Mem. Amer. Entomol. Soc. 31:1–208.

Bentinck, W.C. 1956. Structure and classification. *In* R.L. Usinger (ed.). Aquatic insects of California, pp. 68–73. Univ. Calif. Press, Berkeley.

Bilger, M.D. 1986. A preliminary checklist of the aquatic macroinvertebrates of New England, including New York State. USEPA Region 1. Environ. Serv. Div. Biol. Sec., Lexington, Mass.

Bjarnov, N., and J. Thorup. 1970. A simple method for rearing running-water insects, with some preliminary results. Arch. Hydrobiol. 67:201–209.

*Brigham, A.R., W.U. Brigham, and A. Gnilka (eds.). 1982. Aquatic insects and oligochaetes of North and South Carolina. Midwest Aquatic Enterprises, Mahomet, Ill. 837 pp.

*Chandler, H.P. 1956. Neuroptera. *In* R.L. Usinger (ed.). Aquatic insects of California, pp. 234–236. Univ. Calif. Press, Berkeley.

Cummins, K.W., and M.J. Klug. 1979. Feeding ecology of stream invertebrates. Annu. Rev. Ecol. Syst. 10:147–172.

Cummins, K.W., and R.W. Merritt. 1984a. Ecology and distribution of aquatic insects. *In* R.W. Merritt and K.W. Cummins (eds.). An introduction to the aquatic insects of North America, 2nd ed., pp. 59–65. Kendall/Hunt Publ. Co., Dubuque, Iowa.

——. 1984b. General morphology of aquatic insects. *In* R.W. Merritt and K.W. Cummins (eds.). An introduction to the aquatic insects of North America, 2nd ed., pp. 4–10. Kendall/Hunt Publ. Co., Dubuque, Iowa.

Daly, H.V. 1984. General classification and key to the orders of aquatic and semiaquatic insects. *In* R.W. Merritt and K.W. Cummins (eds.). An introduction to the aquatic insects of North America, 2nd ed., pp. 76–81. Kendall/Hunt Publ. Co., Dubuque, Iowa.

Danks, H.V. 1988. Systematics in support of entomology. Annu. Rev. Entomol. 33:271–296.

Davies, I.J. 1984. Sampling aquatic insect emergence. *In* J.A. Downing and F.H. Rigler (eds.). A manual on methods for the assessment of secondary productivity in fresh waters, 2nd ed., pp. 161–227. IBP handbook 17. Blackwell Scientific, Oxford.

Downing, J.A. 1984. Sampling the benthos of standing waters. *In* J.A. Downing and F.H. Rigler (eds.). A manual on methods for the assessment of secondary productivity in fresh waters, 2nd ed., pp. 87–130. IBP handbook 17. Blackwell Scientific, Oxford.

*Edmondson, W.T. (ed.). 1959. Fresh-water biology, 2nd ed. John Wiley and Sons, New York. 1248 pp.

Eriksen, C.H., V.H. Resh, S.S. Balling, and G.A. Lamberti. 1984. Aquatic insect respiration. *In* R.W. Merritt and K.W. Cummins (eds.). An introduction to the aquatic insects of North America, 2nd ed., pp. 27–37. Kendall/Hunt Publ. Co., Dubuque, Iowa.

Gressit, J.L. 1974. Insect biogeography. Annu. Rev. Entomol. 19:293–321.

*Gurney, A.B., and S. Parfin. 1959. Neuroptera. *In* W.T. Edmondson (ed.). Fresh-water biology, 2nd ed., pp. 973–980. John Wiley and Sons, New York.

Hargeby, A. 1986. A simple trickle chamber for rearing aquatic invertebrates. Hydrobiologia 133:271–274.

Harris, R.A. 1979. A glossary of surface sculpturing. State Calif. Dep. Food Agric. Occas. Pap. Entomol. 28:1–31.

*Hilsenhoff, W.L. 1981. Aquatic insects of Wisconsin, 2nd ed. Nat. Hist. Counc. Publ. no. 2. Univ. Wisc., Madison. 60 pp.

Huggins, D.G., P.M. Liechti, and L.C. Ferrington. 1981. Guide to the freshwater invertebrates of the midwest. Tech. Publ. State Biol. Surv. Kans. 11:1–221.

Hutchinson, G.E. 1981. Thoughts on aquatic insects. BioScience 31:495–500.

Hynes, H.B.N. 1984. The relationship between the taxonomy and ecology of aquatic insects. *In* V.H. Resh and D.M. Rosenberg (eds.). The ecology of aquatic insects, pp. 9–23. Praeger, New York.

Kristensen, N.P. 1981. Phylogeny of insect orders. Annu. Rev. Entomol. 26:135–158.

Lattin, J.D. 1956. Equipment and technique. *In* R.L. Usinger (ed.). Aquatic insects of California, pp. 50–67. Univ. Calif. Press, Berkeley.
*Leech, H.B., and H.P. Chandler. 1956. Coleoptera. *In* R.L. Usinger (ed.). Aquatic insects of California, pp. 293–371. Univ. Calif. Press, Berkeley.
Lehmkuhl, D.L. 1979. How to know the aquatic insects. W.C. Brown Co., Dubuque, Iowa. 168 pp.
Lewis, P.A. 1972. References for the identification of freshwater macroinvertebrates. EPA-R4-72-006.
Macan, T.T., and E.B. Worthington. 1951. Life in lakes and rivers, 3rd ed. Collins, London. 272 pp.
Mason, W.T., Jr., and P.A. Lewis. 1970. Rearing devices for stream insect larvae. Prog. Fish Cult. 32:61–62.
*McCafferty, W.P. 1981. Aquatic entomology. The fishermen's and ecologists' illustrated guide to insects and their relatives. Science Books International, Boston. 448 pp.
*Merritt, R.W., and K.W. Cummins (eds.). 1984. An introduction to the aquatic insects of North America, 2nd ed. Kendall/Hunt Publ. Co., Dubuque, Iowa. 722 pp.
Merritt, R.W., K.W. Cummins, and V.H. Resh. 1984. Collecting, sampling, and rearing methods for aquatic insects. *In* R.W. Merritt and K.W. Cummins (eds.). An introduction to the aquatic insects of North America, 2nd ed., pp. 11–26. Kendall/Hunt Publ. Co., Dubuque, Iowa.
Milne, L., and M. Milne. 1980. The Audubon Society field guide to North American insects and spiders. Knopf, New York. 989 pp.
Needham, J.G., and J.T. Lloyd. 1937. Life of inland waters. Comstock Publ. Co., Ithaca, N.Y. 438 pp.
Needham, J.G., A.D. MacGillivary, O.A. Johannsen, and K.C. Davies. 1903. Aquatic insects in New York State. Bull. N.Y. State Mus. 68:1–517.
*Needham, J.G., and P.R. Needham. 1962. A guide to the study of fresh-water biology, 5th ed. Holden-Day, San Francisco. 107 pp.
Peckarsky, B.L. 1984. Sampling the stream benthos. *In* J.A. Downing and F.H. Rigler (eds.). A manual on methods for the assessment of secondary productivity in fresh waters, 2nd ed., pp. 131–160. IBP handbook 17. Blackwell Scientific, Oxford.
*Pennak, R.W. 1978. Freshwater invertebrates of the United States, 2nd ed. John Wiley and Sons, New York. 803 pp.
*Peterson, B.V. 1981. Simuliidae. *In* J.F. McAlpine, B.V. Peterson, G.E. Shewell, H.J. Teskey, J.R. Vockeroth, and D.M. Wood (coords.). Manual of Nearctic Diptera, vol. 1, pp. 355–392. Res. Branch Agric. Can. Monogr. no. 27. Ottawa.
*Pritchard, A.E., and R.F. Smith. 1956. Odonata. *In* R.L. Usinger (ed). Aquatic insects of California, pp. 106–153. Univ. Calif. Press, Berkeley.
Reid, G.K. 1967. Pond life. Golden Press, New York. 160 pp.
Resh, V.H., and J.O. Solem. 1984. Phylogenetic relationships and evolutionary adaptations of aquatic insects. *In* R.W. Merritt and K.W. Cummins (eds.). An introduction to the aquatic insects of North America, 2nd ed., pp. 66–75. Kendall/Hunt Publ. Co., Dubuque, Iowa.
Rosenberg, D.M., and H.V. Danks. 1987. Aquatic insects of peatlands and marshes in Canada. Mem. Entomol. Soc. Can. 140:1–174.
*Ross, H.H. 1944. The caddis flies, or Trichoptera, of Illinois. Bull. Ill. Nat. Hist. Surv. 23:1–326.
——. 1959. Introduction to aquatic insects. *In* W.T. Edmondson (ed.). Fresh-water biology, 2nd ed., pp. 902–907. John Wiley and Sons, New York.
Schwiebert, E.G. 1973. Nymphs. Winchester Press, New York. 339 pp.
Singh, P., and R.F. Moore (eds.). 1985a. Handbook of insect rearing, vol. 1. Elsevier Sci., New York. 488 pp.
——. 1985b. Handbook of insect rearing, vol. 2. Elsevier Sci., New York. 514 pp.
Smith, M.H. 1984. Laboratory rearing of stream-dwelling insects. Antenna 8:67–68.

Stehr, F.W. (ed.). 1987. Immature insects, vol. 1. Kendall/Hunt Publ. Co., Dubuque, Iowa.

*Unzicker, J.D., V.H. Resh, and J.C. Morse. 1982. Trichoptera. *In* A.R. Brigham, W.U. Brigham, and A. Gnilka (eds.). Aquatic insects and oligochaetes of North and South Carolina, pp. 9.1–9.138. Midwest Aquatic Enterprises, Mahomet, Ill.

*Usinger, R.L. (ed.). 1956. Aquatic insects of California. Univ. Calif. Press, Berkeley. 508 pp.

——. 1967. The life of rivers and streams. McGraw-Hill, New York. 232 pp.

*Waltz, R.D., and W.P. McCafferty. 1979. Freshwater springtails (Hexapoda: Collembola) of North America. Purdue Univ. Agric. Exp. Stn. Res. Bull. 960:1–32.

*Wiggins, G.B. 1977. Larvae of the North American caddisfly genera (Trichoptera). Univ. Toronto Press, Toronto. 401 pp.

Wiley, M., and S.L. Kohler. 1984. Behavioral adaptations of aquatic insects. *In* V.H. Resh and D.M. Rosenberg (eds.). The ecology of aquatic insects, pp. 101–133. Praeger, New York.

Wooton, R.J. 1972. The evolution of insects in freshwater ecosystems. *In* R.B. Clark and R.J. Wooton (eds.). Essays in hydrobiology, pp. 69–82. Univ. Exeter Press, Exeter, England.

——. 1984. Insects of the world. Facts on File Publ., New York.

*Yoon, I.B., and Y.J. Bae. 1984. The classification of Heptageniidae (Ephemeroptera) in Korea. Entomol. Res. Bull. 10:1–34.

3 Semiaquatic Collembola

Classification

The Collembola (springtails) were traditionally placed in the subclass Apterygota, the primitive wingless insects. However, recent taxonomic works generally treat Collembola as a separate class and recognize from one to five orders (Waltz and McCafferty 1979). Three families of Collembola have genera with semiaquatic species.

Life History

Collembolans are ametabolous, with no metamorphosis. They have from 2 to more than 50 instars, and molting continues after the adult stage has been reached (Christiansen and Snider 1984). Patterns of reproduction are quite variable among species; some are viviparous (Pennak 1978), many lack a definitive adult form (Christiansen and Bellinger 1980), and most show little sexual dimorphism, except for *Sminthurides* and *Bourletiella*, which show elaborate behavioral patterns during reproduction (Christiansen and Snider 1984).

Habitat

Collembolans occur on the surface of both lentic and lotic waters, and submerge only accidentally. Their small size and hydrophobic body surface prevent them from breaking through the surface film (Pennak 1978). Thus, they are, strictly speaking, semiaquatic. Waltz and McCafferty (1979) rank the freshwater collembolans on the basis of degree of association with the aquatic habitat. *Primary* aquatic associates are found exclusively on the surface of aquatic habitats and have the highest degree of aquatic specialization; *secondary* aquatic associates are commonly found near or on aquatic habitats or where high humidity exists; and *tertiary aquatic associates* demonstrate an affinity for a variety of habitats other than freshwater and are found only transitorily on temporary rain pools. We consider only the genera with species that are primarily or secondarily associated with freshwater habitats. Other species are found intertidally along the east or west coast of North America (Christiansen and Bellinger 1988), in interstitial habitats, and on the aquatic surface film in caves (Christiansen and Snider 1984).

Feeding

Reported to feed on detritus, algae, fungi, or dead animal matter (Christiansen 1964, Pennak 1978), semiaquatic collembolans typically specialize on the particular foods available on the water surface (Christiansen 1964). The mouthparts of some species are modified for consuming the lipoprotein layer of the surface film or the underlying bacterial populations (Christiansen and Snider 1984).

Respiration

Collembolans are characterized by the presence of a ventral tube (a collophore) on the first abdominal segment that serves as a respiratory, adhesive, and osmoregulatory organ (Christiansen and Snider 1984).

Collection and Preservation

All aquatic species are very small (rarely larger than 3 mm), and they tend to leap off the surface of the water when disturbed. Thus, collecting specimens may be tricky. Although Pennak (1978) suggests using glass sucking tubes, Berlese funnels, or pipettes for collecting collembolans, we generally recommend capturing them by quick sweeps over the surface film with long-handled aquatic D-frame nets. K. Christiansen (pers. comm., 1987) suggests using an aspirator rather than foreceps for removing them from the net. Collembolans may be preserved in 75–95% ethanol. For detailed taxonomic study, Christiansen and Bellinger (1980) describe the most modern and effective mounting techniques.

Identification

All collembolans are characterized by the collophore on the venter of the first abdominal segment. Most aquatic species also possess a jumping organ (furcula), located on the venter of the fourth abdominal segment. The third abdominal segment has a catchlike spring holder (tenaculum) that holds the furcula in place when the animal is at rest. Collembolans are generally distinguished by body shape and segmentation, features of the furcula, coloration, setation, structure of antennae, mouthparts, and eyes.

More-Detailed Information

Scott 1956, Christiansen 1964, Pennak 1978, Waltz and McCafferty 1979, Hilsenhoff 1981, Christiansen and Snider 1984, Christiansen and Bellinger 1988.

Checklist of Semiaquatic Collembola

Isotomidae
- *Agrenia*
- *Isotomurus*

Poduridae
- *Podura*

Sminthuridae
- *Bourletiella*
- *Sminthurides*

Key to Genera of Semiaquatic Collembola

1a. Body somewhat globular; thorax and abdomen indistinctly segmented (Fig. 1) . . . . **Sminthuridae** 2

1b. Body elongate; thorax and abdomen distinctly segmented (Figs. 5, 8) 3

2a (1a). **Sminthuridae**: Female without subanal appendage; males with antennae modified for clasping (Fig. 2); body length 0.5–1.0 mm; last segment of antenna long and not subdivided (Fig. 4); several aquatic and semiaquatic species . . . . *Sminthurides*

2b. Female with subanal appendage (Fig. 3); males without clasping antennae; body length up to 2.2 mm; last segment of antenna subdivided into about 12 subsegments (Fig. 1); olive, yellow, purple, or brown integument . . . . *Bourletiella*

3a (1b). Mouthparts directed downward (Fig. 5); furcula present (Fig. 5), and forks distinctly converge (Fig. 6) . . . . **Poduridae**, *Podura aquatica*

3b. Mouthparts directed forward (Fig. 7); furcula absent or forks do not distinctly converge . . . . **Isotomidae** 4

4a (3b). **Isotomidae**: Bothriotricha present on 1 or more abdominal segments (Fig. 8); mucro quadridentate (Fig. 9), or tridentate (Fig. 10), if tridentate then 4 lamellae (Fig. 11); dens dorsally smooth or crenulate . . . . *Isotomurus*

4b. Bothriotricha absent on abdominal segments; mucro tridentate; dens dorsally tuberculate, never smooth, usually with prominent subapical seta surpassing distal margin of mucro (Fig. 12) . . . . *Agrenia bidenticulata*

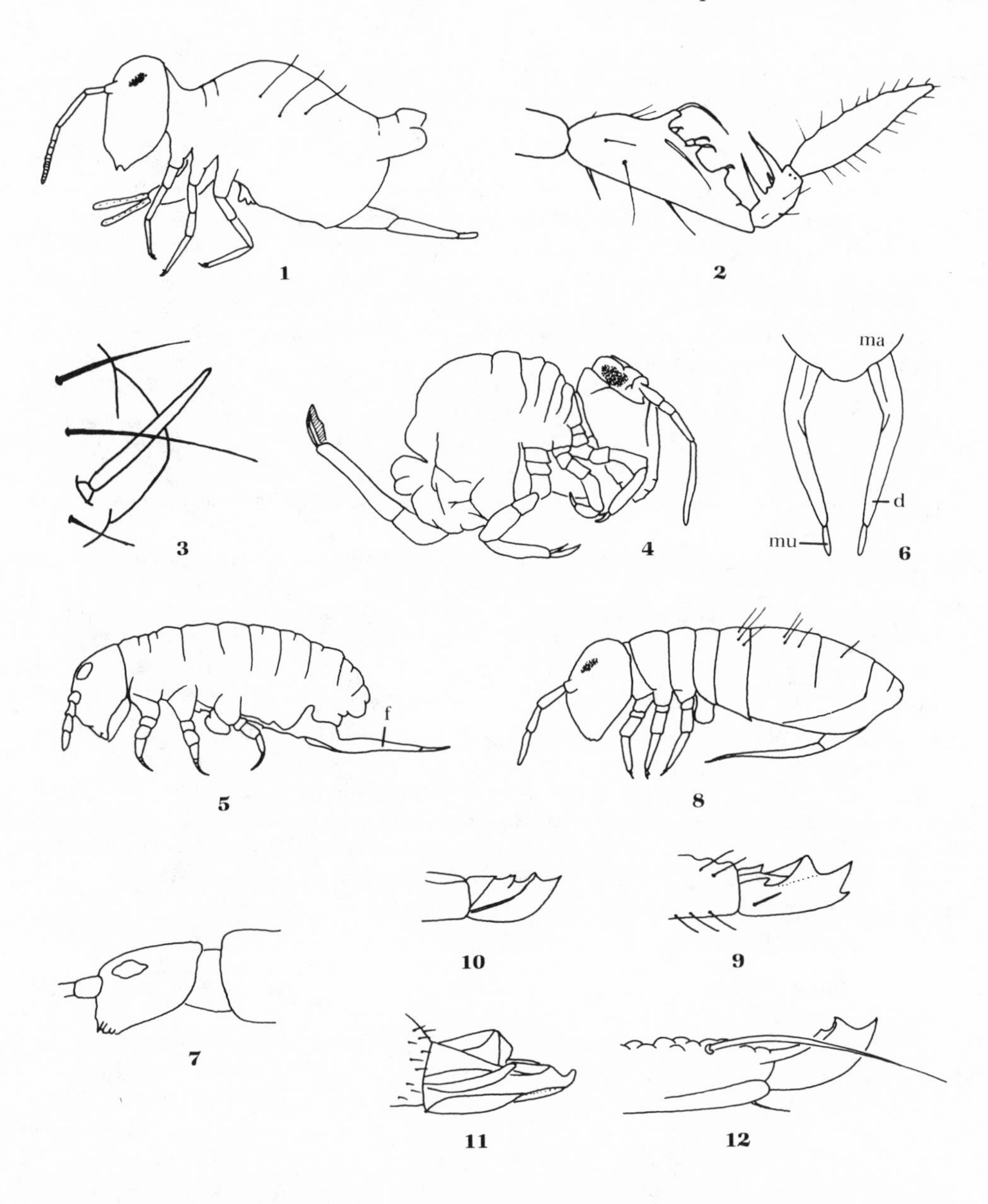

1. Generalized Smiththuridae, lv. **2.** Male antennal organ of *Sminthurides* (Sminthuridae). **3.** Female subanal appendage of *Bourletiella* (Sminthuridae). **4.** *Sminthurides* (Sminthuridae), lv. **5.** *Podura* (Poduridae), lv: f, furcula. **6.** Furculae of *Podura* (Poduridae), dv: ma, manabrium; d, dens; mu, mucro. **7.** Anterior portion of an isotomid, lv. **8.** *Isotomurus* (Isotomidae) showing bothriotricha, lv. **9.** Mucro of *Isotomurus* (Isotomidae), lv. **10.** Mucro of *Isotomurus* (Isotomidae), lv. **11.** Mucro of *Isotomurus* (Isotomidae), lv. **12.** Mucro and dens of *Agrenia* (Isotomidae), lv. dv, dorsal view; lv, lateral view. (1, 2, 5–12 modified or redrawn from Waltz and McCafferty 1979; 3 redrawn from Christiansen and Bellinger 1980; 4 modified from Pennak 1978.)

References on Semiaquatic Collembola Systematics

(*Used in construction of key.)

Christiansen, K. 1964. Bionomics of Collembola. Annu. Rev. Entomol. 9:147–178.

*Christiansen, K., and P. Bellinger. 1980. The Collembola of North America north of the Rio Grande: a taxonomic analysis. Grinnell College, Grinnell, Iowa. 1322 pp.

——. 1988. Marine littoral Collembola of North and Central America. Bull. Mar. Sci. 42(2):215–245.

*Christiansen, K., and R.J. Snider. 1984. Aquatic Collembola. *In* R.W. Merritt and K.W. Cummins (eds.). An introduction to the aquatic insects of North America, 2nd ed., pp. 82–93. Kendall/Hunt Publ. Co., Dubuque, Iowa.

Ellis, W.N., and P.F. Bellinger. 1973. An annotated list of the generic names of Collembola (Insecta) and their type species. Neder. Entomol. Ver., Amsterdam. 74 pp.

Fjellberg, A. 1986. A revision of the genus *Agrenia* Borner 1906 (Collembola: Isotomidae). Entomol. Scand. 17:93–106.

Hilsenhoff, W.L. 1981. Aquatic insects of Wisconsin, 2nd ed. Nat. Hist. Counc. Publ. no. 2. Univ. Wisc., Madison. 60 pp.

*Maynard, E.A. 1951. A monograph of the Collembola of New York State. Comstock Publ. Co., Inc., Ithaca, N.Y. 339 pp.

*Pennak, R.W. 1978. Freshwater invertebrates of the United States, 2nd ed. John Wiley and Sons, New York. 803 pp.

Richards, W.R. 1968. Generic classification, evolution, and biogeography of the Sminthuridae of the world (Collembola). Mem. Entomol. Soc. Can. 53:1–54.

Scott, O.B. 1956. Collembola. *In* R.L. Usinger (ed.). Aquatic insects of California, pp. 74–78. Univ. Calif. Press, Berkeley.

Tomlin, A.D. 1985. *Folsomia candida*. *In* P. Singh and R.F. Moore (eds.). Handbook of insect rearing, vol. 1, pp. 317–320. Elsevier Sci., New York.

*Waltz, R.D., and W.P. McCafferty. 1979. Freshwater springtails (Hexapoda: Collembola) of North America. Purdue Univ. Agric. Exp. Stn. Res. Bull. 960:1–32.

4 Ephemeroptera

Classification

The Ephemeroptera (mayflies) belong to the infraclass Paleoptera; their primitive wings cannot be folded over their backs. They are among the most primitive aquatic insects and probably arose in the Carboniferous period, 280–350 million years ago (Resh and Solem 1984). A number of classification schemes have been proposed for Ephemeroptera above the family level, and there is even some disagreement on family classification (Edmunds et al. 1976, Edmunds 1984, Landa and Soldan 1985). The taxonomy of the immature stages is poorly known because the nymphs of many species have not yet been associated with adult forms. We have incorporated in our keys a number of recent generic-level changes accepted by Edmunds (pers. comm., 1987).

Life History

Mayflies are hemimetabolous insects well known for their short-lived (ephemeral) adult phase, which usually lasts from two hours to three days. However, adults of some species have been reported to live for less than 90 minutes and, in some ovoviviparous species, for several weeks. Mayflies mate in swarms dominated by males. Females fly into a swarm, copulate with a male, then leave the swarm to oviposit (Pennak 1978). They typically oviposit by dipping their abdomens into the surface of the water. Eggs are released a few at a time or all at once, in which case they separate on contact with the water. Females of various species of *Baetis* submerge to oviposit, drop eggs from the air, or sit on stones and dip their ovipositors into the water (Brittain 1982). Eggs are sticky and may have specialized anchoring devices (Edmunds 1984). Mayfly fecundity ranges from 100 to 12,000 eggs per female, with the most common range being between 500 and 3,000 eggs per female. Fecundity increases within species as female size increases (Brittain 1982).

Eggs may hatch immediately upon being laid if they have undergone embryogenic

development while carried by the female. This developmental feature enables some species of mayflies to speed up their life cycles in order to exploit ephemeral habitats. More typically, eggs begin embryonic development after oviposition and hatch synchronously or over staggered time intervals. Eggs of some species have been reported to diapause for up to 11 months (*Parameletus*) (Edmunds 1984). A few species reproduce parthenogenetically (Brittain 1982), with variation from population to population within species (e.g., *Eurylophella funeralis*; Sweeney and Vannote 1987). Nymphs may molt from 12 to 45 times depending on the species and on temperature conditions, completing development in as short a time as about two weeks (*Callibaetis*) or as long as two years (*Hexagenia*). Clifford et al. (1979) have devised a useful scheme for classifying mayfly nymphal developmental stages based on wingpad development, as an alternative to the more difficult instar analysis. Most species are probably *univoltine*, with one flight period each year, but many produce several generations a year (*multivoltine*), while a few are *semivoltine* (take two years to develop but never over their entire geographic range). Mayflies generally overwinter as nymphs.

The final nymphal instar molts to a winged form called the *subimago* or *dun*, which is sexually immature except in the females of a few genera and has translucent wings and a dull body (Edmunds and McCafferty 1988). It is this stage that makes mayflies unique, for the immatures of all other extant insects molt directly into the winged adult stage or pupate before emerging as winged adults. Although the imagoes of most species in northeastern North America are not "subimago-like," some species (e.g., *Caenis*) have imagoes with subimago-like characters. The only sure way to distinguish subimagoes from imagoes is by the presence of sickle-shaped microtrichiae on the wings of subimagoes (Edmunds and McCafferty 1988). Subimagoes often can be found resting on branches overhanging the water, but most go to tree tops and other resting places, where they may remain for a day or two before the final molt to the adult form. In some species, the subimago form lasts only a few minutes, and the molt to the adult stage may be completed in flight. The adult stage, identified by its transparent wings and shiny body, is likewise very brief, usually lasting a few days. The adult stage is strictly for reproduction and dispersal; adults do not feed, having vestigial mouthparts, and sometimes vestigial legs (Edmunds et al. 1976).

Habitat

Mayflies probably evolved in cool streams, a habitat to which nymphs of some of the more primitive groups are currently restricted. They have since radiated to inhabit many lentic habitats as well. They reach peak abundances in cool, clean headwater streams, where they are an important source of food for fish (Edmunds 1984). They have been classified as climbers, sprawlers, burrowers, clingers, and swimmers (Pennak 1978, Edmunds 1984). Some species are common in temporary ponds (e.g., *Callibaetis* spp.) and can tolerate fairly low dissolved-oxygen conditions.

Feeding

Edmunds (1984) classifies mayfly nymphs primarily as grazers and collector-gatherers. Some species are predaceous (Edmunds 1957), and others have interesting adaptations for filter feeding (Oligoneuriidae) (Kondratieff and Voshell 1984). The diet of most mayflies is composed of algae or detritus.

Respiration

Mayfly nymphs have paired tracheal gills on the lateral or dorsal surface of the abdominal segments. Some species move or vibrate their gills to regulate oxygen intake behaviorally, a trait that enables some of them (e.g., *Callibaetis*, *Siphlonurus*) to live in ponds or in stream microhabitats (Ephemerellidae) where dissolved-oxygen content is low. Some burrowing mayflies (*Hexagenia*) use slow gill movements to direct oxygenated water through their burrows, a behavior that may also be involved in suspension feeding. Many mayfly species rely on water current to renew oxygen over gill surfaces and thus are restricted to streams with high dissolved-oxygen content. The integument also may absorb a substantial percentage of the oxygen taken in by mayflies (Pennak 1978). Because the gills of some species are modified for increasing friction with the substrate, they are able to reside in microhabitats with very swift flows. Some have one set of gills modified as a gill cover, or operculum, that protects underlying gills from siltation.

Collection and Preservation

Edmunds et al. (1976) detail various methods for collecting mayfly nymphs. Generally, aquatic D-frame nets are suitable. Edmunds et al. (1976) recommend preservation of nymphs in 95% ethanol because a considerable amount of water is usually collected with them. Specialists often collect specimens in Carnoy or Kahle's fluid and replace this preservative with 80% ethanol after 24 hours. This technique helps preserve color and reduces the loss of legs, antennae, cerci, and gills, common when collections are made in ethanol; it is also recommended for specimens whose internal anatomy will be studied. Collecting vials should be filled completely with preservative to prevent breakage of fragile appendages during transport from the field to the laboratory.

Identification

Figures A and B illustrate the generalized morphological features of nymphal mayflies. The most common features used to identify mayfly genera are gill structure,

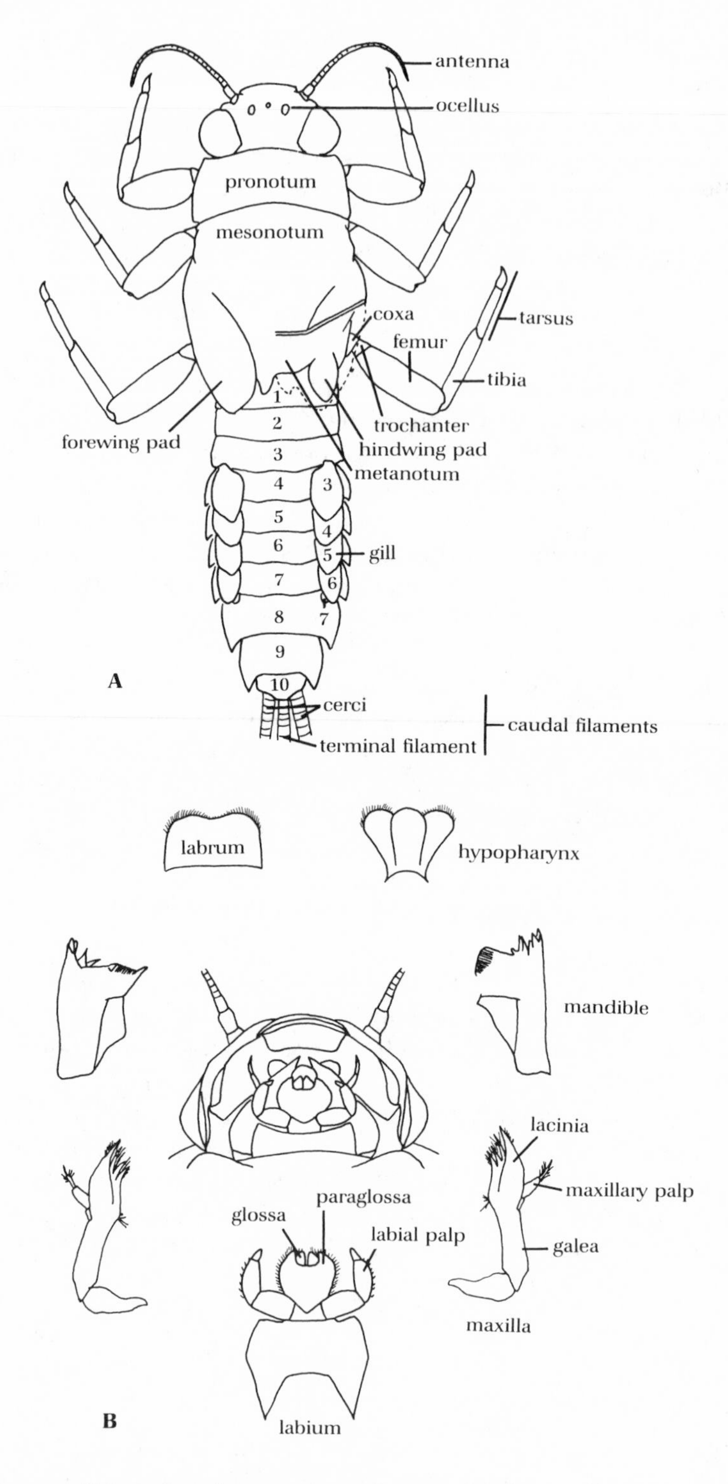

A. *Ephemerella* (Ephemerellidae), dorsal view. **B.** Mouthparts of *Ephemerella* (Ephemerellidae), ventral view. (A, B redrawn from Edmunds et al. 1976.)

caudal filaments, tubercles and spines, claws, mouthparts, and presence or absence of metathoracic wingpads. If specimens are missing legs or gills, definitive identification may not be possible. Identification may also be difficult if early instars are used, because their structures may be poorly developed. We recommend using the key with mature, intact specimens.

More-Detailed Information

Needham et al. 1935, Day 1956, Edmunds et al. 1976, Pennak 1978, Hilsenhoff 1981, McCafferty 1981, Unzicker and Carlson 1982, Edmunds 1984, Edmunds and McCafferty 1988.

Checklist of Ephemeroptera Nymphs

Baetidae
- *Acentrella*
- *Acerpenna*
- *Baetis*
- *Callibaetis*
- *Centroptilum*
- *Cloeon*
- *Diphetor*
- *Heterocloeon*

Baetiscidae
- *Baetisca*

Caenidae
- *Brachycercus*
- *Caenis*

Ephemerellidae
- *Attenella*
- *Dannella*
- *Drunella*
- *Ephemerella*
- *Eurylophella*
- *Serratella*

Ephemeridae
- *Ephemera*
- *Hexagenia*
- *Litobrancha*

Heptageniidae
- *Arthroplea*
- *Cinygmula*
- *Epeorus*
- *Heptagenia*
- *Leucrocuta*
- *Macdunnoa*
- *Nixe*
- *Rhithrogena*
- *Stenacron*
- *Stenonema*

Leptophlebiidae
- *Choroterpes*
- *Habrophlebia*
- *Habrophlebiodes*
- *Leptophlebia*
- *Paraleptophlebia*

Metretopodidae
- *Metretopus*
- *Siphloplecton*

Neoephemeridae
- *Neoephemera*

Oligoneuriidae
- *Isonychia*

Polymitarcyidae
- *Ephoron*

Potamanthidae
- *Potamanthus*

Siphlonuridae
- *Ameletus*
- *Siphlonisca*
- *Siphlonurus*

Tricorythidae
- *Tricorythodes*

Key to Genera of Ephemeroptera Nymphs

1a. Mandibles with large forward-projecting tusks (Figs. 1–5); all gills on abdominal segments 2–7 with fringed margins 2

1b. Mandibles without tusks; some gills on abdominal segments 2–7 may have fringed margins 6

2a (1a). Gills lateral, projecting from sides of abdomen; foretibiae slender, subcylindrical (Fig. 6) . . . . **Potamanthidae**, *Potamanthus*

2b. Gills dorsal, curving up over abdomen; foretibiae fossorial (Figs. 7, 8) 3

3a (2b). Apex of metatibiae rounded (Fig. 7); mandibular tusks curved inward and downward apically, with numerous tubercles on upper surface; extreme tip may point up, resembling a tubercle (Fig. 1) ..... **Polymitarcyidae**, *Ephoron*

3b. Apex of metatibiae projected into acute point ventrally (Fig. 8); mandibular tusks curved upward apically (Fig. 2) .... **Ephemeridae** 4

4a (3b). **Ephemeridae**: Frontal process bifid (Fig. 3) .... *Ephemera*

4b. Frontal process rounded, conical, or truncate (Figs. 4, 5) 5

5a (4b). Gills on abdominal segment 1 bifid (Fig. 9); antennae with whorls of long setae (Fig. 4) .... *Hexagenia*

5b. Gills on abdominal segment 1 single (Fig. 10); antennae with short, scattered setae (Fig. 5) .... *Litobrancha*

6a (1b). Mesonotum modified into a carapacelike structure that covers the gills on abdominal segments 1–6 (Fig. 11) .... **Baetiscidae**, *Baetisca*

6b. Mesonotum not modified into a carapace; gills exposed 7

7a (6b). Gills absent from abdominal segment 2, rudimentary or absent from abdominal segment 1, and present or absent (Fig. 12) from abdominal segment 3; gills on segment 3 or 4 may be operculate (Fig. 13) ...... **Ephemerellidae** 8

7b. Gills present on abdominal segments 1–7 or 2–7 13

8a (7a). **Ephemerellidae**: Lamellate gills present on abdominal segments 3–7; gill on segment 7 may be hidden under gill on segment 6 (Fig. 14) 9

8b. Lamellate gills present on abdominal segments 4–7 (Fig. 12); gills often operculate, not all visible (Fig. 13) 11

9a (8a). Very prominent tubercles or spines present on anterior edge of forefemora (Fig. 15); tarsal claws with 1–4 denticles .... *Drunella*

9b. Tubercles absent from anterior edge of forefemora; a few scattered spines may be present; tarsal claws usually with more than 4 denticles 10

10a (9b). Caudal filaments with whorls of short spines at apex of each segment, and with only sparse intrasegmental setae or none (Fig. 16) (visible only in mature specimens); maxillary palps absent or reduced in size, less than ½ the length of the lacinia (Figs. 17, 18) .... *Serratella*

10b. Caudal filaments with or without whorls of spines at apex of each segment, and with numerous intrasegmental setae (Fig. 19) (visible only in mature specimens); maxillary palps well developed, more than ½ the length of the lacinia (Fig. 20) .... *Ephemerella*

11a (8b). Abdominal terga without paired submedian tubercles .... *Dannella*

11b. Abdominal terga with paired submedian tubercles at least on segments 4–7 (Figs. 12, 13) 12

12a (11b). Middorsal lengths of abdominal terga 2–9 subequal; no operculate gills (Fig. 12) .... *Attenella*

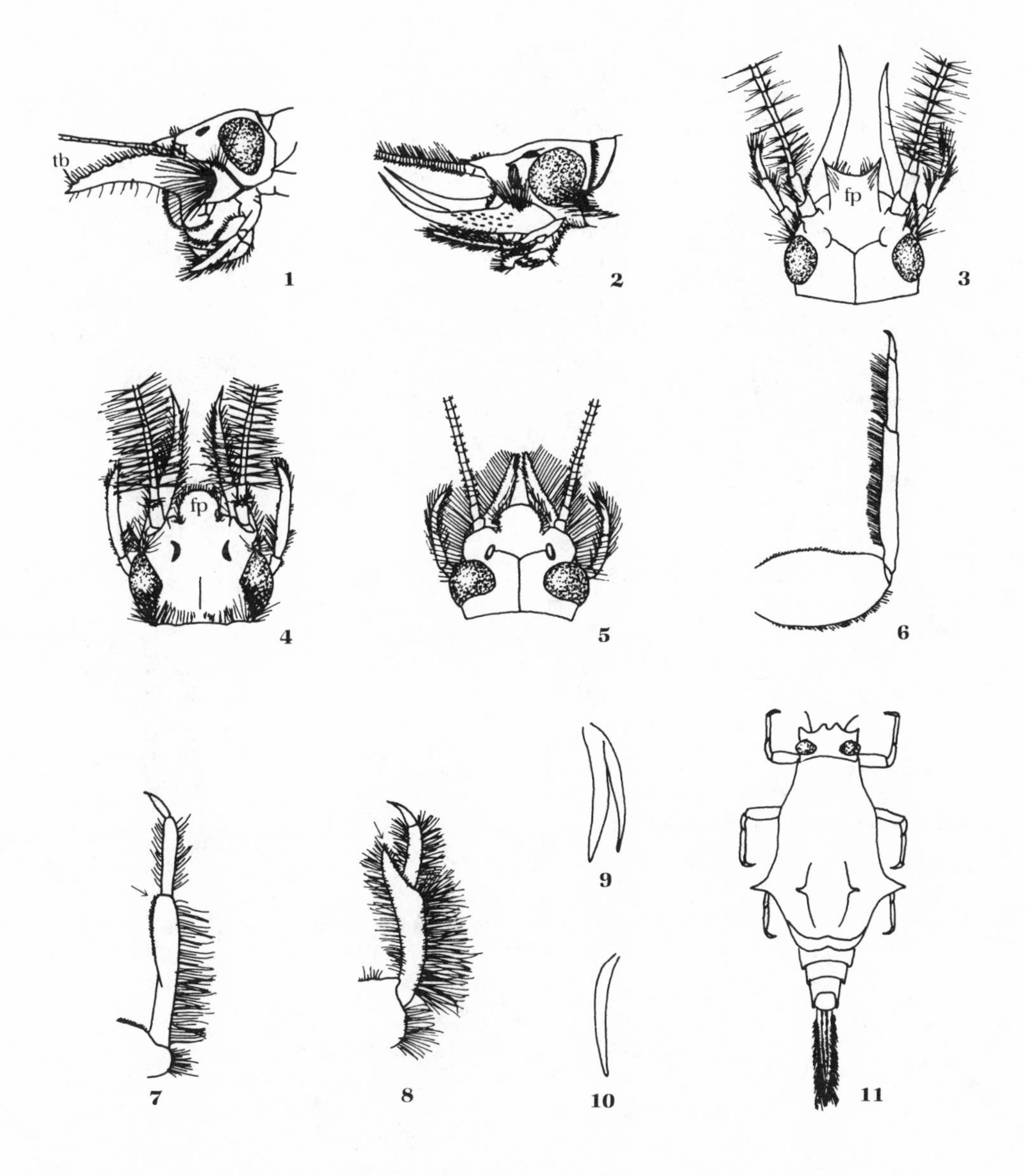

1. Head of *Ephoron* (Polymitarcyidae), lv: tb, tubercle. **2.** Head of *Ephemera* (Ephemeridae), lv. **3.** Head of *Ephemera* (Ephemeridae), dv: fp, frontal process. **4.** Head of *Hexagenia* (Ephemeridae), dv: fp, frontal process. **5.** Head of *Litobrancha* (Ephemeridae), dv. **6.** Prothoracic leg of *Potamanthus* (Potamanthidae), lv. **7.** Hind leg of *Ephoron* (Polymitarcyidae), lv. **8.** Hind leg of *Hexagenia* (Ephemeridae), lv. **9.** Gill 1 of *Hexagenia* (Ephemeridae). **10.** Gill 1 of *Litobrancha* (Ephemeridae). **11.** *Baetisca* (Baetiscidae), dv. dv, dorsal view; lv, lateral view. (1–5, 7–10 modified from Edmunds et al. 1976; 6, 11 redrawn from Hilsenhoff 1981.)

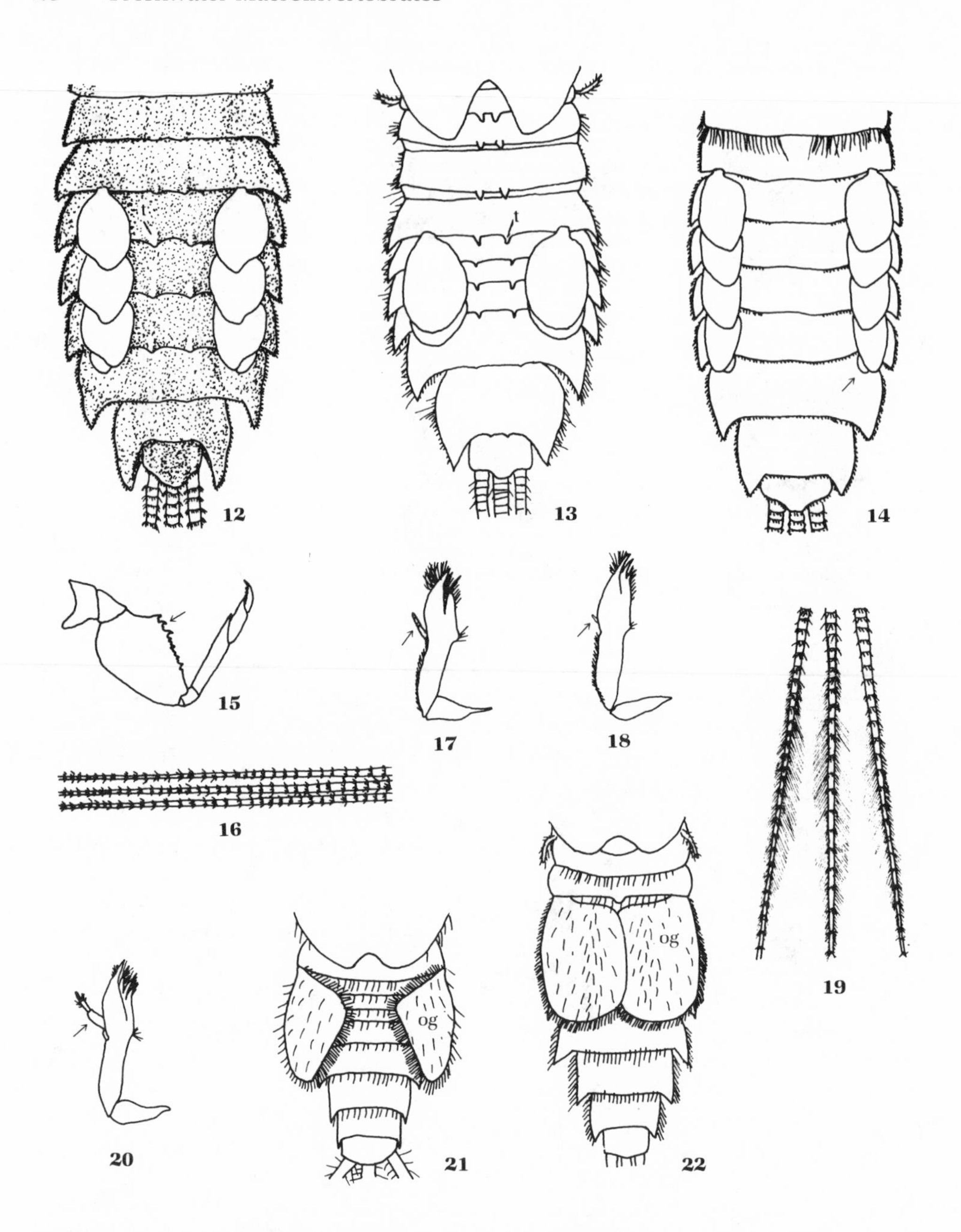

12. Abdomen of *Attenella* (Ephemerellidae), dv: t, submedian tubercle. **13.** Abdomen of *Eurylophella* (Ephemerellidae), dv: t, submedian tubercle. **14.** Abdomen of *Ephemerella* (Ephemerellidae), dv. **15.** Foreleg of *Drunella* (Ephemerellidae), lv. **16.** Caudal filaments of *Serratella* (Ephemerellidae), dv. **17.** Maxilla of *Serratella* (Ephemerellidae), dv. **18.** Maxilla of *Serratella* (Ephemerellidae), dv. **19.** Caudal filaments of *Ephemerella* (Ephemerellidae), dv. **20.** Maxilla of *Ephemerella* (Ephemerellidae), dv. **21.** Abdomen of *Tricorythodes* (Tricorythidae), dv: og, operculate gill. **22.** Abdomen of *Caenis* (Caenidae), dv: og, operculate gill. dv, dorsal view; lv, lateral view. (12–14, 16–20 modified from Edmunds et al. 1976; 15 modified from Morgan 1913; 21, 22 modified or redrawn from Hilsenhoff 1981.)

12b. Abdominal segments 5–7 conspicuously shortened, so that their combined middorsal length is subequal to elongate segment 9; operculate gills on segments 4–7 (Fig. 13) .... *Eurylophella*

13a (7b). Gills on abdominal segment 2 operculate or semioperculate, covering or partially covering the gills on the succeeding segments (Figs. 21, 22) 14
13b. Gills on abdominal segment 2 similar to gills on segments 3–5 17

14a (13a). Operculate gills somewhat triangular and well separated from each other mesally (Fig. 21); succeeding gills without fringed margins .... **Tricorythidae**, *Tricorythodes*
14b. Operculate gills quadrate and overlap mesally (Fig. 22); succeeding gills with fringed margins 15

15a (14b). Mesonotum with distinct rounded lobe on anterolateral corners (Fig. 23); operculate gills joined medially (Fig. 24); developing hind wingpads present .... **Neoephemeridae**, *Neoephemera*
15b. Mesonotum without anterolateral lobes (Fig. 25); operculate gills not joined medially (Fig. 22); developing hind wingpads absent .... **Caenidae** 16

16a (15b). **Caenidae**: Three prominent tubercles on head (Fig. 26); maxillary and labial palps 2-segmented .... *Brachycercus*
16b. No tubercles on head; maxillary and labial palps 3-segmented .... *Caenis*

17a (13b). Head flattened dorsoventrally; eyes and antennae dorsal (Figs. 27–29); gills single lamellae (Figs. 30–32), often with a fibrilliform tuft (Figs. 34, 35) .... **Heptageniidae** 18
17b. Head not flattened dorsoventrally; eyes and antennae lateral; gills variable 27

18a (17a). **Heptageniidae**: Nymphs with only 2 caudal filaments .... *Epeorus*
18b. Nymphs with 3 caudal filaments 19

19a (18b). Gills on abdominal segment 7 vestigial; gills on segment 6 reduced .... *Macdunnoa*
19b. Gills on abdominal segment 7 readily visible, but sometimes reduced (Fig. 33); gills on segment 6 similar to those on segment 5 20

20a (19b). Last pair of gills reduced to a single slender filament with tracheation reduced or absent (Fig. 33) 21
20b. Last pair of gills similar to the pairs that precede it (Figs. 34, 36) 22

21a (20a). Lamellate gills pointed apically (Fig. 30) .... *Stenacron*
21b. Lamellate gills rounded or truncate apically (Figs. 31, 32) .... *Stenonema*

22a (20b). Gill lamellae enlarged on segments 1 and 7; each gill projects to form a ventral disk (Fig. 36) .... *Rhithrogena*

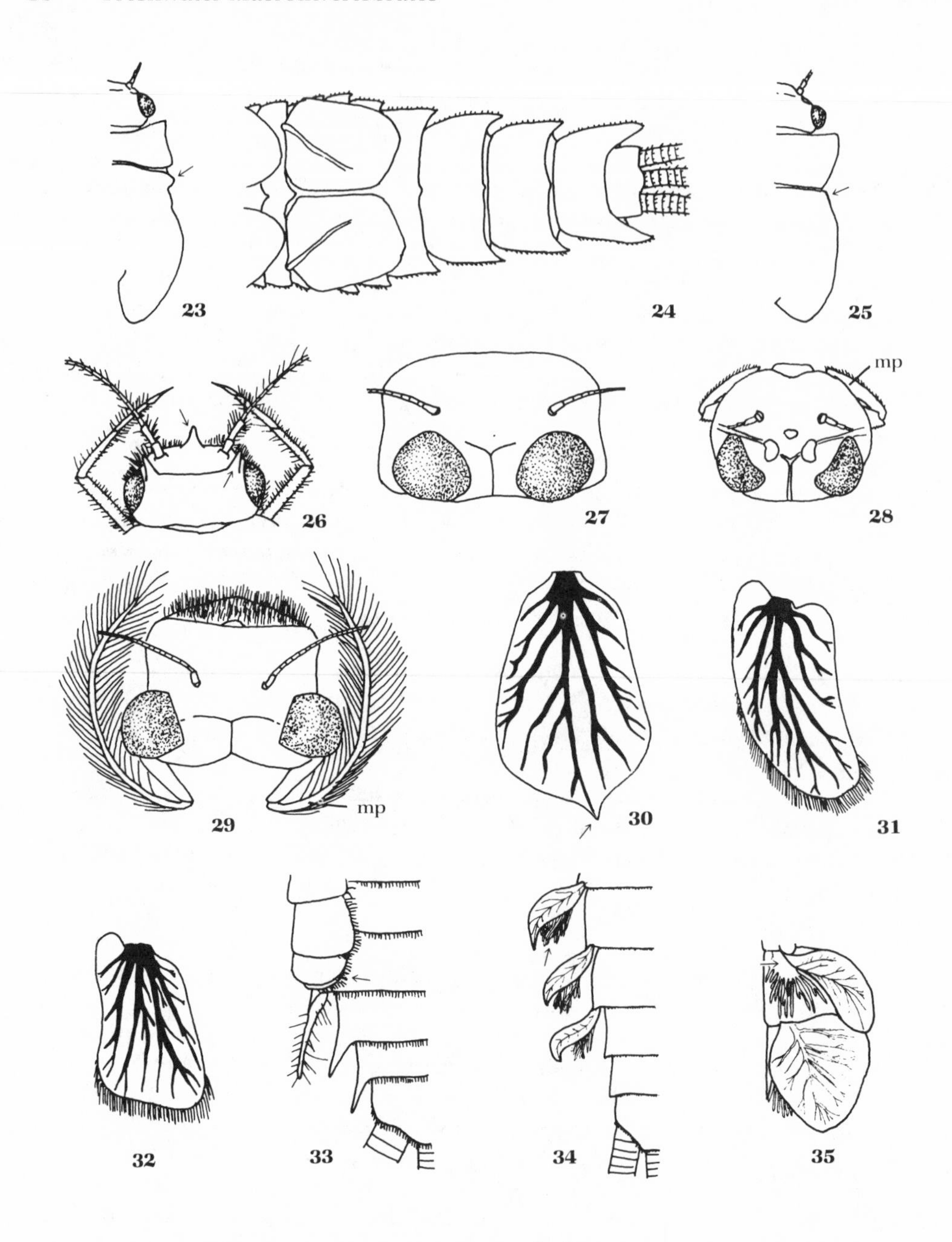

23. Head and thorax of *Neoephemera* (Neoephemeridae), dv. **24.** Abdomen of *Neoephemera* (Neoephemeridae), dv. **25.** Head and thorax of *Caenis* (Caenidae), dv. **26.** Head of *Brachycercus* (Caenidae), dv. **27.** Head of *Stenonema* (Heptageniidae), dv. **28.** Head of *Cinygmula* (Heptageniidae), dv: mp, maxillary palp. **29.** Head of *Arthroplea* (Heptageniidae), dv: mp, maxillary palp. **30.** Gill lamella of *Stenacron* (Heptageniidae). **31.** Gill lamella of *Stenonema* (Heptageniidae). **32.** Gill lamella of *Stenonema* (Heptageniidae). **33.** Left half of abdomen of *Stenonema* (Heptageniidae), dv. **34.** Left half of abdomen of *Heptagenia* (Heptageniidae), dv. **35.** Gills 1 and 2 of *Heptagenia* (Heptageniidae). dv, dorsal view. (23–26, 28, 35 modified from Edmunds et al. 1976; 27, 29–34 redrawn or modified from Hilsenhoff 1981.)

22b. Gill lamellae on segments 1 and 7 smaller than on segments 2–6 23

23a (22b). Distal segment of maxillary palps at least 4 times as long as galea-lacinia with long, conspicuous setae (Fig. 29) (modified for filter-feeding); abundant in mosquito bogs of Massachusetts . . . . *Arthroplea*

23b. Distal segment of maxillary palps much shorter, without long setae (Fig. 28) 24

24a (23b). Front of head distinctly emarginate medially, maxillary palps normally visible at sides of head from dorsal view (Fig. 28); fibrilliform portion of gills absent or reduced to tiny filaments (Fig. 37) . . . . *Cinygmula*

24b. Front of head entire or feebly emarginate; maxillary palps normally not visible from dorsal view; fibrilliform portion of gills distinct at least on segments 1–6 (Fig. 35) 25

25a (24b). Abdominal gill 7 with fibrilliform portion; maxillae with submedian row of ventral setae on galea-lacinia (Fig. 38) . . . . *Heptagenia*

25b. Abdominal gill 7 lacking fibrilliform portion; maxillae with ventral setae on galea-lacinia scattered (Figs. 39, 40) 26

26a (25b). Caudal filaments with whorls of spines at apex and middle of each segment, intrasegmental setae absent (Fig. 41) . . . . *Leucrocuta*

26b. Caudal filaments with intrasegmental setae in addition to whorls of spines (Fig. 42) . . . . *Nixe*

27a (17b). Claws on foretarsi much shorter than those on other tarsi and bifid (Fig. 43); claws of meso- and metatarsi long and slender, about as long as tibiae (Fig. 44) . . . . **Metretopodidae** 28

27b. Claws on all tarsi similar in length and structure 29

28a (27a). **Metretopodidae**: Terminal segment of labial palps expanded and truncate (Fig. 45); distal margin of labrum slightly emarginate (Fig. 46); gills with small setae along outer margin . . . . *Siphloplecton*

28b. Terminal segment of labial palps rounded at its free end (Fig. 47); distal margin of labrum moderately to deeply emarginate (Fig. 48); gills 3–7 with stout spines on outer edge in addition to small setae . . . . *Metretopus*

29a (27b). Prothoracic legs with a dense row of setae along inner surface (Fig. 49) . . . . **Oligoneuriidae**, *Isonychia*

29b. Prothoracic legs without a dense row of setae along inner surface 30

30a (29b). Gills forked (Figs. 50–52), or bilamellate and terminating in a filament or point (Figs. 53, 54), or clusters of filaments (Fig. 55) . . . . . **Leptophlebiidae** 31

30b. Gills single or double lamellae (Figs. 56–58) 35

31a (30a). **Leptophlebiidae**: Each gill on abdominal segments 2–6 consists of 2 clusters of filaments (Fig. 55) . . . . *Habrophlebia*

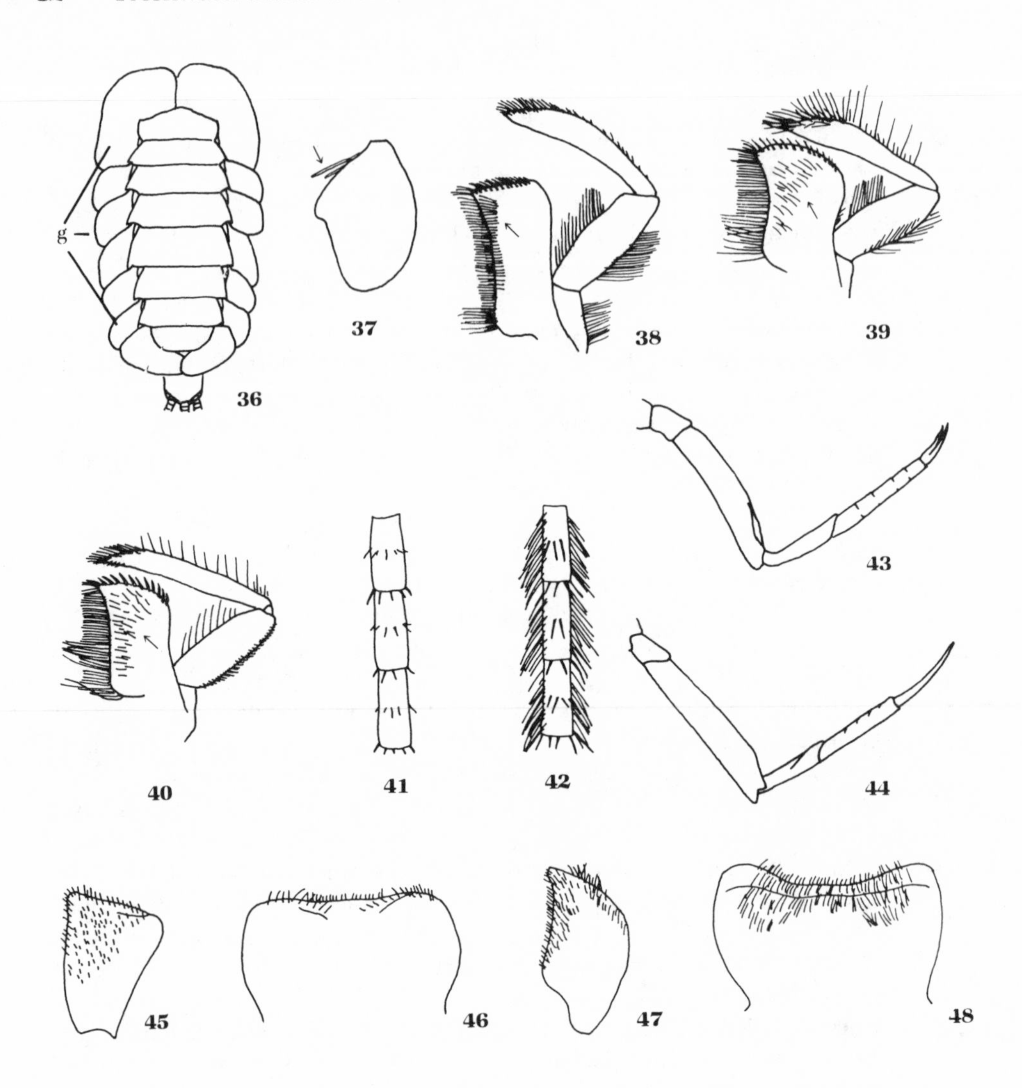

36. Abdomen of *Rhithrogena* (Heptageniidae), vv: g, gills. **37.** Gill 4 of *Cinygmula* (Heptageniidae). **38.** Maxilla of *Heptagenia* (Heptageniidae), vv. **39.** Maxilla of *Leucrocuta* (Heptageniidae), vv. **40.** Maxilla of *Nixe* (Heptageniidae), vv. **41.** Caudal filament of *Leucrocuta* (Heptageniidae). **42.** Caudal filament of *Nixe* (Heptageniidae). **43.** Foreleg of *Siphloplecton* (Metretopodidae), lv. **44.** Hind leg of *Siphloplecton* (Metretopodidae), lv. **45.** Labial palp of *Siphloplecton* (Metretopodidae). **46.** Labrum of *Siphloplecton* (Metretopodidae), dv. **47.** Labial palp of *Metretopus* (Metretopodidae). **48.** Labrum of *Metretopus* (Metretopodidae), dv. dv, dorsal view; lv, lateral view; vv, ventral view. (36 modified from Hilsenhoff 1981; 37, 43, 44 modified from Edmunds et al. 1976; 38–42 modified from Flowers 1980a; 45–48 modified from Berner 1978.)

31b. Gills forked or bilamellate 32

32a (31b). Gills on abdominal segment 1 different in structure from succeeding pairs (Figs. 52–54, 59) 33

32b. Gills on segments 1–7 narrowly lanceolate and bifid (Figs. 50, 51) 34

33a (32a). Gills on segment 1 forked (Fig. 52), remaining gills bilamellate (Fig. 53) . . . . *Leptophlebia*

33b. Gills on segment 1 single, linear lamellae (Fig. 59), remaining gills bilamellate (Fig. 54) . . . . *Choroterpes*

34a (32b). Front of labrum rather deeply emarginate (Fig. 60); posterolateral spines on abdominal segments 8 and 9 . . . . *Habrophlebiodes*

34b. Front of labrum only shallowly emarginate (Fig. 61); posterolateral spines sometimes absent from segment 8 . . . . *Paraleptophlebia*

35a (30b). Median caudal filament well developed, and antennae shorter than twice the width of the head; glossae and paraglossae of labium short and broad (Fig. 62) . . . . **Siphlonuridae** 36

35b. Median caudal filament highly reduced, or if developed then antennae long, more than twice (and usually more than 3 times) as long as the width of the head; glossae and paraglossae long and narrow (Fig. 63) . . . . **Baetidae** 38

36a (35a). **Siphlonuridae**: Gill lamellae double on segments 1 and 2, and sometimes on other segments (Fig. 56) . . . . *Siphlonurus*

36b. Gill lamellae single on all segments (Figs. 57, 58) 37

37a (36b). Gills more or less oval with a sclerotized band along lateral margin and usually with a similar band along mesal margin (Figs. 64, 65); maxillae with a crown of pectinate spines (Fig. 66) . . . . *Ameletus*

37b. Gills more or less heart-shaped (Fig. 67); maxillae without pectinate spines; abdominal segments 5–9 greatly expanded laterally; sternum of mesothorax and metathorax each with a median tubercle; reported from the Adirondacks, Maine, and Newfoundland . . . . *Siphlonisca*

38a (35b). **Baetidae**: Two well-developed cerci; median caudal filament absent or less than ⅛ as long as cerci 39

38b. Three well-developed caudal filaments; although median caudal filament may be shorter and thinner than cerci, it is much longer than ⅛ length of cerci (Fig. 68) 41

39a (38a). Center region of gills with a large pigmented area (Fig. 69); procoxae with a single filamentous gill on inner margin . . . . *Heterocloeon*

39b. Center region of gills without a large pigmented area; procoxae lacking a gill 40

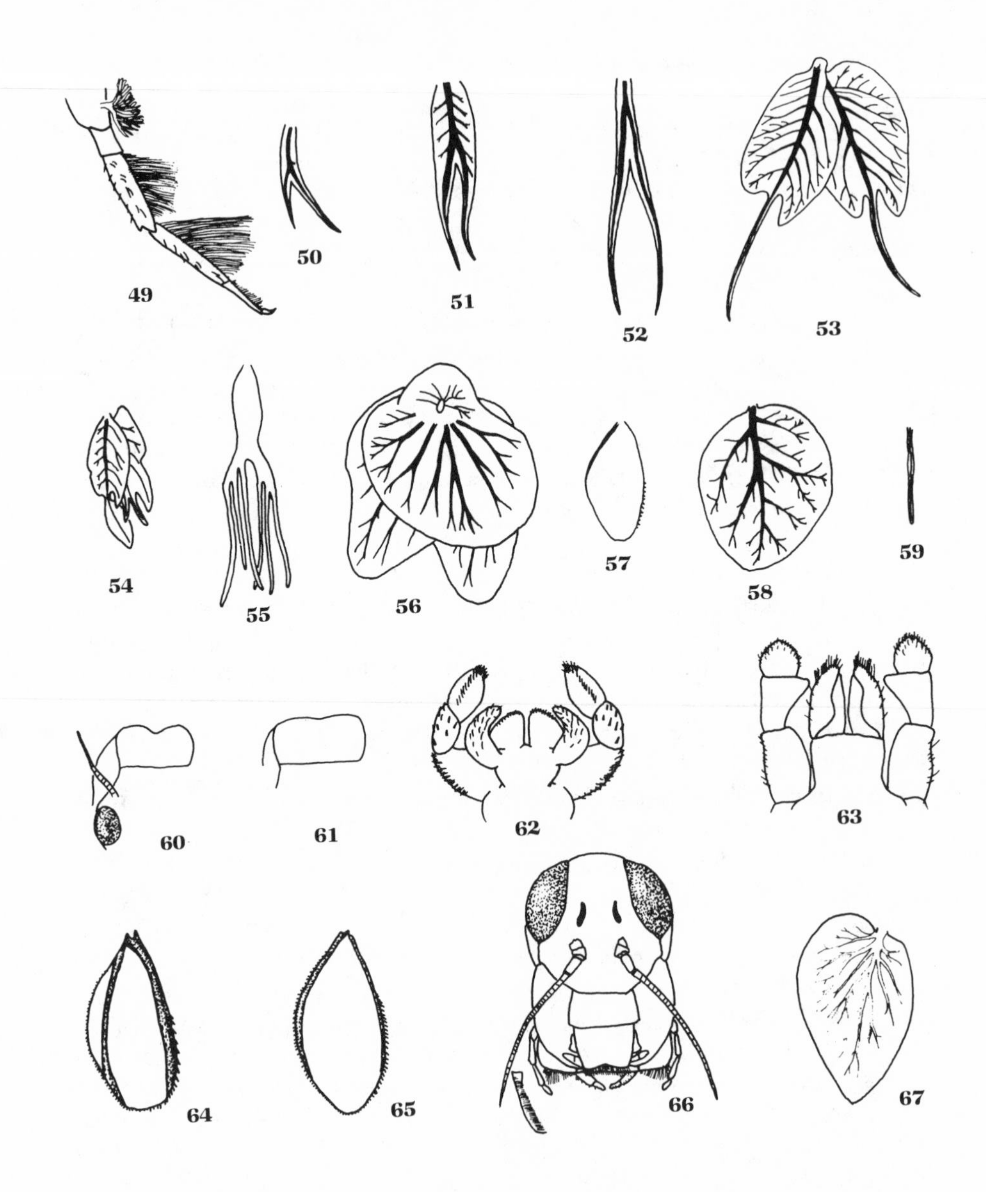

49. Prothoracic leg of *Isonychia* (Oligoneuriidae), lv. **50.** Gill 1 of *Paraleptophlebia* (Leptophlebiidae). **51.** Gill 3 of *Paraleptophlebia* (Leptophlebiidae). **52.** Gill 1 of *Leptophlebia* (Leptophlebiidae). **53.** Gill 3 of *Leptophlebia* (Leptophlebiidae). **54.** Gill 3 of *Choroterpes* (Leptophlebiidae). **55.** Gill 5 of *Habrophlebia* (Leptophlebiidae). **56.** Gill 3 of *Siphlonurus* (Siphlonuridae). **57.** Gill 5 of *Ameletus* (Siphlonuridae). **58.** Gill 3 of *Parameletus* (Siphlonuridae). **59.** Gill 1 of *Choroterpes* (Leptophlebiidae). **60.** Labrum of *Habrophlebiodes* (Leptophlebiidae), dv. **61.** Labrum of *Paraleptophlebia* (Leptophlebiidae), dv. **62.** Labium of *Siphlonurus* (Siphlonuridae), vv. **63.** Labium of *Baetis* (Baetidae), vv. **64.** Gill 4 of *Ameletus* (Siphlonuridae). **65.** Gill 4 of *Ameletus* (Siphlonuridae). **66.** Head of *Ameletus* (Siphlonuridae), av. **67.** Gill 4 of *Siphlonisca* (Siphlonuridae). av, anterior view; dv, dorsal view; lv, lateral view; vv, ventral view. (49–54, 56–61 redrawn or modified from Hilsenhoff 1981; 55, 62–67 modified from Edmunds et al. 1976.)

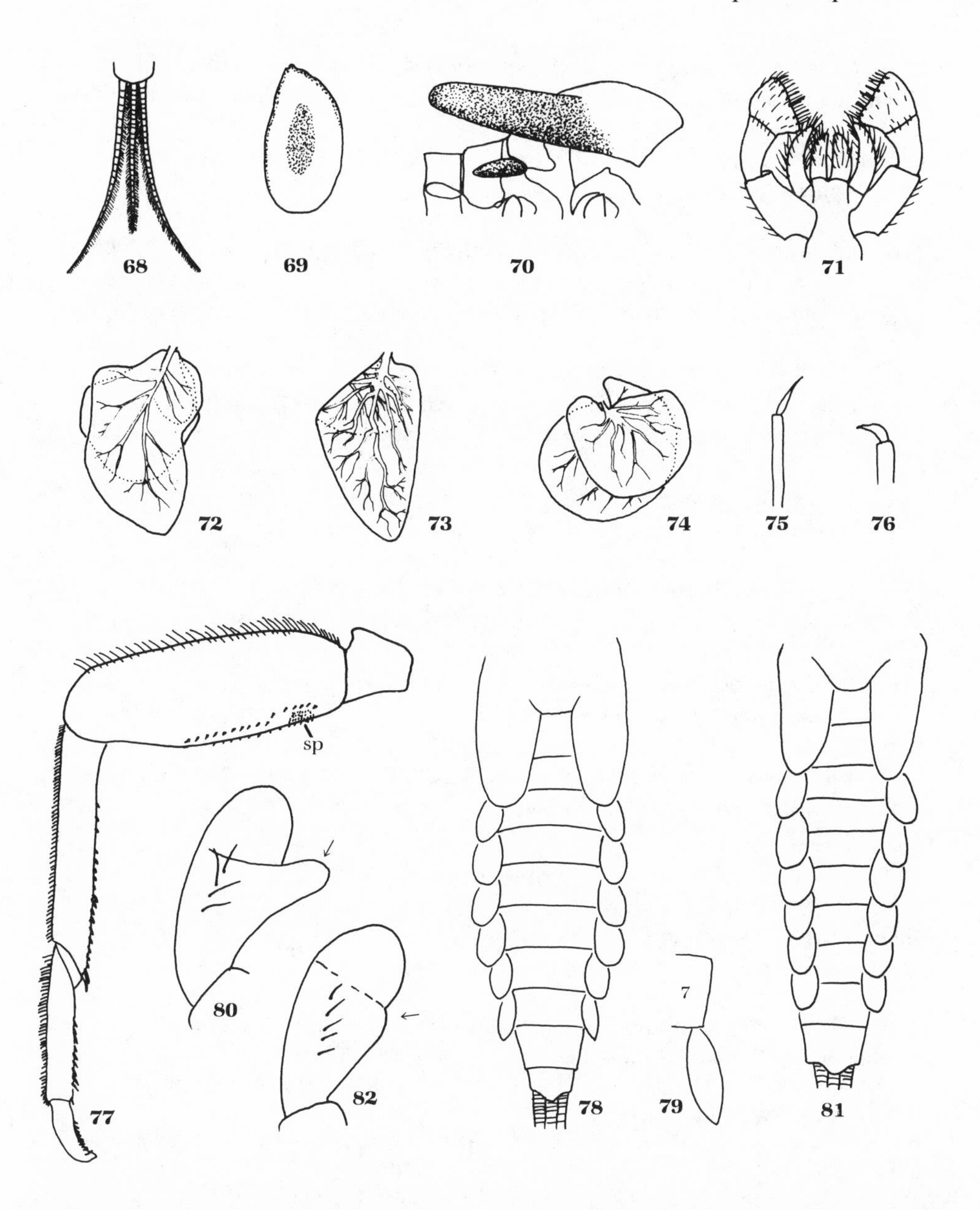

68. Caudal filaments of *Baetis* (Baetidae), dv. **69.** Gill of *Heterocloeon* (Baetidae). **70.** Thorax of *Baetis* (Baetidae), lv. **71.** Labium of *Centroptilum* (Baetidae), vv. **72.** Gill 2 of *Callibaetis* (Baetidae). **73.** Gill 4 of *Callibaetis* (Baetidae). **74.** Gill 1 of *Cloeon* (Baetidae). **75.** Tarsal claw of *Callibaetis* (Baetidae). **76.** Tarsal claw of *Baetis* (Baetidae). **77.** Foreleg of *Heterocloeon* (Baetidae), lv: sp, setal patch. **78.** *Acerpenna* (Baetidae), dv. **79.** 7th gill of *Acerpenna* (Baetidae), dv. **80.** Labial palpus of *Acerpenna* (Baetidae), vv. **81.** *Diphetor* (Baetidae), dv. **82.** Labial palpus of *Diphetor* (Baetidae), vv. dv, dorsal view; lv, lateral view; vv, ventral view. (68–70 redrawn or modified from Hilsenhoff 1981; 71 redrawn from Day 1956; 72–74 modified from Edmunds et al. 1976; 77 modified or redrawn from Waltz and McCafferty 1987a; 78–82 modified or redrawn from Morihara and McCafferty 1979.)

40a (39b). Metathoracic wingpads absent, or reduced to a minute thread . . . *Acentrella*
40b. Metathoracic wingpads present (Fig. 70) . . . . *Baetis* (in part)

41a (38b). Third segment of labial palps truncate (Fig. 71) 42
41b. Third segment of labial palps rounded (Fig. 63) 43

42a (41a). Metathoracic wingpads present (as in Fig. 70); gills 1 and 2 always single, may have recurved flap (as in Fig. 73) . . . . *Centroptilum*
42b. Metathoracic wingpads absent; gills 1 and 2 sometimes double (Fig. 74) . . . . *Cloeon*

43a (41b). Gills on abdominal segments 1–4 with ventral recurved flap (Figs. 72, 73) sometimes double or triple; tarsal claws with an elongate, slender tip (Fig. 75) . . . . *Callibaetis*
43b. Gills all single lamellae, with no ventral recurved flaps; tarsal claws without an elongate, slender tip (Fig. 76) 44

44a (43b). Forelegs with ventral femoral patch of setae (as in Fig. 77) . . . . *Baetis* (in part)
44b. Forelegs without ventral femoral patch of setae 45

45a (44b). Seven pairs of gills, gills 1–6 rounded apically, gill 7 pointed apically (Figs. 78, 79); segment 2 of labial palps with anteromedial thumblike process (Fig. 80) . . . . *Acerpenna*
45b. Gills absent on abdominal segment 1, gills on segments 2–7 rounded apically (Fig. 81); segment 2 of labial palps poorly developed (Fig. 82) . . . . *Diphetor*

References on Ephemeroptera Systematics

(*Used in construction of key.)

Allen, R.K. 1977. A review of *Ephemerella (Dannella)* and the description of a new species (Ephemeroptera: Ephemerellidae). Pan-Pac. Entomol. 53:215–217.
——. 1984. A new classification of the subfamily Ephemerellinae and the description of a new genus. Pan-Pac. Entomol. 60:245–247.
Allen, R.K., and G.F. Edmunds, Jr. 1959. A revision of the genus *Ephemerella* (Ephemeroptera: Ephemerellidae). I. The subgenus *Timpanoga*. Can. Entomol. 91:51–58.
——. 1961a. A revision of the genus *Ephemerella* (Ephemeroptera: Ephemerellidae). II. The subgenus *Caudatella*. Ann. Entomol. Soc. Amer. 54:603–612.
——. 1961b. A revision of the genus *Ephemerella* (Ephemeroptera: Ephemerellidae). III. The subgenus *Attenella*. J. Kans. Entomol. Soc. 34:161–173.
——. 1962a. A revision of the genus *Ephemerella* (Ephemeroptera: Ephemerellidae). IV. The subgenus *Dannella*. J. Kans. Entomol. Soc. 35:333–338.
——. 1962b. A revision of the genus *Ephemerella* (Ephemeroptera: Ephemerellidae). V. The subgenus *Drunella* in North America. Misc. Publ., Entomol. Soc. Amer. 3:147–170.
——. 1963a. A revision of the genus *Ephemerella* (Ephemeroptera: Ephemerellidae). VI. The subgenus *Serratella* in North America. Ann. Entomol. Soc. Amer. 56:583–600.

——. 1963b. A revision of the genus *Ephemerella* (Ephemeroptera: Ephemerellidae). VII. The subgenus *Eurylophella*. Can. Entomol. 95:597–623.

——. 1965. A revision of the genus *Ephemerella* (Ephemeroptera: Ephemerellidae). VIII. The subgenus *Ephemerella* in North America. Misc. Publ., Entomol. Soc. Amer. 4:243–282.

Arbona, F.L., Jr. 1980. Mayflies, the angler, and the trout. Winchester Press, Tulsa, Okla. 188 pp.

Bednarik, A.F., and W.P. McCafferty. 1979. Biosystematic revision of the genus *Stenonema* (Ephemeroptera: Heptageniidae). Can. Bull. Fish. Aquat. Sci. 201:1–73.

Benton, M.J., and G. Pritchard. 1988. New methods for mayfly instar number determination and growth curve estimation. J. Fresh. Ecol. 4:361–368.

Bergman, E.A., and W.L. Hilsenhoff. 1978. *Baetis* (Ephemeroptera: Baetidae) of Wisconsin. Great Lakes Entomol. 11:125–137.

Berner, L. 1956. The genus *Neoephemera* in North America (Ephemeroptera: Neoephemeridae). Ann. Entomol. Soc. Amer. 49:33–42.

——. 1959. A tabular summary of the biology of North American mayfly nymphs (Ephemeroptera). Bull. Fla. State Mus. 4:1–58.

*——. 1978. A review of the mayfly family Metretopodidae. Trans. Amer. Entomol. Soc. 104:91–137.

Berner, L., and M.L. Pescador. 1980. The mayfly family Baetiscidae (Ephemeroptera). Part I. *In* J.F. Flannagan and K.E. Marshall (eds.). Advances in ephemeropteran biology, pp. 511–524. Plenum, New York.

——. 1988. The mayflies of Florida, rev. ed. Univ. Presses of Florida, Gainesville. 415 pp.

Brinck, P. 1957. Reproductive system and mating in Ephemeroptera. Opusc. Entomol. 22:1–37.

Brittain, J.E. 1982. Biology of mayflies. Annu. Rev. Entomol. 27:119–148.

Burks, B.D. 1953. The mayflies or Ephemeroptera of Illinois. Bull. Ill. Nat. Hist. Surv. 26:1–216.

Clifford, H.F., M. Hamilton, and B.A. Killins. 1979. Biology of the mayfly *Leptophlebia cupida* (Say) (Ephemeroptera: Leptophlebiidae). Can. J. Zool. 57:1026–1045.

*Day, W.C. 1956. Ephemeroptera. *In* R.L. Usinger (ed.). Aquatic insects of California, pp. 79–105. Univ. Calif. Press, Berkeley.

Edmunds, G.F., Jr. 1957. The predaceous mayfly nymphs of North America. Proc. Utah Acad. Sci. Arts Lett. 34:23–24.

——. 1972. Biogeography and evolution of Ephemeroptera. Annu. Rev. Entomol. 17:21–42.

——. 1982. Historical and life history factors in the biogeography of mayflies. Amer. Zool. 22:371–374.

*——. 1984. Ephemeroptera. *In* R.W. Merritt and K.W. Cummins (eds.). An introduction to the aquatic insects of North America, 2nd ed., pp. 94–125. Kendall/Hunt Publ. Co., Dubuque, Iowa.

Edmunds, G.F., Jr., R.K. Allen, and W.L. Peters. 1963. An annotated key to the nymphs of the families and subfamilies of mayflies (Ephemeroptera). Univ. Utah. Biol. Ser. 13:1–49.

Edmunds, G.F., Jr., L. Berner, and J.R. Traver. 1958. North American mayflies of the family Oligoneuriidae. Ann. Entomol. Soc. Amer. 51:375–382.

*Edmunds, G.F., Jr., S.L. Jensen, and L. Berner. 1976. The mayflies of North and Central America. Univ. Minn. Press, Minneapolis. 330 pp.

Edmunds, G.F., Jr., and W.P. McCafferty. 1978. A new J.G. Needham device for collecting adult mayflies (and other out-of-reach insects). Entomol. News 89:193–194.

——. 1988. The mayfly subimago. Annu. Rev. Entomol. 33:509–530.

Edmunds, G.F., Jr., and J.R. Traver. 1959. The classification of the Ephemeroptera. I. Ephemeroidea: Behningiidae. Ann. Entomol. Soc. Amer. 52:43–51.

Fink, T.J. 1980. A comparison of mayfly (Ephemeroptera) instar determination methods. *In* J.F. Flannagan and K.E. Marshall (eds.). Advances in ephemeropteran biology, pp. 367–380. Plenum, New York.

Flannagan, J.F., and K.E. Marshall (eds.). 1980. Advances in ephemeropteran biology. Plenum, New York. 552 pp.

*Flowers, R.W. 1980a. Two new genera of Nearctic Heptageniidae (Ephemeroptera). Fla. Entomol. 63:296–306.

——. 1980b. A review of the Nearctic *Heptagenia* (Heptageniidae, Ephemeroptera). *In* J.F. Flannagan and K.E. Marshall (eds.). Advances in ephemeropteran biology, pp. 93–102. Plenum, New York.

*——. 1982. Review of the genus *Macdunnoa* (Ephemeroptera: Heptageniidae) with a description of a new species from Florida. Great Lakes Entomol. 15:25–30.

Flowers, R.W., and W.L. Hilsenhoff. 1975. Heptageniidae (Ephemeroptera) of Wisconsin. Great Lakes Entomol. 8:201–218.

——. 1978. Life cycles and habitats of Wisconsin Heptageniidae (Ephemeroptera). Hydrobiologia 60:159–172.

Harper, F., and P.P. Harper. 1981. Northern Canadian mayflies (Insecta: Ephemeroptera), records and descriptions. Can. J. Zool. 59:1784–1789.

*Hilsenhoff, W.L. 1981. Aquatic insects of Wisconsin. Nat. Hist. Counc. Publ. no. 2. Univ. Wisc., Madison. 60 pp.

——. 1984. Identification and distribution of *Baetisca* nymphs (Ephemeroptera: Baetiscidae) in Wisconsin. Great Lakes Entomol. 17:51–52.

Hubbard, M.D. 1979. The validity of the generic name *Parameletus* Bengtsson (Ephemeroptera: Siphlonuridae). Proc. Entomol. Soc. Wash. 79:409–410.

——. 1982. Catalog of the Ephemeroptera: family-group taxa. Aquat. Insects 4:49–53.

Hubbard, M.D., and J. Kukalova-Peck. 1980. Permian mayfly nymphs: new taxa and systematic characters. *In* J.F. Flannagan and K.E. Marshall (eds.). Advances in ephemeropteran biology, pp. 19–31. Plenum, New York.

Hubbard, M.D., and W.L. Peters. 1978. Environmental requirements and pollution tolerance of Ephemeroptera. EPA-600/4-78-061. 461 pp.

Hubbard, M.D., and H.M. Savage. 1981. The fossil Leptophlebiidae (Ephemeroptera): a systematic and phylogenetic review. J. Paleontol. 55:810–813.

Ide, F.P. 1937. Descriptions of eastern North American species of baetine mayflies with particular reference to the nymphal stages. Can. Entomol. 69:219–231, 235–243.

Jensen, S.L. 1966. The mayflies of Idaho (Ephemeroptera). M.S. thesis. Univ. Utah, Salt Lake City. 365 pp.

——. 1974. A new genus of mayflies from North America (Ephemeroptera: Heptageniidae). Proc. Entomol. Soc. Wash. 76:225–228.

Jensen, S.L., and G.F. Edmunds, Jr. 1973. Some phylogenetic relationships within the family Heptageniidae. *In* W.L. Peters and J.G. Peters (eds.). Proc. First Int. Conf. on Ephemeroptera, August 17–20, 1970, Tallahassee, Fla., pp. 82–87. E.J. Brill, Leiden, Netherlands.

Kondratieff, B.C., and J.R. Voshell, Jr. 1983. A checklist of the mayflies (Ephemeroptera) of Virginia, with a review of the pertinent taxonomic literature. J. Ga. Entomol. Soc. 18:273–279.

——. 1984. The North and Central American species of *Isonychia* (Ephemeroptera: Oligoneuriidae). Trans. Amer. Entomol. Soc. 110:129–244.

Koss, R.W. 1968. Morphology and taxonomic use of Ephemeroptera eggs. Ann. Entomol. Soc. Amer. 61:696–721.

Koss, R.W., and G.F. Edmunds, Jr. 1974. Ephemeroptera eggs and their contribution to phylogenetic studies of the order. Zool. J. Linn. Soc. 55:267–349.

Kukalova, J. 1968. Permian mayfly nymphs. Psyche 75:310–327.

Landa, V. 1969. Comparative anatomy of mayfly larvae (Ephemeroptera). Acta Entomol. Bohemoslov. 66:289–316.

Landa, V., and T. Soldan. 1985. Phylogeny and higher classification of the order Ephemeroptera: a discussion from the comparative anatomical point of view. Studie CSAV 4:1–121.

Lehmkuhl, D.M. 1979b. The North American species of *Cinygma* (Ephemeroptera: Heptageniidae). Can. Entomol. 111:657–680.

Leonard, J.W. 1965. Environmental requirements of Ephemeroptera. *In* C.M. Tarzwell (ed.). Biological problems in water pollution. Third seminar, U.S. Dep. Health, Educ., Welfare Serv., Div. Water Supply Pollut. Control, R.A. Taft Sanit. Engr. Ctr. PHS Publ. no. 999-WP-25, pp. 110–117.

Leonard, J.W., and F.A. Leonard. 1962. Mayflies of Michigan trout streams. Cranbrook Institute of Science, Bloomfield Hills, Mich. 139 pp.

Lewis, P.A. 1978. On the use of pectinate maxillar spines to separate *Stenonema* and *Stenacron* (Ephemeroptera: Heptageniidae). Proc. Entomol. Soc. Wash. 80:655–656.

Martin, I.D., and M.A. Gates. 1984. Patterns of morphological and reproductive variation in *Stenonema* mayflies (Ephemeroptera: Ephemeroidea). Trans. Amer. Entomol. Soc. 101:447–504.

McCafferty, W.P. 1971. A new genus of mayflies from eastern North America (Ephemeroptera: Ephemeridae). J. N.Y. Entomol. Soc. 79:45–51.

——. 1975. The burrowing mayflies of the United States (Ephemeroptera: Ephemeroidea). Trans. Amer. Entomol. Soc. 101:447–504.

*——. 1981a. Distinguishing larvae of North American Baetidae from Siphlonuridae (Ephemeroptera). Entomol. News 92:138–141.

——. 1981b. Aquatic entomology. The fisherman's and ecologists' illustrated guide to insects and their relatives. Science Books International, Boston. 448 pp.

——. 1985. The Ephemeroptera of Alaska. Proc. Entomol. Soc. Wash. 87:381–386.

McCafferty, W.P., and G.F. Edmunds, Jr. 1979. The higher classification of the Ephemeroptera and its evolutionary basis. Ann. Entomol. Soc. Amer. 72:5–12.

Morgan, A.H. 1911. May-flies of Fall Creek. Ann. Entomol. Soc. Amer. 4:93–126.

——. 1912. Homologies in the wing-veins of may-flies. Ann. Entomol. Soc. Amer. 5:89–111.

*——. 1913. A contribution to the biology of may-flies. Ann. Entomol. Soc. Amer. 6:371–413.

*Morihara, D.K., and W.P. McCafferty. 1979. The *Baetis* larvae of North America (Ephemeroptera: Baetidae). Trans. Amer. Entomol. Soc. 105:139–221.

Murphy, H.E. 1922. Notes on the biology of some of our North American species of may-flies. Bull. Lloyd Lib. Bot., Pharmacy Materia Med., Entomol. Ser. 22:1–46.

Needham, J.G. 1917. Burrowing mayflies of our larger lakes and streams. Bull. Bur. Fish. 36:267–292.

Needham, J.G., K.J. Morton, and O. Johannsen. 1905. Mayflies and midges of New York. Bull. N.Y. State Mus. 86:1–352.

Needham, J.G., J.R. Traver, and Y. Hsu. 1935. The biology of mayflies. Comstock Publ. Co., Ithaca, N.Y. 759 pp.

Pasternak, K., and R. Sowa (eds.). 1979. Proc. Second Int. Conf. on Ephemeroptera, August 23–26, 1975. Panstwowe Wydawnictwo Naukowe, Warsaw, Poland. 312 pp.

Pennak, R.W. 1978. Freshwater invertebrates of the United States, 2nd ed. John Wiley and Sons, New York. 803 pp.

Pescador, M.L. 1985. Systematics of the Nearctic genus *Pseudiron* (Ephemeroptera: Heptageniidae: Pseudironinae). Fla. Entomol. 63:432–443.

Pescador, M.L., and L. Berner. 1981. The mayfly family Baetiscidae (Ephemeroptera). Part II. Biosystematics of the genus *Baetisca*. Trans. Amer. Entomol. Soc. 107:163–228.

Peters, W.L. 1980. Phylogeny of the Leptophlebiidae (Ephemeroptera): an introduction. *In* J.F. Flannagan and K.E. Marshall (eds.). Advances in ephemeropteran biology, pp. 33–41. Plenum, New York.

——. 1988. Origins of the North American Ephemeroptera fauna, especially the *Leptophlebiidae*. Mem. Entomol. Soc. Can. 144:13–24.

Peters, W.L., and J.G. Peters (eds.). 1973. Proc. First Int. Conf. on Ephemeroptera, August 17–20, 1970. Tallahassee, Fla. E.J. Brill, Leiden, Netherlands. 312 pp.

Resh, V.H., and J.O. Solem. 1984. Phylogenetic relationships and evolutionary adaptations of aquatic insects. *In* R.W. Merritt and K.W. Cummins (eds.). An introduction to the aquatic insects of North America, 2nd ed., pp. 66–75. Kendall/Hunt Publ. Co., Dubuque, Iowa.

Rick, E.F. 1973. The classification of the Ephemeroptera. *In* W.L. Peters and J.G. Peters (eds.). Proc. First Int. Conf. on Ephemeroptera, August 17–20, 1970, Tallahassee, Fla., pp. 160–178. E.J. Brill, Leiden, Netherlands.

Spieth, H.T. 1947. Taxonomic studies on the Ephemeroptera. IV. The genus *Stenonema*. Ann. Entomol. Soc. Amer. 40:87–122.

Sweeney, B.W., and R.L. Vannote. 1987. Geographic parthenogenesis in the stream mayfly *Eurylophella funeralis* in eastern North America. Holarct. Ecol. 10:52–59.

Traver, J.R. 1940. Collecting mayflies (Ephemeroptera). Compend. Entomol. Methods 1. Wards Natural Science Establishment, Rochester, N.Y. 8 pp.

Tsui, P.T.P. 1973. The use of thoracic morphology in taxonomic and phylogenetic studies of the Leptophlebiidae (Ephemeroptera). *In* W.L. Peters and J.G. Peters (eds.). Proc. First Int. Conf. on Ephemeroptera, August 17–20, 1970. Tallahassee, Fla., pp. 79–81. E.J. Brill, Leiden, Netherlands.

Unzicker, J.D., and P.H. Carlson. 1982. Ephemeroptera. *In* A.R. Brigham, W.U. Brigham, and A. Gnilka (eds.). Aquatic insects and oligochaetes of North and South Carolina, pp. 3.1–3.97. Midwest Aquatic Enterprises, Mahomet, Ill.

*Waltz, R.D., and W.P. McCafferty. 1987a. New genera of Baetidae for some Nearctic species previously included in *Baetis* Leach (Ephemeroptera). Ann. Entomol. Soc. Amer. 80:667–670.

——. 1987b. Systematics of *Pseudocloeon*, *Acentrella*, *Baetiella*, and *Liebebiella*, new genus (Ephemeroptera: Baetidae). J. N.Y. Entomol. Soc. 95:553–568.

Waltz, R.D., W.P. McCafferty, and J.H. Kennedy. 1985. *Barbaetis*: a new genus of eastern Nearctic mayflies (Ephemeroptera: Baetidae). Great Lakes Entomol. 18:161–166.

5 Odonata

Classification

The order Odonata (dragonflies and damselflies) belongs to the primitive infraclass Paleoptera. Odonates are among the most ancient flying insects; their ancestors appeared during the Carboniferous period 280–350 million years ago. Dragonflies of the Carboniferous period reportedly had wingspans of 75 cm, four times that of the largest extant species (Hutchinson 1981). Odonates comprise two suborders, the Anisoptera (dragonflies) and Zygoptera (damselflies). Adult dragonflies hold their wings horizontally when at rest; damselflies rest holding wings vertically over their backs. The taxonomy of this order is relatively well known among the aquatic insects, since nymphs (or naiads) are typically very conspicuous and well studied (Westfall 1984).

Life History

Odonates are hemimetabolous, with relatively long-lived adults (several weeks to several months). Some species oviposit by inserting eggs into incisions made in emergent or submerged vegetation (endophytic). Others deposit eggs on the surface of such vegetation (epiphytic), on sand or silt under the water, at the surface of the water, or in the basins of dry ponds (exophytic). One published report describes an aeshnid ovipositing in holes dug by wild boars in the bottom of dry ponds (Utzeri and Raffi 1983). Male odonates, especially damselflies, may remain in tandem with females after copulation, and accompany them to oviposition sites. This behavior reduces the chances that other males will copulate with mated females and remove sperm deposited by previous copulations, assuring paternity for the mated males (Alcock 1982). Fecundity of odonates ranges from a few hundred to a few thousand eggs per female, with the maximum recorded fecundity being 5,200 eggs in one female (Westfall 1984). Eggs hatch within 12–30 days, depending on the species and on temperature conditions, or they may diapause in dry basins until the basins are reflooded.

Odonates generally overwinter as eggs or as nymphs, which hatch from eggs as *pronymphs* (sometimes termed instar one), then molt almost immediately into what is usually considered the first instar (Westfall 1984). Odonates have been reported to undergo from 10–16 molts and are usually univoltine, but some may be multivoltine (*Enallagma*) or may take up to five years to complete development (Petaluridae) (Westfall 1984). Since instar analysis is difficult, Baker (1986) suggests classifying zygopteran developmental stages on the basis of wingpad length. Nymphs crawl from the water to emerge usually on vertically oriented vegetation.

Mating behavior in odonates has received much attention. Adult males are often territorial, defending regions around ponds or streams that may be prime oviposition sites. Females that fly into male territories are grasped by the head (anisopterans) or the prothorax (zygopterans), and copulation occurs usually near the water. Before copulation, males transfer sperm from the primary genitalia to the secondary genitalia on abdominal segment 2 (translocation). This behavior frees the posterior end of the male for clasping the female during copulation. The female curves her abdomen ventrally to bring her genitalia into contact with the secondary genitalia of the male, creating the characteristic "wheel" copulatory position often photographed by naturalists.

Habitat

Odonate nymphs occur typically in lentic habitats, where they may burrow into the sediments or climb on the aquatic macrophytes. Some are classified as sprawlers. A few species are typical of streams, however, again either as burrowers or climbers (Pennak 1978, Westfall 1984). The lestids are specialized for inhabiting temporary ponds, where they oviposit on sedges above the water surface. If stems are green, females seal the incision made by the ovipositor. During early autumn storms, stems collapse and are insulated in snow. Eggs, containing developed embryos, diapause and hatch when wetted in spring. Species that oviposit in dead, dry, bent stems do not seal the incision, instead producing thicker walled, cold-resistant eggs that diapause early in embryonic development. Wetting of these stems in spring stimulates further development and then hatching.

Feeding

Odonate nymphs are strictly predaceous, using their modified extensible labium or "mask" to capture insect, crustacean, molluscan, or oligochaete prey. They have even been reported to consume small vertebrates (Pennak 1978). Some species burrow in the substrate and ambush prey that they detect by tactile or vibrational cues.

Other species more actively stalk their prey. Odonates are one of the few orders of aquatic insects whose immature stages have eyes well developed for hunting (Peckarsky 1984). Adult odonates have large compound eyes to detect insect prey, which they grasp with their legs while flying (Westfall 1984).

Respiration

Anisopteran nymphs respire through rectal expansion and contraction, which moves water in and out of the anus. The rectal chamber is highly specialized, containing longitudinal rows of minute tracheal gills that project into villi in the cavity. Each gill contains tracheoles that absorb oxygen from the water in the rectum (Pennak 1978). Anisopteran nymphs also use their rectal chambers for jet-propulsion locomotion by rapidly expelling water from the chambers (Westfall 1984). Zygopteran nymphs have three caudal lamellae that serve as tracheal gills as well as swimming structures. These lamellae are supplied with large tracheal trunks. Respiration also occurs cutaneously in the damselflies (Pennak 1978).

Collection and Preservation

Odonate nymphs are easily collected by sweeping the vegetation of ponds with an aquatic D-frame net or by sampling the littoral benthos or stream bottom. They are best preserved by killing them in hot water and then storing them in 70% ethanol.

Identification

Figures A, B, and C illustrate the general morphological characteristics of nymphal zygopterans and anisopterans. Features commonly used to separate genera of odonate nymphs are structure of the terminal appendages, antennal segmentation, shape and tooth structure of the labium, dorsal and lateral abdominal spines, and the length-to-width ratio of abdominal segments. The families Corduliidae and Libellulidae cannot be separated by any single character, differing instead by a suite of features.

More-Detailed Information

Smith and Pritchard 1956, Corbet 1963, Pennak 1978, Corbet 1980, Hilsenhoff 1981, Huggins and Brigham 1982, Westfall 1984, Miller 1987.

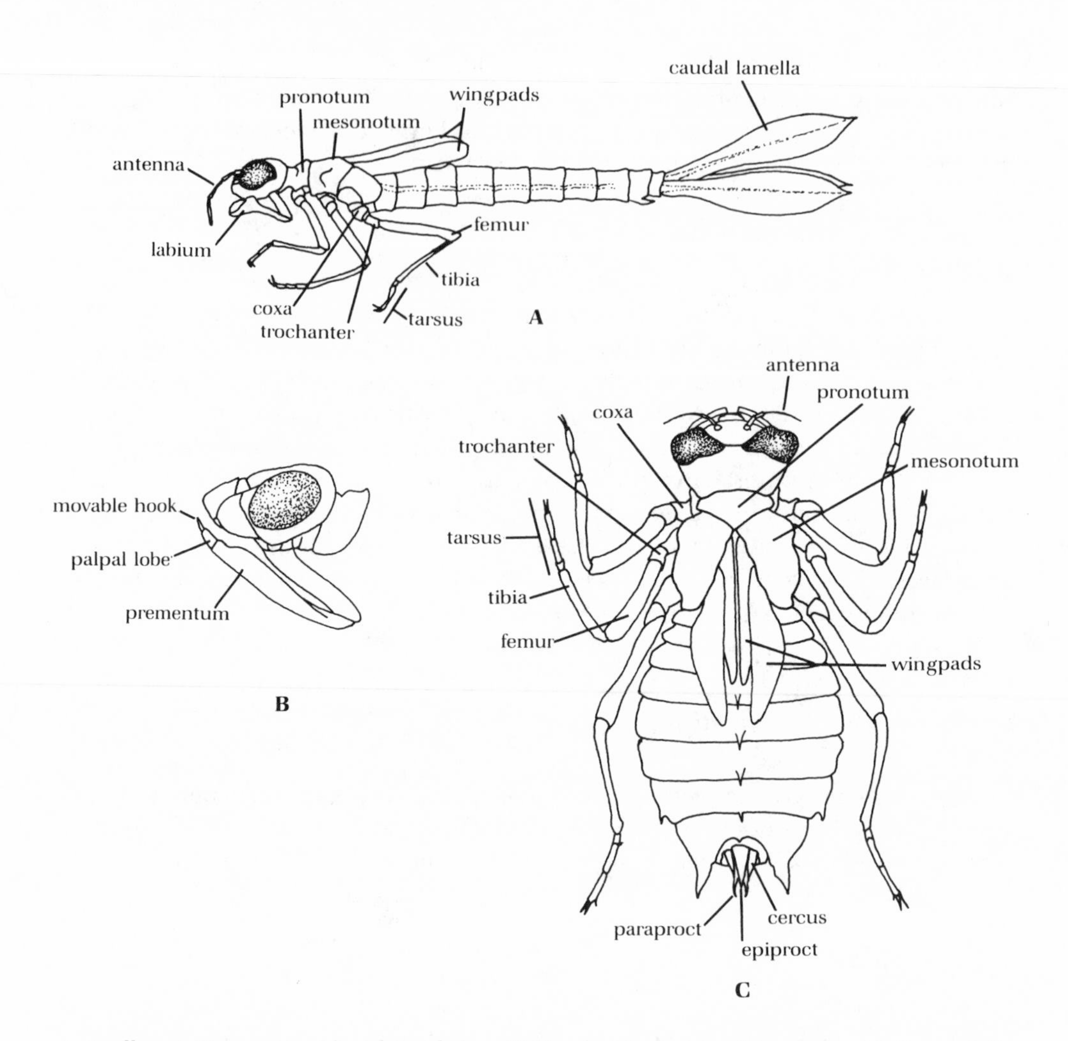

A. *Enallagma* (Coenagrionidae), lateral view. **B.** Head of Aeshnidae, lateral view. **C.** *Epicordulia* (Corduliidae), dorsal view. (A, C redrawn from Smith and Pritchard 1956, B redrawn from Pennak 1978.)

Checklist of Odonata Nymphs

Suborder Anisoptera
- Aeshnidae
  - *Aeshna*
  - *Anax*
  - *Basiaeschna*
  - *Boyeria*
  - *Epiaeschna*
  - *Gomphaeschna*
  - *Nasiaeschna*
- Cordulegastridae
  - *Cordulegaster*[a]
  - *Taeniogaster*[a]
  - *Zoraena*[a]
- Corduliidae
  - *Cordulia*
  - *Dorocordulia*
  - *Epicordulia*
  - *Helocordulia*
  - *Neurocordulia*
  - *Somatochlora*
  - *Tetragoneuria*
  - *Williamsonia*
- Gomphidae
  - *Arigomphus*
  - *Dromogomphus*
  - *Gomphus*[b]
  - *Hagenius*
  - *Lanthus*
  - *Ophiogomphus*
  - *Progomphus*
  - *Stylogomphus*
  - *Stylurus*
- Libellulidae
  - *Celithemis*
  - *Erythemis*
  - *Erythrodiplax*
  - *Ladona*
  - *Leucorrhinia*
  - *Libellula*
  - *Nannothemis*
  - *Pachydiplax*
  - *Pantala*
  - *Perithemis*
  - *Plathemis*
  - *Sympetrum*
  - *Tramea*
- Macromiidae
  - *Didymops*
  - *Macromia*
- Petaluridae
  - *Tachopteryx*

Suborder Zygoptera
- Calopterygidae
  - *Calopteryx*
  - *Hetaerina*
- Coenagrionidae
  - *Amphiagrion*
  - *Anomalagrion*
  - *Argia*
  - *Chromagrion*
  - *Coenagrion*
  - *Enallagma*
  - *Ischnura*
  - *Nehalennia*
- Lestidae
  - *Archilestes*
  - *Lestes*

[a]We have lumped these three genera within *Cordulegaster*, the most common practice among odonate systematists. However, Carle (1983) suggests that the three genera be recognized once again.

[b]Includes subgenera *Gomphurus*, *Gomphus*, and *Hylogomphus*, which some systematists have treated as genera.

Key to Genera of Odonata Nymphs

1a. Abdomen terminating in 3 caudal lamellae, longest more than ⅓ length of abdomen . . . . **Zygoptera** 2

1b. Abdomen terminating in 3–5 stiff, pointed valves, longest less than ⅓ length of abdomen (Fig. 1) . . . . **Anisoptera** 5

2a (1a). **Zygoptera**: First antennal segment as long as, or longer than, remaining segments combined; prementum with deep, median cleft (Figs. 2, 3); lotic . . . . **Calopterygidae** 10

2b. First antennal segment much shorter than others combined; prementum with at most a very small median cleft (Figs. 4, 5) 3

3a (2b). Basal half of prementum greatly narrowed and elongate (Fig. 4); unextended labium reaches back to or past middle coxae . . . . **Lestidae** 4

3b. Basal half of prementum not greatly narrowed (Fig. 5); unextended labium reaches only to fore coxae . . . . **Coenagrionidae** 11

4a (3a). **Lestidae**: Distal margin of each palpal lobe divided into 3 sharp processes, the outermost one much shorter than the movable hook (Fig. 6); caudal gills with 2 well-defined and complete, dark crossbands . . . . *Archilestes*

4b. Distal margin of each palpal lobe divided into 4 processes, 3 sharp hooks, and a short, serrate, truncate projection between the 2 outer hooks, the outermost hook almost as long as the movable hook (Fig. 7); caudal gills without 2 well-defined and complete, dark crossbands . . . . *Lestes*

5a (1b). **Anisoptera**: Prementum flat, or nearly so (Fig. 8), without stout setae 6

5b. Prementum spoon-shaped, covering face to base of antennae (Fig. 9), and armed with stout setae 8

6a (5a). Antennae 4-segmented; mesotarsi 2-segmented . . . . **Gomphidae** 17

6b. Antennae 6- or 7-segmented; mesotarsi 3-segmented 7

7a (6b). Prementum parallel-sided in distal ⅔, abruptly narrowed at base (Fig. 10) . . . . **Petaluridae**, *Tachopteryx*

7b. Prementum widest in distal ½, but not parallel-sided (Fig. 11) . . . . **Aeshnidae** 25

8a (5b). Labium with large irregular teeth on distal edge of palpal lobes (Fig. 12); very hairy nymphs; lotic . . . . **Cordulegastridae**, *Cordulegaster*

8b. Labium with distal edge of palpal lobes entire or with small, even crenulations or teeth (Fig. 13) 9

9a (8b). Head with a prominent, almost erect, thick frontal horn between bases of antennae; legs very long, apex of each metafemur reaching to or beyond apex of abdominal segment 8 (Fig. 14); metasternum with broad, median tubercle (may be obscure) . . . . **Macromiidae** 32

9b. Head without a prominent frontal horn, except in *Neurocordulia molesta*; legs shorter, apex of metafemora usually not reaching apex of abdominal segment 8; metasternum without a median tubercle . . . . **Libellulidae** and **Corduliidae** 33

10a (2a). **Calopterygidae**: Prementum cleft almost halfway to base (Fig. 2) . . . . *Calopteryx*

10b. Prementum cleft only to base of palpal lobes (Fig. 3) . . . . *Hetaerina*

11a (3b). **Coenagrionidae**: Distal margin of each palpal lobe produced into 3 pointed hooks, middle one shorter than end hook and usually about ½ as long as movable hook (Fig. 15); widest part of median caudal lamellae usually ⅓ to ½ as wide as long and in some species quite thick or triangular in cross section . . . . *Argia*

11b. Distal margin of each palpal lobe with a comparatively small end hook; portion between end hook and movable hook more or less truncate and denticulate, middle teeth less than ⅓ as long as movable hook (Fig. 5); widest

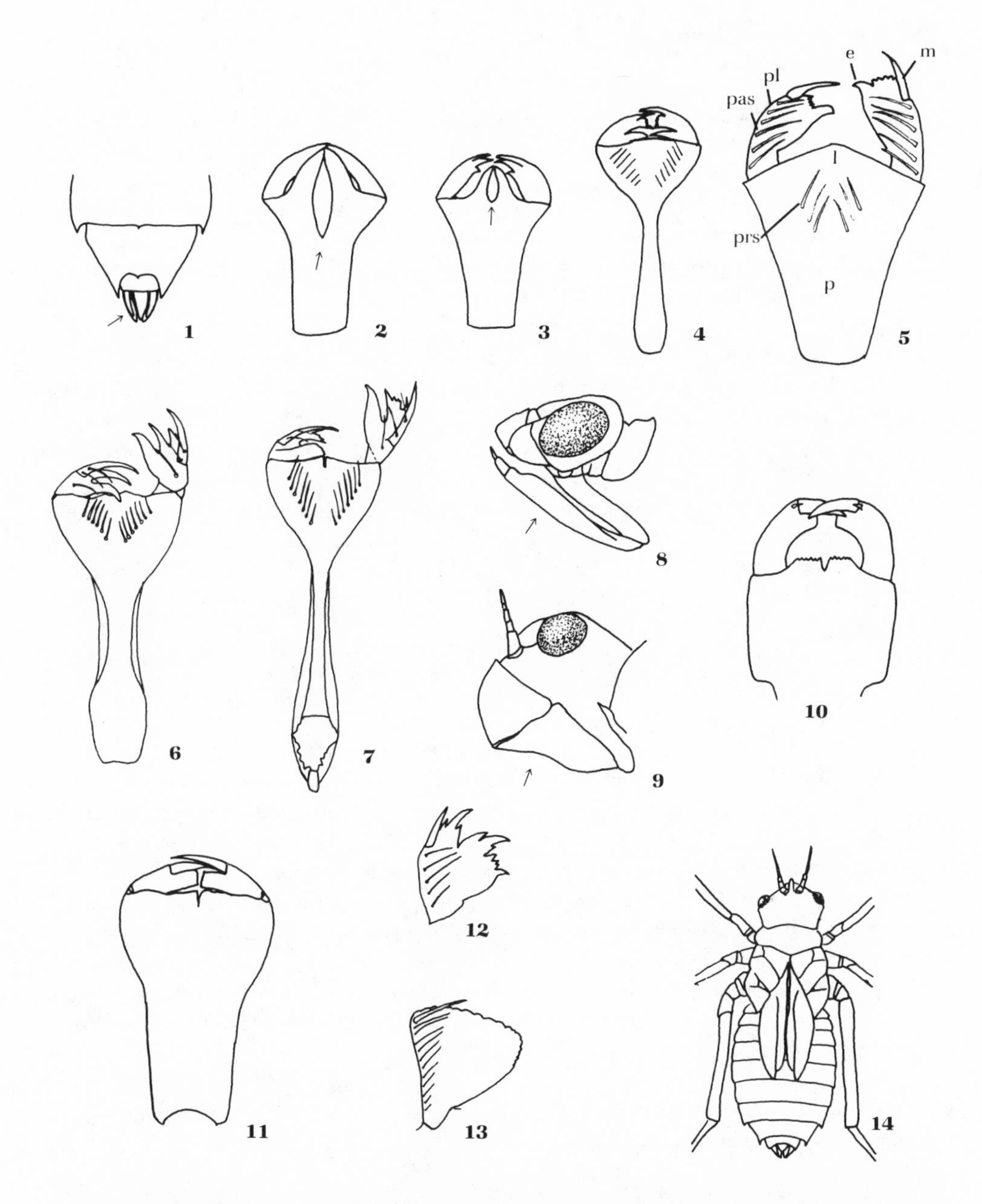

1. Tip of abdomen of *Gomphus* (*Gomphurus*) (Gomphidae), dv. **2.** Prementum of *Calopteryx* (Calopterygidae), vv. **3.** Prementum of *Hetaerina* (Calopterygidae), vv. **4.** Prementum of *Lestes* (Lestidae), dv. **5.** Prementum of *Enallagma* (Coenagrionidae), dv: e, end hook; l, ligula; m, movable hook; p, prementum; pas, palpal setae; pl, palpal lobes; prs, premental setae. **6.** Labium of *Archilestes* (Lestidae), dv. **7.** Labium of *Lestes* (Lestidae), dv. **8.** Head of Aeshnidae, lv. **9.** Head of Libellulidae, lv. **10.** Prementum of *Tachopteryx* (Petaluridae), vv. **11.** Prementum of *Aeshna* (Aeshnidae), vv. **12.** Palpal lobe of *Cordulegaster* (Cordulegastridae), dv. **13.** Palpal lobe of *Tramea* (Libellulidae), dv. **14.** Nymph of *Macromia* (Macromiidae), dv. dv, dorsal view; lv, lateral view; vv, ventral view. (1–5 redrawn from Hilsenhoff 1981; 6, 7 redrawn from Smith and Pritchard 1956; 8–11, 14 redrawn or modified from Pennak 1978; 12, 13 redrawn from Gloyd and Wright 1959.)

part of median caudal lamellae less than ⅓ as wide as long (except in *Amphiagrion*) 12

12a (11b). Posterolateral margin on each side of head angulate, sometimes with angle projecting and forming a blunt tubercle (Fig. 16) 13
12b. Posterolateral margin on each side of head broadly rounded, no blunt tubercle (Fig. 17) 14

13a (12a). Antennae 5- or 6-segmented; caudal lamellae each about ⅓ as broad as long, margins thickly set with setae from base to apex . . . . . *Amphiagrion*
13b. Antennae 7-segmented; caudal lamellae each not more than ⅙ as broad as long, margins with only a few widely separated setae . . . . . *Chromagrion*

14a (12b). Prementum with 1 or 2 dorsal setae on each side of median line, the second, when present, very small . . . . . *Nehalennia*
14b. Prementum with 2–7 well-developed dorsal setae on each side of median line (Fig. 5) 15

15a (14b). Antennae 6-segmented . . . . . *Enallagma*
15b. Antennae 7-segmented (fewer segments in young nymphs) 16

16a (15b). Caudal lamellae terminating in a blunt point (Fig. 18) . . . . . *Coenagrion*
16b. Caudal lamellae terminating in a sharp, tapered point (Fig. 19) . . . . . *Anomalagrion, Ischnura*

17a (6a). **Gomphidae**: Naked antennal segment 4 generally about ¼ as long as hairy segment 3 (Fig. 20); wingpads divergent from longitudinal body axis; mesothoracic legs closer together at base than prothoracic legs (ventral view) . . . . . *Progomphus*
17b. Segment 4 of antennae vestigial or nearly so (Figs. 21–23); wingpads divergent or parallel to body axis; mesothoracic legs not closer together at base than prothoracic legs 18

18a (17b). Wingpads strongly divergent . . . . . *Ophiogomphus*
18b. Wingpads laid parallel along back 19

19a (18b). Body very flat; abdomen nearly circular in dorsal view; paired tubercles on top of head . . . . . *Hagenius*
19b. Abdomen more nearly cylindrical; no tubercles on head 20

20a (19b). Antennal segment 3 flattened and oval, nearly as wide as long (but may be twice as long as wide) (Figs. 22–24) 21
20b. Long antennal segment 3 more or less cylindrical (Fig. 21) 22

21a (20a). Antennal segment 3 widest proximally, mesal borders straight and contiguous in living larvae (may have separated in preserved specimens), with papilliform setae decreasing in length distally, and with a few short scattered setae on edges (Figs. 23, 24); antennal segment 4 flattened and appears embedded in segment 3 (Fig. 23); anterior margin of prementum (ligula) slightly convex . . . . *Stylogomphus*

21b. Antennal segment 3 widest near middle, about 2 times as long as wide and rounded apically (Fig. 22); mesal borders convex and not contiguous and with many long hairs, especially along lateral margins giving segment a very hairy appearance (Fig. 22); antennal segment 4 spherical and not appearing embedded in segment 3 (Fig. 22); ligula convex . . . . *Lanthus*

22a (20b). Middorsal spines on abdominal segments 2–9; distinct but short middorsal carinae on segments 7 and 8, and a longer one on 9, each terminating in a spine (Fig. 25) . . . . *Dromogomphus*

22b. Middorsal spines not present on abdominal segments 2–9, or, if present, the ones on segments 7–9 are not terminations of raised dorsal carinae 23

23a (22b). Abdominal segment 10 longer than wide; end hook on palpal lobe small, about size of palpal teeth (Fig. 27) . . . . *Arigomphus*

23b. Abdominal segment 10 shorter than wide (Fig. 28); end hook on palpal lobe long, strong, curved inward (Fig. 26) 24

24a (23b). Tibial burrowing hooks well developed; palpal lobe of labium with 4–10 regular teeth on inner margin following the longer, curved end hook (Fig. 26) . . . . *Gomphus*

24b. Tibial burrowing hooks absent or vestigial; palpal lobe of labium with 1–4 shallow teeth on inner margin following the longer, curved end hook . . . . *Stylurus*

25a (7b). **Aeshnidae**: Posterolateral margin on each side of head angulate (Fig. 29); lateral spines present on abdominal segments 5–9 26

25b. Posterolateral margin on each side of head rounded (Fig. 30); lateral spines present on abdominal segments 6–9 or 7–9 (in *Aeshna eremita* the hind angles of the head are slightly angulate, and lateral spines are present on abdominal segments 5–9) 29

26a (25a). Blade of palpal lobe of labium wide and squarely truncate on outer end (Fig. 31); tips of paraprocts curved inward; a moundlike protuberance on each side of mesothorax at about midheight . . . . *Boyeria*

26b. Blade of palpal lobe of labium narrowed toward tip (Fig. 32); tips of paraprocts straight 27

27a (26b). Dorsum of abdomen broadly rounded in cross section; epiproct about ⅔ the length of paraprocts . . . . *Basiaeschna*

27b. Dorsum of abdomen with a low median ridge; epiproct about the same length as paraprocts 28

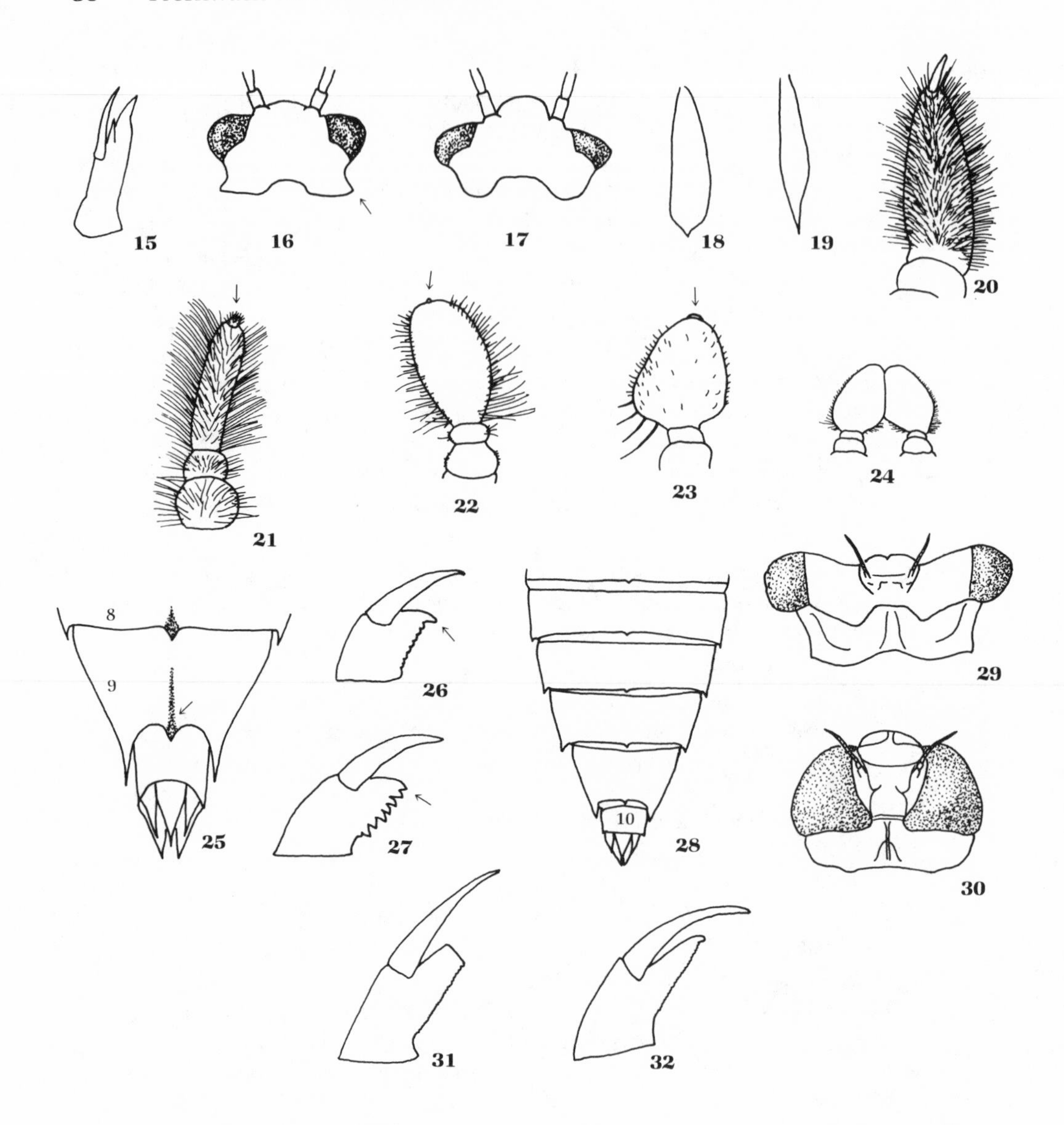

15. Palpal lobe of *Argia* (Coenagrionidae), dv. **16.** Head of *Amphiagrion* (Coenagrionidae), dv. **17.** Head of *Enallagma* (Coenagrionidae), dv. **18.** Lateral caudal lamella of *Coenagrion* (Coenagrionidae), lv. **19.** Lateral caudal lamella of *Ischnura* (Coenagrionidae), lv. **20.** Antenna of *Progomphus* (Gomphidae), dv. **21.** Antenna of *Gomphus* (Gomphidae), dv. **22.** Antenna of *Lanthus* (Gomphidae), dv. **23.** Antenna of *Stylogomphus* (Gomphidae), dv. **24.** Antennae of *Stylogomphus* (Gomphidae), dv. **25.** Tip of abdomen of *Dromogomphus* (Gomphidae), dv. **26.** Palpal lobe of *Gomphus* (Gomphidae), dv. **27.** Palpal lobe of *Arigomphus* (Gomphidae), dv. **28.** Abdomen of *Gomphus* (Gomphidae), dv. **29.** Head of *Basiaeschna, Boyeria, Nasiaeschna* (Aeshnidae), dv. **30.** Head of *Anax* (Aeshnidae), dv. **31.** Palpal lobe of *Boyeria* (Aeshnidae), dv. **32.** Palpal lobe of *Basiaeschna* (Aeshnidae), dv. dv, dorsal view; lv, lateral view. (14, 29, 30 modified from Pennak 1978; 15–19, 21, 23, 26, 27, 31, 32 redrawn from Hilsenhoff 1981; 20, 25 modified from Gloyd and Wright 1959; 22 modified from Carle 1980; 24 and 28 modified from Walker 1958.)

28a (27b). Blunt dorsal hooks on median ridge of abdominal segments 7–9 (Fig. 33); cerci each less than ½ as long as epiproct .... *Nasiaeschna*
28b. No dorsal hooks on median ridge; cerci ¾ length of epiproct .... *Epiaeschna*

29a (25b). Lateral spines present on abdominal segments 6–9 .... *Aeshna* (in part)
29b. Lateral spines present on abdominal segments 7–9 only (rarely, an extremely small one on segment 6) 30

30a (29b). Antenna longer than distance from its base to posterior margin of head .... *Gomphaeschna*
30b. Antenna shorter than distance from its base to posterior margin of head 31

31a (30b). End hook on truncate blade of palpal lobe prominent (Fig. 34) .... *Anax*
31b. End hook not prominent (Fig. 35) .... *Aeshna* (in part)

32a (9a). **Macromiidae**: Labium with 6 palpal setae, 5–6 prominent premental setae, and 3–4 less conspicuous premental setae; lateral spines of abdominal segment 9 short (Fig. 36) .... *Macromia*
32b. Labium with 5 palpal setae, 5 prominent premental setae, and 1–2 less conspicuous premental setae; lateral spines of abdominal segment 9 long, reaching or going beyond tips of abdominal appendages .... *Didymops*

33a (9b). **Libellulidae** and **Corduliidae**: Abdomen with a middorsal hook, spine, or knob on at least segments 6 or 7 34
33b. Abdomen without a middorsal hook, spine, or knob on segments 6 and 7 47

34a (33a). A middorsal hook, spine, or knob on abdominal segment 9 35
34b. No middorsal hook, spine, or knob on abdominal segment 9 41

35a (34a). Lateral spines of abdominal segment 9 reaching almost to tip of epiproct or beyond .... **Corduliidae** (in part) 37
35b. Lateral spines of abdominal segment 9 not reaching beyond midlength of epiproct, usually only to its base 36

36a (35b). Each cercus about as long as epiproct; 6–8 palpal setae .... **Corduliidae**, *Somatochlora* (in part)
36b. Each cercus about half as long as epiproct; 5 palpal setae (Fig. 37) .... **Libellulidae**, *Perithemis*

37a (35a). **Corduliidae** (in part): No lateral spines on segment 8 .... *Williamsonia*
37b. Distinct lateral spines on segment 8 38

38a (37b). Middorsal hooks knoblike, with apices blunt and rounded (Fig. 38); crenulations on distal margin of palpal lobe very deep, each crenula 2 or more times as deep as wide (Fig. 39) .... *Neurocordulia*

38b. Middorsal hooks spinelike, with apices acuminate (Fig. 40); crenulations on distal margin of palpal lobe relatively shallow, each crenula at most as deep as wide (Figs. 41–43) 39

39a (38b). Middorsal hooks absent on abdominal segments 1–5 or 1–6 . . . *Helocordulia*
39b. Middorsal hooks present on abdominal segments 2–9 or 3–9 40

40a (39b). Distal half of dorsal surface of prementum heavily setose (Fig. 44); 4–5 palpal setae . . . . *Epicordulia*
40b. Distal half of dorsal surface of prementum with few, or usually no, setae; 6–8 palpal setae . . . . *Tetragoneuria*

41a (34b). Each cercus ⅔ to equal length of paraprocts; 7 setae on palpal lobe of labium . . . . **Corduliidae**, *Dorocordulia*
41b. Each cercus less than ⅔ length of paraprocts . . . . . . **Libellulidae** (in part) 42

42a (41b). **Libellulidae** (in part): 0–3 premental setae, all fine or inconspicuous . . . . *Ladona*
42b. 7–21 premental setae, all prominent 43

43a (42b). Lateral spines of abdominal segment 9 long and straight, reaching to or beyond tips of paraprocts, and about twice middorsal length of segment 9; no middorsal hook on segment 8 . . . . *Celithemis*
43b. Lateral spines of abdominal segment 9 less than twice middorsal length of that segment; middorsal hook present or absent on segment 8 44

44a (43b). Eyes small, projecting forward from anterolateral margins of head, and less than ½ length of head (excluding labrum and clypeus) (Fig. 45); body with numerous long hairs 45
44b. Eyes larger and more lateral, occupying ½ or more than ½ length of head (excluding labrum and clypeus) (Fig. 46); body with only scattered long hairs 46

45a (44a). Abdominal segments 7–9 with brown or black, shining middorsal ridges; width of head across eyes less than 1.25 times the width of prothorax across dorsolateral ridges; ligula crenulate (Fig. 47) . . . . *Plathemis*
45b. Abdominal segments 7–9 without dark middorsal ridges; width of head across eyes more than 1.25 times the width of prothorax across dorsolateral ridges; ligula evenly contoured, not obviously crenulate . . . . *Libellula* (in part)

46a (44b). Middorsal hook present on segment 3; epiproct and paraprocts about equal in length; dark markings usually present on abdominal sterna . . . . *Leucorrhinia* (in part)
46b. Middorsal hook absent on segment 3; epiproct usually noticeably shorter than paraprocts; no dark markings on abdominal sterna . . . . *Sympetrum* (in part)

47a (33b). Apical third of cerci and paraprocts strongly bowed or curved downward; no lateral spines on abdomen; 7–9 palpal setae .... **Libellulidae**, *Erythemis*

47b. Apical third of cerci and paraprocts straight or only slightly bowed or curved downward; lateral spines may or may not be present on abdomen; number of palpal setae varies 48

48a (47b). Lateral spines prominent on abdominal segment 8, at least as long as middorsal length of that segment .... **Libellulidae** (in part) 49

48b. Lateral spines on abdominal segment 8 absent, or smaller than middorsal length of that segment 50

49a (48a). **Libellulidae** (in part): Epiproct as long as or longer than paraprocts (Fig. 48) .... *Pantala*

49b. Epiproct shorter than paraprocts (Fig. 49) .... *Tramea*

50a (48b). Tips of lateral spines of abdominal segment 9 extending beyond tip of epiproct .... **Libellulidae**, *Pachydiplax*

50b. Tips of lateral spines of abdominal segment 9 not extending beyond tip of epiproct 51

51a (50b). Each cercus not more than ½ length of paraprocts; if eyes large and on lateral margins of head (Fig. 46), then lateral spines on abdominal segment 8 at least ½ the middorsal length of that segment ...... **Libellulidae** (in part) 52

51b. Each cercus more than ⅔ length of paraprocts (except *Sympetrum*, where cerci can be less than ½ length of paraprocts and eyes are large and on lateral margins of head [Fig. 46], but lateral spines on abdominal segment 8 are less than ⅓ middorsal length of that segment) 53

52a (51a). **Libellulidae** (in part): Head with eyes small and on anterolateral corners of head (Fig. 45); lateral spines of abdominal segments 8 and 9 subequal in length; body very hairy; abdominal sterna without dark markings .... *Libellula* (in part)

52b. Head with eyes larger and more laterally located (Fig. 46); lateral spines of abdominal segment 9 longer than those of segment 8; body either less hairy or smooth; dark markings on abdominal sterna .... *Leucorrhinia* (in part)

53a (51b). Abdominal segments 4–9 with a median, dorsal bump bearing a tuft of thick setae; brackish water .... **Libellulidae**, *Erythrodiplax*

53b. Abdominal segments without median bump bearing setae, fresh water ... 54

54a (53b). Paraprocts subequal to epiproct; nymph small, short, < 11 mm; crenulations of palpal lobes shallow (Figs. 41, 43), less than ¼ as deep as broad .... **Libellulidae**, *Nannothemis*

54b. Paraprocts longer than epiproct; nymphs not necessarily small and short; crenulations of palpal lobes usually deep, each crenulation ⅓ to ½ as deep as broad (Fig. 42) (except *Sympetrum*) 55

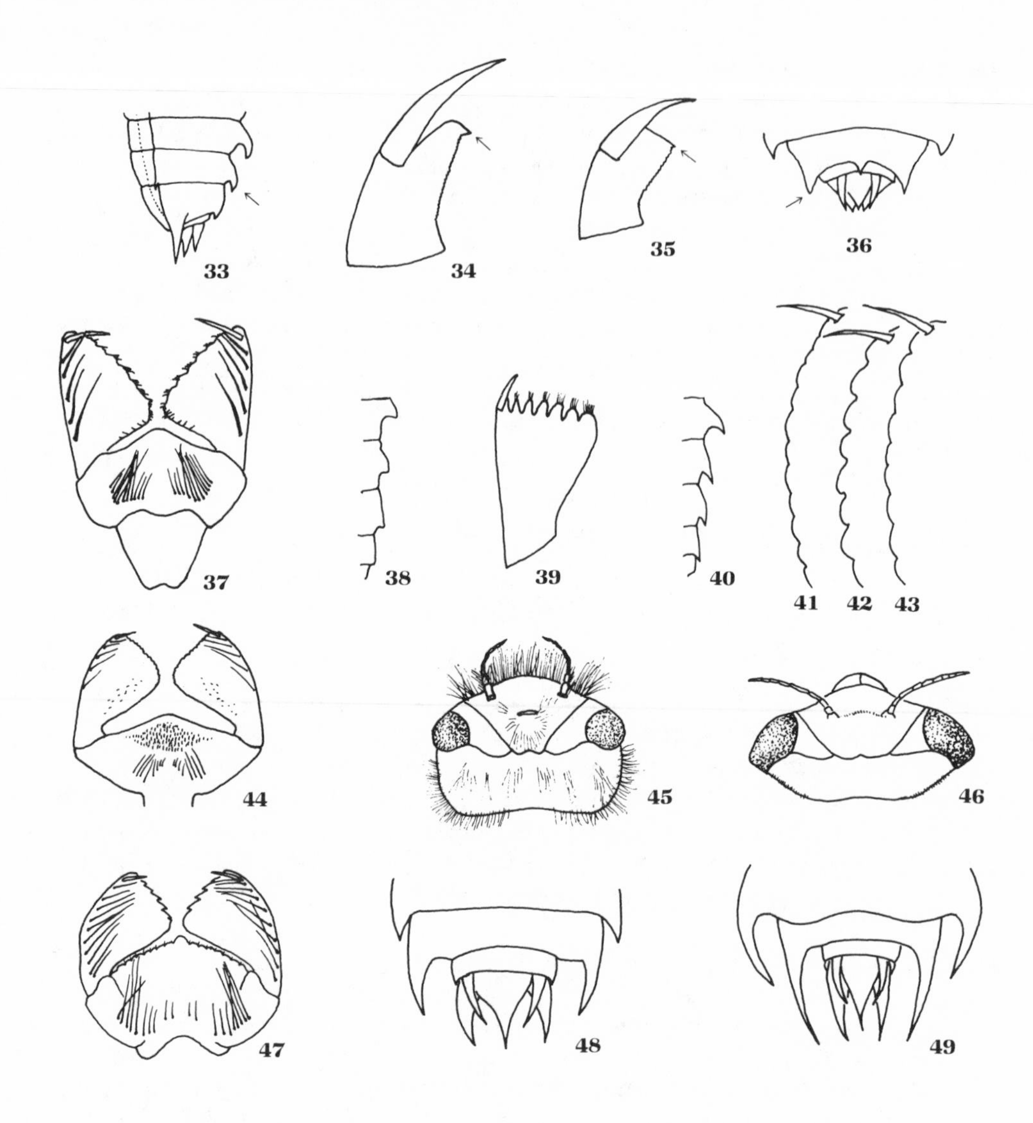

33. Terminal segments of Anisoptera abdomen, lv. **34.** Palpal lobe of *Anax* (Aeshnidae), dv. **35.** Palpal lobe of *Aeshna* (Aeshnidae), dv. **36.** Terminal segments of *Macromia* (Macromiidae), dv. **37.** Prementum of *Perithemis* (Libellulidae), dv. **38.** Middorsal hooks of *Neurocordulia* (Corduliidae), lv. **39.** Palpal lobe of *Neurocordulia* (Corduliidae), dv. **40.** Middorsal hooks of *Tetragoneuria* (Corduliidae), lv. **41.** Distal margin of palpal lobe of *Libellula* (Libellulidae), dv. **42.** Distal margin of palpal lobe of *Dorocordulia* (Corduliidae), dv. **43.** Distal margin of palpal lobe of *Helocordulia* (Corduliidae), dv. **44.** Prementum of *Epicordulia* (Corduliidae), dv. **45.** Head of *Plathemis* (Libellulidae), dv. **46.** Head of *Leucorrhinia* (Libellulidae), dv. **47.** Prementum of *Plathemis* (Libellulidae), dv. **48.** Terminal segments of *Pantala* (Libellulidae), dv. **49.** Terminal segments of *Tramea* (Libellulidae), dv. dv, dorsal view; lv, lateral view. (33, 44 redrawn from Gloyd and Wright 1959; 34, 35, 38–40 modified or redrawn from Hilsenhoff 1981; 36, 37, 41–43, 46–49 redrawn from Pennak 1978; 45 redrawn from Wright and Peterson 1944.)

55a (54b). A dark longitudinal stripe present along the dorsolateral margin of the thorax; 7 palpal setae; crenulations of palpal lobes of medium depth .... **Corduliidae**, *Cordulia*

55b. No dark stripe present on thorax; usually more than 7 palpal setae (may be 7 in *Somatochlora*, then crenulations very deep); crenulations of palpal lobes either very deep or very shallow, not intermediate 56

56a (55b). Crenulations of palpal lobes shallow (Figs. 41, 43); cerci each ⅔ length of epiproct or slightly less; 9 or more palpal setae .... **Libellulidae**, *Sympetrum* (in part)

56b. Crenulations of palpal lobes deep (Fig. 42); cerci each subequal to length of epiproct; 7–8 palpal setae .... **Corduliidae**, *Somatochlora* (in part)

References on Odonata Systematics

(*Used in construction of key.)

Alcock, J. 1982. Post-copulatory mate guarding by males of the damselfly *Hetaerina vulnerata* Selys (Odonata: Calopterygidae). Anim. Behav. 30:99–107.

Baker, R.L. 1986. Developmental stages and the analysis of Zygopteran life histories. J. Freshwat. Ecol. 3:325–332.

Borror, D.J. 1942. A revision of the libelluline genus *Erythrodiplax*. Ohio State Univ. Grad. Sch. Stud., Biol. Ser., Contrib. Zool. Entomol. 4:1–296.

——. 1945. A key to the new world genera of Libellulidae. Ann. Entomol. Soc. Amer. 38:168–194.

Byers, C.F. 1927a. Key to North American species of *Enallagma*, with description of new species (Odonata: Zygoptera). Trans. Amer. Entomol. Soc. 53:249–260.

——. 1927b. The nymph of *Libellula incesta* and a key for the separation of the known nymphs of the genus *Libellula*. Entomol. News 38:113–115.

——. 1939. A study of the dragonflies of the genus *Progomphus* (Gomphoidea) with a description of a new species. Proc. Fla. Acad. Sci. 4:19–86.

Carle, F.L. 1979. Two new *Gomphus* (Odonata: Gomphidae) from eastern North America with adult keys to the subgenus *Hylogomphus*. Ann. Entomol. Soc. Amer. 72:418–426.

*——. 1980. A new *Lanthus* (Odonata: Gomphidae) from eastern North America with adult and nymphal keys to the American octogomphines. Ann. Entomol. Soc. Amer. 73:172–179.

——. 1982. The wing vein homologies and phylogeny of the Odonata: a continuing debate. Soc. Int. Odon. Rapid Commun. 4:1–66.

——. 1983. A new *Zoraena* (Odonata: Cordulegastridae) from eastern North America, with a key to the adult Cordulegastridae of America. Ann. Entomol. Soc. Amer. 76:61–68.

——. 1986. The classification, phylogeny and biogeography of the Gomphidae (Anisoptera). I. Classification. Odonatologica 15:275–326.

Chelmick, D., C. Hammond, N. Moore, and A. Stubbs. 1980. The conservation of dragonflies. Nature Conservancy Council, London. 23 pp.

Corbet, P.S. 1963. A biology of dragonflies. Quadrangle Books, Chicago, Ill. 247 pp.

——. 1980. Biology of Odonata. Annu. Rev. Entomol. 25:189–218.

Corbet, P.S., C. Longfield, and N.W. Moore. 1960. Dragonflies. Collins, London. 260 pp.

Davies, D.A.L. 1981. A synopsis of the extant genera of the Odonata. Soc. Int. Odon. Rapid Commun. no. 3, pp. 1–59.

Davies, D.A.L., and P. Tobin. 1984. The dragonflies of the world: a systematic list of the extant species of Odonata. Vol. 1: Zygoptera, Anisozygoptera. Soc. Int. Odon. Rapid Commun. Suppl. no. 3, pp. 1–127.

Dunkle, S.W. 1977. Larvae of the genus *Gomphaeschna* (Odonata: Aeshnidae). Fla. Entomol. 60:223–225.

Garman, P. 1917. The Zygoptera, or damselflies, of Illinois. Bull. Ill. State Lab. Nat. Hist. 12:411–587.

——. 1927. The Odonata or dragonflies of Connecticut. Guide to the insects of Connecticut. Part V. Bull. Conn. Geol. Nat. Hist. Surv. 39:1–331.

Garrison, R.W. 1981. Description of the larva of *Ischnura gemina* with a key and new characters for separation of sympatric *Ischnura* larvae. Ann. Entomol. Soc. Amer. 74:525–530.

Gibbs, R.H., and S. Preble. 1954. The Odonata of Cape Cod, Massachusetts. J. N.Y. Entomol. Soc. 62:167–184.

Gloyd, L.K. 1973. The status of the generic names *Gomphoides*, *Neogomphoides*, *Progomphus*, and *Ammogomphus* (Odonata: Gomphidae). Occas. Pap. Univ. Mich. Mus. Zool. 668:1–7.

Gloyd, L.K., and M. Wright. 1959. Odonata. *In* W.T. Edmondson (ed.). Fresh-water biology, 2nd ed., pp. 917–940. John Wiley and Sons, New York.

*Hilsenhoff, W.L. 1981. Aquatic insects of Wisconsin. Nat. Hist. Counc. Publ. no. 2. Univ. Wisc., Madison. 60 pp.

House, L.S., III. 1982. The dragonflies and damselflies (Odonata: Anisoptera and Zygoptera) of Otsego County, N.Y. Occas. Pap. no. 10. Biol. Dep. Univ. State N.Y., Oneonta. 135 pp.

Huggins, D.G., and W.U. Brigham. 1982. Odonata. *In* A.R. Brigham, W.U. Brigham, and A. Gnilka (eds.). Aquatic insects and oligochaetes of North and South Carolina, pp. 4.1–4.100. Midwest Aquatic Enterprises, Mahomet, Ill.

Hutchinson, G.E. 1981. Thoughts on aquatic insects. BioScience 31:495–500.

Kiauta, M., and B. Kiauta. 1981. Aspirator for collecting of small zygopterans close above water surface and among dense vegetation. Notul. Odonatol. 1:133–134.

Kormandy, E.J. 1958. Catalogue of the Odonata of Michigan. Misc. Publ. Univ. Mich. Mus. Zool. 104:1–43.

——. 1959a. Review: A reclassification of the order Odonata, by F.C. Fraser. 1957. Entomol. News 70:165–167.

——. 1959b. The systematics of *Tetragoneuria*, based on ecological, life history, and morphological evidence (Odonata: Corduliidae). Misc. Publ. Univ. Mich. Mus. Zool. 107:1–79.

Krull, W.H. 1929. The rearing of dragonflies from eggs. Ann. Entomol. Soc. Amer. 22:651–658.

Levine, H.R. 1957. Anatomy and taxonomy of the mature naiads of the dragonfly genus *Plathemis* (Family Libellulidae). Smithson. Misc. Coll. 134:1–28.

Lyon, M.B. 1915. The ecology of dragonfly nymphs of Cascadilla Creek (Odonata). Entomol. News. 26:1–15.

May, M.L. 1982. Heat exchange and endothermy in Protodonata. Evolution 36:1051–1058.

*McCafferty, W.P. 1981. Aquatic entomology. The fishermen's and ecologists' illustrated guide to insects and their relatives. Science Books International, Boston. 448 pp.

Miller, P.L. 1987. Dragonflies. Naturalists' handbook 7. Cambridge Univ. Press. Cambridge. 84 pp.

Needham, J.G. 1897. Preliminary studies of North American Gomphinae. Can. Entomol. 29:164–186.

——. 1903a. A genealogical study of dragonfly wing venation. Proc. U.S. Natl. Mus. 26:703–764.

——. 1903b. Life histories of Odonata. Suborder Zygoptera. Damsel flies. Bull. N.Y. State Mus. 68:218–279.

——. 1941. Life history studies on *Progomphus* and its nearest allies (Odonata: Aeschnidae). Trans. Amer. Entomol. Soc. 67:221–245.

——. 1948. Studies on the North American species of the genus *Gomphus* (Odonata). Trans. Amer. Entomol. Soc. 73:307–339.

Needham, J.G., and E. Fisher. 1936. The nymphs of North American libelluline dragonflies. Trans. Amer. Entomol. Soc. 62:107–116.

Needham, J.G., and H.B. Heywood. 1929. A handbook of the dragonflies of North America. Charles C Thomas, Springfield, Ill. 378 pp.

*Needham, J.G., and M.J. Westfall, Jr. 1955. A manual of the dragonflies of North America (Anisoptera). Univ. Calif. Press, Los Angeles. 615 pp.

Peckarsky, B.L. 1984. Predator-prey interactions among aquatic insects. *In* V.H. Resh and D.M. Rosenberg (eds.). The ecology of aquatic insects, pp. 198–254. Praeger, New York.

Pellerin, P., and J.G. Pilon. 1975. The life cycle of *Lestes eurinus* Say (Odonata: Lestidae), a rearing technique used in a conditioned environment. Nat. Can. 102:643–652.

*Pennak, R.W. 1978. Freshwater invertebrates of the United States, 2nd ed. John Wiley and Sons, New York. 803 pp.

Ris, F. 1930. A revision of the libelluline genus *Perithemis* (Odonata). Misc. Publ. Univ. Mich. Mus. Zool. 21:1–50.

*Smith, R.F., and A.E. Pritchard. 1956. Odonata. *In* R.L. Usinger (ed.). Aquatic insects of California, pp. 106–153. Univ. Calif. Press, Berkeley.

Snodgrass, R.E. 1954. The dragonfly larva. Smithson. Misc. Coll. 123:1–38.

Utzeri, C., and R. Raffi. 1983. Observations on the behavior of *Aeshna affinis* (VanderLinden) at a dried-up pond (Anisoptera: Aeshnidae). Odonatologica 12:17–26.

Walker, E.M. 1912. The North American dragonflies of the genus *Aeshna*. Univ. Toronto Stud., Biol. Ser. 11:1–23.

——. 1925. The North American dragonflies of the genus *Somatochlora*. Univ. Toronto Stud., Biol. Ser. 26:1–202.

——. 1953. The Odonata of Canada and Alaska. Vol. 1, part 1: General; part 2: The Zygoptera—damselflies. Univ. Toronto Press, Toronto. 292 pp.

*——. 1958. The Odonata of Canada and Alaska. Vol. 2, part 3: The Anisoptera—four families. Univ. Toronto Press, Toronto. 318 pp.

*Walker, E.M., and P.S. Corbet. 1975. The Odonata of Canada and Alaska. Vol. 3, part 4: The Anisoptera—three families. Univ. Toronto Press, Toronto. 304 pp.

Westfall, M.J., Jr. 1984. Odonata. *In* R.W. Merritt and K.W. Cummins (eds.). An introduction to the aquatic insects of North America, 2nd ed., pp. 126–176. Kendall/Hunt Publ. Co., Dubuque, Iowa.

Williamson, E.B. 1909. The North American dragonflies (Odonata) of the genus *Macromia*. Proc. U.S. Natl. Mus. 37:369–398.

*Wright, M., and A. Peterson. 1944. A key to the genera of anisopterous dragonfly nymphs of the United States and Canada (Odonata, suborder Anisoptera). Ohio J. Sci. 44:151–166.

6 Plecoptera

Classification

The order Plecoptera (stoneflies) belongs to the infraclass Neoptera because stoneflies' wings fold over their backs at rest. Wings develop in external wingpads, a characteristic that places Plecoptera in the division Exopterygota. North American stoneflies are generally divided into two groups, Euholognatha and Systellognatha, based on major differences in mouthpart morphology and, hence, feeding biology. The taxonomy of this order, like that of the Ephemeroptera, is poorly known because the larvae of many species have not been associated with adults. Authorities often disagree even on family designations. The classification scheme used here is that of Stewart and Stark (1988).

Life History

Females of this hemimetabolous order generally oviposit by dipping their abdomens into the water as they fly over the water surface. Eggs detach or drop into the water, become entrained in the current, and settle to the bottom. Some females oviposit while submerged (Hynes 1976), and a large flightless stonefly was observed to oviposit as it ran over the surface of the water (Jewett 1963). Stonefly eggs have adhesive disks or gelatinous coverings that enable them to attach to the substrate (Harper and Stewart 1984). One family of stoneflies (Capniidae) has ovoviviparous species, whose eggs complete development within the female and hatch immediately upon oviposition (Hynes 1976). All others develop after oviposition and hatch with or without diapause (Harper and Stewart 1984). A few reports of parthenogenetic species have been published (Harper 1973a). The first nymphal instar usually develops within the egg of plecopterans; thus, it is actually the second instar that hatches from the egg.

Stonefly nymphs undergo between 12 and 36 molts, depending on the species and the water temperature. A few species exhibit an interesting summer diapause (*aesti-*

vation) phase as early nymphal instars (Harper and Hynes 1970). Certain species of the families Taeniopterygidae, Nemouridae, and Capniidae avoid high water temperatures by burrowing into the interstices and entering diapause near the groundwater table. Stonefly nymphs are cold stenotherms, as a rule, and many undergo their major growth during colder seasons. Most species are univoltine. Some of the larger species may take two to four years to complete their development; for example, members of the family Perlidae often have representatives of two- or three-year classes coexisting at all times of the year.

As in the odonates, the last-instar stonefly nymphs crawl out of the water and rest on vertical objects or on the undersides of leaves to emerge. Adults live from one to four weeks. Species of two stonefly families (Taeniopterygidae and Capniidae) emerge on snow and ice in midwinter, and on sunny days they may be seen mating on the surface of the snow (Pennak 1978). Stonefly males, and in some species females, communicate with acoustical signaling by beating their abdomens on the surface of the substrate or vibrating their abdomens at high frequency (Harper and Stewart 1984). Drumming patterns are species-specific and vary with temperature. This behavior may act as a reproductive isolating mechanism for coexisting stonefly species.

Habitat

Plecopteran nymphs are restricted to cool, clean streams with high dissolved-oxygen content. Some species, however, may be found along the wave-swept shores of large oligotrophic lakes. Most stoneflies are classified as clingers or sprawlers, as they are closely associated with the substrate or leaf litter. A few species have been reported from the hyporheic zone (Stanford and Gaufin 1974, Harper and Stewart 1984).

Feeding

Generally, stonefly nymphs are either shredders or predators. Some groups that are predaceous as late instars have been reported to be herbivorous or detritivorous in early instars, while late instars of large detritivores (Pteronarcyidae) may consume some prey. Predators are engulfers, that is, they swallow their prey whole or bite off and swallow parts of prey. They are active search or pursuit predators, using their long filamentous antennae to locate prey using tactile, wave disturbance, and chemical cues (Martínez and Peckarsky, submitted; Peckarsky and Wilcox 1989). Many species are opportunistic feeders, consuming prey in proportion to their relative abundance. Other species are selective for prey species or sizes (Allan et al. 1987; Peckarsky and Penton 1989). In some families adults feed, and in others they do not (Hitchcock 1974).

Respiration

Nymphs of many species of stoneflies respire through tufts of filamentous tracheal gills. Others, however, do not have gills and rely entirely on cutaneous oxygen exchange. These species require highly oxygenated habitats, a fact that may explain why they are restricted to cold, swift streams (Hitchcock 1974, Pennak 1978). Perlidae and Perlodidae can often be observed in collecting pans of still water doing "push-ups" in response to oxygen stress.

Collection and Preservation

Stonefly nymphs can be collected by disturbing the substrate or leaf packs upstream of an aquatic D-frame net, by using a hand screen, or by hand collecting nymphs from the undersides of stones (Pennak 1978). The latter method is not as thorough as the former two because species that do not cling to the substrate will release their hold as stones are lifted from the stream. Preservation in 70% ethanol is adequate.

Identification

Figures A and B illustrate the general morphological features of nymphal stoneflies. Features commonly used to identify stonefly nymphs are presence and location of gills, shape of the wingpads, mouthpart morphology, and body patterning. The keys are most effective with late-instar nymphs, since many of the identifying features (e.g., shape of the wingpads) cannot be distinguished in early instars. We have attempted to include some characters that are valid even in early instars so that smaller specimens can be identified. The recent generic revisions accepted by Stewart and Stark (1988) are incorporated in this key.

More-Detailed Information

Jewett 1956, Brinck 1957, Hitchcock 1974, Hynes 1976, Baumann et al. 1977, Pennak 1978, Hilsenhoff 1981, Unzicker and McCaskill 1982, Harper and Stewart 1984, Stewart and Stark 1988.

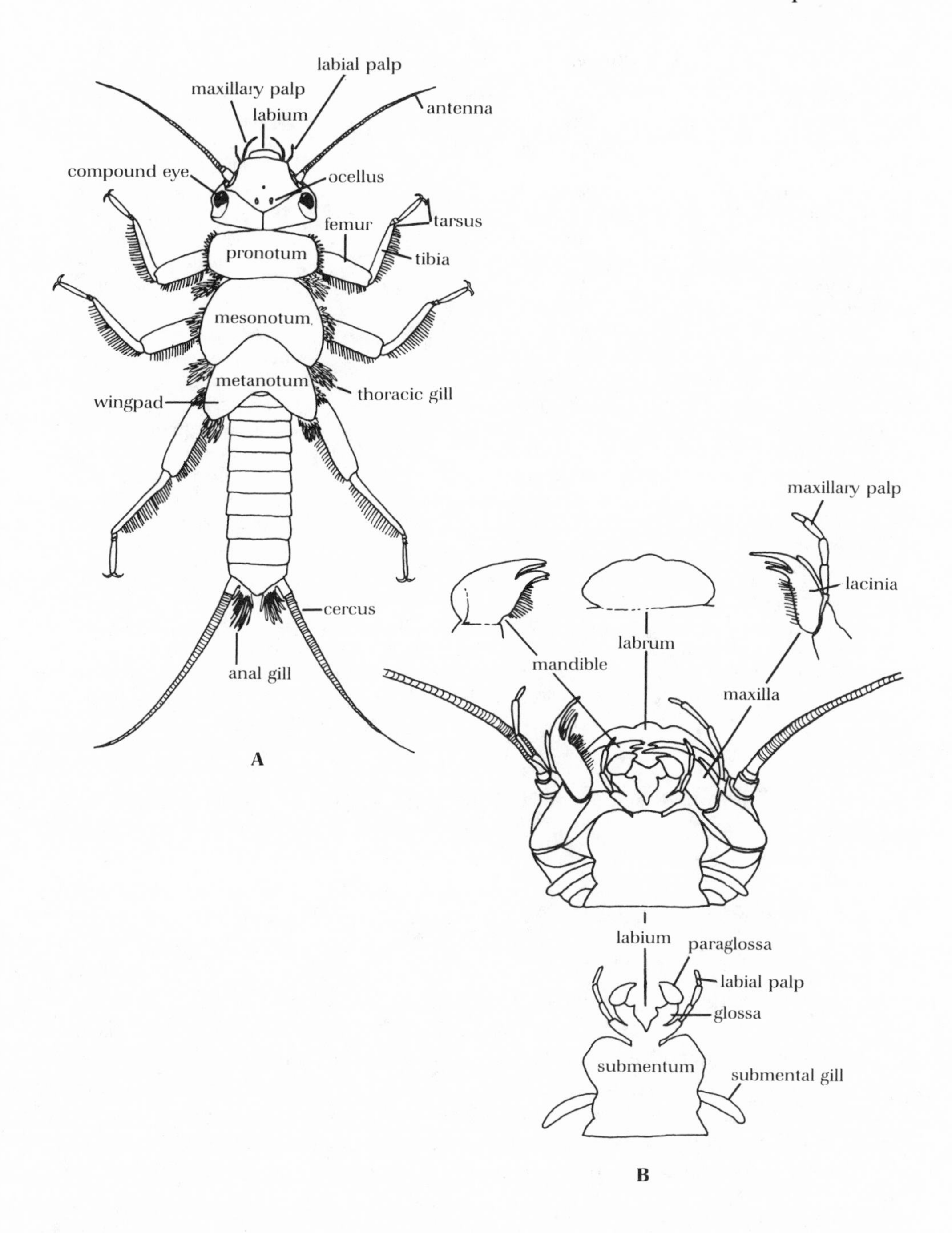

A. Generalized stonefly nymph, dorsal view. **B.** Head of a generalized stonefly nymph with detail of mouthparts, ventral view. (A redrawn from Baumann et al. 1977; B redrawn from Ward 1985.)

Checklist of Plecoptera Nymphs

Capniidae
- *Allocapnia*
- *Capnia*
- *Nemocapnia*
- *Paracapnia*
- *Utacapnia*

Chloroperlidae
- *Alloperla*
- *Haploperla*
- *Rasvena*
- *Suwallia*
- *Sweltsa*
- *Utaperla*

Leuctridae
- *Leuctra*
- *Paraleuctra*
- *Zealeuctra*

Nemouridae
- *Amphinemura*
- *Nemoura*
- *Ostrocerca*
- *Paranemoura*
- *Podmosta*
- *Prostoia*
- *Shipsa*
- *Soyedina*
- *Zapada*

Peltoperlidae
- *Peltoperla*[a]
- *Tallaperla*[a]

Perlidae
- *Acroneuria*
- *Agnetina*
- *Attaneuria*
- *Beloneuria*
- *Eccoptura*
- *Hansonoperla*
- *Neoperla*
- *Paragnetina*
- *Perlesta*
- *Perlinella*

Perlodidae
- *Arcynopteryx*
- *Clioperla*
- *Cultus*
- *Diploperla*
- *Diura*
- *Helopicus*
- *Hydroperla*
- *Isogenoides*
- *Isoperla*
- *Malirekus*
- *Remenus*

Pteronarcyidae
- *Pteronarcys*

Taeniopterygidae
- *Bolotoperla*
- *Oemopteryx*
- *Strophopteryx*
- *Taenionema*
- *Taeniopteryx*

[a]Stewart and Stark (1988) distinguish nymphs of these two genera on the basis of presence (*Peltoperla*) or absence (*Talloperla*) of solid dark pigment spots on the meso- and metanota. However, we follow Unzicker and McCaskill (1982), who suggest that nymphs of these genera cannot be reliably separated. Stark and Stewart (1981) separate some species of peltoperlids on the basis of configuration of mandibular teeth.

Key to Genera of Plecoptera Nymphs

1a. Finely branched gills present on sides and venter of all thoracic segments 2

1b. Gills absent, restricted to cervical or coxal area, or fingerlike, without numerous branches 3

2a (1a). Finely branched gill tufts present on abdominal sterna 1 and 2 .... **Pteronarcyidae**, *Pteronarcys*

2b. Finely branched gill tufts absent from abdominal sterna .... **Perlidae** 27

3a (1b). Thoracic sternal plates shieldlike, overlapping succeeding segment (Fig. 1); body roachlike; head bent under ...... **Peltoperlidae**, *Peltoperla, Tallaperla*

3b. Thoracic sterna not shieldlike or overlapping; body not roachlike; head not bent under 4

4a (3b). Tips of glossae extend nearly as far forward as tips of paraglossae (Fig. 2) 5

4b. Tips of glossae situated far behind tips of paraglossae (Fig. 3) 8

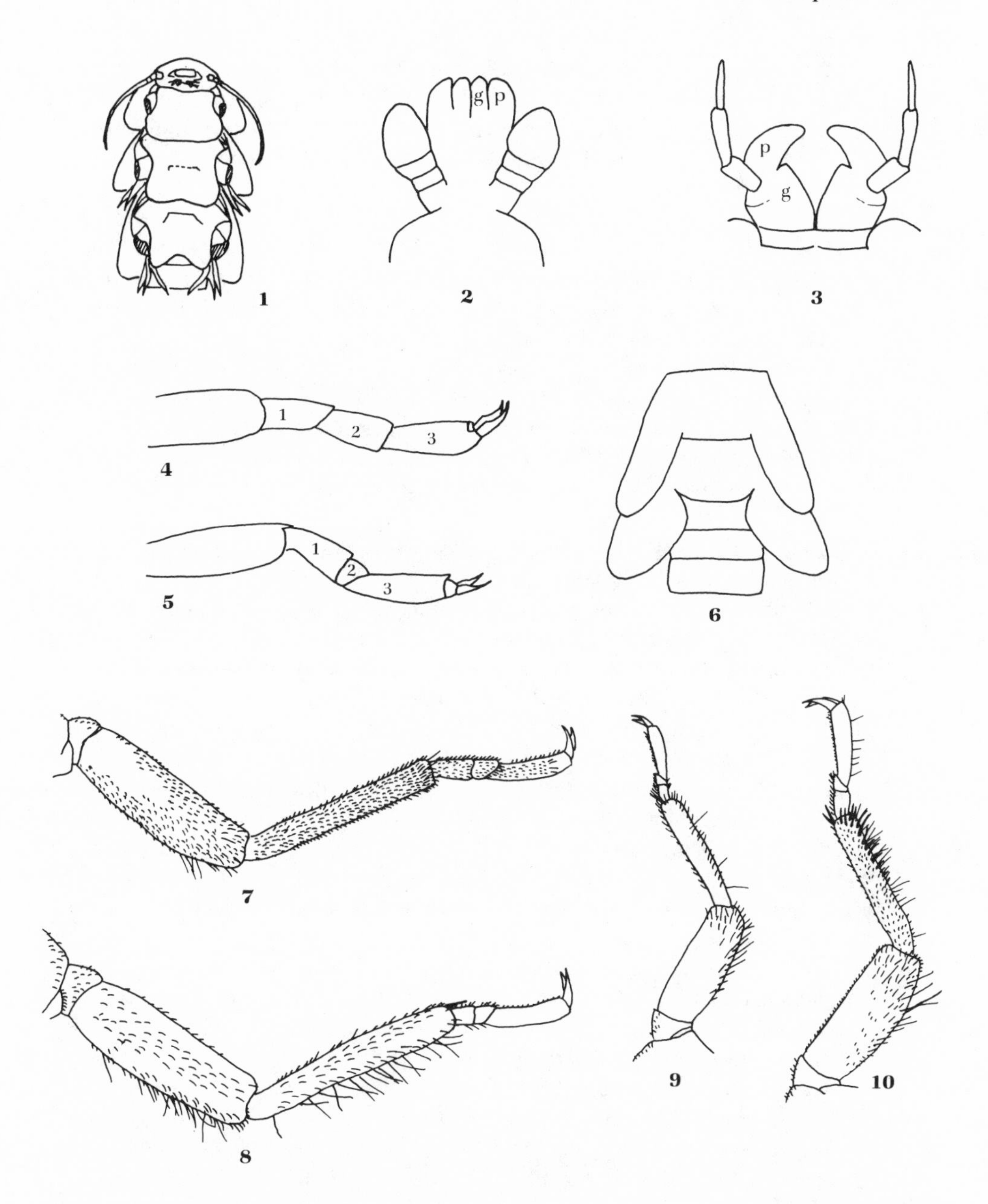

1. Thorax of *Peltoperla* (Peltoperlidae), vv. **2.** Labium of *Nemoura* (Nemouridae), vv: g, glossa; p, paraglossa. **3.** Labium of *Isoperla* (Perlodidae), vv: g, glossa; p, paraglossa. **4.** Tarsal segments (1,2,3) of *Taeniopteryx* (Taeniopterygidae), lv. **5.** Tarsal segments (1,2,3) of *Nemoura* (Nemouridae), lv. **6.** Wingpads of *Nemoura* (Nemouridae), dv. **7.** Foreleg of *Paranemoura* (Nemouridae), lv. **8.** Foreleg of *Shipsa* (Nemouridae), lv. **9.** Foreleg of *Ostrocerca* (Nemouridae), lv. **10.** Foreleg of *Prostoia* (Nemouridae), lv. dv, dorsal view; lv, lateral view; vv, ventral view. (1 modified from Pennak 1978; 2, 4–6 redrawn or modified from Hilsenhoff 1981; 3 redrawn from Frison 1942; 7–10 redrawn from Baumann 1975.)

5a (4a). Coxae with single, telescoping gills (Fig. 28), or abdomen with large, triangular, ventroapical plate (Figs. 29–31); 1st and 2nd tarsal segments subequal in length (Fig. 4) . . . . **Taeniopterygidae** 15

5b. Coxal gills and ventroapical plate absent; 2nd tarsal segment wedge-shaped and shorter than 1st (Fig. 5) 6

6a (5b). Metathoracic wingpads divergent from long axis of body in mature nymphs (Fig. 6); cervical gills sometimes present (Figs. 33, 34); body short; extended, hindlegs reach approximately to tip of abdomen; body usually hairy, with many long and short hairs and/or spines, especially on legs (Figs. 7–10) . . . . **Nemouridae** 19

6b. Metathoracic wingpads parallel to long axis of body in mature nymphs (Figs. 11–14); cervical gills absent; body slender and elongate; extended, hindlegs reach far short of tip of abdomen; body not very hairy, short hairs and spines few or lacking, long silky hairs only on legs 7

7a (6b). Abdomen cylindrical, segments ringlike in cross section; abdomen with middle segments only slightly wider than basal and terminal segments; lateral margins of abdominal segments essentially parallel; no more than segments 1–6 divided by a membranous fold ventrolaterally (Fig. 15), other segments a complete sclerotized ring; body usually lightly pigmented yellow, almost transparent at times, rarely darkly pigmented when mature . . . **Leuctridae** 9

7b. Abdomen with ventral side flattened, segments semicircular in cross section; abdomen with segments 5–7 distinctly wider than basal and terminal segments; lateral margins of abdominal segments rounded; segments 1–9 divided by a membranous fold ventrolaterally (Fig. 16); body usually highly pigmented brown . . . . **Capniidae** 11

8a (4b). Metathoracic wingpads strongly divergent from long axis of body in mature nymphs (Fig. 17); cerci as long as or longer than abdomen; thorax and head with a contrasting light and dark pattern in most mature specimens; wingpads usually lack setae (if wingpads are not developed, then metanotum 1½–2 times as wide as long, elliptical, posterior margins may or may not be sharply rounded and diverging); terminal segment of maxillary palps long and tapering, diameter of base of last segment not much thinner than diameter of next-to-last segment (Fig. 18) . . . . **Perlodidae** 36

8b. Metathoracic wingpads parallel to or weakly divergent from long axis of body (Figs. 19, 20) in mature nymphs; cerci shorter, usually $\leq$ ¾ length of abdomen; thorax and head usually without pigmented pattern; wingpads usually with many setae (if wingpads are not developed, then metanotum nearly as long as wide, not elliptical, lateral margins never sharply rounded and diverging, usually broadly rounded); terminal segment of maxillary palps small and fingerlike, diameter of base of last segment much thinner than diameter of next-to-last segment (Fig. 21) . . . . **Chloroperlidae** 46

9a (7a). **Leuctridae**: Labial palps extending well beyond paraglossae (Fig. 22); abdomen thickly bristled; subanal lobes of abdomen bearing many short and a

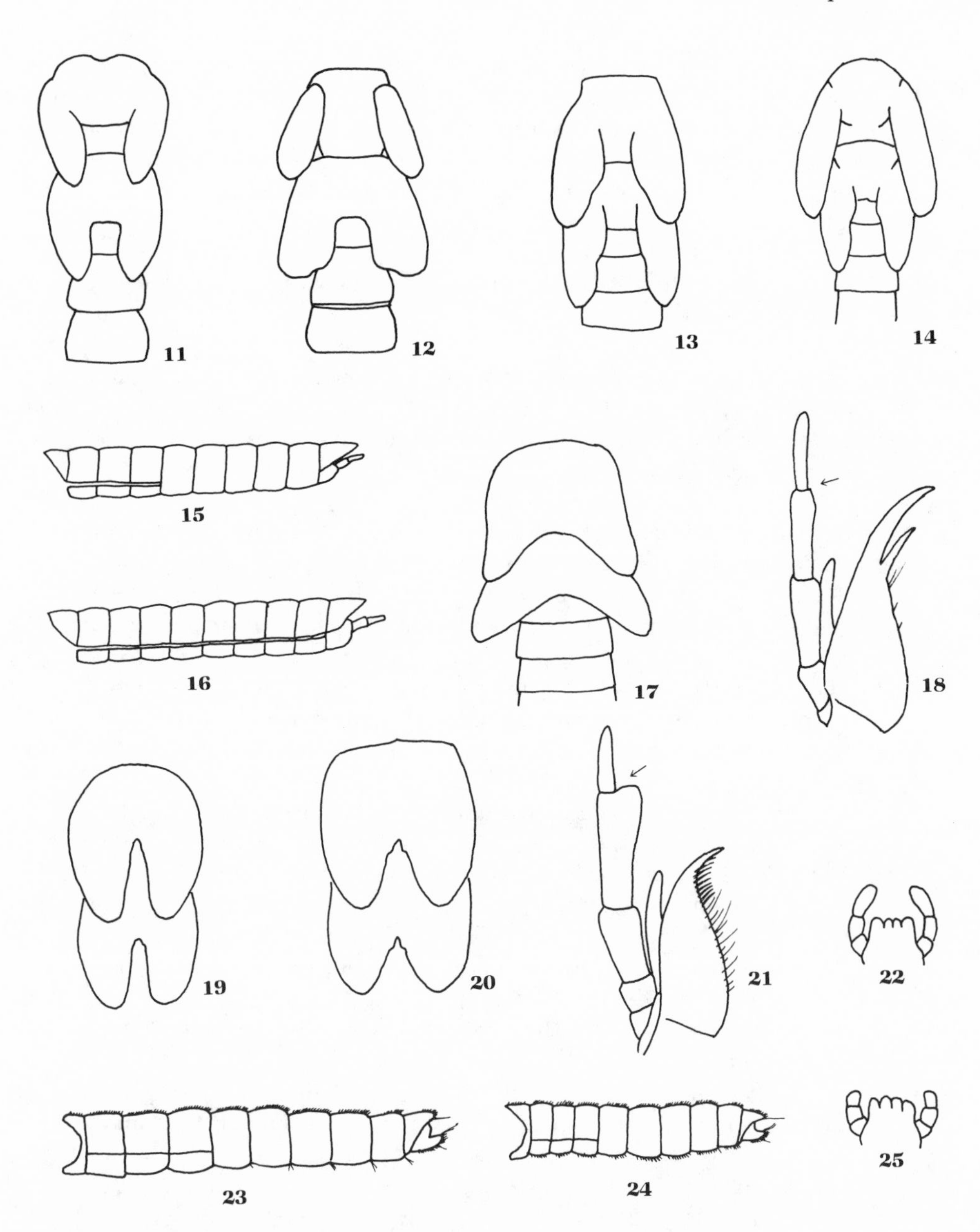

11. Wingpads of *Paracapnia* (Capniidae), dv. **12.** Wingpads of *Allocapnia* (Capniidae), dv. **13.** Wingpads of *Capnia* (Capniidae), dv. **14.** Wingpads of *Leuctra* (Leuctridae), female, dv. **15.** Abdomen of *Allocapnia* (Capniidae), lv. **16.** Abdomen of *Leuctra* (Leuctridae), lv. **17.** Wingpads of *Isoperla* (Perlodidae), dv. **18.** Maxilla of *Hydroperla* (Perlodidae), vv. **19.** Wingpads of *Haploperla* (Chloroperlidae), dv. **20.** Wingpads of *Alloperla* (Chloroperlidae), dv. **21.** Maxilla of *Alloperla* (Chloroperlidae), vv. **22.** Labium of *Leuctra* (Leuctridae), vv. **23.** Abdomen of *Leuctra sibleyi* (Leuctridae), lv. **24.** Abdomen of *Leuctra duplicata* (Leuctridae), lv. **25.** Labium of *Paraleuctra* (Leuctridae), vv. dv, dorsal view; lv, lateral view; vv, ventral view. (11–13, 15–17, 19 redrawn or modified from Hilsenhoff 1981; 14 modified from Frison 1935; 18, 21 redrawn from Frison 1942; 20 modified from Hilsenhoff 1975; 22–25 redrawn from Harper and Hynes 1971a.)

couple of long bristles (Figs. 23, 24); only 1st 4 abdominal segments divided by ventrolateral fold (Fig. 15) . . . . *Leuctra*

9b. Labial palps extending to approximate tip of paraglossae (Fig. 25); abdomen sparsely bristled; subanal lobes of abdomen sometimes bearing short bristles, but no long bristles; first 6 abdominal segments divided by ventrolateral fold 10

10a (9b). Body covered with distinct short hairs and/or bristles . . . . *Paraleuctra*

10b. Body nearly naked . . . . *Zealeuctra*

11a (7b). **Capniidae**: Distal ½ of cerci with a well-developed vertical fringe consisting of several long bristles from each segment (Fig. 26) . . . . *Nemocapnia*

11b. Distal ½ of cerci without a well-developed fringe; bristles principally at joints of segments (Fig. 27) 12

12a (11b). Numerous conspicuous bristles, mostly along posterior margins of abdominal terga and bordering pronotum and wingpads, approximately ½ as long as middorsal length of an abdominal tergum; head with a dorsal, purple-brown, reticulate pattern; meso- and metathoracic wingpads rounded (Fig. 11) . . . . *Paracapnia*

12b. Abdominal bristles, if conspicuous, not more than ⅓ middorsal length of an abdominal tergum; head without a dorsal purple-brown, reticulate pattern; metathoracic wingpads truncate (Fig. 12) or rounded (Fig. 13) 13

13a (12b). Metathoracic wingpads usually truncate and notched near tip (Fig. 12) . . . . *Allocapnia*

13b. Metathoracic wingpads similar to those on mesothorax and notched on inner margin halfway to tip (Fig. 13) 14

14a (13b). Femur and tibia with > 50 fine short hairs, femur with a few long dorsal fringe hairs . . . . *Utacapnia*

14b. Femur and tibia with < 20 short hairs, femur usually with no long dorsal fringe hairs . . . . *Capnia*

15a (5a). **Taeniopterygidae**: Single gill present on inner side of each coxa, may be small in early instars (Fig. 28) . . . . *Taeniopteryx*

15b. No gills on coxae 16

16a (15b). Proximal 8–10 cercal segments with short, fine dorsal hair fringe (Fig. 29); cercal and antennal segments distinctly ringed with narrow brown pigment bands at joints . . . . *Bolotoperla*

16b. Proximal cercal segments without dorsal hair fringe, or a single long hair present on each segment (Fig. 30); cercal and antennal segments may or may not be banded 17

17a (16b). Proximal cercal segments each with a single, long, fine dorsal hair (Fig. 30) . . . . *Oemopteryx*

17b. Proximal cercal segments without long dorsal hairs (Fig. 31), but may have short bristles dorsally 18

18a (17b). Body dark brown, pattern indistinct; legs uniformly dark brown . . . . *Taenionema*

18b. Body light brown or yellow, with distinct darker pattern on head and thorax (Fig. 32), abdomen distinctly banded; legs light brown, femur darker brown apically . . . . *Strophopteryx*

19a (6a). **Nemouridae**: Gills present on prosternum near cervical region (Figs. 33, 34) 20

19b. No gills on prosternum 21

20a (19a). Gills simple, 2 on each side of midline (Fig. 33) . . . . *Zapada*

20b. Gills highly branched (Fig. 34) . . . . *Amphinemura*

21a (19b). Pronotum with a distinct, single lateral fringe of spines of equal length, sometimes longer on posterolateral margins (Figs. 35, 36); mesonotum and metanotum with fringes of spines along lateral margins 22

21b. Pronotum without a distinct lateral fringe of spines (Fig. 38), or fringe irregular with long and short spines interspersed (Fig. 37); mesonotum and metanotum without fringes of spines or with very sparse fringes mostly visible near posterolateral margins 23

22a (21a). Pronotum with shallow notch laterally (Fig. 35); sometimes a longer, thinner seta in lateral fringe at each anterolateral angle and near each posterolateral angle (Fig. 35) . . . . *Soyedina*

22b. Pronotum rounded laterally (Fig. 36); longer, thinner setae absent from lateral fringe . . . . *Nemoura*

23a (21b). Foreleg with a few large spines on tibia and femur; outer margin of tibia with spines of similar length to those on surface and without fringe of long hairs (Fig. 7), or with only occasional hairs . . . . *Paranemoura*

23b. Foreleg with numerous large spines on tibia and femur; outer margin of tibia with row of large spines and/or fringe of long hairs (Figs. 8–10, 39) 24

24a (23b). Foretibiae without outer fringe of long hairs, but with occasional single hairs (Figs. 9, 39) 25

24b. Foretibiae with outer fringe of long hairs, with or without stout outer spines (Figs. 8, 10) 26

25a (24a). Pronotum with irregular lateral fringe of spines (Fig. 37) . . . . *Ostrocerca*

25b. Pronotum without lateral fringe (Fig. 40) . . . . *Podmosta*

26a (24b). Dorsal and ventral bristles of cercal whorls longer than lateral bristles (Fig. 41); foretibia with hair fringe and row of stout spines along outer margin (Fig. 10); legs not banded . . . . *Prostoia*

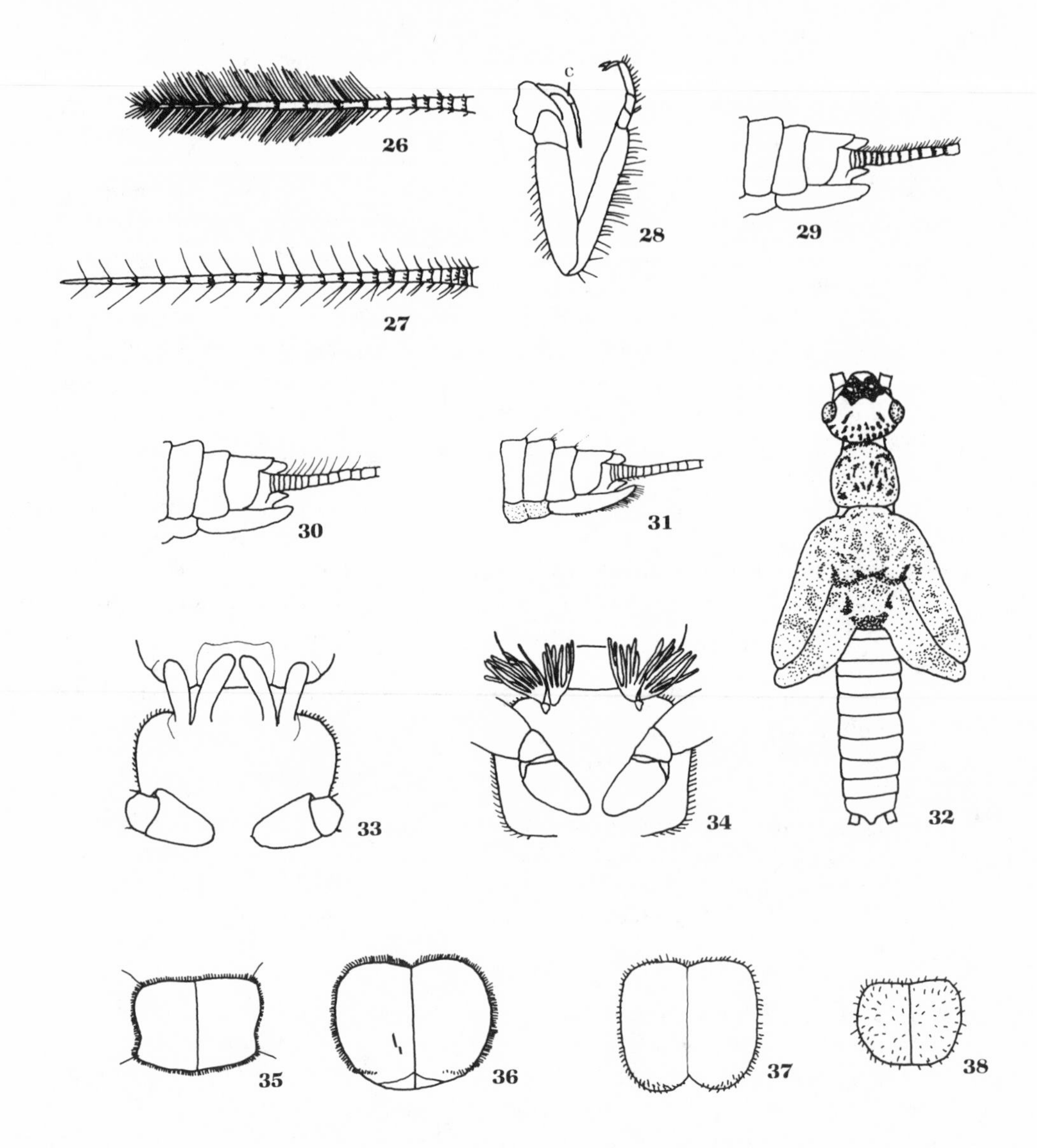

26. Cercus of *Nemocapnia* (Capniidae), lv. **27.** Cercus of *Paracapnia* (Capniidae), lv. **28.** Hind leg of *Taeniopteryx* (Taeniopterygidae), lv: c, coxal gill. **29.** Terminal abdominal segments of *Bolotoperla* (Taeniopterygidae), lv. **30.** Terminal abdominal segments of *Oemopteryx* (Taeniopterygidae), lv. **31.** Terminal abdominal segments of *Taenionema* (Taeniopterygidae), lv. **32.** *Strophopteryx* (Taeniopterygidae), dv. **33.** Cervical area of *Zapada* (Nemouridae), vv. **34.** Cervical area of *Amphinemura* (Nemouridae), vv. **35.** Pronotum of *Soyedina* (Nemouridae), dv. **36.** Pronotum of *Nemoura* (Nemouridae), dv. **37.** Pronotum of *Ostrocerca* (Nemouridae), dv. **38.** Pronotum of *Prostoia* (Nemouridae), dv. dv, dorsal view; lv, lateral view; vv, ventral view. (26, 27 redrawn from Harper and Hynes 1971b; 28 modified from Pennak 1978; 31, 32 redrawn or modified from Harper and Hynes 1971c; 33, 34 modified from Baumann et al. 1977; 35, 36 redrawn from Hilsenhoff 1981; 38 redrawn from Hilsenhoff 1975.)

26b. Only ventral bristles of cercal whorls longer than lateral bristles (Fig. 42); foretibia with hair fringe, but without stout spines along outer margin (Fig. 8); legs indistinctly banded . . . . *Shipsa*

27a (2b). **Perlidae**: Eyes situated much anterior to hind margin of head (Fig. 43) . . . 28
27b. Eyes situated close to hind margin of head (Figs. 44–46) 29

28a (27a). Both front and hind margins of legs bearing many long hairs; branched subanal gills present (as in Fig. 47) . . . . *Perlinella*
28b. Only hind margins of legs bearing many long hairs; branched subanal gills absent (as in Fig. 48) . . . . *Hansonoperla*

29a (27b). Two ocelli, with anterior ocellus absent; distinct transverse occipital ridge across back of head (Fig. 44); subanal gills present (as in Fig. 47) . . . . *Neoperla*
29b. Three ocelli present (Figs. 43, 45, 46); distinct transverse occipital ridge may be present or absent; subanal gills present or absent 30

30a (29b). Occipital ridge present; a closely set regular row of spinules completely across back of head inserted on occipital ridge (Fig. 44) 31
30b. Occipital ridge absent; spinules on back of head present mainly near eyes, or arranged in a transverse row of varying completeness, but always at least a little wavy or irregular (Fig. 45) 32

31a (30a). Branched subanal gills present (Fig. 47); basal cercal segments without an inner fringe of long hairs (Fig. 47) . . . . *Agnetina*
31b. Branched subanal gills absent (Fig. 48); basal cercal segments with an inner fringe of long hairs (Fig. 48) . . . . *Paragnetina*

32a (30b). Back of head without spinules, except near eyes (Fig. 46) 33
32b. Back of head with an irregular row of large spinules (Fig. 45) 35

33a (32a). Head with large areas of yellow in front of anterior ocellus (Fig. 49) . . . . *Eccoptura*
33b. Head mostly brown, often with yellow M-shaped marking in front of anterior ocellus (Fig. 50) 34

34a (33b). Cerci with fringe of long silky setae, at least on basal segments . . . *Acroneuria*
34b. Cerci without fringe of silky setae . . . . *Beloneuria*

35a (32b). Branched subanal gills present (as in Fig. 47); head patterned . . . . . . *Perlesta*
35b. Branched subanal gills absent (as in Fig. 48); head monochromatic brown . . . . *Attaneuria*

36a (8a). **Perlodidae**: Anterior ends of arms of Y ridge of mesosternum meet anterior corners of furcal pits (Fig. 52); submental gills present, about twice as long as greatest width (as in Fig. 51) . . . . *Arcynopteryx*

39. Foreleg of *Podmosta* (Nemouridae), lv. **40.** Pronotum of *Podmosta* (Nemouridae), dv. **41.** Terminal segments of cercus of *Prostoia* (Nemouridae), lv. **42.** Terminal segments of cercus of *Shipsa* (Nemouridae), lv. **43.** Head of *Perlinella* (Perlidae), dv. **44.** Head of *Neoperla* (Perlidae), dv. **45.** Head of *Perlesta* (Perlidae), dv. **46.** Head of *Acroneuria* (Perlidae), dv. **47.** Terminal abdominal segments of *Agnetina* (Perlidae), dv. **48.** Terminal abdominal segments of *Paragentina* (Perlidae), dv. **49.** Head of *Eccoptura* (Perlidae), dv. **50.** Head of *Acroneuria* (Perlidae), dv. **51.** Labium of *Isogenoides* (Perlodidae), vv. dv, dorsal view; lv, lateral view; vv, ventral view. (39 redrawn from Baumann 1975; 41–48, 51 redrawn from Hilsenhoff 1981; 49 modified from Surdick and Kim 1976; 50 modified from Frison 1942.)

36b. Anterior ends of arms of Y ridge of mesosternum meet posterior corners of furcal pits (Figs. 53, 54) or posterior ends of arms meet posterior edge of mesosternum separately (Fig. 55); submental gills may be present or absent 37

37a (36b). Submental gills at least twice as long as greatest width (Fig.51) 38

37b. Submental gills less than twice as long as greatest width, or absent (Fig. 56); posterolateral swelling of submentum sometimes fleshy, resembling a short gill (Fig. 57) 40

38a (37a). Median ridge of mesosternum extends anteriorly beyond fork of Y to transverse ridge (Fig. 54) . . . . *Isogenoides*

38b. Median ridge of mesosternum does not extend anteriorly beyond fork of Y; transverse ridge absent (Fig. 53) or indistinct 39

39a (38b). Head with broad, dark transverse band between eyes (Fig. 58) . . . . . *Helopicus*

39b. Head with dark M-shaped band between eyes, with light M-shaped band anterior to it (Fig. 59) . . . . *Hydroperla*

40a (37b). Lacinia of maxilla terminating in a single tooth lacking spinules or hairs on the mesal margin . . . . *Remenus*

40b. Lacinia with a shorter spine mesal to major spine and commonly with additional spinules or hairs (Figs. 60–63) 41

41a (40b). Lacinia with a sharp angle just below second, smaller tooth sometimes in the form of a knob with tufts or spinules of hairs (Fig. 62) 42

41b. Lacinia without a knob, rounded (Fig. 63) or tapering (Figs. 60, 61) from the smaller spine to base 43

42a (41a). Submental gills present; abdominal tergites each with 6–8 light spots arranged more or less in transverse rows . . . . *Malirekus*

42b. Submental gills absent; abdominal tergites each with broad lateral longitudinal band . . . . *Diura*

43a (41b). Posterior ends of arms of mesosternal ridge meet posterior edge of mesosternum separately (Fig. 55); terminal lacinial spine long, ½ as long as lacinia (Fig. 60) . . . . *Diploperla*

43b. Posterior ends of arms of mesosternal ridge meet posteriorly, forming a forked-Y pattern (Fig. 53) 44

44a (43b). Lacinia tapering from 2nd tooth to base, with apical lacinial tooth ½ length of lacinia (Fig. 61); abdominal segments with anterior, transverse dark bands . . . . *Cultus*

44b. Lacinia rounded from 2nd tooth to base (Fig. 63) 45

45a (44b). Dorsal abdominal segments with alternating transverse or longitudinal light and dark stripes or bands . . . . *Isoperla*

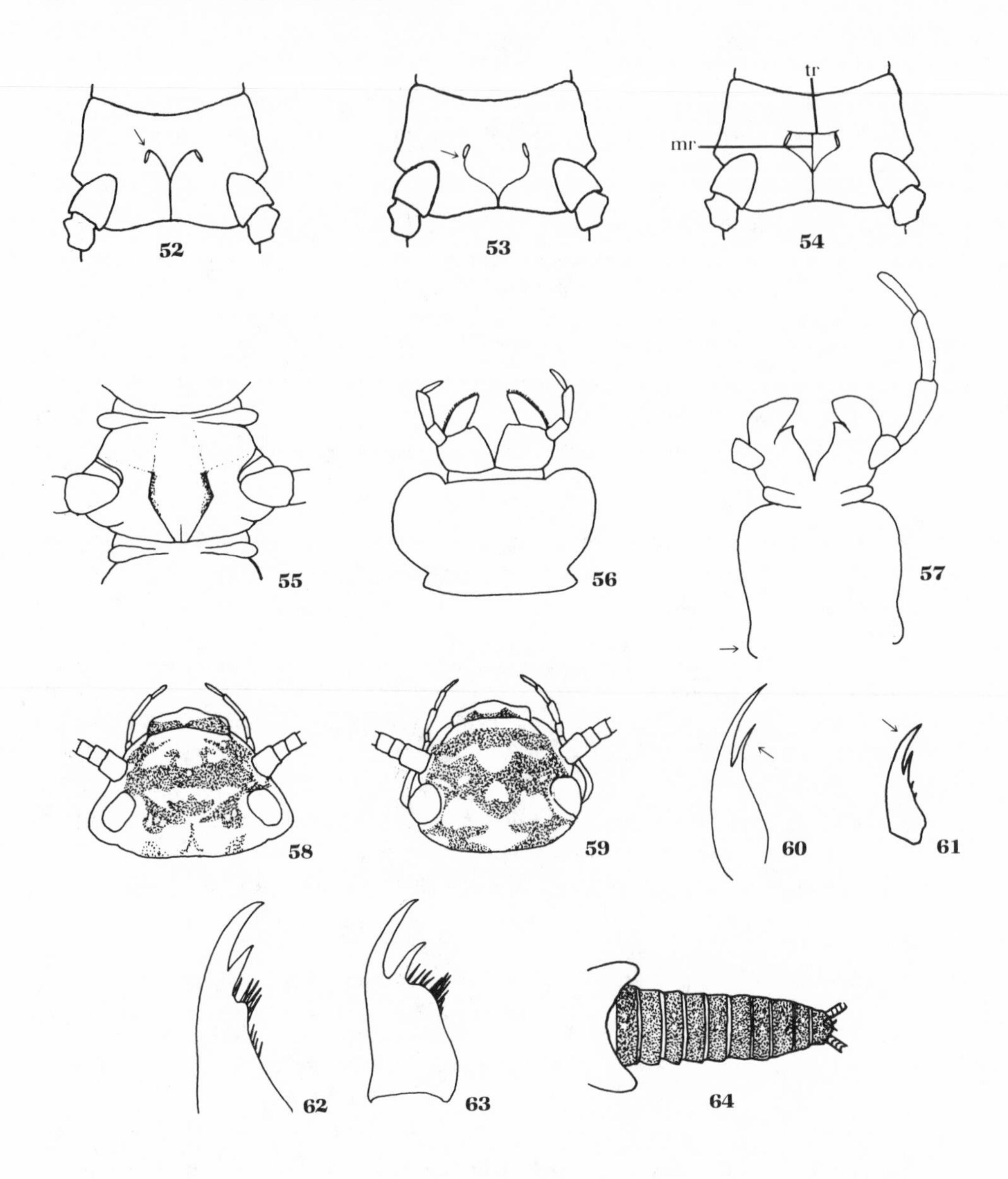

52. Mesosternal ridge pattern of *Arcynopteryx* (Perlodidae), vv. **53.** Mesosternal ridge pattern of *Isoperla* (Perlodidae), vv. **54.** Mesosternal ridge pattern of *Isogenoides* (Perlodidae), vv: mr, median ridge, tr, transverse ridge. **55.** Mesosternal ridge pattern of *Diploperla* (Perlodidae), vv. **56.** Labium of *Isoperla* (Perlodidae), vv. **57.** Labium of *Diploperla* (Perlodidae), vv. **58.** Head of *Helopicus* (Perlodidae), dv. **59.** Head of *Hydroperla* (Perlodidae), dv. **60.** Lacinia of *Diploperla* (Perlodidae), vv. **61.** Lacinia of *Cultus* (Perlodidae), vv. **62.** Lacinia of *Isogenoides* (Perlodidae), vv. **63.** Lacinia of *Clioperla* (Perlodidae), vv. **64.** Abdomen of *Clioperla* (Perlodidae), dv. dv, dorsal view; vv, ventral view. (55, 58, 59 modified from Stewart and Stark 1984; 56 redrawn from Hilsenhoff 1981; 61 redrawn from Ricker 1959; 62, 63 redrawn from Hilsenhoff and Billmyer 1973; 64 modified from Szczytko and Stewart 1981.)

45b. Dorsal abdominal segments uniform, brownish except for a few small light spots (which may be in longitudinal rows) (Fig. 64) (caution: small *Isoperla* nymphs may have uniform color) . . . . *Clioperla*

46a (8b). **Chloroperlidae**: Outside surface of hindleg lacking a well-developed fringe of long, fine setae (Fig. 65); basal cercal segments having many long setae in apical corona visible in lateral view (Fig. 66); body about 10 times as long as width of abdomen . . . . *Utaperla*

46b. Outside surface of hindtibia and often hindfemur with a well-developed fringe of long, fine setae (Figs. 67, 71, 77, 82); basal cercal segments having few long setae in apical corona visible in lateral view (Figs. 68, 72, 78, 83); body less than 10 times as long as width of abdomen 47

47a (46b). Pronotum with few or no setae on front and, especially, hind margins; setae present only on corners (Fig. 69); apical 7–10 cercal segments with numerous vertical long setae between apical coronas forming a feathery fringe visible in lateral view (Fig. 68); setae absent on 8th tergite at mesal posterior margin (Fig. 70); integument yellow-gold . . . . *Alloperla*

47b. Pronotum with variable setation, but always with some setae along hind margin, and usually on front margin as well (Figs. 73, 79, 84); cerci entirely lacking setae or with scattered long setae between apical coronas (Figs. 72, 78, 83); setae present on 8th tergite over entire posterior margin (Figs. 74, 80, 85); integument gold-brown 48

48a (47b). Inner margin of hind wingpads parallel to body axis (Fig. 76); fewer than 6 short, coarse setae between compound eye and hind margin of head (Fig. 75); abdominal tergites lightly setose; pronotum having only sparse, fine setae on dorsal surface . . . . *Haploperla*

48b. Inner margin of hind wingpads angled away from body axis (Fig. 81); more than 6 short, coarse setae between compound eye and hind margin of head (Fig. 86); abdominal tergites heavily setose; pronotum with coarse, closely appressed setae on dorsal surface (Figs. 79, 84) 49

49a (48b). Dorsum of abdomen with 4 longitudinal dark stripes in mature larvae; 8th abdominal tergite with setae absent from a band comprising the proximal ¼ to ⅓ of the tergite (Fig. 80) . . . . *Rasvena*

49b. Dorsum of abdomen monochromatic except when adult coloration shows through, appearing as a median longitudinal band; 8th abdominal tergite with setae absent from a proximal band at most ⅕ the length of the tergite 50

50a (49b). Thick, black, depressed hairs present laterally on all thoracic sterna . . . . *Sweltsa*

50b. Thick, black, depressed hairs absent from lateral thoracic sterna; sternal hairs erect, light-colored . . . . *Suwallia*

Surdick (1981) and Harper and Stewart (1984) suggest that mouthparts need to be mounted for the most reliable separation of *Sweltsa* and *Suwallia*.

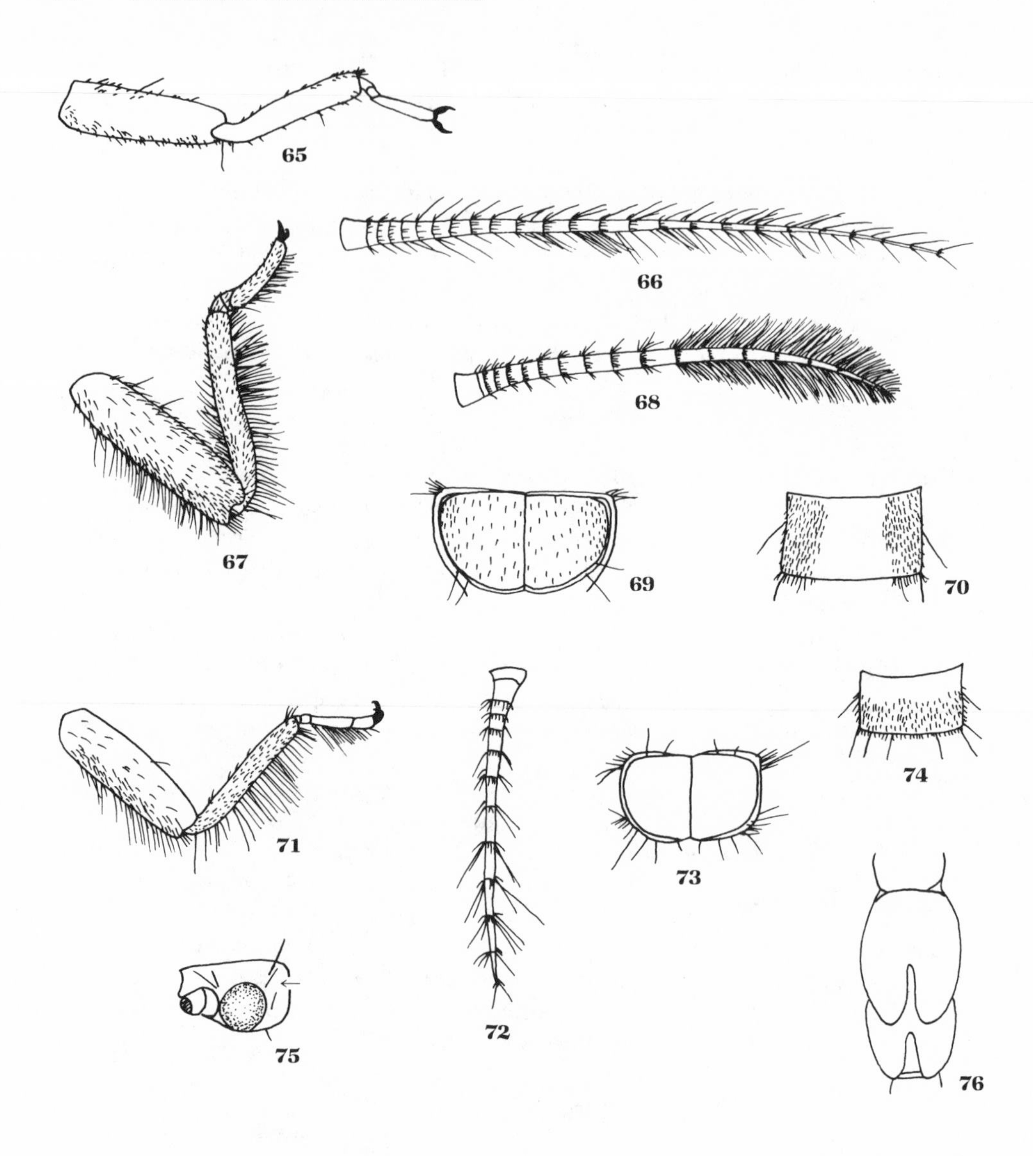

65. Right hindleg of *Utaperla* (Chloroperlidae), lv. **66.** Cercus of *Utaperla* (Chloroperlidae), lv. **67.** Right hindleg of *Alloperla* (Chloroperlidae), lv. **68.** Cercus of *Alloperla* (Chloroperlidae), lv. **69.** Pronotum of *Alloperla* (Chloroperlidae), dv. **70.** Eighth tergite of *Alloperla* (Chloroperlidae), dv. **71.** Right hindleg of *Haploperla* (Chloroperlidae), lv. **72.** Cercus of *Haploperla* (Chloroperlidae), lv. **73.** Pronotum of *Haploperla* (Chloroperlidae), dv. **74.** Eighth tergite of *Haploperla* (Chloroperlidae), dv. **75.** Left side of head of *Haploperla* (Chloroperlidae), lv. **76.** Wingpads of *Haploperla* (Chloroperlidae), dv. dv, dorsal view; lv, lateral view. (65–76 redrawn or modified from Fiance 1977.)

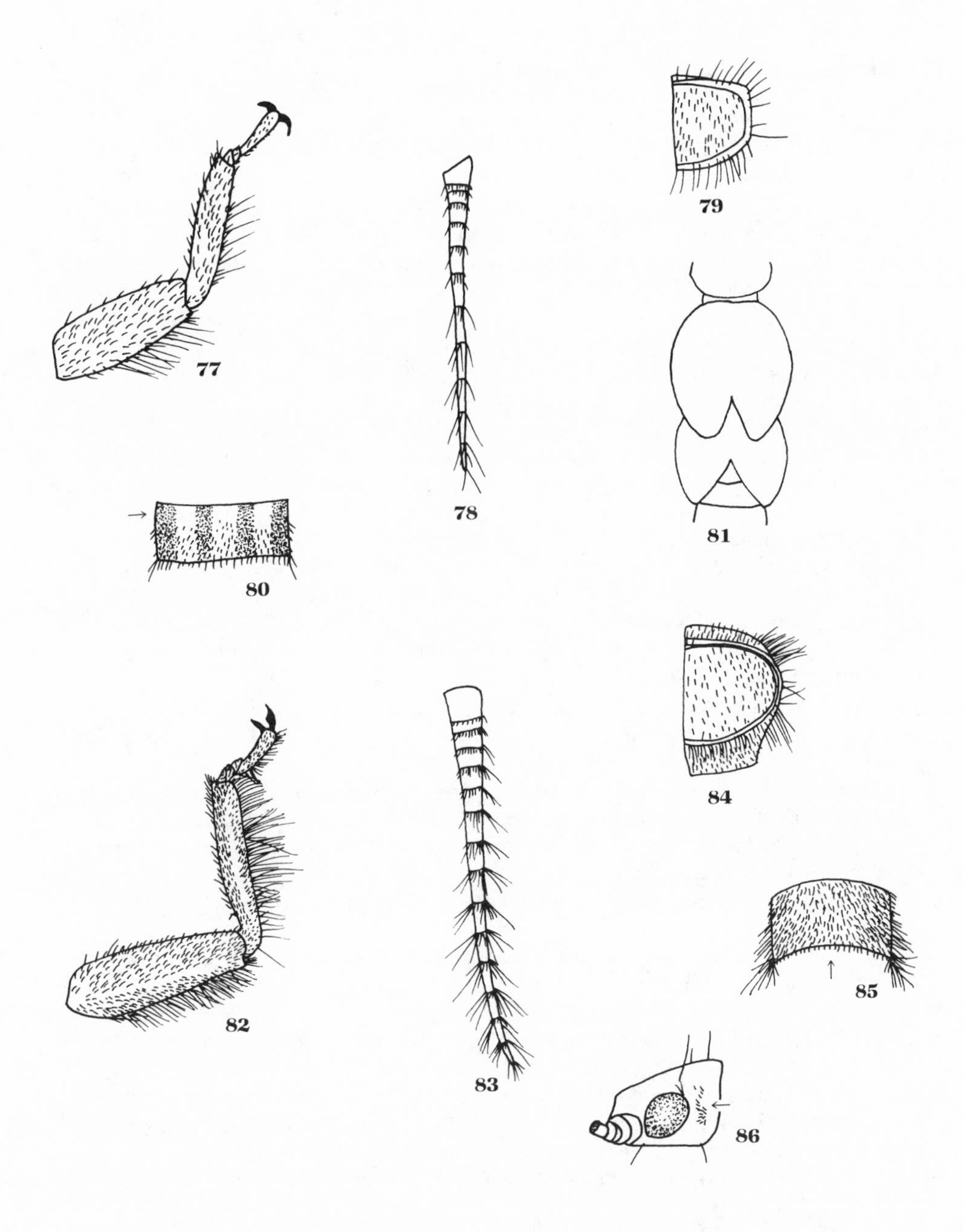

77. Right hindleg of *Rasvena* (Chloroperlidae), lv. **78.** Cercus of *Rasvena* (Chloroperlidae), lv. **79.** Right side of pronotum of *Rasvena* (Chloroperlidae), dv. **80.** Eighth tergite of *Rasvena* (Chloroperlidae), dv. **81.** Wingpads of *Rasvena* (Chloroperlidae), dv. **82.** Right hindleg of *Sweltsa* (Chloroperlidae), lv. **83.** Cercus of *Sweltsa* (Chloroperlidae), lv. **84.** Right side of pronotum of *Sweltsa* (Chloroperlidae), dv. **85.** Eighth tergite of *Sweltsa* (Chloroperlidae), dv. **86.** Left side of head of *Sweltsa* (Chloroperlidae), lv. dv, dorsal view; lv, lateral view. (77–86 redrawn or modified from Fiance 1977.)

References on Plecoptera Systematics

(*Used in construction of key.)

Allan, J.D., A.S. Flecker, and N.L. McClintock. 1987. Prey preference in stoneflies: sedentary vs. mobile prey. Oikos 49:323–331.

*Baumann, R.W. 1975. Revision of the stonefly family Nemouridae: a study of the world fauna at the generic level. Smithson. Contrib. Zool. 211:1–74.

——. 1976. A report on the fifth international symposium on Plecoptera. Proc. Biol. Soc. Wash. 88:399–428.

——. 1979. Nearctic stonefly genera as indicators of ecological parameters (Plecoptera: Insecta). Great Basin Nat. 39:241–244.

*Baumann, R.W., A.R. Gaufin, and R.F. Surdick. 1977. The stoneflies (Plecoptera) of the Rocky Mountains. Mem. Amer. Entomol. Soc. 31:1–208.

Brinck, P. 1956. Reproductive system and mating in Plecoptera. Opusc. Entomol. 21:57–127.

Brodsky, A.K. 1982. Evolution of wing apparatus in the stoneflies (Plecoptera). Part IV. The kinematics of the wings and general conclusions. Rev. d'Entomol. de l'URSS 61:485–490.

Claassen, P.W. 1931. Plecoptera nymphs of America (north of Mexico). Thomas Say Found. Entomol. Soc. Amer. 3:1–199.

——. 1940. A catalogue of the Plecoptera of the world. Mem. Cornell Univ. Agric. Exp. Stn. 232:1–235.

*Cummins, K.W., and R.W. Merritt. 1978. General morphology of aquatic insects. *In* R.W. Merritt and K.W. Cummins (eds.). An introduction to the aquatic insects of North America, pp. 5–12. Kendall/Hunt Publ. Co., Dubuque, Iowa.

*——. 1984. General morphology of aquatic insects. *In* R.W. Merritt and K.W. Cummins (eds.). An introduction to the aquatic insects of North America, 2nd ed., pp. 4–10. Kendall/Hunt Publ. Co., Dubuque, Iowa.

Ernst, M.R., and K.W. Stewart. 1985. Emergence and an assessment of collecting methods for adult stoneflies (Plecoptera) in an Ozark foothills stream. Can. J. Zool. 63:2962–2968.

*Fiance, S.B. 1977. The genera of eastern North American Chloroperlidae (Plecoptera): key to larval stages. Psyche 84:308–318.

Frison, T.H. 1929. Fall and winter stoneflies, or Plecoptera, of Illinois. Bull. Ill. Nat. Hist. Surv. 18:345–409.

*——. 1935. The stoneflies or Plecoptera of Illinois. Bull. Ill. Nat. Hist. Surv. 20:281–471.

*——. 1942. Studies of North American Plecoptera, with special reference to the fauna of Illinois. Bull. Ill. Nat. Hist. Surv. 22:231–355.

Fullington, K.E., and K.E. Stewart. 1980. Nymphs of the stonefly genus *Taeniopteryx* (Plecoptera: Taeniopterygidae) of North America. J. Kans. Entomol. Soc. 53:237–259.

Gaufin, A.R. 1962. Environmental requirements of Plecoptera. *In* C.M. Tarzwell (ed.). Biological problems in water pollution. Third seminar, U.S. Dep. Health, Educ., Welfare Serv., Div. Water Supply Pollut. Control, R.A. Taft Sanit. Engr. Ctr. PHS Publ. no. 999-WP-25, pp. 105–110.

Hanson, J.F. 1946. Comparative morphology and taxonomy of the Capniidae (Plecoptera). Amer. Midl. Nat. 35:193–249.

*Harden, P.H. 1942. The immature stages of some Minnesota Plecoptera. Ann. Entomol. Soc. Amer. 35:318–331.

*Harden, P.H., and C.E. Mickel. 1952. The stoneflies of Minnesota (Plecoptera). Univ. Minn. Agric. Exp. Stn. Tech. Bull. 201:1–84.

Harper, P.P. 1973a. Emergence, reproduction, and growth of setipalpian Plecoptera in southern Ontario. Oikos 24:94–107.

——. 1973b. Life histories of Nemouridae and Leuctridae in southern Ontario (Plecoptera). Hydrobiologia 41:309–356.

*——. 1978. Plecoptera. *In* R.W. Merritt and K.W. Cummins (eds.). An introduction to the aquatic insects of North America, pp. 105–118. Kendall/Hunt Publ. Co., Dubuque, Iowa.

Harper, P.P., and F. Harper. 1984. Biogeography and associations of winter stoneflies in southern Québec (Plecoptera). Can. Entomol. 115:1465–1476.

Harper, P.P., and H.B.N. Hynes. 1970. Diapause in the nymphs of Canadian winter stoneflies. Ecology 51:925–927.

*——. 1971a. The Leuctridae of eastern Canada (Insecta: Plecoptera). Can. J. Zool. 49:915–920.

*——. 1971b. The Capniidae of eastern Canada (Insecta: Plecoptera). Can. J. Zool. 49:921–940.

*——. 1971c. The nymphs of the Taeniopterygidae of eastern Canada (Insecta: Plecoptera). Can. J. Zool. 49:941–947.

*——. 1971d. The nymphs of the Nemouridae of eastern Canada (Insecta: Plecoptera). Can. J. Zool. 49:1129–1142.

*Harper, P.P., and K.W. Stewart. 1984. Plecoptera. *In* R.W. Merritt and K.W. Cummins (eds.). An introduction to the aquatic insects of North America, 2nd ed., pp. 182–230. Kendall/Hunt Publ. Co., Dubuque, Iowa.

*Hilsenhoff, W.L. 1975. Aquatic insects of Wisconsin, with generic keys and notes on biology, ecology and distribution. Tech. Bull. Wisc. Dep. Nat. Resour. 89:1–53.

*——. 1981. Aquatic insects of Wisconsin. Nat. Hist. Counc. Publ. no. 2, Univ. Wisc., Madison. 60 pp.

*Hilsenhoff, W.L., and S.J. Billmyer. 1973. Perlodidae (Plecoptera) of Wisconsin. Great Lakes Entomol. 6:1–14.

*Hitchcock, S.W. 1974. Guide to the insects of Connecticut. Part VII. The Plecoptera or stoneflies of Connecticut. Bull. Conn. State Geol. Nat. Hist. Surv. 107:1–262.

Hynes, H.B.N. 1976. Biology of Plecoptera. Annu. Rev. Entomol. 21:135–153.

——. 1988. Biogeography and origins of the North American stoneflies (Plecoptera). Mem. Entomol. Soc. Can. 144:31–38.

Illies, J. 1965. Phylogeny and zoogeography of the Plecoptera. Annu. Rev. Entomol. 10:117–140.

——. 1966. Katalog der rezenten Plecoptera. Das Tierreich. Lieferung 82. Walter de Gruyter and Co., Berlin. 632 pp.

Illies, J., and P. Zwick (eds.). 1977. Proc. Sixth Int. Symp. on Plecoptera, August 3–6, 1977. Schlitz, Germany.

Jewett, S.J., Jr. 1956. Plecoptera. *In* R.L. Usinger (ed.). Aquatic insects of California, pp. 155–181. Univ. Calif. Press, Berkeley.

——. 1963. A stonefly aquatic in the adult stage. Science 139:484–485.

Kapoor, N.N. 1972. Rearing and maintenance of plecopteran nymphs. Hydrobiologia 40:51–53.

Kawai, T. (ed.). 1981. Proc. Seventh Int. Symp. on Plecoptera, August 19–22, 1980, Women's Univ., Nara, Japan. Jap. Biol. Inland Wat. 2:19–43.

Kondratieff, B.C., and R.F. Kirchner. 1982. Notes on the winter stonefly genus *Allocapnia* (Plecoptera: Capniidae). Proc. Entomol. Soc. Wash. 84:240–244.

——. 1987. Additions, taxonomic corrections, and faunal affinities of the stoneflies (Plecoptera) of Viriginia, USA. Proc. Entomol. Soc. Wash. 89:24–30.

Kondratieff, B.C., R.F. Kirchner, and K.W. Stewart. 1988. A review of *Perlinella* Banks (Plecoptera: Perlidae). Ann. Entomol. Soc. Amer. 81:19–27.

Kondratieff, B.C., R.F. Kirchner, and J.R. Voshell, Jr. 1981. Nymphs of *Diploperla*. Ann. Entomol. Soc. Amer. 74:428–430.

Kondratieff, B.C., and J.R. Voshell, Jr. 1982. The Perlodinae of Virginia, USA (Plecoptera: Perlodidae). Proc. Entomol. Soc. Wash. 84:761–774.

Kuusela, K., and H. Pulkkinen. 1978. A simple trap for collecting newly emerged stoneflies (Plecoptera). Oikos 31:323–325.

Martínez, L.A., and B.L. Peckarsky (submitted). Behavioral responses of stonefly and mayfly nymphs to chemical stimuli: a case for chemoreception by stream-dwelling insects. Can. J. Zool.

Mingo, T.M. 1983. Annotated checklist of stoneflies of Maine (Plecoptera). Entomol. News 94:65–72.

Needham, J.H., and P.W. Claassen. 1925. A monograph of the Plecoptera or stoneflies of America north of Mexico. Thomas Say Found. Entomol. Soc. Amer. 2:1–397.

Nelson, C.H. 1979. *Hansonoperla appalachia*, a new genus and a new species of eastern Nearctic Acroneuriini (Plecoptera: Perlidae), with a phenetic analysis of the genera of the tribe. Ann. Entomol. Soc. Amer. 72:735–739.

——. 1988. Note on the phylogenetic systematics of the family Pteronarcyidae (Plecoptera), with a description of the eggs and nymphs of the Asian species. Ann. Entomol. Soc. Amer. 81:560–576.

Nelson, C.H., and J.F. Hanson. 1971. Contribution to the anatomy and phylogeny of the family Pteronarcidae (Plecoptera). Trans. Amer. Entomol. Soc. 97:123–200.

Nelson, C.R., and R.W. Baumann. 1987. Scanning electron microscopy for the study of the winter stonefly genus *Capnia* (Plecoptera: Capniidae). Proc. Entomol. Soc. Wash. 89:51–56.

Peckarsky, B.L. 1979. A review of the distribution, ecology, and evolution of the North American species of *Acroneuria* and six related genera (Plecoptera: Perlidae). J. Kans. Entomol. Soc. 52:787–809.

Peckarsky, B.L., and M.A. Penton. 1989. Mechanisms of prey selection by stream-dwelling stoneflies. Ecology (in press).

Peckarsky, B.L., and R.S. Wilcox. 1989. Stonefly use of hydrodynamic cues to distinguish prey from non-prey. Oecologia 79:265–270.

*Pennak, R.W. 1978. Freshwater invertebrates of the United States, 2nd ed. John Wiley and Sons, New York. 803 pp.

Ricker, W.E. 1950. Some evolutionary trends in Plecoptera. Proc. Ind. Acad. Sci. 59:197–209.

——. 1952. Systematic studies in Plecoptera. Ind. Univ. Publ. Sci. Ser. 18:1–200.

*——. 1959. Plecoptera. *In* W.T. Edmondson (ed.). Fresh-water biology, 2nd ed., pp. 941–957. John Wiley and Sons, New York.

Ricker, W.E., R. Malouin, P. Harper, and H.H. Ross. 1968a. Distribution of Quebec stoneflies (Plecoptera). Nat. Can. 95:1085–1123.

——. 1968b. North American species of *Taeniopteryx* (Plecoptera, Insecta). J. Fish. Res. Bd. Can. 25:1423–1439.

Ricker, W.E., and H.H. Ross. 1969. The genus *Zealeuctra* and its position in the family Leuctridae. Can. J. Zool. 47:1113–1127.

——. 1975. Synopsis of the Brachypterinae (Insecta: Plecoptera: Taeniopterygidae). Can. J. Zool. 53:132–153.

Ross, H.H., and W.E. Ricker. 1971. The classification, evolution, and dispersal of the winter stonefly genus *Allocapnia*. Univ. Ill. Biol. Monogr. 45:1–116.

Shepard, W.D., and K.W. Stewart. 1983. Comparative study of nymphal gills in North American stonefly (Plecoptera) genera and a new, proposed paradigm of Plecoptera gill evolution. Misc. Publ. Entomol. Soc. Amer. 55:1–57.

Stanford, J.A., and A.R. Gaufin. 1974. Hyporheic communities of two Montana Rivers. Science 185:700–702.

*Stark, B.P. 1983. The *Tallaperla maria* complex of eastern North America (Plecoptera: Peltoperlidae). J. Kans. Entomol. Soc. 56:398–410.

——. 1986. The Nearctic species of *Agnetina*. J. Kans. Entomol. Soc. 59:437–445.

Stark, B.P., and R.W. Baumann. 1978. New species of nearctic *Neoperla* (Plecoptera: Perlidae) with notes on the genus. Great Basin Nat. 38:97–114.

Stark, B.P., and A.R. Gaufin. 1976a. The Nearctic genera of Perlidae (Plecoptera). Misc. Publ. Entomol. Soc. Amer. 10:1–80.

——. 1976b. The Nearctic species of *Acroneuria* (Plecoptera: Perlidae). J. Kans. Entomol. Soc. 49:221–253.

Stark, B.P., and D.H. Ray. 1983. A revision of the genus *Helopicus* (Plecoptera: Perlodidae). Freshwat. Invert. Biol. 2:16–27.

Stark, B.P., and K.W. Stewart. 1981. The Nearctic genera of Peltoperlidae (Plecoptera). J. Kans. Entomol. Soc. 54:285–311.

Stark, B.P., and S.W. Szczytko. 1976. The genus *Beloneuria* (Plecoptera: Perlidae). Ann. Entomol. Soc. Amer. 69:1120–1124.

——. 1981. Contributions to the systematics of *Paragnetina* (Plecoptera: Perlidae). J. Kans. Entomol. Soc. 54:625–648.

——. 1982. Egg morphology and phylogeny in Pteronarcyidae (Plecoptera). Ann. Entomol. Soc. Amer. 75:519–529.

——. 1984. Egg morphology and classification of Perlodinae (Plecoptera: Perlodidae). Ann. Limnol. 20:99–104.

——. 1988. Egg morphology and phylogeny in Arcynopterygini (Plecoptera: Perlodidae). J. Kans. Entomol. Soc. 61:143–160.

Stark, B.P., S.W. Szczytko, and R.W. Baumann. 1986. North American stoneflies (Plecoptera): systematics, distribution, and taxonomic references. Great Basin Nat. 46:383–397.

Stewart, K.W., R.W. Bauman, and B.P. Stark. 1974. The distribution and past dispersal of southwestern United States Plecoptera. Trans. Amer. Entomol. Soc. 99:507–546.

*Stewart, K.W., and B.P. Stark. 1984. Nymphs of North American Perlodinae genera (Plecoptera: Perlodidae). Great Basin Nat. 44:373–415.

*——. 1988. Nymphs of North American stonefly genera (Plecoptera). Thomas Say Found. Entomol. Soc. Amer. 12:1–460.

Stewart, K.W., and D.D. Ziegler. 1984. The use of larval morphology and drumming in Plecoptera systematics, and further studies of drumming. Ann. Limnol. 20:105–114.

Steyskal, G.C. 1976. Notes on the nomenclature and taxonomic growth of the Plecoptera. Proc. Biol. Soc. Wash. 88:408–410.

Surdick, R.F. 1981. New Nearctic Chloroperlidae (Plecoptera). Great Basin Nat. 41:349–359.

*——. 1985. Nearctic genera of Chloroperlinae (Plecoptera: Chloroperlidae). Ill. Biol. Monogr. 54:1–146.

Surdick, R.F., and A.R. Gaufin. 1978. Environmental requirements and pollution tolerance of Plecoptera. EPA-600/4-78-062. 417 pp.

*Surdick, R.F., and K.C. Kim. 1976. Stoneflies (Plecoptera) of Pennsylvania. Bull. Penn. State Univ. Coll. Agric. 808:1–73.

*Szczytko, S.W., and K.W. Stewart. 1977. The stoneflies of Texas. Trans. Amer. Entomol. Soc. 103:327–378.

*——. 1981. Reevaluation of the genus *Clioperla*. Ann. Entomol. Soc. Amer. 74:563–569.

Unzicker, J.D., and V.H. McCaskill. 1982. Plecoptera. *In* A.R. Brigham, W.U. Brigham, and A. Gnilka (eds.). Aquatic insects and oligochaetes of North and South Carolina, pp. 5.1–5.50. Midwest Aquatic Enterprises, Mahomet, Ill.

*Ward, J.V. 1985. An illustrated guide to the mountain stream insects of Colorado. Kinko's Copies: Professor Publishing, Fort Collins, Colo.

Zwick, P. 1973. Insecta: Plecoptera, phylogenetisches System und Katalog. Das Tierreich. Lieferung 94. Walter de Gruyter and Co., Berlin. 465 pp.

——. 1980. Plecoptera. Handb. Zool. 4:1–111.

——. 1984. Notes on the genus *Agnetina* (= *Phasganophora*) (Plecoptera: Perlidae). Aquat. Insects 6:71–79.

7 Aquatic and Semiaquatic Hemiptera

Classification

Hemiptera, or true bugs, belong to the infraclass Neoptera, division Exopterygota; their wings develop externally and can be folded over the dorsum. Only about 10% of all species of Hemiptera are associated with water, and these are representatives of 15 families of the suborder Heteroptera, 14 of which occur in northeastern North America (Polhemus 1984). The taxonomy of this group is well known and, therefore, relatively stable.

Life History

Hemipterans are paurometabolous, undergoing incomplete, gradual metamorphosis from egg to nymph to adult (Hilsenhoff 1981). Females typically oviposit on floating or submerged vegetation, although some hemipterans have very unusual oviposition habits. For example, some species of Corixidae have been observed attempting to oviposit on blue automobile roofs (Schaefer and Schaefer 1979), and one corixid species oviposits exclusively on the exoskeletons of crayfish (Griffith 1945). The most unusual oviposition behavior in the hemipterans occurs in the subfamily Belostomatinae of the Belostomatidae, in which females oviposit on the backs of the males. The males brood and oxygenate the eggs until they hatch, behavior that has been shown to improve brooder fitness (Smith 1979).

Eggs hatch after one to four weeks of embryonic development, then the nymphs undergo five molts before molting to the sexually mature adults. Hemipterans generally overwinter as adults, and in regions where ponds freeze many species fly to streams to overwinter. Adults are strong fliers, although the wing muscles may atrophy after flight. Some species (gerrids) exhibit wing polymorphism: in stable or predictable habitats *apterous* (wingless) or *brachypterous* (short-winged) morphs are more common, and in temporary or unpredictable habitats selection pressure is greater for *macropterous* (long-winged) forms (Harrison 1980). However, wing polymorphism does not necessarily reflect dispersal polymorphism (Fairbairn 1986). In fact, macropterous forms may be less successful than apterous individuals at moving overland or "skating" against a current.

Adults of some families of Hemiptera signal acoustically or by surface wave production to facilitate mating. For example, corixids rub pegs on their forefemora against sharp edges on the sides of their heads, producing species specific signals, although these signals are not obligatory for mating success (Aiken 1985). Gerrid males vibrate the water surface by moving their forefemora at specific frequencies producing courtship or calling signals, which females may answer with receptive signals. Gerrids are able to code into these surface waves all the information necessary for successful copulation (Wilcox 1979) such as location and receptiveness. Both corixids and gerrids can also use signals defensively against predators or competitors (Aiken 1985, Wilcox and Spence 1986). Open-ocean water striders (*Halobates*) have been shown to mate in swarms, or "flotillas," which are first formed by males and, as females enter the swarms, gradually become composed of mating pairs and unmated males. Since females not in copula, whether they have copulated previously or not, are readily grabbed by free males, males prevent takeover of their mated females by prolonging copulation (up to nine hours). One case of a 47-hour copulation was reported by Foster and Treherne (1982). This behavior, like the brooding behavior of male Belostomatinae, assures paternity and thus improves male fitness.

Habitat

Aquatic and semiaquatic hemipterans are found in three general habitats. Six families are fully aquatic in all life history stages (true water bugs: Belostomatidae, Corixidae, Naucoridae, Nepidae, Notonectidae, and Pleidae). These families are classified as swimmers, clingers, or climbers. Two of them (Notonectidae and Pleidae) swim upside down (venter up) and have been given the common name *backswimmers*. Four families are surface dwellers or skaters (Gerridae, Hydrometridae, Mesoveliidae, and Veliidae) and are thus semiaquatic. Four families are shore bugs and live along the edges of ponds or streams. Gelastocoridae, Hebridae, Ochteridae, and Saldidae contain species classified as skaters, climbers, clingers, burrowers, or sprawlers. Hemipterans are generally found in lentic habitats or in backwater or pool areas of streams to which they may have flown to overwinter (Polhemus 1984).

Feeding

Adult and nymphal hemipterans are predaceous, having mouthparts specialized for piercing and sucking the contents of their prey. Because of their unusual triangular beak and scooplike foretarsi, corixids had been thought to be strictly detritivorous or herbivorous (Hungerford 1919); however, recent observations demonstrate that some species can be predaceous (Lauck 1979). Aquatic and semiaquatic hemipterans prey on a variety of aquatic insects and crustaceans. Forelegs are generally specialized for seizing and holding prey while their contents are sucked up through the piercing mouthparts (Pennak 1978). There are even accounts of large aquatic hemipterans consuming vertebrates; for example, Matheson (1907) reported observing a *Lethocerus americanus* attacking a woodpecker, and Wilson (1958) observed *L. uhleri* feeding on a 30-cm-long banded water snake! Giant water bugs can be important pests in fish hatcheries, since they feed on fish up to 7.5 cm long (Harvey 1907a, b; Wilson 1958).

Respiration

Surface-dwelling hemipterans breathe through open spiracles. True water bugs carry an air store (gas gill) dorsally between the front wings (*hemelytra*) and the abdomen, and some have a physical gill or plastron that is structurally supported by hydrofuge hairs on the ventral surface (Naucoridae, Pleidae) (Polhemus 1984). True water bugs replace their air stores at the pond surface using a variety of structures (tubes, air straps, the pronotum, or the tip of the abdomen).

Collection and Preservation

Aquatic D-frame nets are suitable for collecting aquatic or surface-dwelling hemipterans. Some of the surface dwellers are quick, making their capture challenging. Aquatic entomologists should not handle hemipterans with their fingers because some species can give a painful bite. Specimens can be preserved in 70% ethanol or pinned as are terrestrial hemipterans.

Identification

Characters used to identify families and genera of Hemiptera are structure of the antennae and rostrum, body shape and size (for adults), position of tarsal claws, hemelytral pattern, shape or position of eyes, and leg segmentation. Corixids are the most difficult to identify, requiring differentiation of males and females. Although the taxonomy of the suborder Heteroptera is well known, and adults can be accurately identified to species, nymphs of many genera cannot be keyed (Hilsenhoff 1981). We therefore recommend using the keys only for adults.

More-Detailed Information

Hungerford 1919, Usinger 1956, Pennak 1978, Menke 1979, Hilsenhoff 1981, Sanderson 1982, Polhemus 1984, Jansson 1986.

Checklist of Aquatic and Semiaquatic Hemiptera

Belostomatidae
Belostoma
Lethocerus
Corixidae
Arctocorisa
Callicorixa
Cenocorixa
Corisella
Cymatia
Dasycorixa
Hesperocorixa
Palmacorixa
Ramphocorixa
Sigara
Trichocorixa
Gelastocoridae
Gelastocoris
Gerridae
Gerris
Limnogonus
Limnoporus
Metrobates
Neogerris
Rheumatobates
Trepobates
Hebridae
Hebrus
Merragata
Hydrometridae
Hydrometra
Mesoveliidae
Mesovelia
Naucoridae
Pelocoris
Nepidae
Nepa
Ranatra
Notonectidae
Buenoa
Notonecta
Ochteridae
Ochterus
Pleidae
Neoplea
Saldidae
Veliidae
Microvelia
Paravelia
Rhagovelia

Key to Genera of Aquatic and Semiaquatic Hemiptera

1a. Antennae shorter than head, concealed in groove beneath eye; aquatic; (except Ochteridae: antennae shorter than head, but exposed; riparian) 2
1b. Antennae as long as head or longer, usually plainly visible; semiaquatic (including water striders) 9

2a (1a). Ocelli present; metathoracic legs without swimming hairs (may have very small cilia); semiaquatic; riparian 3
2b. Ocelli absent; metathoracic legs with swimming hairs (Fig. 1) (except Nepidae); aquatic 4

3a (2a). Antennae concealed; front legs raptorial; eyes protuberant; body length 7–9 mm . . . . **Gelastocoridae**, *Gelastocoris*
3b. Antennae exposed; front legs not raptorial; eyes not protuberant; body length 3.5–5 mm . . . . **Ochteridae**, *Ochterus*

4a (2b). Rostrum (beak) short, blunt, and triangular, not distinctly segmented, although often with transverse grooves (Fig. 2); front tarsus (pala) a 1-segmented scoop (Fig. 3); foretibia short (Fig. 3) . . . . **Corixidae** 14
4b. Rostrum cylindrical or cone-shaped and relatively long, distinctly 3- or 4-segmented; front tarsus not scooplike; foretibia long 5

5a (4b). Abdomen with long, slender, tubelike respiratory appendages (Fig. 4) . . . . **Nepidae** 25
5b. Apical respiratory appendages, if present, short and flat 6

6a (5b). Body length ≥ 18 mm; abdomen with a pair of short, flat, retractile air straps (Fig. 5) . . . . **Belostomatidae** 24
6b. Body length < 16 mm; apical abdominal respiratory appendages absent . . . 7

7a (6b). Forefemora almost as wide as long (Fig. 6); body broad and flat (Fig. 7), body length 10–12 mm; climber or swimmer .... **Naucoridae**, *Pelocoris*
7b. Forefemora elongate; body elongate or hemispherical; backswimmer 8

8a (7b). Body shape hemispherical (Fig. 8), body length 2.0–2.5 mm; rostrum 3-segmented .... **Pleidae**, *Neoplea*
8b. Body elongate, > 5 mm long; rostrum 4-segmented .... **Notonectidae** 26

9a (1b). Claws of at least protarsi inserted before apex (Fig. 9) 10
9b. Claws of all tarsi at apex (Fig. 10) 11

10a (9a). Hind femur long, distal end greatly surpassing end of abdomen; midlegs inserted closer to hind legs than to forelegs (Fig. 11); metasternum or metasternal area with a single median scent-gland opening (Fig. 12), lateral channels absent; dorsum of head usually without median groove or line .... **Gerridae** 27
10b. Hind femur short, distal end scarcely, if at all, surpassing end of abdomen; midlegs inserted about midway between front and hind pair (except in *Rhagovelia*); metasternum with a pair of groovelike or suturelike channels extended laterally, ending on pleura in front of hind coxa (Fig. 13); dorsum of head usually with a median, longitudinal groove or shiny glabrous line .... **Veliidae** 33

11a (9b). Head as long as entire thorax; very slender; eyes set about halfway to base (Fig. 14); body length 7.5–10.0 mm .... **Hydrometridae**, *Hydrometra*
11b. Head short and stout, rounded body shape; eyes near posterior margin (Figs. 16, 19) 12

12a (11b). Winged, with veins in the membrane of the hemelytra (Fig. 15) .... **Saldidae** (riparian, not keyed further)
12b. Wingless, or if winged, without veins in the membrane (Figs. 16, 19) 13

13a (12b). Tarsi 3-segmented; head without deep longitudinal ventral groove to receive rostrum; rostrum completely visible from lateral view (Fig. 17); legs with scattered, stiff, black bristles (Fig. 18); body length 2.5–4.0 mm .... **Mesoveliidae**, *Mesovelia*
13b. Tarsi 2-segmented; head with deep longitudinal ventral groove to receive rostrum; rostrum only partially visible from lateral view (Fig. 20); legs without bristles; body length < 2.5 mm .... **Hebridae** 35

14a (4a). **Corixidae**: Rostrum without transverse grooves; pronotum without transverse bands; maximum width of membrane of hemelytra is < ⅓ length of membrane (Fig. 21); body length 5.9–8.3 mm .... *Cymatia*
14b. Rostrum with transverse grooves (Fig. 23); pronotum with transverse dark bands, although they may be indistinct; maximum width of membrane of hemelytra is greater than ⅓ of length of membrane (except in *Palmacorixa*) (Fig. 22) 15

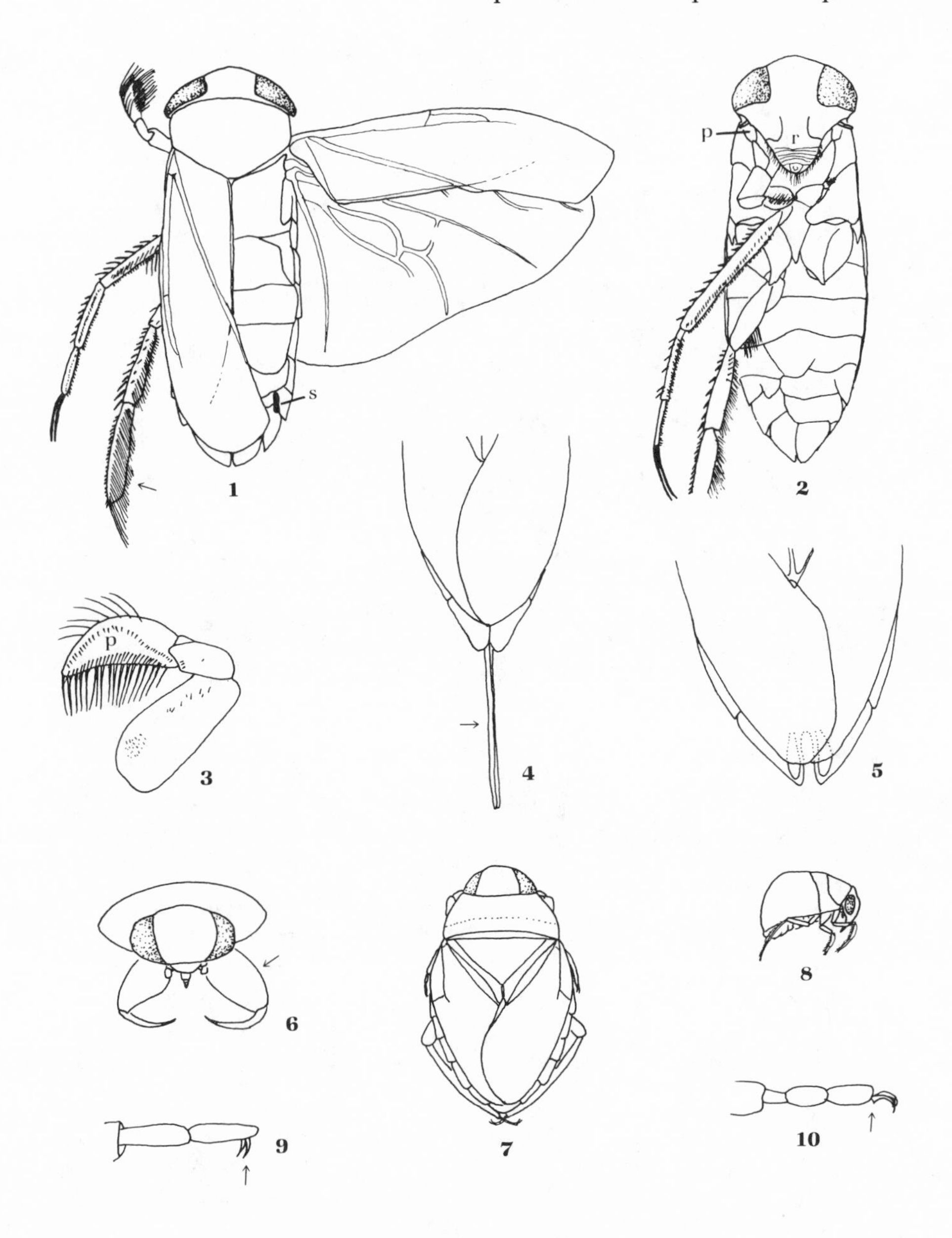

1. Male *Hesperocorixa* (Corixidae), dv: s, strigil. **2.** *Hesperocorixa* (Corixidae), vv: p, prothoracic lobe; r, rostrum. **3.** Front leg of male *Cenocorixa* (Corixidae), dv: p, pala. **4.** End of abdomen of *Nepa* (Nepidae), dv. **5.** End of abdomen of *Belostoma* (Belostomatidae), dv. **6.** *Pelocoris* (Naucoridae), av. **7.** *Pelocoris* (Naucoridae), dv. **8.** *Neoplea* (Pleidae), lv. **9.** Tibia and tarsal segments of prothoracic leg of *Gerris* (Gerridae), lv. **10.** Tibia and tarsal segments of prothoracic leg of *Mesovelia* (Mesoveliidae), lv. av, anterior view; dv, dorsal view; lv, lateral view; vv, ventral view. (1, 2 redrawn from Usinger 1956; 3 modified from Menke 1979; 4, 5, 8–10 modified or redrawn from Hilsenhoff 1981; 6, 7 modified from Pennak 1978.)

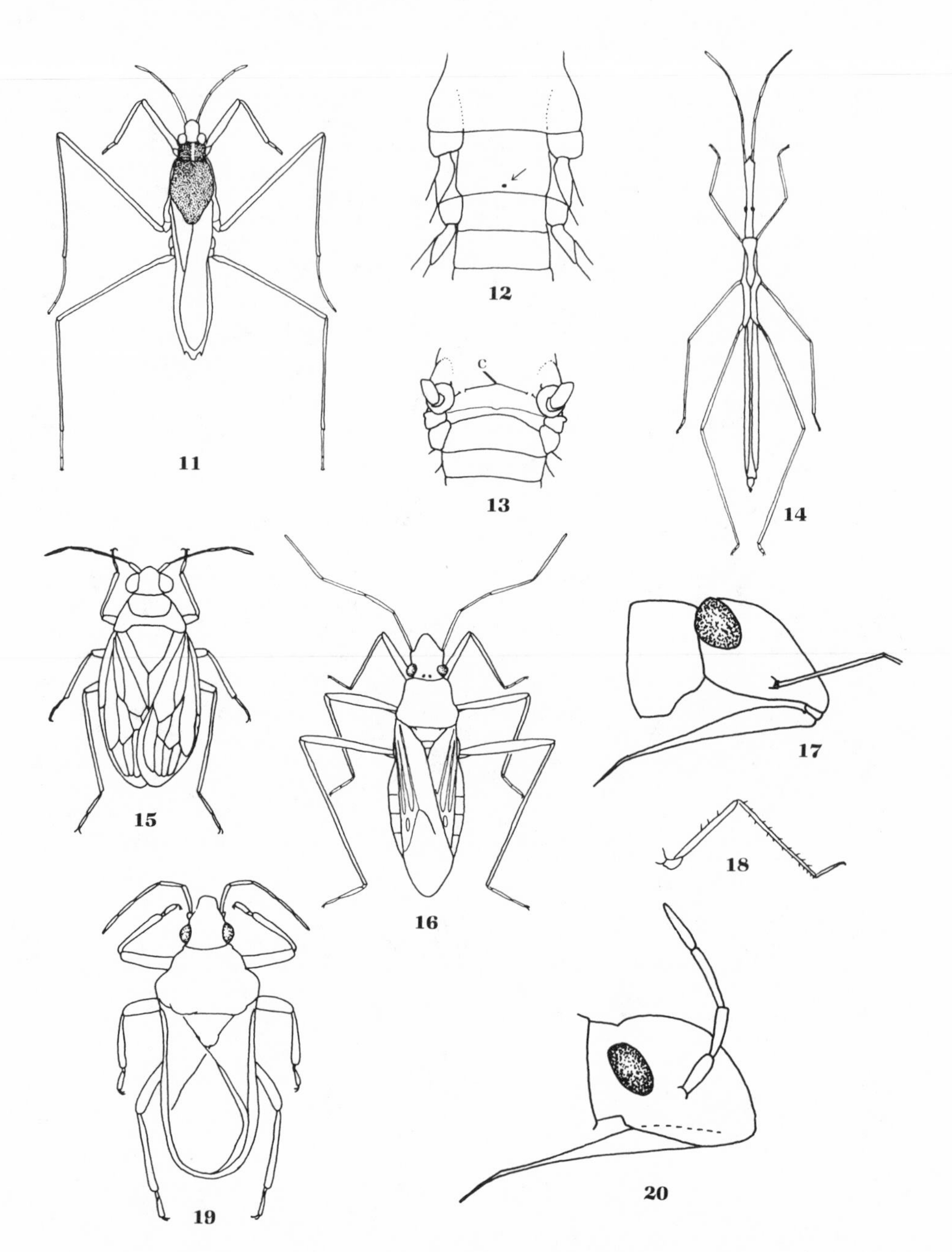

11. *Gerris* (Gerridae), dv. **12.** Meso- and metathorax, and abdominal segments 2–3 of *Gerris* (Gerridae), vv. **13.** Meso- and metathorax, and abdominal segments 2–3 of *Rhagovelia* (Veliidae), vv: c, scent-gland channel. **14.** *Hydrometra* (Hydrometridae), dv. **15.** *Salda* (Saldidae), dv. **16.** *Mesovelia* (Mesoveliidae), dv. **17.** Head of *Mesovelia* (Mesoveliidae), lv. **18.** Mesothoracic leg of *Mesovelia* (Mesoveliidae), lv. **19.** *Hebrus* (Hebridae), dv. **20.** Head of Hebridae, lv. dv, dorsal view; lv, lateral view; vv, ventral view. (11, 15, 19 modified from Usinger 1956; 12, 13 redrawn from Menke 1979; 14, 16 modified from Pennak 1978; 18 redrawn from Hilsenhoff 1981.)

15a (14b). Frons of both sexes flat in lateral view and densely covered with hair; eyes protuberant with inner anterior angles broadly rounded (Fig. 23) .... *Dasycorixa*

15b. Frons of female convex in lateral view (may be flattened in male) and hair sparse if present; eyes not protuberant, inner anterior angles not broadly rounded (Fig. 2) 16

16a (15b). Entire hemelytral pattern usually obliterated; upper surface of pala of male deeply incised (Fig. 24); vertex of male acuminate and carinate; both sexes with terminal claw of pala serrate at ventral margin; body length 5.0–5.5 mm .... *Ramphocorixa*

16b. Hemelytral pattern distinct, although limited areas may be obliterated in some species; male pala not deeply incised; vertex of male usually not acuminate (except *Arctocorisa* spp., *Cenocorixa* spp., *Sigara conocephala*), but if acuminate, not carinate; terminal claw of pala not serrate 17

17a (16b). Posterior apex of clavus not, or scarcely, surpassing a transverse line drawn through nodal furrows (Fig. 25); in males, strigil on left side as seen dorsally; body length 2.8–4.6 mm .... *Trichocorixa*

17b. Posterior apex of clavus plainly surpassing a transverse line drawn through nodal furrows (Fig. 26); in males, strigil on right side as seen dorsally (Fig. 1) 18

18a (17b). Rear margin of head angulate at middle; pronotum shorter than the length of the head (Fig. 27); interocular space much narrower than the width of an eye; body length 4.0–6.0 mm .... *Palmacorixa*

18b. Rear margin of head gently curved; pronotum longer than length of head (Fig. 28); interocular space about equal to the width of an eye 19

19a (18b). Prothoracic lobe (see Fig. 2) quadrate or trapezoidal, more or less truncate at end (Fig. 29); pruinose area at base of claval suture usually about ⅔ as long as postnodal pruinose areas and broadly rounded at apex (Fig. 30); body length 6.3–11.4 mm .... *Hesperocorixa*

19b. Prothoracic lobe rounded (Fig. 31); pruinose area at base of claval suture almost as long as postnodal pruinose area and narrowly rounded or pointed at apex (Fig. 32) 20

20a (19b). Markings on clavus and corium transverse, on corium linear or coarsely reticulate (Fig. 33), never finely reticulate 21

20b. Markings on clavus and corium narrow and broken, on corium finely reticulate (Fig. 34) 22

21a (20a). Strigil absent in male; pala of male with 2 rows of pegs (Fig.35); hemelytra dark olive and brown with little contrast; coriopruina not exceeding ½ the length of the clavopruina (Figs. 22, 25); body length 6.9–8.1 mm . . . *Callicorixa*

21b. Strigil present in male (Fig. 1); pala of male usually with 1 row of pegs (two

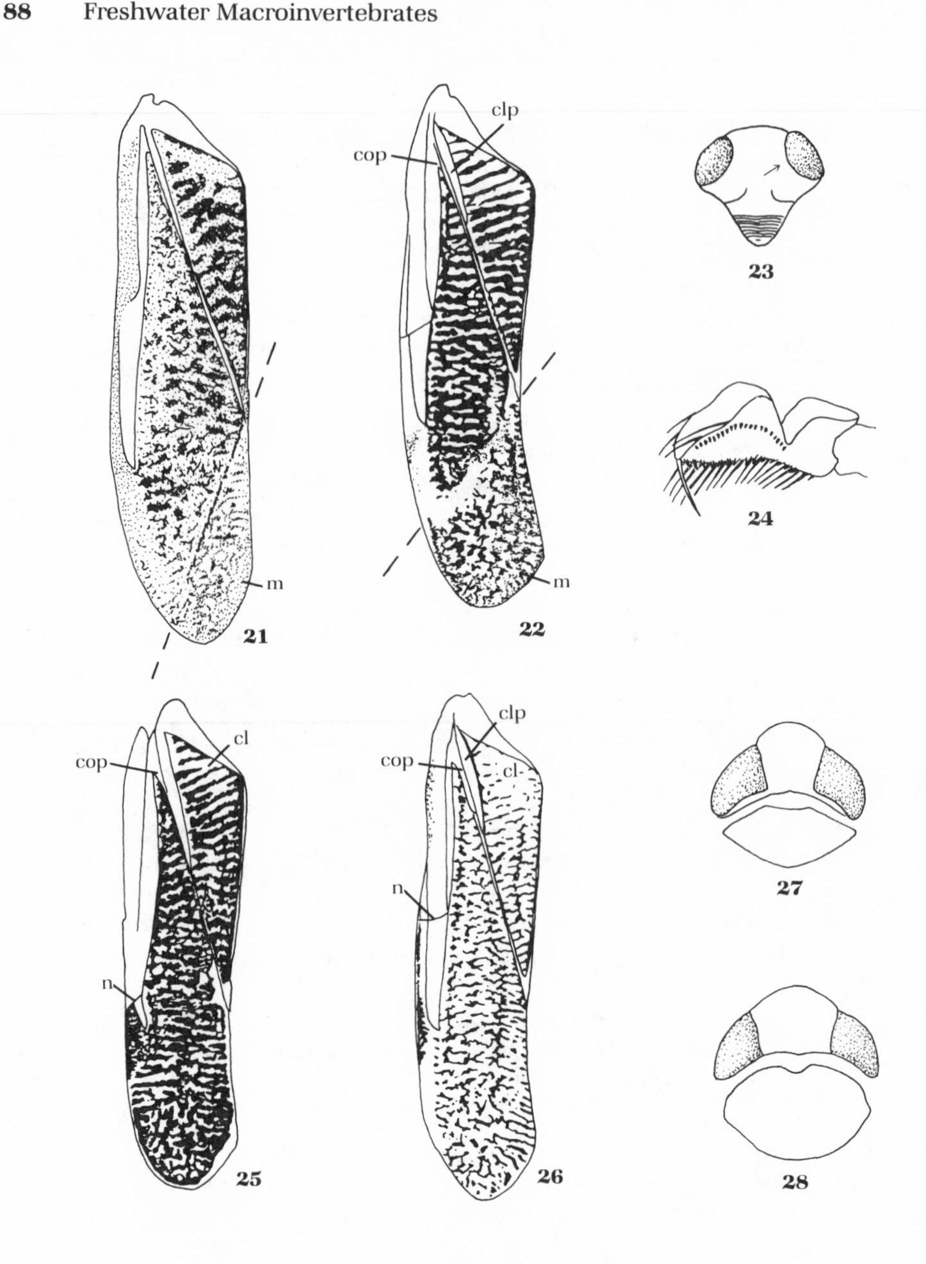

21. Hemelytron of *Cymatia* (Corixidae), dv: m, membrane. **22.** Hemelytron of *Callicorixa* (Corixidae), dv: clp, clavopruina; cop, coriopruina; m, membrane. **23.** Head of *Dasycorixa* (Corixidae), vv. **24.** First tarsus of *Ramphocorixa* (Corixidae), dv. **25.** Hemelytron of *Trichocorixa* (Corixidae), dv: cl, clavus; cop, coriopruina; n, nodal furrow. **26.** Hemelytron of *Corisella* (Corixidae), dv: cl, clavus; clp, clavopruina; cop, coriopruina; n, nodal furrow. **27.** Head and pronotum of *Palmacorixa* (Corixidae), dv. **28.** Head and pronotum of *Corisella* (Corixidae), dv. dv, dorsal view; vv, ventral view. (21 modified from Brooks and Kelton 1967; 22, 25, 26 modified from Menke 1979; 24 modified from Pennak 1978; 27, 28 redrawn from Hilsenhoff 1981.)

exceptions), but never as in Fig. 35; hemelytra contrasting yellow and brown; coriopruina usually extending half the length of clavopruina or more (Fig. 26) (four exceptions); body length 3.6–9.2 mm . . . . *Sigara*

22a (20b). Pala of male triangular (Fig. 36); frons and hemelytra without hair; prothoracic lobe tapering to a narrowly rounded apex (Fig. 37); body length 5.3–8.0 mm . . . . *Corisella*

22b. Pala of male elongate (Fig. 38) or arcuate (Fig. 3); frons and hemelytra with sparse, fine, long hair; prothoracic lobe not tapering to a narrowly rounded apex; body length 5.2–10.0 mm 23

23a (22b). Pronotum with a well-defined longitudinal median carina extending from anterior to posterior margin; body length 7.1–10.0 mm . . . . *Arctocorisa*

23b. Pronotum with a poorly defined longitudinal median carina confined to the anterior ⅓; body length 5.2–7.8 mm . . . . *Cenocorixa*

24a (6a). **Belostomatidae**: Tibia and tarsus of hindleg and middle similar (Fig. 39); basal segment of rostrum longer than or nearly equal to the 2nd segment; body length 18–25 mm . . . . *Belostoma*

24b. Tibia and tarsus of hindleg strongly compressed, thin, much broader than tibia and tarsus of middle leg (Fig. 40); basal segment of rostrum shorter than 2nd segment; body length > 40 mm . . . . *Lethocerus*

25a (5a). **Nepidae**: Body shape oval, more than ⅓ as wide as long; body length 18–20 mm . . . . *Nepa*

25b. Body slender subcylindrical, sticklike; body length 23–42 mm . . . . . . *Ranatra*

26a (8b). **Notonectidae**: Hemelytral commissure with an elliptical pit at anterior end (Fig. 41); body slender; antennae 3-segmented, 3rd segment longer than 2nd; body length 4.1–8.3 mm . . . . *Buenoa*

26b. Hemelytral commissure without a pit (Fig. 42); body thicker; antennae 4-segmented, 4th segment much smaller than 3rd; body length 8.5–15.5 mm . . . . *Notonecta*

27a (10a). **Gerridae**: Inner margin of eyes emarginate (Fig. 43); abdomen at least 4 times as long as wide 28

27b. Inner margin of eyes convexly rounded; abdomen less than 4 times as long as wide (Fig. 44) 31

28a (27a). Pronotum shiny 29

28b. Pronotum dull 30

29a (28a). Pronotum with median dorsal and lateral yellow stripes that sometimes continue to posterior lobe . . . . *Limnogonus*

29b. Pronotum with single, median pale spot . . . . *Neogerris*

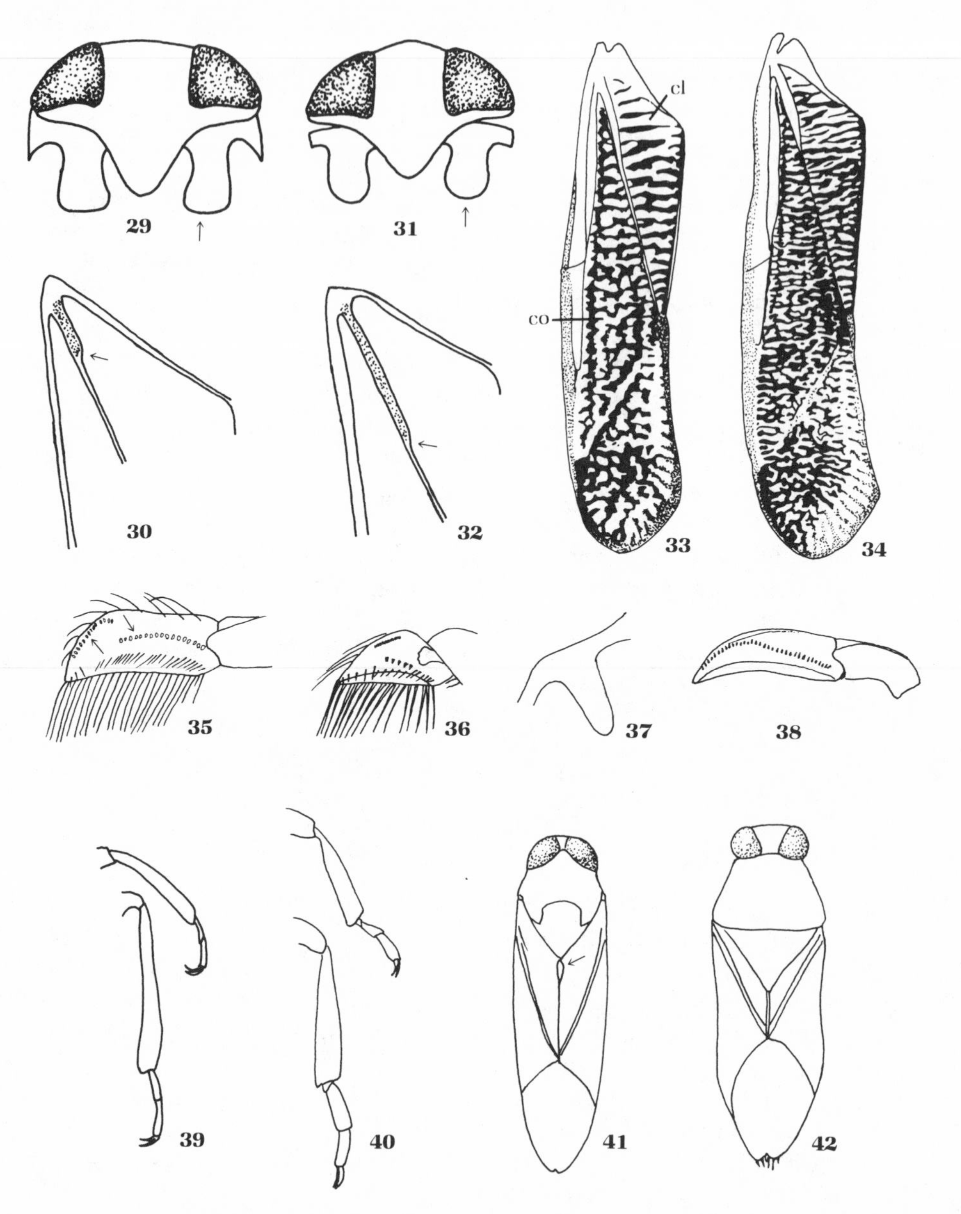

29. Head and prothoracic lobes of *Hesperocorixa* (Corixidae), vv. **30.** Claval suture of wing of *Hesperocorixa* (Corixidae), dv. **31.** Head and prothoracic lobes of *Sigara* (Corixidae), vv. **32.** Claval suture of wing of *Sigara* (Corixidae), dv. **33.** Hemelytron of *Sigara* (Corixidae), dv: cl, clavus; co, corium. **34.** Hemelytron of *Arctocorisa* (Corixidae), dv. **35.** Pala of male *Callicorixa* (Corixidae), dv. **36.** Pala of male *Corisella* (Corixidae), dv. **37.** Prothoracic lobe of *Corisella* (Corixidae), vlv. **38.** Pala of male *Arctocorisa* (Corixidae), dv. **39.** Middle and hindlegs of *Belostoma* (Belostomatidae), dv. **40.** Middle and hindlegs of *Lethocerus* (Belostomatidae), dv. **41.** *Buenoa* (Notonectidae) without legs, dv. **42.** *Notonecta* (Notonectidae) without legs, dv. dv, dorsal view; lv, lateral view; vlv, ventrolateral view; vv, ventral view. (33, 34, 38 modified from Brooks and Kelton 1967; 35, 42 redrawn or modified from Pennak 1978; 36, 40 modified from Menke 1979; 37 redrawn from Hilsenhoff 1981; 39, 41 modified from Usinger 1956.)

30a (28b). First antennal segment considerably shorter than 2nd and 3rd segments together .... *Limnoporus*

30b. First antennal segment longer than, equal to, or slightly shorter than 2nd and 3rd segments together .... *Gerris*

31a (27b). First antennal segment subequal in length to remaining 3 together; body length 3.0–5.0 mm .... *Metrobates*

31b. First antennal segment much shorter than remaining 3 together 32

32a (31b). Third antennal segment with several stiff bristles (Fig. 45); abdomen not much shorter than rest of body (Fig. 44); body length 2.3–3.5 mm .... *Rheumatobates*

32b. Third antennal segment with fine pubescence only; abdomen much shorter than rest of body; body length 3.0–4.3 mm .... *Trepobates*

33a (10b). **Veliidae**: Mesotarsi with plumose hairs and leaflike claws (Fig. 46); body length 3.4–4.6 mm .... *Rhagovelia*

33b. Mesotarsi without plumose hairs or leaflike claws 34

34a (33b). Tarsal formula 1:2:2; body length 1.5–3.0 mm .... *Microvelia*

34b. Tarsal formula 3:3:3; body length 4.0–6.2 mm .... *Paravelia*

35a (13b). **Hebridae**: Antennae consisting of 4 large segments; body length 1.7–2.2 mm .... *Merragata*

35b. Antennae consisting of 5 large segments (as in Fig. 19); body length 1.8–2.2 mm; riparian .... *Hebrus*

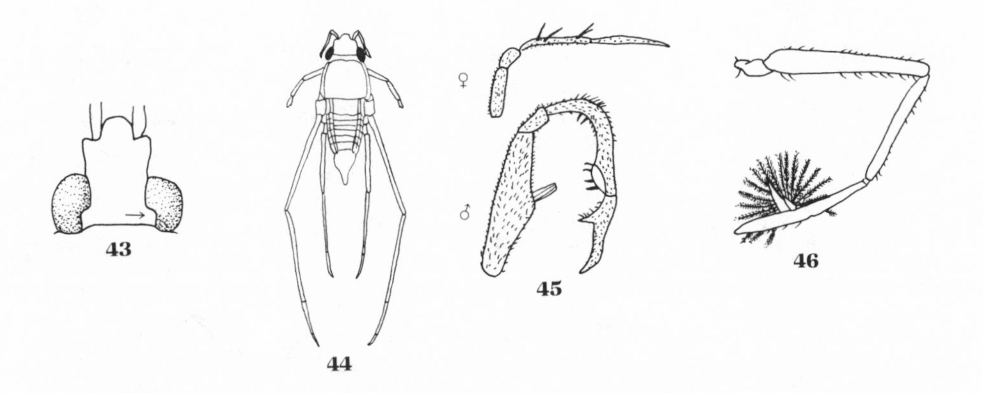

43. Head of *Gerris* (Gerridae), dv. **44.** *Rheumatobates* (Gerridae), dv. **45.** Antennae of *Rheumatobates* (Gerridae), lv. **46.** Mesotarsus of *Rhagovelia* (Veliidae), lv. dv, dorsal view; lv, lateral view. (43, 45 modified or redrawn from Hilsenhoff 1981; 44 modified from Pennak 1978; 46 modified from Menke 1979.)

References on Aquatic and Semiaquatic Hemiptera Systematics

(*Used in construction of key.)

Aiken, R.B. 1985. Sound production by aquatic insects. Biol. Rev. 60:163–212.

Anderson, L.D. 1932. A monograph of the genus *Metrobates* (Hemiptera: Gerridae). Univ. Kans. Sci. Bull. 20:297–311.

Anderson, N.M. 1979. Phylogenetic inference as applied to the study of evolutionary diversification of semiaquatic bugs (Hemiptera: Gerromorpha). Syst. Zool. 28:554–578.

——. 1981. Semiaquatic bugs: phylogeny and classification of the Hebridae (Heteroptera: Gerromorpha) with revisions of *Timasius, Neotimasius* and *Hyrcanus*. Syst. Entomol. 6:377–412.

——. 1982. The semiaquatic bugs (Hemiptera, Gerromorpha). Scand. Sci. Press, Klamperborg, Denmark. 455 pp.

Anderson, N.M., and J.T. Polhemus. 1980. Four new genera of Mesoveliidae (Hemiptera: Gerromorpha) and the phylogeny and classification of the family. Entomol. Scand. 11:369–392.

Bacon, J.A. 1956. A taxonomic study of the genus *Rhagovelia* of the Western Hemisphere. Univ. Kans. Sci. Bull. 38:695–913.

Blatchley, W.S. 1926. Heteroptera or true bugs of eastern North America with especial reference to the faunas of Indiana and Florida. Nature Publ. Co., Indianapolis, Ind. 1,116 pp.

Bobb, M.L. 1974. The insects of Virginia. No. 7: The aquatic and semiaquatic Hemiptera of Virginia. Va. Polytech. Inst. Res. Div. Bull. 87:1–195.

Britton, W.E. (ed.). 1923. Guide to the insects of Connecticut. Part IV. The Hemiptera or sucking insects of Connecticut. Bull. Conn. State Geol. Nat. Hist. Surv. 34:1–807.

*Brooks, A.R., and L.A. Kelton. 1967. Aquatic and semiaquatic Heteroptera of Alberta, Saskatchewan, and Manitoba (Hemiptera). Mem. Entomol. Soc. Can. 51:1–92.

Calabrese, D. 1974. Keys to the adults and nymphs of the genus *Gerris* Fabricius occurring in Connecticut. *In* R.L. Beard (ed.). 25th Anniv. Mem. Conn. Entomol. Soc., pp. 227–266. New Haven, Conn.

——. 1977. Vicariance biogeography of the insect family Gerridae. Amer. Zool. 17:950.

*——. 1980. Zoogeography and cladistic analysis of the Gerridae (Hemiptera: Heteroptera). Misc. Publ. Entomol. Soc. Amer. 11:1–119.

——. 1985. The Gerridae (Hemiptera: Heteroptera): a discussion of two reconstructed phylogenies. Ann. Entomol. Soc. Amer. 78:137–140.

Cheng, L. 1975. Insecta, Hemiptera: Heteroptera, Gerridae, genus *Halobates*. Fiches Identif. Zooplankton 147:1–4.

——. 1985. Biology of *Halobates* (Heteroptera: Gerridae). Annu. Rev. Entomol. 30:111–136.

Cheng, L., and C.H. Fernando. 1970. The waterstriders of Ontario. R. Ont. Mus. Life Sci. Misc. Publ. 23 pp.

China, W.E. 1955. The evolution of the water bugs. Bull. Natl. Inst. Sci. India 7:91–103.

China, W.E., and R.L. Usinger. 1949. Classification of the Veliidae (Hemiptera) with a new genus from South Africa. Ann. Mag. Nat. Hist. 12:343–354.

Deay, H.O., and G.E. Gould. 1936. The Hemiptera of Indiana. I. Family Gerridae. Amer. Midl. Nat. 17:753–769.

Drake, C.J. 1948. Two new Mesoveliidae with check list of American species (Hemiptera). Bol. Entomol. Venez. 7:145–147.

Drake, C.J., and H.C. Chapman. 1953. A preliminary report on the Pleidae of the Americas. Proc. Biol. Soc. Wash. 66:53–60.

Drake, C.J., and H.M. Harris. 1932a. IV. A synopsis of the genus *Metrobates* Uhler (Hemiptera: Gerridae). Ann. Carnegie Mus. 21:83–88.

——. 1932b. A survey of the species of *Trepobates* Uhler (Hemiptera: Gerridae). Bull. Brooklyn Entomol. Soc. 27:113–123.

——. 1934. The Gerrinae of the Western Hemisphere (Hemiptera). Ann. Carnegie Mus. 23:179–240.

Drake, C.J., and L. Hoberlandt. 1950. Catalog of genera and species of Saldidae. Acta Entomol. Mus. Natn. Prague 26(376):1–12.

Drake, C.J., and F.C. Hottes. 1951. Notes on the genus *Rheumatobates*. Proc. Biol. Soc. Wash. 64:147–155.

——. 1954. New American waterstriders. Fla. Entomol. 37:151–155.

Drake, C.J., and R.F. Hussey. 1955. Concerning the genus *Microvelia* Westwood with descriptions of two new species and a check-list of the American forms (Hemiptera: Veliidae). Fla. Entomol. 38:95–115.

Fairbairn, D.J. 1986. Does alary dimorphism imply dispersal dimorphism in the waterstrider *Gerris remigis*? Ecol. Entomol. 11:358–368.

Foster, W.A., and J.E. Treherne. 1982. Reproductive behavior of the ocean skater *Halobates robustus* (Hemiptera: Gerridae) in the Galapagos Islands. Oecologia 55:202–207.

Griffith, M.E. 1945. The environment, life history, and structure of the water boatman, *Ramphocorixa acuminata* (Uhler). Univ. Kans. Sci. Bull. 30:241–365.

Harrison, R.G. 1980. Dispersal polymorphisms in insects. Annu. Rev. Ecol. Syst. 1:95–118.

Harvey, G.W. 1907a. A ferocious water bug. Can. Entomol. 39:17–21.

——. 1907b. A ferocious water bug. Proc. Entomol. Soc. Wash. 8:72–75.

Hilsenhoff, W.L. 1970. Corixidae of Wisconsin. Trans. Wisc. Acad. Sci. Arts Lett. 58:203–235.

*——. 1981. Aquatic insects of Wisconsin. Nat. Hist. Counc. Publ. no. 2. Univ. Wisc., Madison. 60 pp.

*——. 1984. Aquatic Hemiptera of Wisconsin. Great Lakes Entomol. 17:29–50.

*——. 1986. Semiaquatic Hemiptera of Wisconsin. Great Lakes Entomol. 19:7–20.

*Hungerford, H.B. 1919. The biology and ecology of aquatic and semi-aquatic Hemiptera. Univ. Kans. Sci. Bull. 11:3–328.

——. 1922a. The life history of the toad bug. Univ. Kans. Sci. Bull. 14:145–171.

——. 1922b. The Nepidae in North America north of Mexico (Heteroptera). Univ. Kans. Sci. Bull. 14:432–469.

——. 1924. Stridulation of *Buenoa limnocastoris* Hungerford and systematic notes on the *Buenoa* of the Douglas Lake region of Michigan, with the description of a new form. Ann. Entomol. Soc. Amer. 17:223–236.

——. 1933. The genus *Notonecta* of the world. Univ. Kans. Sci. Bull. 21:5–196.

*——. 1948. The Corixidae of the Western Hemisphere (Hemiptera). Univ. Kans. Sci. Bull. 32:1–827.

——. 1954. The genus *Rheumatobates* Bergoth (Hemiptera: Gerridae). Univ. Kans. Sci. Bull. 36:529–588.

——. 1958. Some interesting aspects of the world distribution and classification of aquatic and semiaquatic Hemiptera. *In* E.C. Becker (ed.). Proc. 10th Int. Cong. of Entomol., vol. 1, August 17–25, 1956, pp. 337–348. Montreal.

*——. 1959. Hemiptera. *In* W.T. Edmondson (ed.). Fresh-water biology, 2nd ed., pp. 958–972. John Wiley and Sons, New York.

Hungerford, H.B., and N.E. Evans. 1934. The Hydrometridae of the Hungarian National Museum and other studies in the family (Hemiptera). Ann. Mus. Natl. Hungaricici 28:31–113.

Hungerford, H.B., and R. Matsuda. 1960. Keys to the subfamilies, tribes, genera, and subgenera of the Gerridae of the world. Univ. Kans. Sci. Bull. 41:3–23.

Hutchinson, G.E. 1945. On the species of *Notonecta* (Hemiptera: Heteroptera) inhabiting New England. Trans. Conn. Acad. Arts Sci. 36:599–605.

Jaczewski, T. 1930. Notes on the American species of the genus *Mesovelia*. Muls. Ann. Mus. Zool. Polonici 9:3–12.

Jansson, A. 1986. The Corixidae of Europe and some adjacent regions. Acta Entomol. Fenn. 47:1–94.

Jarvinen, O., M. Nummelin, and K. Vepsalainen. 1977. A method for estimating population densities of water-striders (*Gerris*). Notul. Entomol. 57:25–28.

La Rivers, I. 1948. A new species of *Pelocoris* from Nevada, with notes on the genus in the United States (Hemiptera: Naucoridae). Ann. Entomol. Soc. Amer. 41:371–376.

Lauck, D.R. 1979. Family Corixidae/water boatmen. *In* A.S. Menke (ed.). The semiaquatic Hemiptera of California, pp. 87–123. Bull. Calif. Insect Surv. 21:1–166.

Lauck, D.R., and A.S. Menke. 1961. The higher classification of the Belostomatidae. Ann. Entomol. Soc. Amer. 54:644–657.

Matheson, R. 1907. *Belostoma* eating a bird. Entomol. News 18:452.

*Matsuda, R. 1960. Morphology, evolution and a classification of the Gerridae. Univ. Kans. Sci. Bull. 41:25–632.

Menke, A.S. 1958. A synopsis of the genus *Belostoma* Latreille, of America north of Mexico, with the description of a new species. Bull. S. Calif. Acad. Sci. 57:154–174.

——. 1963. A review of the genus *Lethocerus* in North and Central America, including the West Indies. Ann. Entomol. Soc. Amer. 56:261–267.

*——. 1979. The semiaquatic and aquatic Hemiptera of California. Bull. Calif. Insect Surv. 21:1–166.

*Miller, N.C.E. 1956. Biology of the Heteroptera. Leonard Hill, London. 172 pp.

Pajunen, V.I. 1972. Evaluation of a removal method for estimating the numbers of rock pool corixids (Hemiptera: Corixidae). Ann. Zool. Fenn. 9:152–155.

Papacek, M., and P. Stys. 1985. Fossil Clypostemmatinae compared with modern Notonectidae (Heteroptera). Acta Entomol. Bohemoslov. 82:28–48.

Parshley, H.M. 1921. On the genus *Microvelia* Westwood. Bull. Brooklyn Entomol. Soc. 16:87–93.

——. 1925. A bibliography of the North American Hemiptera-Heteroptera. Smith College, Northampton, Mass. 252 pp.

*Pennak, R.W. 1978. Freshwater invertebrates of the United States, 2nd ed. John Wiley and Sons, New York. 803 pp.

Polhemus, J.T. 1976. A reconsideration of the status of the genus *Paravelia* Breddin, with other notes and a check-list of species (Veliidae: Heteroptera). J. Kans. Entomol. Soc. 49:509–513.

——. 1982. Marine Hemiptera of the Northern Territory. J. Austral. Entomol. Soc. 21:5–11.

——. 1984. Aquatic and semiaquatic Hemiptera. *In* R.W. Merritt and K.W. Cummins (eds.). An introduction to the aquatic insects of North America, 2nd ed., pp. 231–260. Kendall/Hunt Publ. Co., Dubuque, Iowa.

——. 1985. Nomenclatural changes for North American Saldidae. Proc. Entomol. Soc. Wash. 87:893.

Polhemus, J.T., and C.N. McKinnon. 1983. Notes on the Hebridae of the western hemisphere with descriptions of two new species (Heteroptera: Hemiptera). Proc. Entomol. Soc. Wash. 85:110–115.

Rao, T.K.R. 1976. Bioecological studies on some aquatic Hemiptera: Nepidae. Entomon. 1:123–132.

Ruhoff, F.A. 1968. Bibliography and index to scientific contributions of Carl J. Drake for the years 1914–1967. U.S. Natl. Mus. Bull. 267:1–81.

Sailer, R.I. 1948. The Genus *Trichocorixa*. Univ. Kans. Sci. Bull. 32:289–407.

Sanderson, M.W. 1982. Aquatic and semiaquatic Heteroptera. *In* A.R. Brigham, W.U. Brigham, and A. Gnilka (eds.). Aquatic insects and oligochaetes of North and South Carolina, pp. 6.1–6.94. Midwest Aquatic Enterprise, Mahomet, Ill.

Schaefer, C.W., and M.I. Schaefer. 1979. Corixids (Hemiptera: Heteroptera) attracted to automobile roof. Entomol. News 90:230.

Schell, D.V. 1943. The Ochteridae of the Western Hemisphere. J. Kans. Entomol. Soc. 16:29–47.

Schuh, R.T. 1967. The shore bugs of the Great Lakes region. Contrib. Amer. Entomol. Inst. 2:1–35.

Slater, J.A. 1974. A preliminary analysis of the derivation of the Heteroptera fauna of the northeastern United States with special reference to the fauna of Connecticut. *In* R.L. Beard (ed.). 25th Anniv. Mem. Conn. Entomol. Soc., pp. 145–213. New Haven, Conn.

*Smith, C.L., and J.T. Polhemus. 1978. The Veliidae of America north of Mexico: keys and a check list. Proc. Entomol. Soc. Wash. 60:56–68.

Smith, R.L. 1979. Paternity assurance and altered roles in the mating behaviour of a giant water bug, *Abedus herberti* (Heteroptera, Belostomatidae). Anim. Behav. 27:716–725.

Spence, J.R. 1980. Density estimation for water-striders (Heteroptera: Gerridae). Freshwat. Biol. 10:563–570.

Todd, E.L. 1955. A taxonomic revision of the family Gelastocoridae. Univ. Kans. Sci. Bull. 37:277–475.

Torre-Bueno, J.R. de la. 1924. A preliminary survey of the species of *Microvelia* Westwood (Veliidae: Heteroptera) of the western world, with description of a new species from the southern United States. Bull. Brooklyn Entomol. Soc. 19:186–194.

——. 1926. The family Hydrometridae in the Western Hemisphere. Entomol. Amer. 7:83–138.

Truxal, F.S. 1953. A revision of the genus *Buenoa* (Hemiptera: Notonectidae). Univ. Kans. Sci. Bull. 35:1351–1523.

Usinger, R.L. 1941. Key to the subfamilies of Naucoridae with a generic synopsis of the new subfamily Ambrysinae (Hemiptera). Ann. Entomol. Soc. Amer. 34:5–16.

*——. 1956. Aquatic Hemiptera. *In* R.L. Usinger (ed.). Aquatic insects of California, pp. 182–228. Univ. Calif. Press, Berkeley.

*Van Duzee, E.P. 1917. Catalogue of the Hemiptera of America north of Mexico, excepting the Aphididae, Coccidae, and Aleyrodidae. Univ. Calif. Publ., Tech. Bull. Coll. Agric. Agric. Exp. Stn. Entomol. 2:1–902.

Wilcox, R.S. 1979. Sex discrimination in *Gerris remigis*: role of a surface wave signal. Science 206:1325–1327.

Wilcox, R.S., and J.R. Spence. 1986. The mating system of two hybridizing species of water striders (Gerridae). I. Ripple signal functions. Behav. Ecol. Sociobiol. 19:79–86.

Wilson, C.A. 1958. Aquatic and semi-aquatic Hemiptera of Mississippi. Tulane Stud. Zool. 6:115–170.

Woodruff, L.C. 1956. Biographical notes and bibliography of Herbert Barker Hungerford. Univ. Kans. Sci. Bull. 38:ii–xiv.

8 Trichoptera

Classification

The Trichoptera (caddisflies) belong to the infraclass Neoptera, division Endopterygota; their wings develop internally instead of externally in wingpads like those of the previous orders. They are closely related to the Lepidoptera, an insect order with very few aquatic species. The larvae of all the species of Trichoptera are aquatic, except for a few cases (Erman 1981). Three superfamilies of Trichoptera have been distinguished, which include five groups based on case-building behavior. Hydropsychoidea (Hydropsychidae, Philopotamidae, Polycentropodidae, and Psychomyiidae) are the net spinners and retreat makers; Rhyacophiloidea include the free-living forms (Rhyacophilidae), saddle-case makers (Glossosomatidae), and purse-case makers (Hydroptilidae); and Limnephiloidea are the tube-case makers (all other families; see Checklist) (Wiggins 1977, 1984). A new classification for Trichoptera has been proposed (Weaver 1984, Weaver and Morse 1986), but it is still controversial and will need some modification before it is widely accepted. We therefore follow the classification using three superfamilies presented above.

Life History

Caddisflies are holometabolous insects whose eggs are deposited in gelatinous matrices in or out of water. These matrices minimize water loss and maximize respiratory exchange. Wiggins et al. (1976) demonstrated that the gelatin of a limnephilid egg mass is composed of fibrils resembling the chorion of terrestrial insect eggs, enabling eggs of inhabitants of temporary ponds to survive a dry period. Some species drop their eggs into the water during flight, while others actually descend beneath the water surface to deposit their eggs on the substrate. Some (Rhyacophilidae) place their eggs singly, but most deposit them in sheets attached to the rocks (Balduf 1939). Many species that inhabit temporary ponds lay eggs in the basins of dry pools in the fall, when the soil surface is beginning to become more moist. After larvae hatch, they may remain within the gelatinous matrix for months until the pool basin is reflooded (Wiggins 1973c). Moisture is the apparent stimulus for larvae to break out of the gelatinous matrix and begin case building or net spinning.

Larvae go through five instars (rarely six or seven). Caddisfly larvae of the superfamilies Limnephiloidea and Rhyacophiloidea (excluding the Rhyacophilidae) and are well known for their diversity in case making: they use silk to cement portable cases made of substrate materials in shapes that may be species-specific. Case-building behavior may be stereotypic, although many species are flexible in the materials they use for case construction. Some species add pieces of substrate to the anterior end of their cases as they grow; others abandon cases and construct new ones after each molt (Pennak 1978). Cases function in respiration (see Respiration) and as ballast, camouflage, and protection from biting predators. Larvae of the superfamily Hydropsychoidea spin silken nets for filter feeding and may reside in adjacent fixed retreats that they construct with silk and/or substrate materials. They periodically emerge from these retreats to obtain food from catch nets (see Feeding).

Most species of caddisflies are univoltine, but some complete more than one generation per year, whereas others require two years for development (Wiggins 1984). After completing larval development, case makers and retreat makers seal off the ends of their cases or retreats and attach them with silk to the substrate. Free-living forms spin a silken cocoon. The pupal phase is generally two to three weeks, although some species may overwinter as pupae.

Many caddisflies undergo a diapause phase, which, depending on the species, occurs in one of several life history stages. Diapausing eggs may overwinter in dry pool basins; larvae may diapause as terminal instars within the sealed pupal case, a stage termed the *prepupa*; and adults may undergo an ovarian diapause, delaying sexual maturity until late summer. The latter behavior prevents species of temporary ponds from ovipositing into pool basins that are excessively dry, and it fosters deposition of eggs into the moister environment characteristic of early autumn (Wiggins 1977, Wiggins et al. 1980).

Caddisfly pupae have functional mandibles that they use to chew their way out of the pupal case once they are ready to emerge as adults (Wiggins 1977). Adults are mostly crepuscular, limiting their activity to dusk or darkness. Some species feed on plant nectar, while others do not feed at all. Adults live from a few weeks to several months, depending on the species and the nature of the habitat. Caddisflies mate in flight, on the ground, or on vegetation near aquatic habitats (Unzicker et al. 1982). Mass emergences of some species from large rivers are considered a nuisance by residents, since the insects are attracted to outdoor lights; human allergies to the scales on their wings have also been reported (Fremling 1960).

Habitat

Like mayflies and stoneflies, caddisflies probably evolved in cold, fast-flowing streams, since families with more primitive characteristics (e.g., Rhyacophilidae) are restricted to those habitats. Wiggins (1977) hypothesizes that the use of silk for case construction enabled the Trichoptera to become more diverse ecologically, providing a respiratory mechanism whereby habitats with higher temperatures and lower dissolved-oxygen levels could be exploited (Mackay and Wiggins 1979). At present, caddisflies inhabit a wide range of habitats from the ancestral cool streams to warm streams, permanent lakes and marshes, and permanent and temporary ponds (Wiggins 1977). One species has been found in tide pools off the coast of New Zealand; the females oviposit through the papillar pores of starfishes (Winterbourn and Anderson 1980). Caddisflies have been generally classified as clingers, sprawlers, or climbers, although a few are burrowers (Wiggins 1984).

Feeding

Caddisfly larvae occupy every conceivable trophic level or functional feeding group. Many Limnephiloidea are shredders or grazers, and Hydropsychoidea are characteristically filter feeders or predators, using silken nets to collect seston or catch prey. Mesh sizes of these nets vary, a phenomenon that was once thought to enable coexisting species to capture food particles of different sizes (Wallace et al. 1977, Wallace and Merritt 1980). However, recent evidence suggests that caddisfly nets are not simple sieves, since a correlation does not necessarily exist between the size of the mesh and the sizes of the particles entrained in the nets. Caddisflies can also be selective feeders, preferentially removing more-nutritious foods from their nets (algae, animals) (Alstad 1982). Some Limnephiloidea filter feed (Brachycentridae: *Brachycentrus*) by orienting their legs into the current to trap particles suspended in the water column. The saddle- and purse-case makers are specialized algal grazers and can defend territories of rich algal resources or depress algal densities, thereby outcompeting other grazers (McAuliffe 1984, Hart 1985). The free-living forms and a few of the tube-case makers feed on other insect larvae, crustaceans, or annelids (Pennak 1978).

Respiration

The evolution of case-building behavior greatly increased the respiratory efficiency of the Trichoptera, enabling them to invade lentic habitats where dissolved oxygen was less plentiful. The case is generally held in place by two or three (depending on the species) spacing tubercles on the first abdominal segment of the larva. These humps maintain a space between the body of the larva and the case through which a current is created by dorsoventral abdominal undulations. In this way, larvae can replenish the dissolved-oxygen supply around their gills and can survive in conditions of low oxygen. Lotic non-case-bearing caddisflies have numerous branched abdominal gills (Hydropsychidae) or rely largely on cuticular respiration (Rhyacophilidae).

Collection and Preservation

Simple D-frame dip nets are adequate for collecting most caddisfly larvae. Eighty percent ethanol is a suitable preservative, but before they are preserved, Wiggins (1977) suggests killing them by dropping them in Kahle's solution or in boiling water.

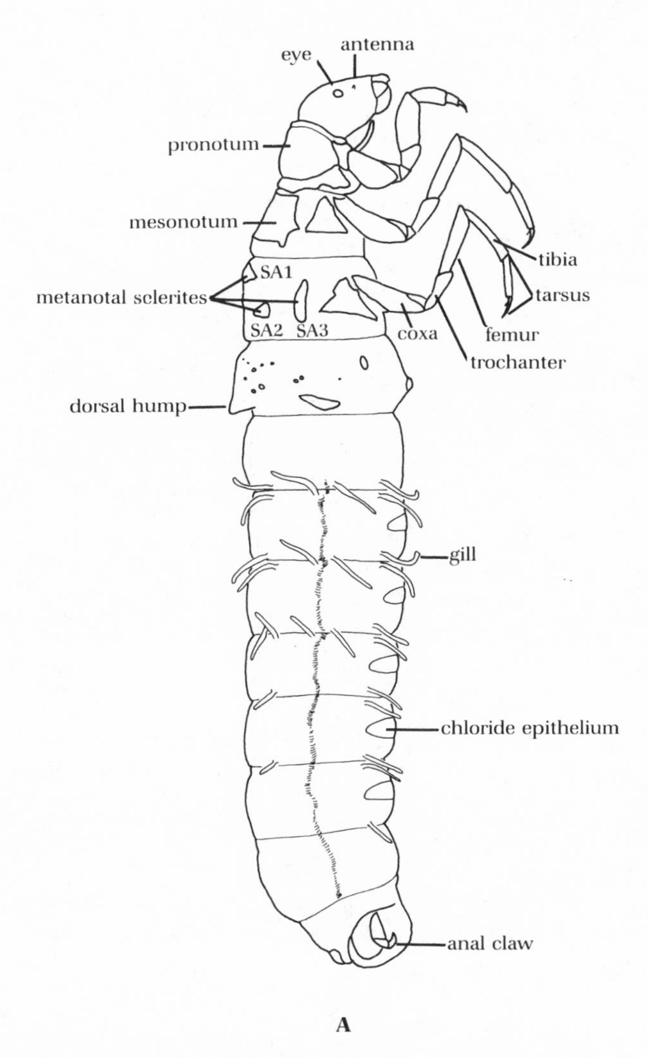

A. *Hydatophylax* (Limnephilidae), lateral view. (Modified from Wiggins 1977.)

Identification

Wiggins's (1977) definitive work has made it possible for students to identify larvae of most genera of caddisflies. However, the larvae of only one-third of the caddisfly species known to exist in North America have been identified. Clearly, larval association is needed. Use of the laboratory rearing techniques described by Wiggins (1977) may yield this information. Figure A illustrates the general morphological features of larval caddisflies. Features commonly used in our keys to separate genera of larval caddisflies are head and thoracic sclerotization or setation, gill development, thoracic leg and tarsal claw structure, shape of the fore trochantin, features of the anal claw, head or prothorax coloration, antennal length and location, presence of abdominal chloride epithelia, and structure of cases.

More-Detailed Information

Balduf 1939, Denning 1956, Wallace et al. 1977, Wiggins 1977, Pennak 1978, Wallace and Merritt 1980, Hilsenhoff 1981, Unzicker et al. 1982, Wiggins 1984.

Checklist of Trichoptera Larvae

Superfamily Hydropsychoidea
- Hydropsychidae
  - *Arctopsyche*[a]
  - *Cheumatopsyche*
  - *Diplectrona*
  - *Homoplectra*
  - *Hydropsyche*
  - *Macrostemum*
  - *Parapsyche*[a]
  - *Potamyia*
- Philopotamidae
  - *Chimarra*
  - *Dolophilodes*
  - *Wormaldia*
- Polycentropodidae
  - *Cernotina*
  - *Cyrnellus*
  - *Neureclipsis*
  - *Nyctiophylax*
  - *Phylocentropus*[b]
  - *Polycentropus*
- Psychomyiidae
  - *Lype*
  - *Psychomyia*

Superfamily Limnephiloidea
- Beraeidae
  - *Beraea*
- Brachycentridae
  - *Adicrophleps*
  - *Brachycentrus*
  - *Micrasema*
- Calamoceratidae
  - *Heteroplectron*
- Helicopsychidae
  - *Helicopsyche*
- Lepidostomatidae
  - *Lepidostoma*
  - *Theliopsyche*
- Leptoceridae
  - *Ceraclea*
  - *Leptocerus*
  - *Mystacides*
  - *Nectopsyche*
  - *Oecetis*
  - *Setodes*
  - *Triaenodes*
- Limnephilidae
  - *Anabolia*
  - *Apatania*
  - *Arctopora*
  - *Asynarchus*
  - *Chyranda*
  - *Frenesia*
  - *Glyphopsyche*
  - *Goera*
  - *Goerita*
  - *Grammotaulius*
  - *Hesperophylax*
  - *Hydatophylax*
  - *Ironoquia*
  - *Lenarchus*
  - *Leptophylax*[c]
  - *Limnephilus*
  - *Madeophylax*
  - *Nemotaulius*
  - *Onocosmoecus*
  - *Phanocelia*[d]
  - *Platycentropus*
  - *Pseudostenophylax*
  - *Psychoglypha*
  - *Pycnopsyche*
- Molannidae
  - *Molanna*
- Odontoceridae
  - *Marilia*
  - *Psilotreta*
- Phryganeidae
  - *Agrypnia*
  - *Banksiola*
  - *Fabria*
  - *Hagenella*
  - *Oligostomis*
  - *Phryganea*
  - *Ptilostomis*
- Sericostomatidae
  - *Agarodes*
- Uenoidae
  - *Neophylax*

Superfamily Rhyacophiloidea
- Glossosomatidae
  - *Agapetus*
  - *Culoptila*
  - *Glossosoma*
  - *Protoptila*
- Hydroptilidae
  - *Agraylea*
  - *Dibusa*
  - *Hydroptila*
  - *Ithytrichia*
  - *Leucotrichia*
  - *Mayatrichia*
  - *Neotrichia*
  - *Ochrotrichia*
  - *Orthotrichia*
  - *Oxyethira*
  - *Palaeagapetus*
  - *Stactobiella*
- Rhyacophilidae
  - *Rhyacophila*

[a]Schmid (1968) and Nimmo (1987) place these two genera in the family Arctopsychidae, but we follow Wiggins (1977) in placing them in the family Hydropsychidae.

[b]Schmid (1980a) places this genus in the family Hyalopsychidae, but a distinct family status is questionable.

[c]Adults have been described, but larvae are still unknown; not keyed.

[d]Fairchild and Wiggins (1989) described the previously unknown larva of *Phanocelia* from sphagnum bogs as this book was in press. It is therefore not keyed here, and the reader is advised to refer to their paper when collecting in sphagnum bogs.

Key to Genera of Trichoptera Larvae

1a. Case spiral and made of sand grains or tiny stones so that it resembles a snail shell (Fig. 1); anal claw with many short teeth forming a comb (Fig. 2) . . . . **Helicopsychidae**, *Helicopsyche*

1b. Case not like a snail shell, made of varying materials; anal claw not as above 2

2a (1b). Each thoracic segment covered with a single dorsal plate, which may have a mesal or transverse fracture line 3

2b. Metanotum mostly membranous, having only scattered hairs or small plates, or divided into 2 or more sclerites 22

3a (2a). Abdomen with ventral rows of branched gills; no portable case . . . . **Hydropsychidae** 4

3b. Abdomen without gills, and usually much enlarged (Fig. 3); larvae < 5 mm long and usually in a barrel- or purselike case that may be attached to the substrate . . . . **Hydroptilidae** 11

4a (3a). **Hydropsychidae**: Head with a broad, depressed, flat, dorsal area surrounded by an extensive arcuate carina (Figs. 4, 5); anterior margin of tibia and tarsus of foreleg with a dense brush of pale setae (Fig. 6) . . . . *Macrostemum*

4b. Head somewhat flat dorsally, but lacking carina; tibia and tarsus of foreleg without setal brush 5

5a (4b). Genae completely separated by an elongate gula (Figs. 7, 8) 6

5b. Genae fused for most of their length dividing gula into an anterior and posterior sclerite, the latter often reduced or absent (Figs. 9, 10) 7

6a (5a). Gula with sides nearly parallel (Fig. 7); abdomen with short, dark, scalelike setae on dorsum and arranged in tufts along posterior margin (Fig. 11) . . . . *Parapsyche*

6b. Gula narrowed posteriorly (Fig. 8); abdomen with only coarse hairs of varying lengths, never in tufts (Fig. 12) . . . . *Arctopsyche*

7a (5b). Posterior gular sclerite triangular and elongate (Fig. 9); meso- and metanotum divided by transverse fracture line in posterior third; fore trochantin simple (Fig. 13) 8

7b. Posterior gular sclerite triangular and minute or absent (Fig. 10); meso- and metanotum entire; fore trochantin forked (Fig. 14) or with a dorsal spur 9

8a (7a). Pronotum with transverse groove behind which the posterior ⅓ or so of pronotum is constricted from anterior ⅔ (Fig. 15); anterior margin of frontoclypeal apotome asymmetrical, notched on left side (Fig. 16) . . . *Homoplectra*

8b. Pronotum lacking transverse groove as above, constricted only slightly at posterior border; anterior margin of frontoclypeal apotome symmetrical and

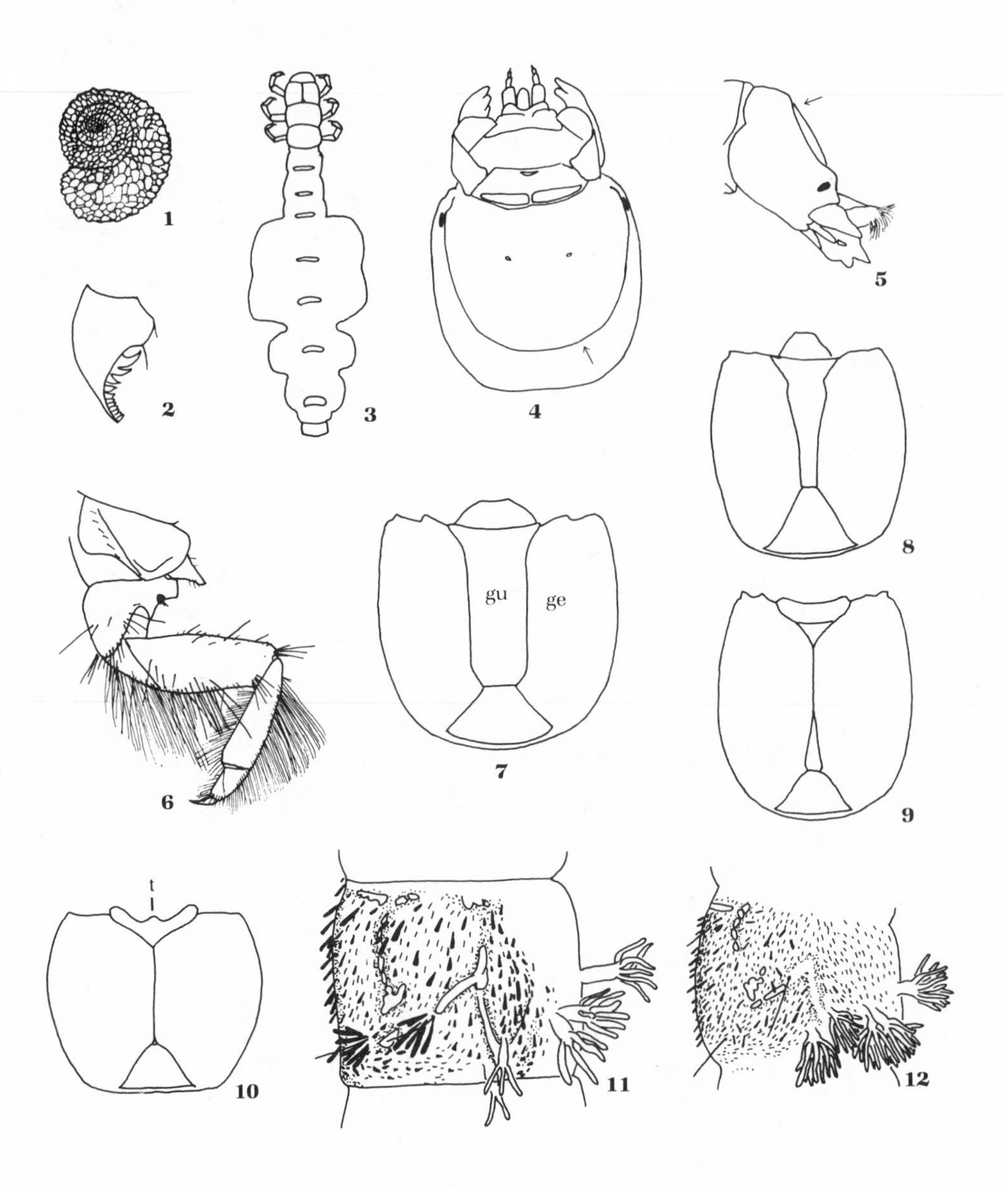

1. Case of *Helicopsyche* (Helicopsychidae), dlv. **2.** Anal claw of *Helicopsyche* (Helicopsychidae), lv. **3.** *Leucotrichia* (Hydroptilidae), dv. **4.** Head of *Macrostemum* (Hydropsychidae), dv. **5.** Head of *Macrostemum* (Hydropsychidae), lv. **6.** Foreleg of *Macrostemum* (Hydropsychidae), lv. **7.** Head of *Parapsyche* (Hydropsychidae), vv: ge, gena; gu, gula. **8.** Head of *Arctopsyche* (Hydropsychidae), vv. **9.** Head of *Diplectrona* (Hydropsychidae), vv. **10.** Head of *Potamyia* (Hydropsychidae), vv: t, tubercle. **11.** Abdominal segment of *Parapsyche* (Hydropsychidae), lv. **12.** Abdominal segment of *Arctopsyche* (Hydropsychidae), lv. dlv, dorsolateral view; dv, dorsal view; lv, lateral view; vv, ventral view. (1–3 modified from Hilsenhoff 1981; 4–12 modified from Wiggins 1977.)

entire (Fig. 17), or if notched then notch is more nearly median and the left mandible bears a large dorsal protuberance (Fig. 18) . . . . *Diplectrona*

9a (7b). Prosternal plate with a pair of detached, moderate-sized, posterior sclerites (often under folded intersegmental membrane) (Fig. 19) . . . . *Hydropsyche*

9b. Prosternal plate with at most a pair of detached, very minute, sclerotized dots (Fig. 20) 10

10a (9b). Anterior gular sclerite with a prominent anteromedian tubercle (Fig. 10); abdominal sternum 9 with posterior margin of setate sclerites entire (Fig. 21); lateral border of mandibles flanged (Fig. 22); fore trochantin simple or forked (Figs. 13, 14) . . . . *Potamyia*

10b. Anterior gular sclerite lacking a prominent tubercle (Fig. 23); abdominal sternum 9 with posterior margin of setate sclerites notched (Fig. 24); lateral border of mandibles not flanged (Fig. 25): fore trochantin forked (Fig. 14) . . . . *Cheumatopsyche*

11a (3b). **Hydroptilidae** (Key is for instars III or greater; early instars are free-living.): Meso- and metanotal sclerotized plates usually lacking middorsal ecdysial line, but line present on pronotum (Fig. 3); tarsal claws short and stout, tarsi approximately twice as long as claws (Fig. 26); segments 5 and 6 of abdomen usually abruptly broader than others in dorsal view (Fig. 3); case translucent, ovoid and flattened (Fig. 27) . . . . *Leucotrichia*

11b. Pro-, meso-, and metanotal sclerotized plates each with middorsal ecdysial line (Fig. 33); tarsal claws variable, but tarsi approximately twice as long as claws only when claws elongate and slender (Fig. 34) 12

12a (11b). Larva dorsoventrally compressed, abdominal segments 1–8 with truncate, fleshy tubercles on each side (Fig. 35); case of 2 flattened, elliptical valves covered with liverwort pieces (Fig. 36) . . . . *Palaeagapetus*

12b. Larva more or less laterally compressed, segments 1–8 without lateral tubercles 13

13a (12b). Abdominal segments with conspicuous dorsal and ventral projections (Fig. 37); transparent case made entirely of silken secretion, a flat pouch, open posteriorly and with a small anterior hole . . . . *Ithytrichia*

13b. Abdominal segments without dorsal and ventral projections 14

14a (13b). Middle and hindlegs almost 3 times as long as forelegs (Fig. 38); case flattened and bottle-shaped, made entirely of silk . . . . *Oxyethira*

14b. Middle and hindlegs not more than 1½ times as long as forelegs (Figs. 39, 40) 15

15a (14b). Tarsal claws of middle and hindlegs about same length or longer than short, thick tarsi (Figs. 39–41); case purselike (Figs. 28–30) 16

15b. Tarsal claws of middle and hindlegs usually much shorter than slender tarsi (Fig. 42); case not purselike or cylindrical 20

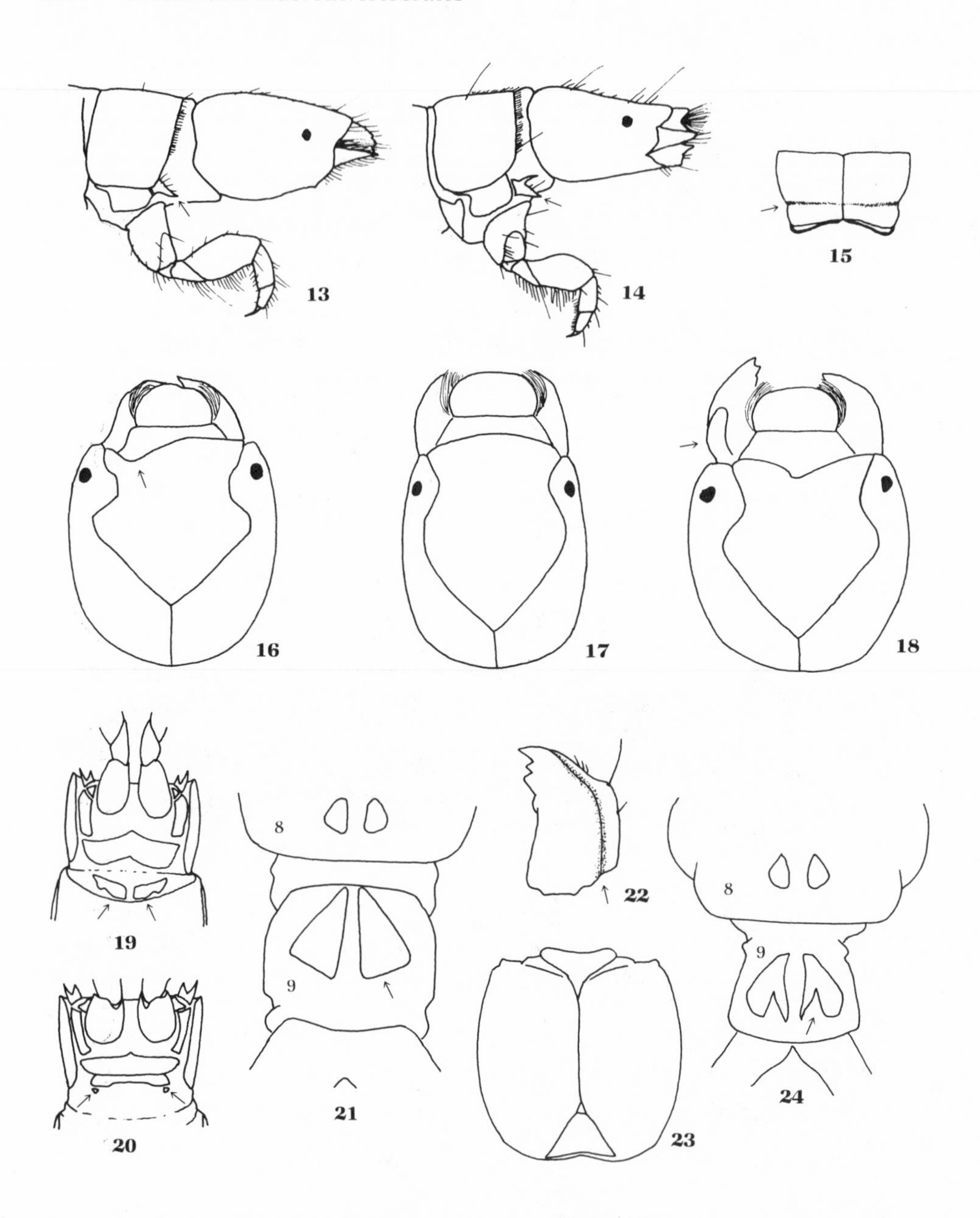

13. Head and prothorax of *Parapsyche* (Hydropsychidae), lv. **14.** Head and prothorax of *Cheumatopsyche* (Hydropsychidae), lv. **15.** Pronotum of *Homoplectra* (Hydropsychidae), dv. **16.** Head of *Homoplectra* (Hydropsychidae), dv. **17.** Head of *Diplectrona* (Hydropsychidae), dv. **18.** Head of *Diplectrona* (Hydropsychidae), dv. **19.** Prothorax of *Hydropsyche* (Hydropsychidae), vv. **20.** Prothorax of *Cheumatopsyche* (Hydropsychidae), vv. **21.** Abdominal segments 8 and 9 of *Potamyia* (Hydropsychidae), vv. **22.** Mandible of *Potamyia* (Hydropsychidae), dv. **23.** Head of *Cheumatopsyche* (Hydropsychidae), vv. **24.** Abdominal segments 8 and 9 of *Cheumatopsyche* (Hydropsychidae), vv. dv, dorsal view; lv, lateral view; vv, ventral view. (13–24 modified or redrawn from Wiggins 1977.)

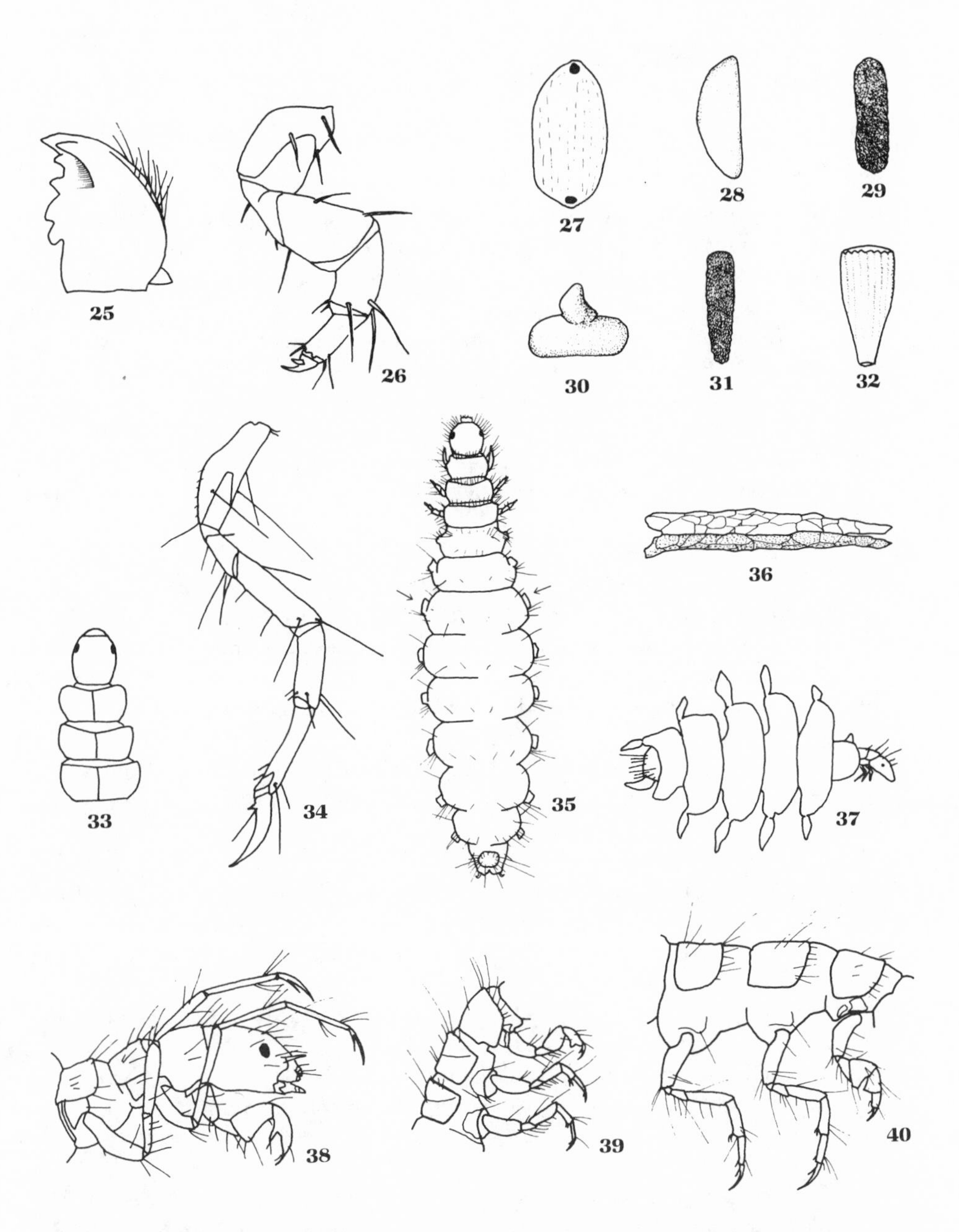

25. Mandible of *Cheumatopsyche* (Hydropsychidae), dv. **26.** Middle leg of *Leucotrichia* (Hydroptilidae), lv. **27.** Case of *Leucotrichia* (Hydroptilidae). **28.** Case of *Stactobiella* (Hydroptilidae). **29.** Case of *Hydroptila* (Hydroptilidae). **30.** Case of *Ochrotrichia* (Hydroptilidae). **31.** Case of *Neotrichia* (Hydroptilidae). **32.** Case of *Mayatrichia* (Hydroptilidae). **33.** Head and thorax of *Palaeagapetus* (Hydroptilidae), dv. **34.** Middle leg of *Orthotrichia* (Hydroptilidae), lv. **35.** *Palaeagapetus* (Hydroptilidae), dv. **36.** Case of *Palaeagapetus* (Hydroptilidae), lv. **37.** *Ithytrichia* (Hydroptilidae), lv. **38.** Head and thorax of *Oxyethira* (Hydroptilidae), lv. **39.** Thorax of *Hydroptila* (Hydroptilidae), lv. **40.** Thorax of *Agraylea* (Hydroptilidae), lv. dv, dorsal view; lv, lateral view. (25, 26, 34–36, 38–40 modified from Wiggins 1977; 27–32, 37 modified from Hilsenhoff 1981.)

16a (15a). Tarsal claws stout and abruptly curved, with a thick, blunt spur at base (Fig. 41) 17

16b. Tarsal claws slender, smoothly curved, with a thin, pointed spur at base (Fig. 43) 18

17a (16a). Dorsal abdominal setae stout, each with small sclerotized area around base; dorsal rings of abdominal segments clearly delineated (Fig. 44); larvae found on a red alga (*Lemanea*), case of 2 symmetrical valves made of this alga (Fig. 45) . . . . *Dibusa*

17b. Dorsal abdominal setae slender, bases without sclerotized area; dorsal rings indistinct (Fig. 46); case 2 matched valves of silk (Fig. 28); larvae found on stones in small, rapid streams . . . . *Stactobiella*

18a (16b). Tibiae of hindlegs twice as long as wide (Fig. 40); case of silk with algal filaments incorporated concentrically . . . . *Agraylea*

18b. Tibiae of hindlegs about as long as wide (Fig. 39) 19

19a (18b). Anterolateral angle of metanotum not produced; one or more setae at or near anterolateral angle of metanotum (Fig. 47); case laterally compressed, with 2 silken valves covered with sand grains or occasionally diatoms or algae . . . . *Hydroptila*

19b. Anterolateral angle of metanotum produced to form a small lobe; all setae on anterior of metanotum well dorsad of small lobe (Fig. 48); case laterally compressed, with 2 silken valves covered with sand grains or occasionally filamentous algae . . . . *Ochrotrichia*

20a (15b). Anal prolegs apparently combined with body mass (Fig. 49); abdominal tergum 8 with only 1 or 2 pairs of weak setae (Fig. 50) . . . . *Orthotrichia*

20b. Anal prolegs distinctly projecting from body mass (Fig. 51); abdominal tergum 8 with many setae (Fig. 52) 21

21a (20b). Thoracic terga clothed with slender, erect, fine setae (Fig. 53); case of sand grains and evenly tapered (Fig. 31) . . . . *Neotrichia*

21b. Thoracic terga clothed with stout, black setae (Fig. 54); case of silk, evenly tapered, semitranslucent, and dorsal side fluted with raised ridges (Fig. 32) . . . . *Mayatrichia*

22a (2b). Meso- and metanotum entirely membranous, or with only weak sclerites on mesonotum at SA1 covering less than half the mesonotum (for location of SA1 and other setal areas see Figs. 55, 57) 23

22b. Meso- and often metanotum with some conspicuous sclerotized plates; mesonotal plates covering more than half the mesonotum 47

23a (22a). Abdominal segment 9 with dorsum entirely membranous; no portable case 24

23b. Abdominal segment 9 bearing a sclerotized dorsal plate; with or without case 36

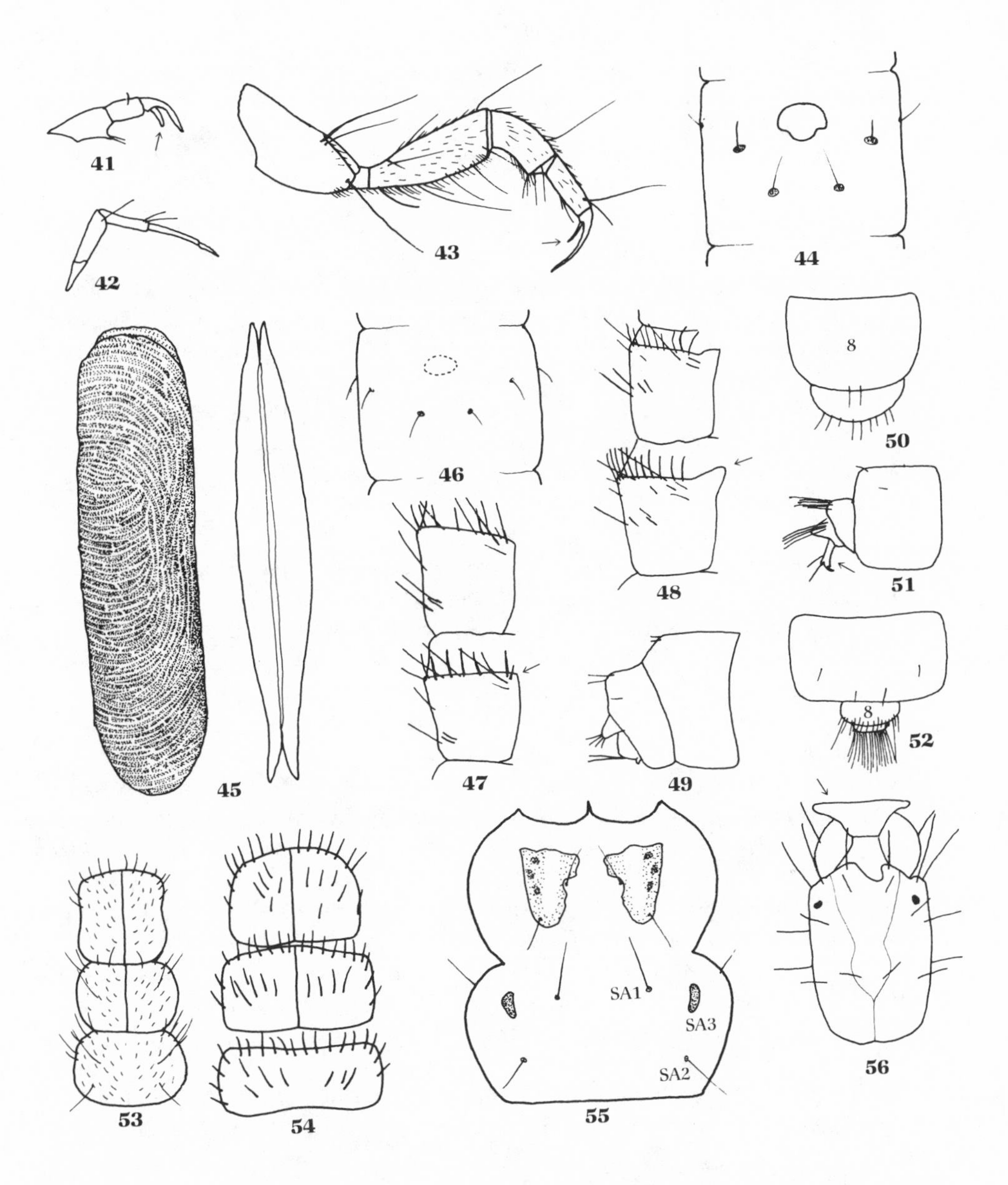

41. Metatibia and tarsus of *Stactobiella* (Hydroptilidae), lv. **42.** Metathoracic leg of *Orthotrichia* (Hydroptilidae), lv. **43.** Mesothoracic leg of *Hydroptila* (Hydroptilidae), lv. **44.** Abdominal segment of *Dibusa* (Hydroptilidae), dv. **45.** Case of *Dibusa* (Hydroptilidae). **46.** Abdominal segment of *Stactobiella* (Hydroptilidae), dv. **47.** Meso- and metanota of *Hydroptila* (Hydroptilidae), lv. **48.** Meso- and metanota of *Ochrotrichia* (Hydroptilidae), lv. **49.** Abdominal segments 8–10 of *Orthotrichia* (Hydroptilidae), lv. **50.** Abdominal segments 8–9 of *Orthotrichia* (Hydroptilidae), dv. **51.** Abdominal segments 8–10 of *Neotrichia* (Hydroptilidae), lv. **52.** Abdominal segments 7–9 of *Neotrichia* (Hydroptilidae), dv. **53.** Thorax of *Neotrichia* (Hydroptilidae), dv. **54.** Thorax of *Mayatrichia* (Hydroptilidae), dv. **55.** Meso- and metathorax of *Agapetus* (Glossosomatidae), dv. **56.** Head of *Chimarra* (Philopotamidae), dv. dv, dorsal view; lv, lateral view. (41, 42, 49–54, 56 modified or redrawn from Hilsenhoff 1981; 43, 45, 47, 48, 55 modified from Wiggins 1977.)

24a (23a). Labrum membranous and T-shaped (Fig. 56), often withdrawn in preserved specimens; "case" a fixed sack-shaped net of silk (Fig. 58); head without muscle scars . . . . **Philopotamidae** 27

24b. Labrum sclerotized and widest near base (Fig. 59); head often with muscle scars (Fig. 59) 25

25a (24b). Fore trochantin broad, hatchet-shaped, separated from episternum by dark suture line (Fig. 60); head usually without muscle scars . . **Psychomyiidae** 26

25b. Fore trochantin pointed, fused with episternum without separating suture (Fig. 61); head usually with muscle scars (Fig. 59) . . . . **Polycentropodidae** 29

26a (25a). **Psychomyiidae**: Anal claw with several long teeth ventrally (Fig. 62); mentum with a pair of high, quadrangular sclerites (Fig. 63) . . . . *Psychomyia*

26b. Anal claw lacking ventral teeth (Fig. 64); mentum with a pair of wide, short sclerites (Fig. 65) . . . . *Lype*

27a (24a). **Philopotamidae**: Apex of frontoclypeus deeply emarginate, often with a large or pointed left lobe and a smaller right one (Fig. 56) . . . . *Chimarra*

27b. Apex of frontoclypeus nearly symmetrical (Figs. 66, 67) 28

28a (27b). Fore trochantin projecting anteriorly only a short distance (Fig. 68); frontoclypeus almost perfectly symmetrical, widened abruptly near anterior margin (Fig. 66) . . . . *Wormaldia*

28b. Fore trochantin projecting anteriorly to form a long fingerlike process (Fig. 69); frontoclypeus slightly asymmetrical, widened gradually toward anterior margin (Fig. 67) . . . . *Dolophilodes*

29a (25b). **Polycentropodidae**: Tarsi broad, flat, and densely pilose (Fig. 70); mandibles short and triangular, each with a large, thick mesal brush (Fig. 71) . . . . *Phylocentropus*

29b. Tarsi not broad and flat, with little or no pile (Fig. 72); mandibles elongate, without a mesal brush (Fig. 73) 30

30a (29b). Anal claw with well-developed ventral teeth (Fig. 74) . . . . *Nyctiophylax*

30b. Anal claw without well-developed ventral teeth (Figs. 75–78) 31

31a (30b). Basal segment of anal proleg subequal to distal segment and without setae, except sometimes a few distally (Fig. 76); ventral margin of anal claw with a row of many tiny spines (visible only under high power) . . . . *Neureclipsis*

31b. Basal segment of anal proleg about 1½ times length of distal segment and with numerous setae (Fig. 78); ventral margin of anal claw without tiny spines 32

32a (31b). Suture on anal proleg segment black, dorsal, not X-shaped; muscle scars on head lighter than surroundings . . . . *Cyrnellus*

32b. Suture on anal proleg segment black, dorsal, X-shaped (Figs. 75, 79); muscle scars on head darker than surroundings (Fig. 59) or indistinct 33

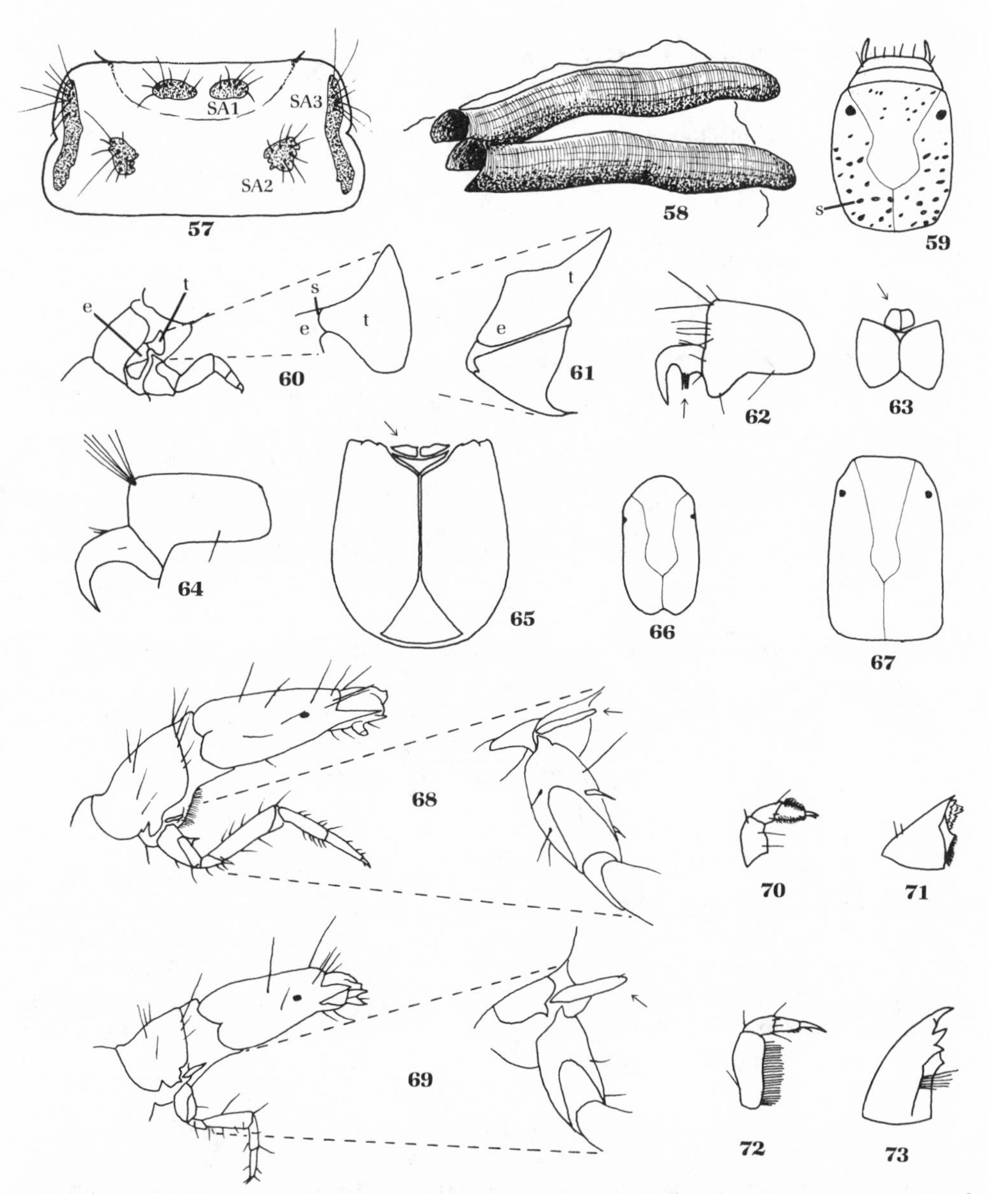

57. Metathorax of *Chyranda* (Limnephilidae), showing setal areas, dv. **58.** Silk cases of *Dolophilodes* (Philopotamidae), vv. **59.** Head of *Polycentropus* (Polycentropodidae), dv: s, scars. **60.** Prothorax of *Lype* (Psychomyiidae), with enlargement of fore trochantin, lv: e, episternum; s, suture; t, trochantin. **61.** Lateral prothoracic sclerites of *Neureclipsis* (Polycentropodidae), lv: e, episternum; t, trochantin. **62.** Anal claw and last segment of proleg of *Psychomyia* (Psychomyiidae), lv. **63.** Genae and mentum of *Psychomyia* (Psychomyiidae), vv. **64.** Anal claw and last segment of *Lype* (Psychomyiidae), lv. **65.** Genae and mentum of *Lype* (Psychomyiidae), vv. **66.** Head of *Wormaldia* (Philopotamidae) (mouthparts omitted), dv. **67.** Head of *Dolophilodes* (Philopotamidae) (mouthparts omitted), dv. **68.** Head and prothorax of *Wormaldia* (Philopotamidae) with enlargement of fore trochantin, lv. **69.** Head and prothorax of *Dolophilodes* (Philopotamidae) with enlargement of fore trochantin, lv. **70.** Femur, tibia, and tarsus of *Phylocentropus* (Polycentropodidae), lv. **71.** Mandible of *Phylocentropus* (Polycentropodidae), dv. **72.** Femur, tibia, and tarsus of *Polycentropus* (Polycentropodidae), lv. **73.** Mandible of *Polycentropus* (Polycentropodidae), dv. dv, dorsal view; lv, lateral view; vv, ventral view. (57, 58, 60, 61, 65, 68, 69 modified from Wiggins 1977; 59, 62–64, 66, 67, 70–73 modified or redrawn from Hilsenhoff 1981.)

33a (32b). Forelegs with tarsi broad and only ½ as long as tibiae (Fig. 80) . . . . *Polycentropus* (in part)
33b. Forelegs with tarsi narrow and at least ⅔ as long as tibiae (Fig. 81) 34

34a (33b). Anal claws obtusely curved . . . . *Polycentropus* (in part)
34b. Anal claws curved approximately 90° 35

35a (34b). Anal claws each with 2 or 3 dorsal accessory spines . . *Polycentropus* (in part)
35b. Anal claws each with only 1 accessory spine (Fig. 79) . . . . *Cernotina*

36a (23b). SA3 on meso- and metanotum consisting of a cluster of setae (Figs. 82–85); prosternal horn present; case of vegetable matter is readily vacated . . . . **Phryganeidae** 41
36b. SA3 on meso- and metanotum consisting of a single seta (Fig. 86); prosternal horn absent 37

37a (36b). Anal claw long, about as long as elongate sclerite on anal proleg (Fig. 88); fore trochantin conspicuous; no portable case . . . . **Rhyacophilidae**, *Rhyacophila*
37b. Anal claw small, much shorter than elongate sclerite on anal proleg (Fig. 89); fore trochantin obscure; saddle-shaped or turtlelike case (Fig. 90) . . . . **Glossosomatidae** 38

38a (37b). **Glossosomatidae**: Mesonotum without sclerites (Fig. 86) . . . . . . *Glossosoma*
38b. Mesonotum with 2 or 3 sclerites (Figs. 55, 87) 39

39a (38b). Mesonotum with 2 sclerites (Fig. 55); anal claw with 1 large tooth and a smaller ventral tooth . . . . *Agapetus*
39b. Mesonotum with a large median sclerite and 2 smaller lateral sclerites (Fig. 87); anal claw with many teeth (Fig. 91); body length, < 4 mm 40

40a (39b). Basal seta of each tarsal claw long and thin, arising from side of stout basal process (Fig. 92); case constructed of large stones . . . . *Protoptila*
40b. Basal seta of each tarsal claw short and stout, larger than basal process (Fig. 93); case of uniformly small stones . . . . *Culoptila*

41a (36a). **Phryganeidae**: SA1 of mesonotum with 2 large brownish yellow sclerites (Fig. 82); case a series of rings (Fig. 94) . . . . *Oligostomis*
41b. SA1 of mesonotum membranous (Fig. 86), or (in *Hagenella* and *Ptilostomis*) SA1 sclerite very small, confined to base of seta 42

42a (41b). Head and pronotum uniformly light brown, without contrasting light and dark stripes . . . . *Hagenella*
42b. Head and pronotum with contrasting dark stripes on light background . . . 43

43a (42b). Ventral combs of prothoracic coxae conspicuous, their teeth evident at a magnification of 50× (Fig. 95) 44

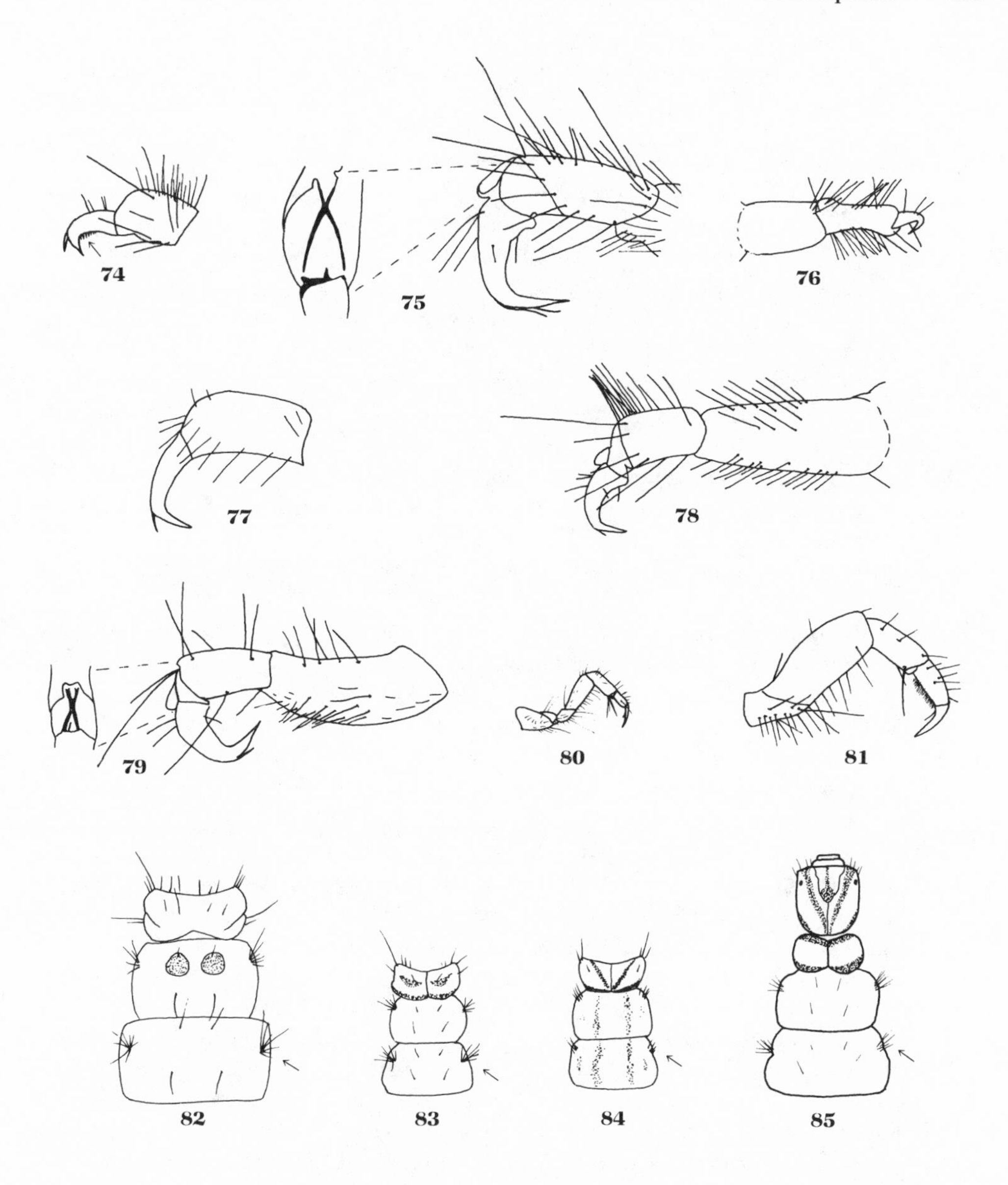

74. Anal claw of *Nyctiophylax* (Polycentropodidae), lv. **75.** Anal proleg of *Polycentropus* (Polycentropodidae), lv. **76.** Anal proleg of *Neureclipsis* (Polycentropodidae), lv. **77.** Anal claw and last segment of proleg of *Cyrnellus* (Polycentropodidae), lv. **78.** Anal proleg of *Cyrnellus* (Polycentropodidae), lv. **79.** Anal proleg of *Cernotina* (Polycentropodidae), lv. **80.** Prothoracic leg of *Polycentropus* (Polycentropodidae), lv. **81.** Prothoracic leg of *Cernotina* (Polycentropodidae), lv. **82.** Thorax of *Oligostomis* (Phryganeidae), dv. **83.** Thorax of *Ptilostomis* (Phryganeidae), dv. **84.** Thorax of *Banksiola* (Phryganeidae), dv. **85.** Thorax of *Phryganea* (Phryganeidae), dv. dv, dorsal view; lv, lateral view. (74, 82–85 modified from Hilsenhoff 1981; 75, 76, 78, 80 modified from Wiggins 1977; 77 modified from Hilsenhoff 1975; 79, 81 modified from Hudson et al. 1981.)

43b. Ventral combs of prothoracic coxae small, each comb appearing as a tiny raised point at 50× (Fig. 96) 45

44a (43a). Small pigmented sclerite (sternellum) usually present between prothoracic coxae (Fig. 98); frontoclypeus with or without a dark longitudinal stripe; pronotum either with diagonal dark stripes or a uniformly dark anterior margin (Fig. 97); if pronotum has an anterior black border, meso- and metanota have dark blotches; spiral case smooth . . . . *Agrypnia*

44b. No pigmented sclerite between prothoracic coxae; a dark longitudinal stripe on frontoclypeus (Fig. 85); pronotum with black border at anterior margin, followed by a dark brown band of variable width (Fig. 85); meso- and metanota without dark blotches; spiral case rough . . . . *Phryganea*

45a (43b). Meso- and metanota with 2 irregular, longitudinal dark bands, separated by a pale area (Fig. 84); pronotum with dark stripes converging posteriorly (Fig. 84) . . . . *Banksiola*

45b. Meso- and metanota with blotches or of fairly uniform pigmentation 46

46a (45b). Pronotum with a dark line along anterior margin, without dark, central transverse markings (Fig. 99); case of spiral construction, with trailing ends of plant fragments giving it a bushy appearance . . . . *Fabria*

46b. Pronotum without dark line along anterior margin, with dark, central transverse markings near center of each sclerite (Fig. 83); case of ring construction, without trailing ends . . . . *Ptilostomis*

47a (22b). Claws of hindlegs very small, those of middle and forelegs long (Fig. 100); case of sand with lateral flanges (Fig. 101) . . . . **Molannidae**, *Molanna*

47b. Claws of hindlegs as long as those of middle legs 48

48a (47b). Anal proleg with lateral sclerite much reduced in size and produced posteriorly as a lobe around base of stout apical seta (Fig. 102); base of anal claw with mesal surface membranous, with prominent brush of 25–30 fine setae (Fig. 103); case of sand grains (Fig. 104); in wet soil of spring seepage areas . . . . **Beraeidae**, *Beraea*

48b. Anal proleg with lateral sclerite not reduced in size or produced posteriorly as a lobe around base of apical seta; base of anal claw with mesal surface largely sclerotized and lacking prominent brush of fine setae 49

49a (48b). Labrum with a transverse row of approximately 16 long setae across central part (Fig. 105); case of plant materials variously arranged . . . . **Calamoceratidae**, *Heteroplectron*

49b. Labrum with only 6 or fewer long setae across central part 50

50a (49b). Mesonotum weakly sclerotized, except for a pair of dark, narrow, curved or angled bars (Fig. 106); cases ovate or convex (Fig. 107) . . . . **Leptoceridae** (in part), *Ceraclea*

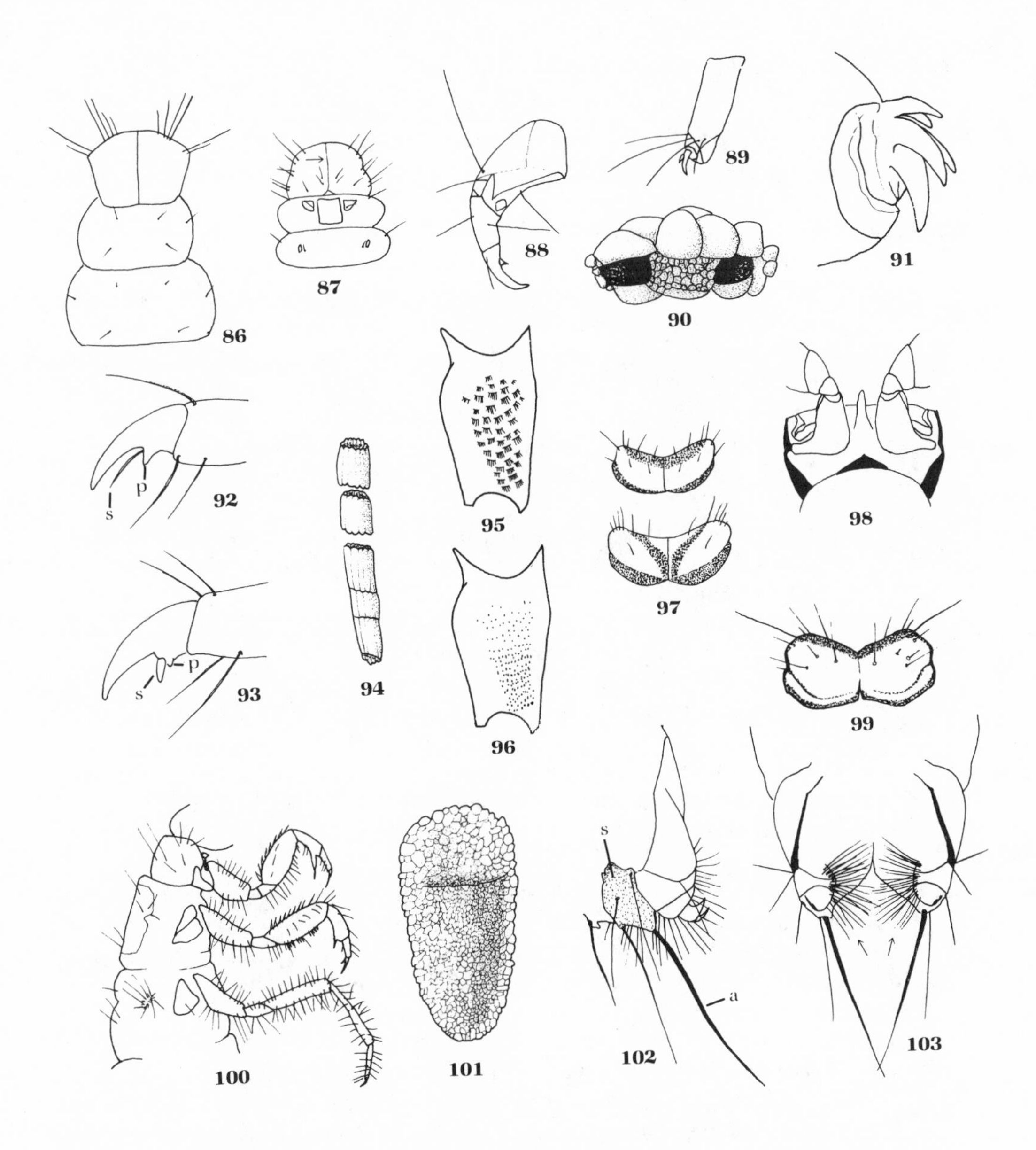

86. Thorax of *Glossosoma* (Glossosomatidae), dv. **87.** Thorax of *Protoptila* (Glossosomatidae), dv. **88.** Anal proleg of *Rhyacophila* (Rhyacophilidae), lv. **89.** Anal proleg of *Glossosoma* (Glossosomatidae), lv. **90.** Case of *Glossosoma* (Glossosomatidae), vlv. **91.** Anal claw of *Protoptila* (Glossosomatidae), lv. **92.** Mesothoracic tarsus of *Protoptila* (Glossosomatidae), lv: p, process; s, seta. **93.** Mesothoracic tarsus of *Culoptila* (Glossosomatidae), lv: p, process; s, seta. **94.** Case of *Ptilostomis* (Phryganeidae), vv. **95.** Prothoracic coxa of *Phryganea* (Phryganeidae), vv. **96.** Prothoracic coxa of *Banksiola* (Phryganeidae), vv. **97.** Pronota of 2 species of *Agrypnia* (Phryganeidae), dv. **98.** Prothorax of *Agrypnia* (Phryganeidae), vv. **99.** Pronotum of *Fabria* (Phryganeidae), dv. **100.** Thorax of *Molanna* (Molannidae), lv. **101.** Case of *Molanna* (Molannidae), vv. **102.** Anal proleg of *Beraea* (Beraeidae), lv: a, apical seta; s, sclerite. **103.** Anal prolegs of *Beraea* (Beraeidae), vv. dv, dorsal view; lv, lateral view; vlv, ventrolateral view; vv, ventral view. (86–90, 94, 97, 100, 101 modified or redrawn from Hilsenhoff 1981; 91–93, 95, 98, 99, 102, 103 modified or redrawn from Wiggins 1977.)

50b. Mesonotum strongly or weakly sclerotized, but without such a pair of sclerotized bars 51

51a (50b). Antennae long, at least 6 times as long as wide, and arising near base of mandibles (Fig. 112) .... **Leptoceridae** (in part) 52
51b. Antennae very short, not more than 3 times as long as wide, often inconspicuous and arising at various points (Figs. 113, 114) 57

52a (51a). **Leptoceridae** (in part): Mesothoracic legs with claw stout and hook-shaped, tarsus bent (Fig. 115); case slender and transparent (Fig. 108) ..... *Leptocerus*
52b. Mesothoracic legs with claw slender, slightly curved, tarsus straight (Fig. 116); case may or may not be slender, but seldom transparent 53

53a (52b). Mandibles long, sharp at apex, with medial teeth well below apex (Fig. 117); maxillary palps extending far beyond anterior edge of labrum (Figs. 117, 118); case robust, often shaped like horn of plenty .... *Oecetis*
53b. Mandibles shorter, blunt at apex, with medial teeth near apex (Fig. 119); maxillary palps extending little, if at all, beyond labrum; case slender 54

54a (53b). Anal prolegs with sclerotized, concave plate on each side of the anal opening, each plate with marginal spines, and extended into ventral lobe (Fig. 120); case cylindrical, of small stones .... *Setodes*
54b. Anal prolegs without sclerotized plates, although patches of spines or setae may be present (Fig. 121) 55

55a (54b). Tibiae of hindlegs without a fracture near middle; gula abbreviated posteriorly and formed into a rounded triangle (Fig. 122); case of various material, often with projecting sticks (Fig. 109) .... *Nectopsyche*
55b. Tibiae of hindlegs with a fracture near middle, which appears to divide tibiae into 2 segments (Figs. 123, 124); gula rectangular 56

56a (55b). Tibiae of hindlegs with a regular fringe of long hair (Fig. 123); case of spirally arranged bits of vegetation (Fig. 110) .... *Triaenodes*
56b. Tibiae of hindlegs with only irregularly placed hairs of varying lengths (Fig. 124); case of sand, stones, or vegetation, often with projecting pieces (Fig. 111) .... *Mystacides*

57a (51b). No seta or sclerites at SA1 of metanotum; dorsal and lateral humps lacking on first abdominal segment; pronotum divided by a sharp transverse furrow across middle, area in front of furrow depressed (Fig. 113) .... **Brachycentridae** 58
57b. Setae or sclerites at SA1 of metanotum (Figs. 55, 82, 86); at least lateral humps present on 1st abdominal segment; pronotum with furrow absent or shallow (Fig. 114) (may be deep in *Ironoquia*) 60

58a (57a). **Brachycentridae**: Middle and hindlegs long, their femora approximately same length as head capsule (Fig. 125), their tibiae produced distally into

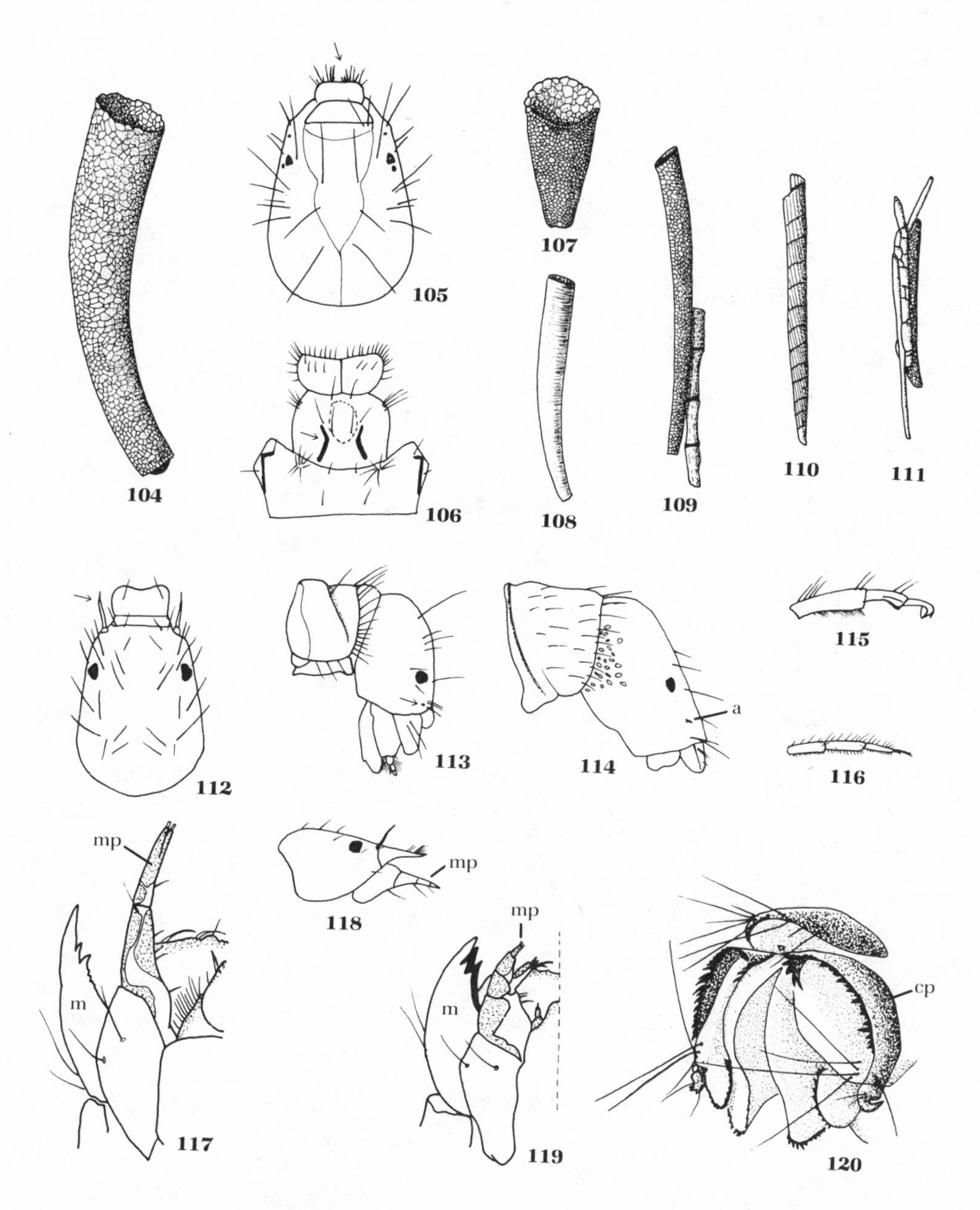

104. Case of *Beraea* (Beraeidae). **105.** Head of *Heteroplectron* (Calamoceratidae), dv. **106.** Thorax of *Ceraclea* (Leptoceridae), dv. **107.** Case of *Ceraclea* (Leptoceridae). **108.** Case of *Leptocerus* (Leptoceridae). **109.** Case of *Nectopsyche* (Leptoceridae). **110.** Case of *Triaenodes* (Leptoceridae). **111.** Case of *Mystacides* (Leptoceridae). **112.** Head of *Leptocerus* (Leptoceridae), dv. **113.** Pronotum and head of *Brachycentrus* (Brachycentridae), lv. **114.** Pronotum and head of *Pseudostenophylax* (Limnephilidae), lv: a, antenna. **115.** Mesothoracic tibia and tarsus of *Leptocerus* (Leptoceridae), lv. **116.** Mesothoracic femur, tibia, and tarsus of *Nectopsyche* (Leptoceridae), lv. **117.** Mandible (m) and maxillary palp (mp) of *Oecetis* (Leptoceridae), vv. **118.** Head of *Oecetis* (Leptoceridae), lv: mp, maxillary palp. **119.** Mandible (m) and maxillary palp (mp) of *Mystacides* (Leptoceridae), dv. **120.** Anal prolegs of *Setodes* (Leptoceridae), plv: cp, concave plate. dv, dorsal view; lv, lateral view; plv, posterolateral view; vv, ventral view. (104, 105, 112, 117, 119, 120 modified from Wiggins 1977; 106–111, 113–116, 118 modified or redrawn from Hilsenhoff 1981.)

prominent processes from which stout spurs arise (Fig. 126); mesonotum with 4 elongate sclerites; plates of metanotum heavily sclerotized (Fig. 127); case often 4-sided and tapered . . . . *Brachycentrus*

58b. Middle and hindlegs shorter, their femora much shorter than head capsule (Fig. 128), their tibiae not produced distally into prominent processes, although spurs arise from about same point (Fig. 128); case 4-sided or rounded 59

59a (58b). Gula longer than wide, somewhat narrowed at posterior end (Fig. 129); larva with very short prosternal horn; mesonotum with 6 sclerotized plates; case 4-sided, composed of small lengths of plant material placed crosswise with loose ends often protruding (Fig. 130) . . . . *Adicrophleps*

59b. Gula usually wider than long, sometimes squarish (Fig. 131); larva without prosternal horn; mesonotum with 2 very wide sclerites that may be longitudinally divided near lateral margins; plates of metanotum only lightly sclerotized; case cylindrical, composed of lengths of plant material wound around the circumference (Fig. 132) or of silk and sand . . . . *Micrasema*

60a (57b). Antennae situated very close to anterior margin of eyes (Fig. 133); median dorsal hump on abdominal segment 1 lacking; case tapered, of bits of vegetable matter or sand (Figs. 134–137) . . . . **Lepidostomatidae** 61

60b. Antennae situated at least as close (Fig. 114) or closer to anterior margin of head capsule as to eyes; dorsal hump usually prominent on abdominal segment 1; case of sand or rock fragments 62

61a (60a). **Lepidostomatidae**: Gula as long as or longer than median ventral ecdysial line (Fig. 138); case of plant pieces, usually 4-sided but may be constructed in other ways, sometimes case of sand grains (Figs. 134–137) or sand and plant pieces . . . . *Lepidostoma*

61b. Gula shorter than median ventral ecdysial line (Fig. 139); case of sand grains . . . . *Theliopsyche*

62a (60b). Antennae situated at or close to anterior margin of head capsule (as in Fig. 113); prosternal horn and abdominal chloride epithelia never present; cases only of mineral materials 63

62b. Antennae situated approximately midway between anterior margin of head capsule and eyes (Fig. 114); prosternal horn present (Fig. 184), although sometimes short; chloride epithelia usually present on some abdominal segments (Figs. 166, 168); case of wide range of materials and construction . . . 65

63a (62a). Anal proleg with dorsal cluster of approximately 30 or more setae posteromesad of lateral sclerite (Fig. 141); fore trochantin relatively large, the apex hook-shaped (Fig. 140); case of fine sand grains, easily crushed . . . . **Sericostomatidae**, *Agarodes*

63b. Anal proleg with approximately 3–5 dorsal setae posteromesad of lateral sclerite, sometimes with short spines; fore trochantin small, the apex not

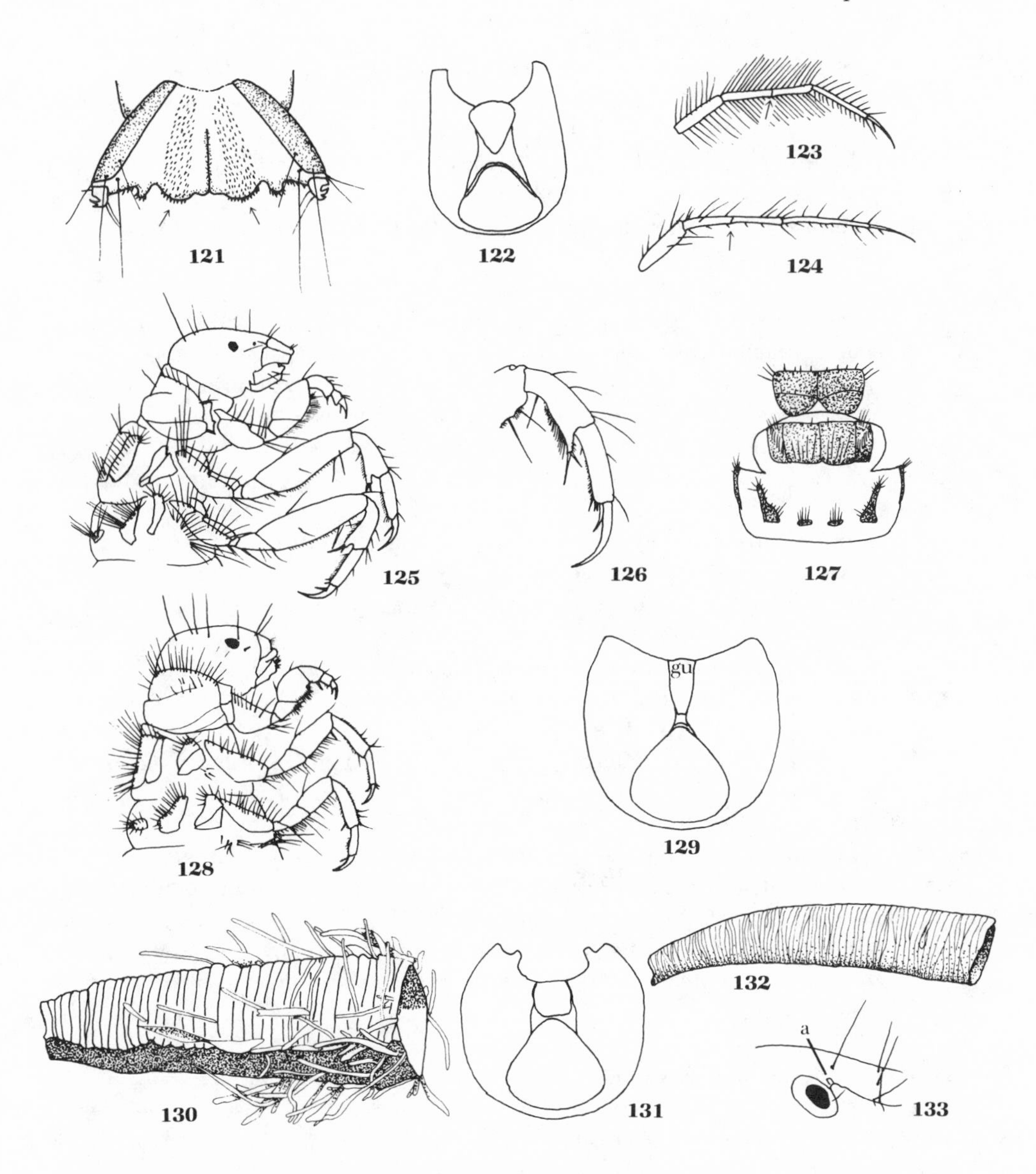

121. Anal prolegs of *Triaenodes* (Leptoceridae), vv. **122.** Head of *Nectopsyche* (Leptoceridae), vv. **123.** Metathoracic leg of *Triaenodes* (Leptoceridae), lv. **124.** Metathoracic leg of *Mystacides* (Leptoceridae). **125.** Anterior of *Brachycentrus* (Brachycentridae), lv. **126.** Hindleg of *Brachycentrus* (Brachycentridae), lv. **127.** Thorax of *Brachycentrus* (Brachycentridae), dv. **128.** Anterior of *Adicrophleps* (Brachycentridae), lv. **129.** Head of *Adicrophleps* (Brachycentridae), vv: gu, gula. **130.** Case of *Adicrophleps* (Brachycentridae). **131.** Head of *Micrasema* (Brachycentridae), vv. **132.** Case of *Micrasema* (Brachycentridae). **133.** Head of *Lepidostoma* (Lepidostomatidae), lv. dv, dorsal view; lv, lateral view; pv, posterior view; vv, ventral view. (121, 122, 125, 126, 128–133 modified or redrawn from Wiggins 1977; 123, 124, 127 modified or redrawn from Hilsenhoff 1981.)

hook-shaped (Fig. 142); case of small rock fragments, extremely hard .... **Odontoceridae** 64

64a (63b). **Odontoceridae**: Anterolateral margins of pronotum produced into long, sharp, forward-projecting points (Fig. 143) .... *Psilotreta*

64b. Anterolateral margins of pronotum rounded and not produced (Fig. 144) .... *Marilia*

65a (62b). Anterior margin of mesonotum with anteromedian emargination (Fig. 145); head elongated; case of sand grains and tiny stones (Fig. 146) .... **Uenoidae**, *Neophylax*

65b. Anterior margin of mesonotum without anteromedian emargination; head nearly ovoid; case of wide variety of materials; very widespread .... **Limnephilidae** 66

66a (65b). **Limnephilidae**: Anterolateral margins of pronotum produced into long, forward-projecting points (Fig. 147) 67

66b. Anterolateral margins of pronotum not produced into long points 68

67a (66a). (Note: gills of immature limnephilid larvae may not be well developed.) Gills mostly 3-branched (Fig. 148); case tubular, of sand or small rock fragments, with larger pebbles along each side (Fig. 149) .... *Goera*

67b. Gills absent; case of small rock fragments, tapered and slightly curved, outline smooth .... *Goerita*

68a (66b). Abdominal gills absent; venter of abdominal segment 1 with irregular ovoid translucent sclerite anteriorly (Fig. 150); case of rock fragments, often with plant materials fastened to dorsum .... *Madeophylax*

68b. Abdominal gills present, although may be poorly developed in early instars; abdomen without ventral translucent sclerite on segment 1; case of various materials 69

69a (68b). Most abdominal gills single 70

69b. Most abdominal gills of dorsal and ventral rows multiple (Fig. 151), although lateral gills sometimes single 75

70a (69a). Large single sclerite at base of lateral hump enclosing posterior half of hump and extending posterodorsad as irregular lobe (Fig. 152); case of leaves or bark formed into flattened tube with lateral seam and narrow flange along each side (Fig. 153) .... *Chyranda*

70b. Sclerites at base of lateral hump not as above 71

71a (70b). One rounded posterior sclerite and 1 elongate dorsal sclerite at base of lateral hump (Fig. 154); case a straight tube of rock and wood fragments (Fig. 155); femora, tibiae, and tarsi usually annulate with black (Fig. 154) .... *Psychoglypha*

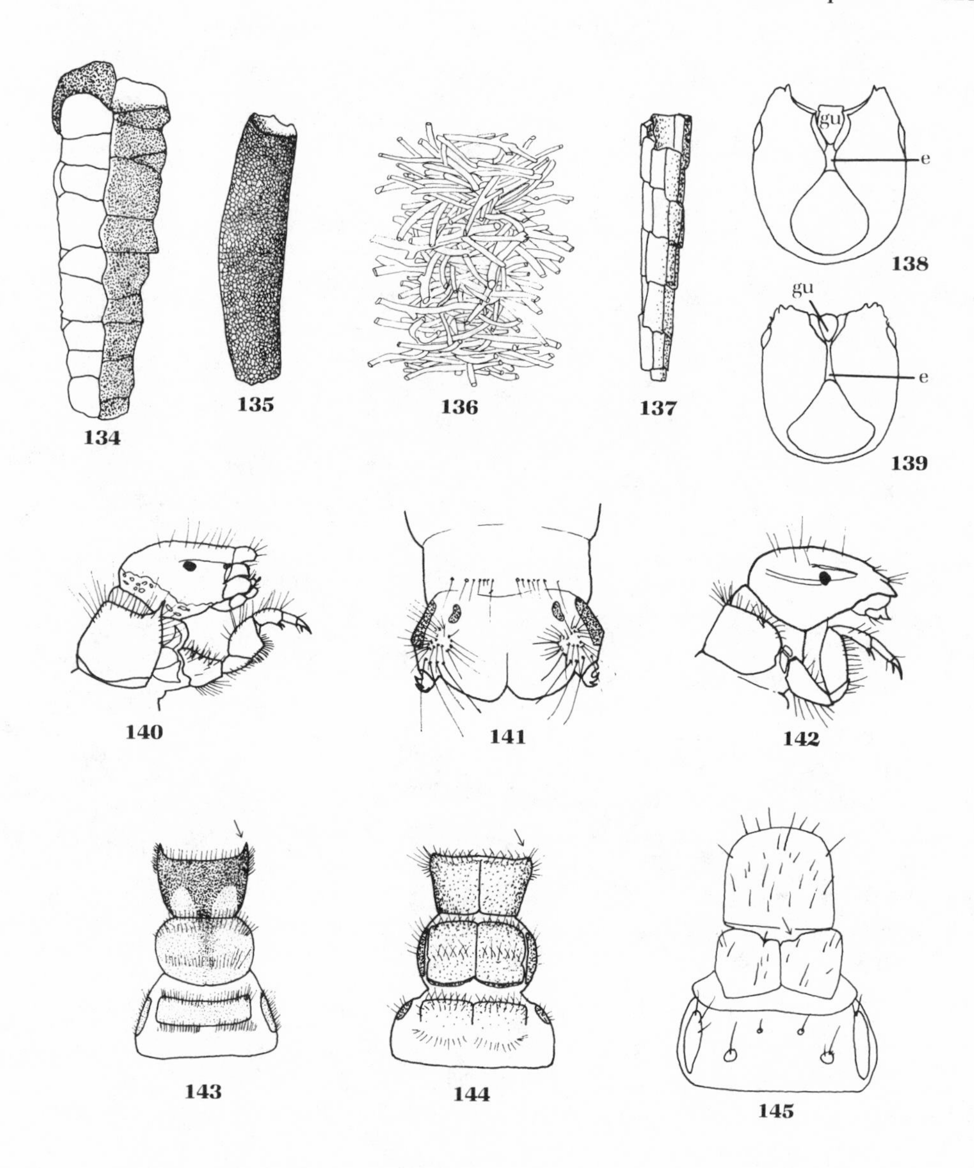

134–137. Cases of *Lepidostoma* (Lepidostomatidae). **138.** Head of *Lepidostoma* (Lepidostomatidae), vv: e, ecdysial line; gu, gula. **139.** Head of *Theliopsyche* (Lepidostomatidae), vv: e, ecdysial line; gu gula. **140.** Head and prothorax of *Agarodes* (Sericostomatidae), lv. **141.** Segment 9 and anal prolegs of *Agarodes* (Sericostomatidae), dv. **142.** Head and prothorax of *Marilia* (Odontoceridae), lv. **143.** Thorax of *Psilotreta* (Odontoceridae), dv. **144.** Thorax of *Marilia* (Odontoceridae), dv. **145.** Thorax of *Neophylax* (Uenoidae), dv. dv, dorsal view; lv, lateral view; vv, ventral view. (134–142, 144 modified or redrawn from Wiggins 1977; 143, 145 modified or redrawn from Hilsenhoff 1981.)

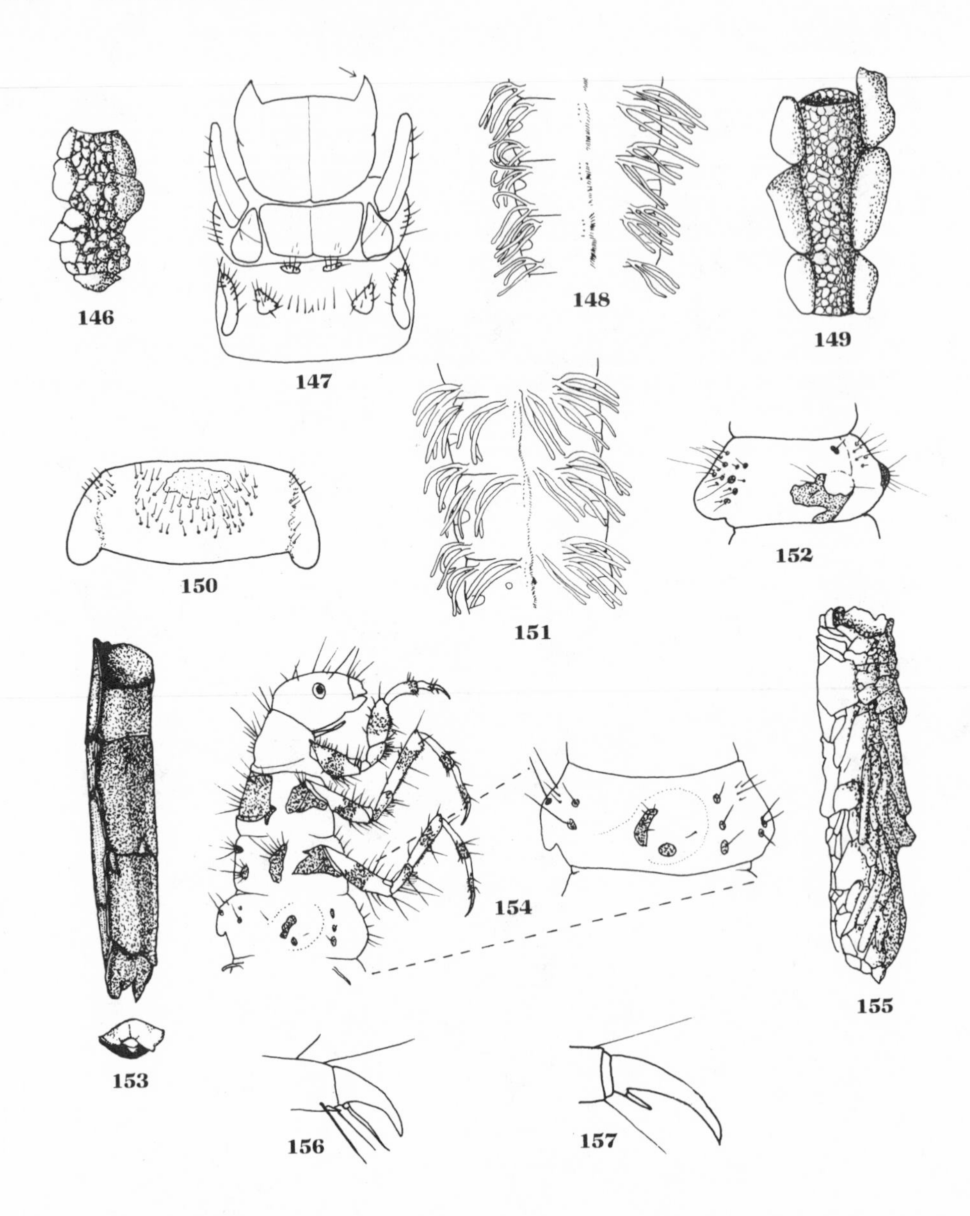

146. Case of *Neophylax* (Uenoidae). **147.** Thorax of *Goera* (Limnephilidae), dv. **148.** Abdominal segments of *Goera* (Limnephilidae), lv. **149.** Case of *Goera* (Limnephilidae). **150.** Abdominal segment 1 of *Madeophylax* (Limnephilidae), vv. **151.** Abdominal segments of *Anabolia* (Limnephilidae), lv. **152.** Abdominal segment 1 of *Chyranda* (Limnephilidae), lv. **153.** Case of *Chyranda* (Limnephilidae). **154.** Head, thorax, and 1st abdominal segment of *Psychoglypha* (Limnephilidae), with enlargement of abdominal segment 1, lv. **155.** Case of *Psychoglypha* (Limnephilidae). **156.** Tarsal claw of *Apatania* (Limnephilidae), lv. **157.** Tarsal claw of *Hydatophylax* (Limnephilidae), lv. dv, dorsal view; lv, lateral view; vv, ventral view. (146–149 modified or redrawn from Hilsenhoff 1981; 150 redrawn from Huryn and Wallace 1984; 151–156 modified or redrawn from Wiggins 1977.)

71b. Sclerotization on lateral humps not as above; legs lacking contrasting annuli 72

72a (71b). Basal seta of tarsal claws extending almost to tip of claw (Fig. 156); anterior metathoracic plates (SA1) replaced by a transverse row of setae (Fig. 158); case of sand grains, slightly tapered posteriorly . . . . . *Apatania*

72b. Basal seta of tarsal claws extending far short of tip (Fig. 157); anterior metathoracic plates (SA1) present 73

73a (72b). Head brown with lighter inconspicuous muscle scars posteriorly (Fig. 114); lateral humps without a sclerite; case of sand grains, slightly tapered and curved (Fig. 159) . . . . . *Pseudostenophylax*

73b. Head pale with dark scars and blotches; a small sclerite at posterior of each lateral hump; case of vegetable matter or sand 74

74a (73b). Metanotal SA1 sclerites fused medially, with no visible suture (Fig. 165); abdominal sterna 2–7 with chloride epithelia (Fig. 166) . . . . . *Hydatophylax*

74b. Metanotal SA1 sclerites not fused medially, although may be close together (Fig. 167) or joined by a suture; abdominal sterna 3–7 with chloride epithelia (Fig. 168) . . . . . *Pycnopsyche*

75a (69b). Some gills in clusters of 4 or more on basal abdominal segments 76

75b. No gills in clusters of more than 3 80

76a (75a). Gills on basal abdominal segments never with more than 4 branches 77

76b. Most gills on basal segments in clusters of 6 or more 78

77a (76a). Femur of hindleg with two major setae arising from ventral edge (Fig. 169); case of sedge or similar leaves arranged longitudinally to form cylinder . . . . . *Grammotaulius*

77b. Femur of hindleg with more than 2 major setae arising from ventral edge (Fig. 170); case usually of pieces of wood and bark (Fig. 163) . . *Onocosmoecus*

78a (76b). Femur of middle and hindlegs with about 5 long setae along ventral edge (Fig. 171); case of bark and leaves (Fig. 161) or sand; in temporary ponds or streams . . . . . *Ironoquia*

78b. Femur of middle and usually hindlegs with about 2 long setae along ventral edge (Fig. 172) 79

79a (78b). Metanotum with all setae confined to sclerites (Fig. 173); case of lengths of sedge leaves arranged longitudinally to form a cylinder (Fig. 174) or of fragments of bark and leaves; lentic, also marshes and temporary ponds . . . . . *Lenarchus*

79b. Metanotum with at least a few setae arising from membrane between sclerites (Fig. 175); case variable, mostly of small rock fragments with some plant pieces (Fig. 160); in permanent streams . . . . . *Hesperophylax*

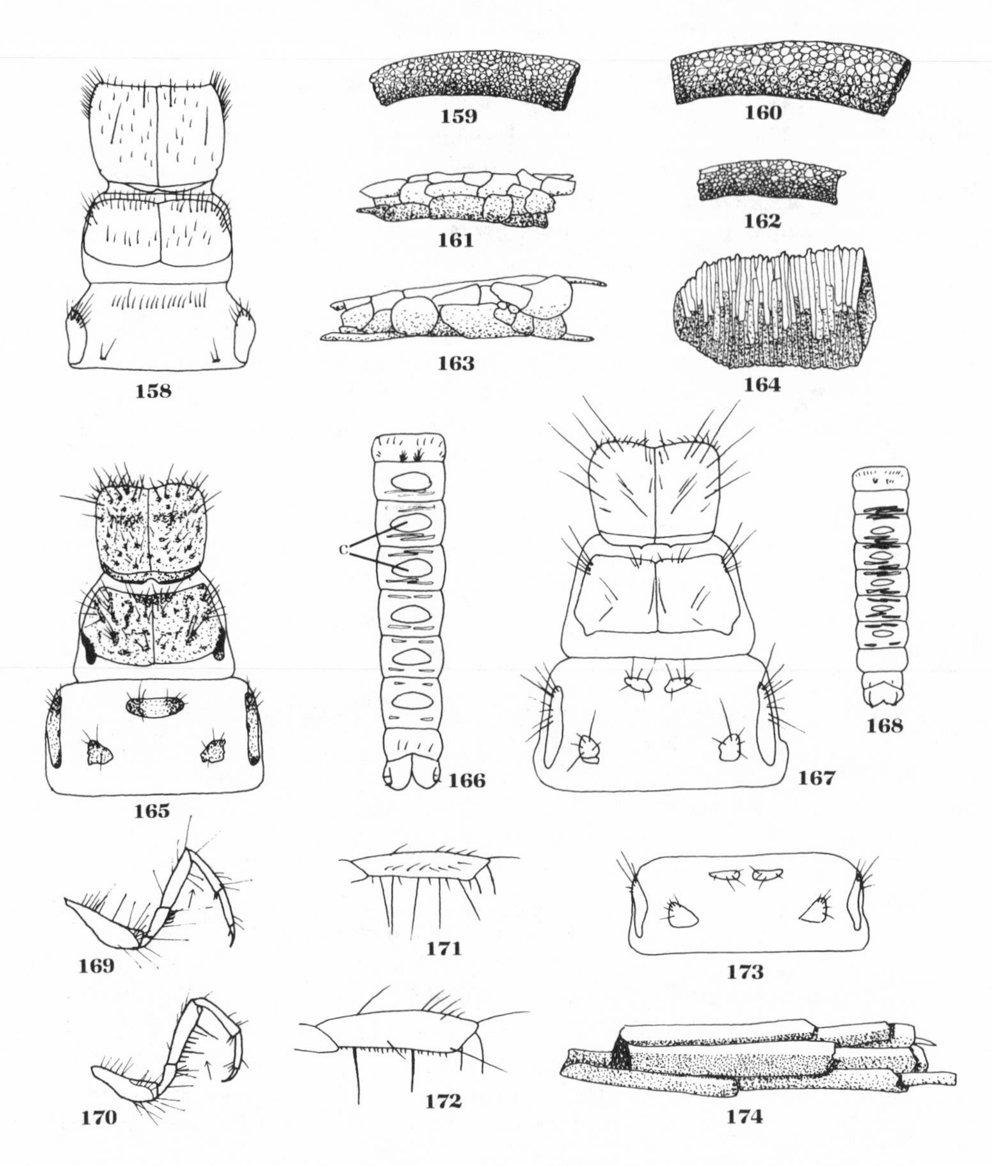

158. Thorax of *Apatania* (Limnephilidae), dv. **159.** Case of *Pseudostenophylax* (Limnephilidae). **160.** Case of *Hesperophylax* (Limnephilidae). **161.** Case of *Ironoquia* (Limnephilidae). **162.** Case of *Frenesia* (Limnephilidae). **163.** Case of *Onocosmoecus* (Limnephilidae). **164.** Case of *Platycentropus* (Limnephilidae). **165.** Thorax of *Hydatophylax* (Limnephilidae), dv. **166.** Abdomen of *Hydatophylax* (Limnephilidae), vv: c, chloride epithelia. **167.** Thorax of *Pycnopsyche* (Limnephilidae), dv. **168.** Abdomen of *Pycnopsyche* (Limnephilidae), vv. **169.** Metathoracic leg of *Grammotaulius* (Limnephilidae), lv. **170.** Metathoracic leg of *Onocosmoecus* (Limnephilidae), lv. **171.** Mesofemur of *Ironoquia* (Limnephilidae), lv. **172.** Mesofemur of *Hesperophylax* (Limnephilidae), lv. **173.** Metathorax of *Lenarchus* (Limnephilidae), dv. **174.** Case of *Lenarchus* (Limnephilidae). dv, dorsal view; lv, lateral view; vv, ventral view. (158–164, 166, 168 modified from Hilsenhoff 1981; 165, 167, 169–174 modified from Wiggins 1977.)

80a (75b). Tibiae and tarsi with contrasting black annuli; case of sticks . . . *Glyphopsyche*
80b. Tibiae and tarsi not annulate 81

81a (80b). Dorsum of head with dark stripes (Fig. 176) and/or posterior portion of frontoclypeus with 3 light areas, 1 along each side and 1 at posterior edge (Figs. 177, 178) 82
81b. Dorsum of head uniformly colored except for some muscle scars (Figs. 179, 180) 84

82a (81a). SA1 of mesonotum consists of a single seta; case of leaf pieces . . *Nemotaulius*
82b. SA1 of mesonotum consists of more than 1 seta 83

83a (82b). Chloride epithelia both dorsally and ventrally on most abdominal segments; case of plant and rock materials . . . . *Asynarchus* (in part)
83b. Chloride epithelia absent dorsally, but present ventrally on most abdominal segments; case of a wide variety of material . . . . *Limnephilus* (in part)

84a (81b). Pronotum with numerous stout, pale setae along anterior margin (Fig. 181); setae on bulbous ventral portion of anal prolegs (Figs. 182, 183); head uniformly brown with light muscle scars posteriorly (Fig. 179); case of sand and small stones (Fig. 162) . . . . *Frenesia*
84b. Pronotum lacking numerous stout, pale setae along anterior margin; no setae on bulbous ventral portion of anal prolegs 85

85a (84b). SA2 on metanotum with few (usually 2) setae and no sclerite; case of elongate leaf pieces arranged in a smooth cylinder . . . . *Arctopora*
85b. SA2 on metanotum with more than 2 setae (visible under high power) and with a sclerite; case variable 86

86a (85b). Head with numerous dark spots, coalescing especially on frontoclypeal apotome to form diffuse blotches (Fig. 180); a small patch of minute spines at anterolateral corner of pronotum; case of plant material arranged longitudinally . . . . *Anabolia*
86b. Head often with numerous dark spots, but spots usually not coalescing to form blotches on frontoclypeal apotome; if blotches are present then case of plant material arranged transversely (Fig. 164) 87

87a (86b). Prosternal horn extending beyond apices of coxae of forelegs (Fig. 184); case of transversely placed vegetation (Fig. 164) . . . . *Platycentropus*
87b. Prosternal horn at most reaching apices of coxae of forelegs 88

88a (87b). Chloride epithelia both dorsally and ventrally on most abdominal segments; case of plant and rock materials . . . . *Asynarchus* (in part)
88b. Chloride epithelia absent dorsally, but present ventrally on most abdominal segments; case of a wide variety of materials . . . . *Limnephilus* (in part)

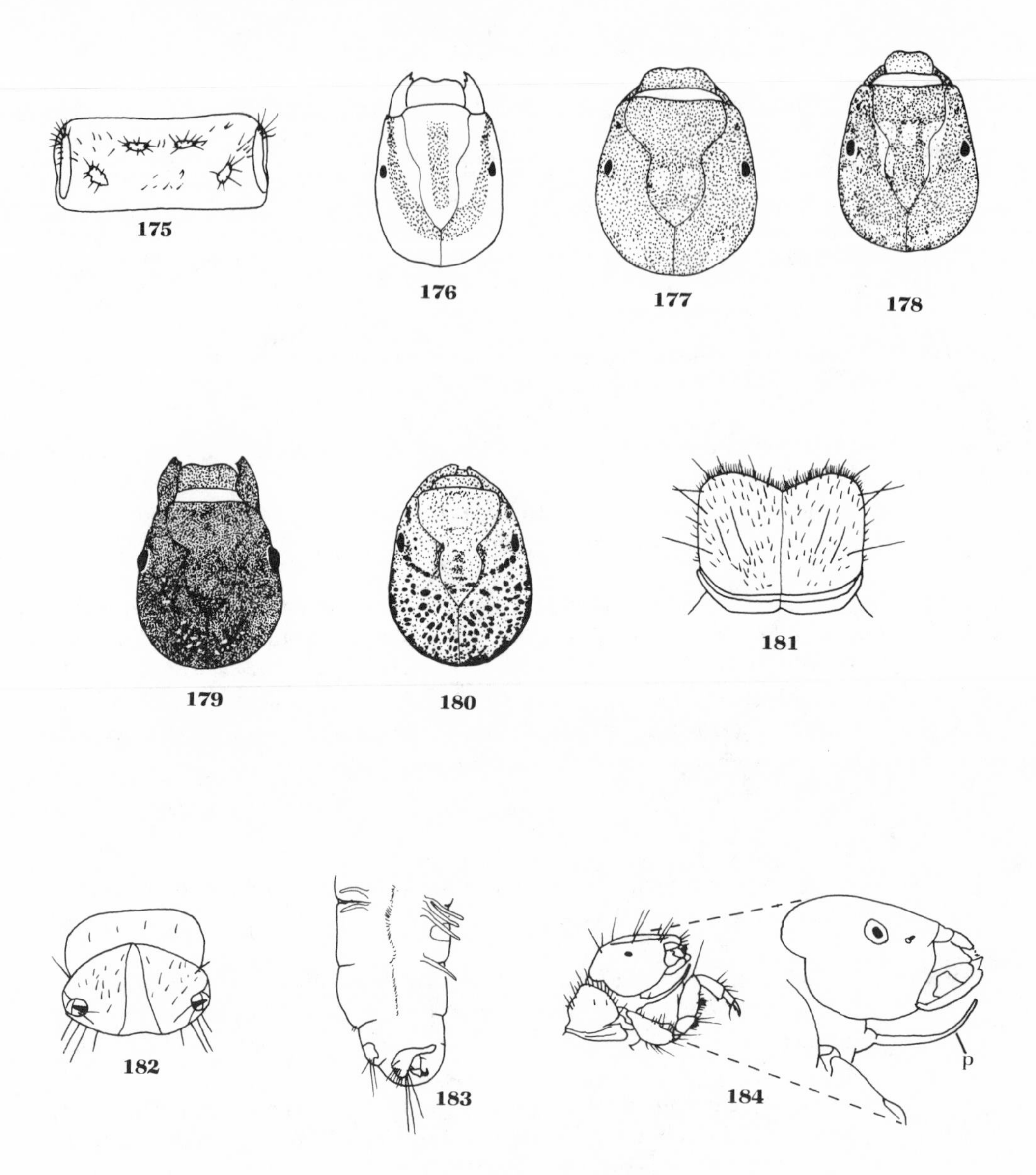

175. Metathorax of *Hesperophylax* (Limnephilidae), dv. **176.** Head of *Nemotaulius* (Limnephilidae), dv. **177.** Head of *Asynarchus* (Limnephilidae), dv. **178.** Head of *Limnephilus* (Limnephilidae), dv. **179.** Head of *Frenesia* (Limnephilidae), dv. **180.** Head of *Anabolia* (Limnephilidae), dv. **181.** Pronotum of *Frenesia* (Limnephilidae), dv. **182.** Abdominal segments 9–10 of *Frenesia* (Limnephilidae), vv. **183.** Last abdominal segments of *Frenesia* (Limnephilidae), lv. **184.** Prothorax and head of *Platycentropus* (Limnephilidae), lv: p, prosternal horn. dv, dorsal view; lv, lateral view; vv, ventral view. (175, 177–181, 183, 184 modified from Wiggins 1977; 176, 182 modified from Hilsenhoff 1981.)

References on Trichoptera Systematics

(*Used in construction of key.)

Alstad, D.N. 1982. Current speed and filtration rate link caddisfly phylogeny and distributional patterns on a stream gradient. Science 216:533–534.

Anderson, N.H. 1978. Continuous rearing of the limnephilid caddisfly, *Clistoronia magnifica* (Banks). *In* M.I. Crichton (ed.). Proc. Second Int. Symp. on Trichoptera, Univ. Reading, Engl., July 25–29, 1977, Dr. D.W. Junk, pp. 317–330. The Hague, Netherlands.

Balduf, W.V. 1939. The bionomics of entomophagous insects. Part II. J.S. Swift Co., St. Louis. 384 pp.

Betten, C. 1901. Aquatic insects in the Adirondacks. Order Trichoptera. Caddis flies. Bull. N.Y. State Mus. 47:561–573, pl. 13, 30–34.

——. 1934. The caddisflies or Trichoptera of New York State. Bull. N.Y. State Mus. 292:1–576.

——. 1950. The genus *Pycnopsyche* (Trichoptera). Ann. Entomol. Soc. Amer. 43:508–522.

Blickle, R.L. 1964. Hydroptilidae (Trichoptera) of Maine. Entomol. News 75:159–162.

——. 1979. Hydroptilidae (Trichoptera) of America north of Mexico. N.H. Agric. Exp. Stn. Bull., Univ. New Hamp. 509:1–97.

Bournaud, M., and H. Tachet (eds.). 1987. Proc. Fifth Int. Symp. on Trichoptera, Lyon, France, July 21–26, 1986. Dr. D.W. Junk, The Hague, Netherlands. 397 pp.

Branch, H.E. 1922. A contribution to the knowledge of the internal anatomy of Trichoptera. Ann. Entomol. Soc. Amer. 15:256–280.

Crichton, M.I. (ed.). 1978. Proc. Second Int. Symp. on Trichoptera, Univ. Reading, Engl., July 25–29, 1977. Dr. D.W. Junk, The Hague, Netherlands. 359 pp.

Denning, D.G. 1948. Review of the Rhyacophilidae (Trichoptera). Can. Entomol. 80:97–117.

——. 1956. Trichoptera. *In* R.L. Usinger (ed.). Aquatic insects of California, pp. 237–270. Univ. Calif. Press, Berkeley.

——. 1982. A review of the Goeridae (Trichoptera). Can. Entomol. 114:637–642.

Erman, N.A. 1981. Terrestrial feeding migration and life history of the stream-dwelling caddisfly, *Desmona bethula* (Trichoptera: Limnephilidae). Can. J. Zool. 59:1658–1665.

Fairchild, W.L., and G.B. Wiggins. 1989. Immature stages and biology of the North American caddisfly genus *Phanocelia* Banks (Trichoptera: Limnephilidae). Can. Entomol. 121:515–519.

Flint, O.S., Jr. 1956. The life history and biology of the genus *Frenesia* (Trichoptera: Limnephilidae). Bull. Brooklyn Entomol. Soc. 51:93–108.

*——. 1960. Taxonomy and biology of Nearctic limnephilid larvae (Trichoptera) with special reference to species in the eastern United States. Entomol. Amer. 40:1–117.

——. 1962. Larvae of the caddis fly genus *Rhyacophila* in eastern North America (Trichoptera: Rhyacophilidae). Proc. U.S. Natl. Mus. 113:465–493.

——. 1964. Notes on some Nearctic Psychomyiidae with special reference to their larvae (Trichoptera). Proc. U.S. Natl. Mus. 115:467–481.

——. 1984. The genus *Brachycentrus* in North America, with a proposed phylogeny of the genera of Brachycentridae (Trichoptera). Smithson. Contrib. Zool. 398:1–58.

Flint, O.S., Jr., and J. Bueno-Soria. 1982. Studies of neotropical caddisflies. XXXII. The immature stages of *Macronema variipenne* Flint and Bueno, with the division of *Macronema* by the resurrection of *Macrostemum* (Trichoptera: Hydropsychidae). Proc. Biol. Soc. Wash. 95:358–370.

Flint, O.S., Jr., and G.B. Wiggins. 1961. Records and descriptions of North American species in the genus *Lepidostoma*, with a revision of the *vernalis* group (Trichoptera: Lepidostomatidae). Can. Entomol. 93:279–297.

Fremling, C.R. 1960. Biology and possible control of nuisance caddisflies of the upper Mississippi River. Res. Bull. Cult. Home Econ. Exp. Stn., State Univ. Sci. Technol., Ames, Iowa 483:856–879.

Haddock, J.D. 1977. The biosystematics of the caddisfly genus *Nectopsyche* in North America with emphasis on the aquatic stages. Amer. Midl. Nat. 98:382–421.

Harris, T.L., and T.M. Lawrence. 1978. Environmental requirements and pollution tolerance of Trichoptera. EPA-600/4-78-063. 310 pp.

Hart, D.D. 1985. Causes and consequences of territoriality in a grazing stream insect. Ecology 66:404–414.

*Hilsenhoff, W.L. 1975. Aquatic insects of Wisconsin, with generic keys and notes on biology, ecology and distribution. Tech. Bull. Wisc. Dep. Nat. Resour. 89:1–53.

*——. 1981. Aquatic insects of Wisconsin. Nat. Hist. Counc. Publ. no. 2. Univ. Wisc., Madison. 60 pp.

——. 1985. The Brachycentridae (Trichoptera) of Wisconsin. Great Lakes Entomol. 18:149–154.

Holzenthal, R.W. 1982. The caddisfly genus *Setodes* in North America (Trichoptera: Leptoceridae). J. Kans. Entomol. Soc. 55:253–271.

*Hudson, P.L., J.C. Morse, and J.R. Voshell, Jr. 1981. Larva and pupa of *Cernotina spicata*. Ann. Entomol. Soc. Amer. 74:516–519.

*Huryn, A.D., and J.B. Wallace. 1984. New eastern Nearctic limnephilid (Trichoptera) with unusual zoogeographical affinities. Ann. Entomol. Soc. Amer. 77:284–292.

Hutchins, R.E. 1966. Caddis insects, nature's carpenters and stonemasons. Dodd, Mead and Co., New York. 80 pp.

Kelley, R.W. 1984. Phylogeny, morphology and classification of the microcaddisfly genus *Oxyethira* Eaton (Trichoptera: Hydroptilidae). Trans. Amer. Entomol. Soc. 110:435–463.

——. 1985. Revision of the micro-caddisfly genus *Oxyethira* (Trichoptera: Hydroptilidae). Part II. Subgenus *Oxyethira*. Trans. Amer. Entomol. Soc. 111:223–254.

——. 1986. Revision of the micro-caddisfly genus *Oxyethira* (Trichoptera: Hydroptilidae). Part III. Subgenus *Holarctotrichia*. Proc. Entomol. Soc. Wash. 88:777–785.

Krafka, J., Jr. 1915. A key to the families of trichopterous larvae. Can. Entomol. 47:217–225.

——. 1923. Morphology of the head of trichopterous larvae as a basis for the revision of the family relationships. J. N.Y. Entomol. Soc. 31:31–52.

LaFontaine, G. 1981. Caddisflies. Nick Lyons Books, New York. 336 pp.

Lloyd, J.T. 1915. Notes on the immature stages of some New York Trichoptera. J. N.Y. Entomol. Soc. 23:201–210.

*——. 1921. The biology of the North American caddisfly larvae. Bull. Lloyd Lib. Bot., Pharmacy Materia Med., Entomol. Ser. 21:1–124.

Mackay, R.J. 1978. Larval identification and instar association in some species of *Hydropsyche* and *Cheumatopsyche* (Trichoptera: Hydropsychidae). Ann. Entomol. Soc. Amer. 71:499–509.

Mackay, R.J., and G.B. Wiggins. 1978. Concepts of evolutionary ecology in Nearctic Trichoptera. *In* M.I. Crichton (ed.). Proc. Second Int. Symp. on Trichoptera, Univ. Reading, Engl., July 25–29, 1977, pp. 267–268. Dr. D.W. Junk, The Hague, Netherlands.

——. 1979. Ecological diversity in Trichoptera. Annu. Rev. Entomol. 24:185–208.

Malicky, H. (ed.). 1976. Proc. First Int. Symp. on Trichoptera, Lunz, Austria, Sept. 16–20, 1975. Dr. D.W. Junk, The Hague, Netherlands. 213 pp.

Marshall, J.E. 1979. A review of the genera of Hydroptilidae. Bull. Br. Mus. (Nat. Hist.) Entomol. Ser. 39:135–239.

McAuliffe, J.R. 1984. Resource depression by a stream herbivore: effects on distributions and abundances of other grazers. Oikos 42:327–333.

Merrill, D., and G.B. Wiggins. 1971. The larva and pupa of the caddisfly genus *Setodes* in North America (Trichoptera: Leptoceridae). R. Ont. Mus. Life Sci. Occas. Pap. 19:1–12.

Milne, L.J., and M.J. Milne. 1938. The Arctopsychidae of continental America north of Mexico (Trichoptera). Bull. Brooklyn Entomol. Soc. 33:97–110.

Milne, M.J. 1939. Immature North American Trichoptera. Psyche 46:9–19.

Moretti, G.P. (ed.). 1981. Proc. Third Int. Symp. on Trichoptera, Perugia, Italy, July 28–Aug. 2, 1980. Series Entomol., vol. 20. Dr. D.W. Junk, The Hague, Netherlands. 472 pp.

Morse, J.C. 1978. Evolution of caddisfly genus *Ceraclea* in Africa—implications for age of Leptoceridae (Trichoptera). *In* M.I. Crichton (ed.). Proc. Second Int. Symp. on Trichoptera, Univ. Reading, Engl., July 25–29, 1977, pp. 199–206. Dr. D.W. Junk, The Hague, Netherlands.

—— (ed.). 1984. Proc. Fourth Int. Symp. on Trichoptera, Clemson, S.C., July 11–16, 1983. Dr. D.W. Junk, The Hague, Netherlands. 486 pp.

Morse, J.C., and R.W. Holzenthal. 1984. Trichoptera genera. *In* R.W. Merritt and K.W. Cummins (eds.). An introduction to the aquatic insects of North America, 2nd ed., pp. 312–347. Kendall/Hunt Publ. Co., Dubuque, Iowa.

Neves, R.J. 1979. Checklist of caddisflies (Trichoptera) from Massachusetts. Entomol. News 90:167–175.

Nielsen, A. 1948. Postembryonic development and biology of the Hydroptilidae. A contribution to the phylogeny of the caddis flies and to the question of the origin of the case-building instinct. Dan. Vidensk. Selsk. Biol. Skr. 5:1–200.

——. 1957. A comparative study of the genital segments and their appendages in male Trichoptera. Dan. Vidensk. Selsk. Biol. Skr. 8:1–159.

Nimmo, A.P. 1981. Francis Walker types of and new synonymies for North American *Hydropsyche* species (Trichoptera: Hydropsychidae). Psyche 88:259–264.

——. 1986. The adult Polycentropodidae of Canada and adjacent United States. Quaest. Entomol. 22:143–252.

——. 1987. The adult Arctopsychidae and Hydropsychidae (Trichoptera) of Canada and adjacent United States. Quaest. Entomol. 23:1–189.

Noyes, A.A. 1914. The biology of the net-spinning Trichoptera of Cascadilla Creek. Ann. Entomol. Soc. Amer. 7:251–272.

Parker, C.R., and G.B. Wiggins. 1985. The Nearctic caddisfly genus *Hesperophylax* Banks (Trichoptera: Limnephilidae). Can. J. Zool. 63:2443–2472.

——. 1987. Revision of the caddisfly genus *Psilotreta* (Trichoptera: Odontoceridae). R. Ont. Mus. Zool. Paleontol. Life Sci. Contrib. 144:1–55.

Pennak, R.W. 1978. Freshwater invertebrates of the United States, 2nd ed. John Wiley and Sons, New York. 803 pp.

Resh, V.H. 1972. A technique for rearing caddisflies. Can. Entomol. 104:1959–1961.

——. 1976. The biology and immature stages of the caddisfly genus *Ceraclea* in eastern North America (Trichoptera: Leptoceridae). Ann. Entomol. Soc. Amer. 69:1039–1061.

Ross, H.H. 1944. The caddis flies or Trichoptera of Illinois. Bull. Ill. Nat. Hist. Surv. 23:1–326.

——. 1946. A review of the Nearctic Lepidostomatidae (Trichoptera). Ann. Entomol. Soc. Amer. 39:265–290.

——. 1947. Descriptions and records of North American Trichoptera with synoptic notes. Trans. Amer. Entomol. Soc. 73:125–168.

——. 1948. New Nearctic Rhyacophilidae and Philopotamidae (Trichoptera). Ann. Entomol. Soc. Amer. 41:17–26.

——. 1950. Synoptic notes on some Nearctic limnephilid caddisflies (Trichoptera, Limnephilidae). Amer. Midl. Nat. 43:410–429.

——. 1951a. Phylogeny and biogeography of the caddisflies of the genera *Agapetus* and *Electragapetus* (Trichoptera: Rhyacophilidae). J. Wash. Acad. Sci. 41:347–356.

——. 1951b. The origin and dispersal of a group of primitive caddisflies. Evolution 5:102–115.

——. 1956. Evolution and classification of the mountain caddisflies. Univ. Ill. Press, Urbana. 213 pp.

*——. 1959. Trichoptera. *In* W.T. Edmondson (ed.). Fresh-water biology, 2nd ed., pp. 1024–1029. John Wiley and Sons, New York.

——. 1964. Evolution of caddisworm cases and nets. Amer. Zool. 4:209–220.

——. 1965. The evolutionary history of *Phylocentropus* (Trichoptera: Polycentropodidae). J. Kans. Entomol. Soc. 38:398–400.

——. 1967. The evolution and past dispersal of the Trichoptera. Annu. Rev. Entomol. 12:169–206.

Ross, H.H., and D.G. Gibbs. 1952. An annotated key to the Nearctic males of *Limnephilus* (Trichoptera, Limnephilidae). Amer. Midl. Nat. 47:435–455.

——. 1973. The subfamily relationships of the Dipseudopsinae (Trichoptera, Polycentropodidae). J. Ga. Entomol. Soc. 8:312–316.

Ross, H.H., and D.C. Scott. 1974. A review of the caddisfly genus *Agarodes* with descriptions of new species (Trichoptera: Sericostomatidae). J. Ga. Entomol. Soc. 9:149–155.

Ross, H.H., and J.D. Unzicker. 1977. The relationship of the genera of American Hydropsychinae as indicated by phallic structures (Trichoptera, Hydropsychidae). J. Ga. Entomol. Soc. 12:298–312.

Ross, H.H., and J.B. Wallace. 1974. The North American genera of the family Sericostomatidae (Trichoptera). J. Ga. Entomol. Soc. 9:42–48.

Rutherford, J.E. 1985. An illustrated key to the pupae of six species of *Hydropsyche* (Trichoptera: Hydropsychidae) common in southern Ontario streams. Great Lakes Entomol. 18:123–132.

——. 1986. Mortality in reared hydropsychid pupae (Trichoptera: Hydropsychidae). Hydrobiologia 131:97–112.

Sattler, W. 1963. Über den Körperban, die Ökologie und Ethologie der Larva und Puppe von *Macronema* Pict. (Hydropsychidae), ein als Larva sich von "micro-drift" ernährendes Trichopter aus dem Amazonosgebiet. Arch. Hydrobiol. 59:29–60.

*Schefter, P.W., G.B. Wiggins, and J.D. Unzicker. 1986. A proposal for assignment of *Ceratopsyche* as a subgenus of *Hydropsyche*, with new synonyms and a new species (Trichoptera: Hydropsychidae). J. N. Amer. Benthol. Soc. 5:67–84.

Schmid, F. 1968. La famille des Arctopsychides. Mem. Soc. Entomol. Québec. no. 1. 84 pp.

——. 1980a. Les insectes et arachnides du Canada. Partie 7. Genera des Trichoptères du Canada et des Etats adjacents. Agric. Can. Publ. no. 1692. Ottawa. 296 pp.

——. 1980b. Esquisse pour une classification et une phylogénie des Goerides (Trichoptera). Nat. Can. 107:185–194.

——. 1981. Revisions des Trichoptères Canadiens. I. La famille des Rhyacophilidae (Annulipalpia). Mem. Entomol. Soc. Can. 116:1–86.

——. 1982. Revision des Trichoptères Canadiens. II. Les Glossosomatidae et Philopotamidae (Annulipalpia). Mem. Entomol. Soc. Can. 122:1–76.

——. 1983. Revisions des Trichoptères Canadiens. III. Les Hyalopsychidae, Psychomyiidae, Goeridae, Brachycentridae, Odontoceridae, Calamoceratidae et Molannidae. Mem. Entomol. Soc. Can. 125:1–109.

Schmude, K.L., and W.L. Hilsenhoff. 1986. Biology, ecology, larval taxonomy, and distribution of Hydropsychidae (Trichoptera) in Wisconsin. Great Lakes Entomol. 19:123–146.

Schuster, G.A. 1984. *Hydropsyche? Symphitopsyche? Ceratopsyche?*: a taxonomic enigma. *In* J.C. Morse (ed.). Proc. Fourth Int. Symp. on Trichoptera, Clemson, S.C., July 11–16, 1983, pp. 339–345. Dr. D.W. Junk, The Hague, Netherlands.

*Schuster, G.A., and D.A. Etnier. 1978. A manual for the identification of the larvae of the caddisfly genera *Hydropsyche* Pictet and *Symphitopsyche* Ulmer in eastern and central North America (Trichoptera: Hydropsychidae). EPA-600/4-78-060. 129 pp.

Smith, D.H., and D.M. Lehmkuhl. 1981. The larvae of four *Hydropsyche* species with the checkerboard head pattern (Trichoptera: Hydropsychidae). Quaest. Entomol. 16:621–634.

Unzicker, J.D., V.H. Resh, and J.C. Morse. 1982. Trichoptera. *In* A.R. Brigham, W.U. Brigham, and A. Gnilka (eds.). Aquatic insects and oligochaetes of North and South Carolina, pp. 9.1–9.138. Midwest Aquatic Enterprises, Mahomet, Ill.

*Vineyard, R.N., and G.B. Wiggins. 1988. Further revision of the caddisfly family Uenoidae (Trichoptera): evidence for the inclusion of Neophylacinae and Thremmatidae. Syst. Entomol. 13:361–372.

Wallace, J.B., and R.W. Merritt. 1980. Filter feeding ecology of aquatic insects. Annu. Rev. Entomol. 25:103–132.

Wallace, J.B., J.R. Webster, and W.R. Woodall. 1977. The role of filter feeders in flowing water. Arch. Hydrobiol. 79:506–532.

Waltz, R.D., and W.P. McCafferty. 1983. The caddisflies of Indiana (Insecta: Trichoptera). Purdue Univ. Agric. Exp. Stn. Res. Stn. Bull. 978:1–25.

Weaver, J.S. 1984. The evolution and classification of Trichoptera. Part I. The groundplan of Trichoptera. *In* J.C. Morse (ed.). Proc. Fourth Int. Symp. on Trichoptera, Clemson, S.C., July 11–16, 1983, pp. 413–419. Dr. D.W. Junk, The Hague, Netherlands.

——. 1985. New species and new generic synonym of Nearctic caddisfly genus *Homoplectra* (Trichoptera: Hydropsychidae). Entomol. News 96:71–77.

Weaver, J.S., and J.C. Morse. 1986. Evolution of feeding and case-making behavior in Trichoptera. J. N. Amer. Benth. Soc. 5:150–158.

Weaver, J.S., and J.L. Sykora. 1979. The *Rhyacophila* of Pennsylvania, with larval descriptions of *R. banksi* and *R. carpenteri* (Trichoptera: Rhyacophilidae). Ann. Carnegie Mus. 48:403–423.

Wiggins, G.B. 1954. The caddisfly genus *Beraea* in North America (Trichoptera). R. Ont. Mus. Zool. Paleontol. Life Sci. Contrib. 39:1–14.

——. 1956. A revision of the North American caddisfly genus *Banksiola* (Trichoptera: Phryganeidae). R. Ont. Mus. Div. Zool. Paleontol. Life Sci. Contrib. 43:1–12.

——. 1959. A method of rearing caddisflies (Trichoptera). Can. Entomol. 91:402–405.

——. 1960. A preliminary systematic study of the North American larvae of the caddisfly family Phryganeidae (Trichoptera). Can. J. Zool. 38:1153–1170.

——. 1962. A new subfamily of phryganeid caddisflies from western North America (Trichoptera: Phryganeidae). Can. J. Zool. 40:879–891.

——. 1963. Larvae and pupae of two North American limnephilid caddisfly genera (Trichoptera: Limnephilidae). Bull. Brooklyn Entomol. Soc. 63:103–112.

——. 1965. Additions and revisions to the genera of North American caddisflies of the family Brachycentridae with special reference to the larval stages (Trichoptera). Can. Entomol. 97:1089–1106.

——. 1973a. Contributions to the systematics of the caddisfly family Limnephilidae (Trichoptera). I. R. Ont. Mus. Life Sci. Contrib. 94:1–31.

——. 1973b. New systematic data for the North American caddisfly genera *Lepania*, *Goeracea*, and *Goerita* (Trichoptera: Limnephilidae). R. Ont. Mus. Life Sci. Contrib. 91:1–33.

——. 1973c. A contribution to the biology of caddisflies in temporary pools. R. Ont. Mus. Life Sci. Contrib. 88:1–28.

——. 1975. Contributions to the systematics of the caddisfly family Limnephilidae (Trichoptera). II. Can. Entomol. 107:325–336.

——. 1976. Contributions to the systematics of the caddisfly family Limnephilidae (Trichoptera). III. The genus *Goereilla*. *In* H. Malicky (ed.). Proc. First Int. Symp. on Trichoptera, Lunz, Austria, Sept. 16–20, 1975, pp. 7–19. Dr. D.W. Junk, The Hague, Netherlands.

*——. 1977. Larvae of the North American caddisfly genera (Trichoptera). Univ. Toronto Press, Toronto, Can. 401 pp.

*——. 1978. Trichoptera. *In* R.W. Merritt and K.W. Cummins (eds.). An introduction to the aquatic insects of North America, pp. 147–186. Kendall/Hunt Publ. Co., Dubuque, Iowa.

*——. 1984. Trichoptera. *In* R.W. Merritt and K.W. Cummins (eds.). An introduction to the aquatic insects of North America, 2nd ed., pp. 271–311. Kendall/Hunt Publ. Co., Dubuque, Iowa.

Wiggins, G.B., and N.H. Anderson. 1968. Contributions to the systematics of the caddisfly genera *Pseudostenophylax* and *Philocasca* with special reference to the immature stages (Trichoptera: Limnephilidae). Can. J. Zool. 46:61–75.

Wiggins, G.B., E.Y.C. Lin, and K.E. Chua. 1976. Preliminary SEM investigation of an aqueous carbohydrate material, the gelatinous matrix of caddisfly eggs (Insecta: Trichoptera). Scan. Electron Microsc. 1976:605–610.

Wiggins, G.B., R.J. Mackay, and I.M. Smith. 1980. Evolutionary and ecological strategies of animals in annual temporary pools. Arch. Hydrobiol. Suppl. 58:97–206.

Wiggins, G.B., J.S. Weaver III, and J.D. Unzicker. 1985. Revision of the caddisfly family Uenoidae (Trichoptera). Can. Entomol. 117:763–800.

Winterbourn, M.J., and N.H. Anderson. 1980. The life history of *Philanisus plebius* Walker (Trichoptera: Chathamidae), a caddisfly whose eggs were found in a starfish. Ecol. Entomol. 5:293–303.

Wood, J.R., V.H. Resh, and E.M. McEwan. 1982. Egg masses of Nearctic sericostomatid caddisfly genera (Trichoptera). Ann. Entomol. Soc. Amer. 75:430–434.

Yamamoto, T., and H.H. Ross. 1966. A phylogenetic outline of the caddisfly genus *Mystacides* (Trichoptera: Leptoceridae). Can. Entomol. 98:627–632.

Yamamoto, T., and G.B. Wiggins. 1964. A comparative study of the North American species in the caddisfly genus *Mystacides* (Trichoptera: Leptoceridae). Can. J. Zool. 42:1105–1126.

9 | Aquatic Lepidoptera

Classification

Lepidoptera (butterflies and moths) are closely related to Trichoptera, having diverged from the Trichoptera probably in the early Mesozoic. They belong to the infraclass Neoptera, division Endopterygota. The larvae of only one family of moths, the Pyralidae, have become truly aquatic. A few other families occur on emergent or floating aquatic macrophytes, but their larvae do not actually submerge (Lange 1984). We consider only the truly aquatic forms in this chapter.

Life History

Lepidopterans are holometabolous and have one to three generations per year. Aquatic moths lay egg clusters or single eggs, often in overlapping layers on a mineral or plant substrate. Females of some species fly over the surface of the water, repeatedly dipping their abdomens to lay single eggs or small egg masses (25 eggs or more), which sink to the bottom. Females of other species walk completely into the water or actively swim using their wings. The submerged female respires by means of a plastron while laying eggs in one or two clusters (totaling several hundred eggs) beneath the stones; she usually dies while still submerged (Brigham and Herlong 1982). Generally, egg incubation lasts from 6 to 15 days. Larvae of most species go through four or five instars before pupating underwater in silken cocoons spun inside either the larval case or the oval pupal case. Adults emerge by swimming or floating to the surface (Lange 1984) and live for at most two weeks. They are crepuscular, usually mating within two hours past sunset. No courtship behavior has been observed (Brigham and Herlong 1982). The females of some species are larger than the males (Munroe 1972).

Habitat

Aquatic pyralid caterpillars are generally found in two habitats. One group (e.g., *Munroessa, Synclita*) is associated with lentic macrophytes. In some areas in the southern United States, these caterpillars may become so abundant on lily pads as to be considered pests. Using bits of aquatic vegetation, they generally build cases that are either portable or attached to plants. Lange (1984) classifies lentic pyralids as climbers or swimmers. Species of the second group of aquatic pyralids are clingers (Lange 1984) and are characteristic of lentic (*Eoparargyractis*) and lotic (*Petrophila*) habitats. They are free-living but usually construct protective silken nets or feltlike canopies on submerged plants or stony substrates (Fiance and Moeller 1978, Pennak 1978) from which they extend to graze on periphyton (McAuliffe 1984).

Feeding

Like their terrestrial counterparts, aquatic caterpillars are strictly herbivorous. The lentic species are leaf miners, shredders, or stem borers. Some lotic species have mandibles specialized for scraping periphyton from stones and have been reported to compete with purse-case-making caddisfly larvae for algae (McAuliffe 1984). As adults many aquatic moths are known to feed on plant nectar and will drink water.

Respiration

The first instars of all known aquatic pyralid species lack gills and often constitute the dispersal phase of the larval life stage. Subsequent instars of the lentic species may or may not have tracheal gills; if gills are present, they typically appear on the second instar and become progressively more abundant (up to 400). The gills of one genus of pyralids (*Parapoynx*) are branched. If gills are absent, larvae respire through open spiracles on the abdomen. Older instars may retain an air bubble inside their cases for oxygen exchange (Brigham and Herlong 1982); young instars often inhabit water-filled cases. Gills are always present in the lotic species. Pupal cases or chambers generally contain an air space in which the pupa develops.

Collection and Preservation

Lentic aquatic caterpillars can be located most effectively by examining submerged macrophytes for tissue damage. Lotic species do not readily detach from stones or submerged plants and can be overlooked when a generalized stream sampling technique is used. Submerged substrates should be inspected for silken canopies, and larvae collected therein. Unsuspecting students often inadvertently preserve terrestrial caterpillars that have dropped from overhanging vegetation into collecting pans. These specimens can be a source of great frustration later in the laboratory. One safeguard against preserving terrestrial caterpillars is to discard any living specimens that refuse to submerge (are *hydrophobic*). Such behavior is often a good clue to their true habitat; however, late instars of *Synclita* and *Munroessa* possess a water-resistant cuticle that often causes them to "float" on the water surface. Brigham and Herlong (1982) recommend fixing larvae in hot water of KAAD (kerosene, acetic acid, ethanol, and dioxane) before storing them in 80–95% ethanol. Preservation in 80% ethanol without prior fixation is also acceptable, although some larvae may lose their color.

Identification

Only those aquatic caterpillars possessing gills and those found in cases made of aquatic plants can be identified with our keys. Although a larva may be aquatic, if it is missing its case or does not have long filamentous gills, it will key out to "terrestrial." Features used to identify genera of truly aquatic caterpillars—that is, those associated with submerged substrates—are gill structure, spiracles, and case morphology.

More-Detailed Information

Welch 1916, Berg 1950, McGaha 1954, Welch 1959, Lange 1956a, Monroe 1972, Pennak 1978, Brigham and Herlong 1982, Lange 1984.

Checklist of Aquatic Lepidoptera Larvae

Pyralidae
- *Acentria*
- *Eoparargyractis*
- *Langessa*
- *Munroessa*
- *Nymphula*
- *Nymphuliella*
- *Parapoynx*
- *Petrophila*
- *Synclita*

Key to Genera of Aquatic Lepidoptera Larvae

1a. Long filamentous gills present, longer than ½ the width of abdominal segments 2
1b. Long filamentous gills absent 4

2a (1a). All gills unbranched (Fig. 1), with about 120 gill filaments; larva free-living on rocks in lotic habitats . . . . *Petrophila*
2b. Some gills branched (Fig. 2) 3

3a (2b). Branched gills on all abdominal segments; 2nd, 3rd, and 4th abdominal spiracles enlarged, others rudimentary; larva in a case of material cut from the food plant (*Nuphar, Potamogeton, Vallisneria,* and others) . . . . *Parapoynx*

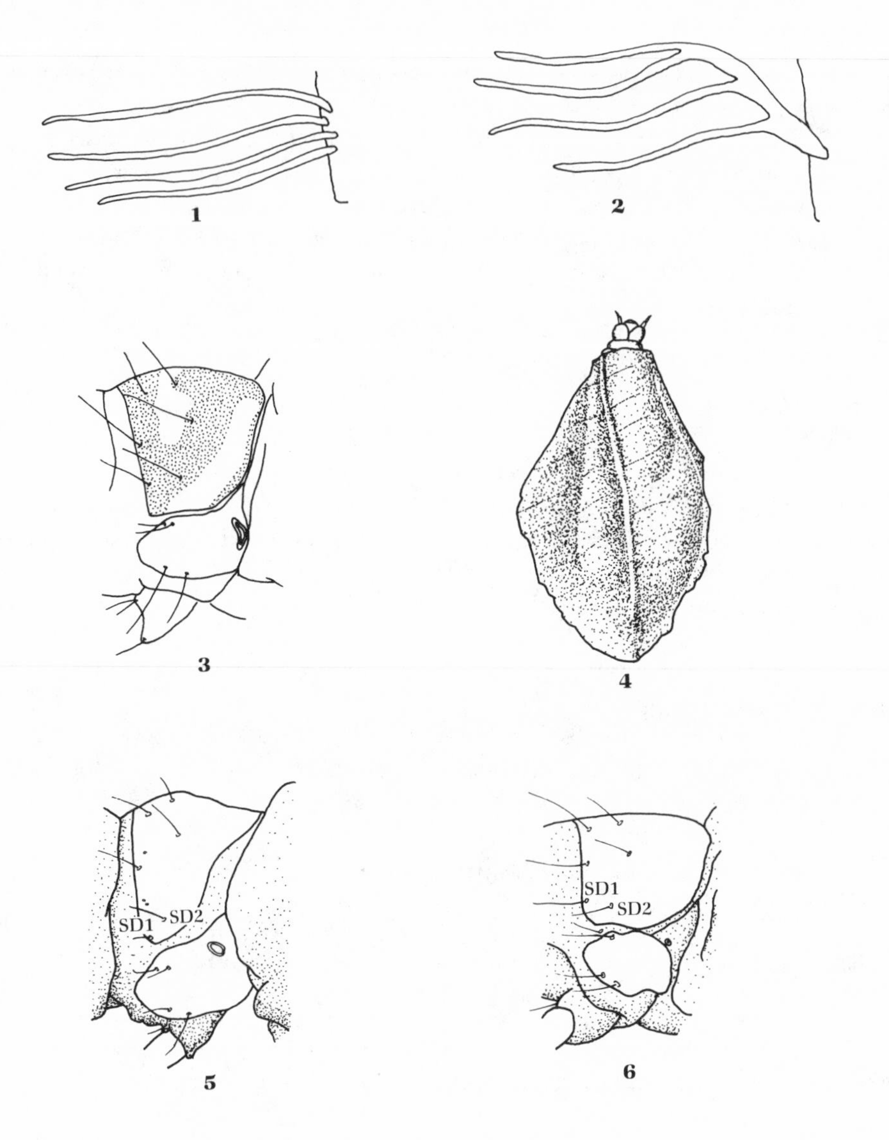

1. Gill of *Petrophila*. **2.** Gill of *Parapoynx*. **3.** Prothorax of *Langessa*, lv. **4.** Case of *Synclita*. **5.** Prothorax of *Munroessa*, lv. **6.** Prothorax of *Synclita*, lv. lv, lateral view. (3, 5, 6 redrawn from Brigham and Herlong 1982; 4 modified from Lange 1956a.)

3b. Branched gills only on 10th abdominal segment; all abdominal spiracles of uniform size; larva without a case . . . . *Eoparargyractis*

4a (1b). Head and cephalic shield of prothorax with extensive pattern of brown bands on cream-colored background (in live specimens only) (Fig. 3) . . . . *Langessa*

4b. Head and cephalic shield of prothorax usually unpatterned, or only head patterned 5

5a (4b). Larva in a case constructed from its food plant (Fig. 4) 6

5b. Larva free-living, without a case . . . . terrestrial (not keyed)

6a (5a). Prothorax with a very large blackish brown shield dorsally, and with vestigial spiracles; restricted to sphagnum bogs; case of *Cladopodiella* (= *Cephalozia*) . . . . *Nymphuliella*

6b. Prothorax without blackish brown shield, and with functional spiracles; case not of *Cladopodiella* 7

7a (6b). Crochets biordinal, in an incomplete ellipse with almost parallel sides and open on the mesal side; case not flat, made of 2 or more bits of *Ceratophyllum* or other aquatic plants (only in late instars; early instars are stem borers) . . . . *Acentria*

7b. Crochets uniordinal or biordinal in circles; case flat or biconvex, usually made of 2 bits of leaves of various aquatic plants 8

8a (7b). Spiracles on abdominal segments 2–4 twice as large as those on 1, 5–8; case of *Lemna* or fragments of other plants . . . . *Nymphula*

8b. Spiracles on abdominal segments 1–4 larger than other spiracles; case of *Potamogeton, Nymphaea, Lemna,* or other plants 9

9a (8b). Seta SD2 distinctly posterodorsad of SD1 on prothorax (Fig. 5); cephalic row of crochets same size as caudal row on prolegs . . . . *Monroessa*

9b. Seta SD2 frequently on same horizontal level as SD1, but occasionally posteroventrad (Fig. 6); cephalic row of crochets distinctly larger than caudal row on prolegs . . . . *Synclita*

References on Aquatic Lepidoptera Systematics

(*Used in construction of key.)

Berg, C.O. 1950. Biology of certain aquatic caterpillars (Pyralidae: *Nymphula* spp.) which feed on *Pomatogeton.* Trans. Amer. Microscop. Soc. 69:254–266.

*Brigham, A.R., and D.D. Herlong. 1982. Aquatic and semiaquatic Lepidoptera. *In* A.R. Brigham, W.U. Brigham, and A. Gnilka (eds.). Aquatic insects and oligochaetes of North and South Carolina, pp. 12.1–12.36. Midwest Aquatic Enterprises, Mahomet, Ill.

Fiance, S.B., and R.E. Moeller. 1978. Immature stages and ecological observations of *Eoparargyractis plevie* (Pyralidae: Nymphulinae). J. Lepidopt. Soc. 31:81–88.

Forbes, W.T.M. 1910. The aquatic caterpillars of Lake Quinsigamond. Psyche 17:219–227.
——. 1911. Another aquatic caterpillar (*Elophila*). Psyche 18:120–121.
——. 1938. *Acentropus* in America (Lepidoptera, Pyralidae). J. N.Y. Entomol. Soc. 46:338.
*Hilsenhoff, W.L. 1981. Aquatic insects of Wisconsin. Nat. Hist. Counc. Publ. no. 2. Univ. Wisc., Madison. 60 pp.
*Lange, W.H., Jr. 1956a. Aquatic Lepidoptera. *In* R.L. Usinger (ed.). Aquatic insects of California, pp. 271–288. Univ. Calif. Press, Berkeley.
*——. 1956b. A generic revision of aquatic moths of North America (Lepidoptera: Pyralidae, Nymphulinae). Wash. J. Biol. 14:59–144.
*——. 1984. Aquatic and semi-aquatic Lepidoptera. *In* R.W. Merritt and K.W. Cummins (eds.). An introduction to the aquatic insects of North America, 2nd ed., pp. 348–357. Kendall/Hunt Publ. Co., Dubuque, Iowa.
Lavery, M.A., and R.R. Costa. 1972. Reliability of the Surber sampler in estimating *Parargyractis fulicalis* (Clemens) (Lepidoptera: Pyralidae) populations. Can. J. Zool. 50:1335–1336.
——. 1973. Geographic distribution of the genus *Parargyractis* Lange (Lepidoptera: Pyralidae) throughout the Lake Erie and Lake Ontario watersheds. J. N.Y. Entomol. Soc. 81:42–49.
Lloyd, J.T. 1914. Lepidopterous larvae from rapid streams. J. N.Y. Entomol. Soc. 22:145–152.
McAuliffe, J.R. 1984. Competition for space, disturbance, and the structure of a benthic stream community. Ecology 65:894–908.
McCafferty, W.P., and M.C. Minno. 1979. The aquatic and semiaquatic Lepidoptera of Indiana and adjacent areas. Greak Lakes Entomol. 12:179–188.
McGaha, Y.T. 1954. Contribution to the biology of some Lepidoptera which feed on certain aquatic flowering plants. Trans. Amer. Microscop. Soc. 73:167–177.
Munroe, E. 1972. Pyraloidea Pyralidae comprising subfamilies Scopariinae, Nymphulinae, Fasc. 13.1A. *In* R.B. Dominick et al. (eds.). The moths of America north of Mexico, including Greenland, pp. 1–134. E.W. Classey Ltd., London.
*Pennak, R.W. 1978. Freshwater invertebrates of the United States, 2nd ed. John Wiley and Sons, New York. 803 pp.
Welch, P.S. 1916. Contribution to the biology of certain aquatic Lepidoptera. Ann. Entomol. Soc. Amer. 9:159–187.
——. 1959. Lepidoptera. *In* W.T. Edmondson (ed.). Fresh-water biology, 2nd ed., pp. 1050–1056. John Wiley and Sons, New York.

10 Aquatic Coleoptera

Classification

The order Coleoptera (beetles) is the largest order of insects. It belongs to the infraclass Neoptera, division Endopterygota. Members of this order have an anterior pair of wings (the *elytra*) that are hard and leathery and not used in flight; the membranous hindwings, which are used for flight, are concealed under the elytra when the animals are at rest. Only 10% of the 350,000 described species of beetles are aquatic. Aquatic adaptations have arisen along many distinct phylogenetic lines, and aquatic species occur in two major suborders, the Adephaga and the Polyphaga (Brigham et al. 1982). Both larvae and adults of six beetle families are aquatic (Dytiscidae, Elmidae, Gyrinidae, Haliplidae, Hydrophilidae, and Noteridae). Five families (Chrysomelidae, Limnichidae, Psephenidae, Ptilodactylidae, and Scirtidae) have aquatic larvae and terrestrial adults, as do most of the other orders of aquatic insects; adult limnichids, however, readily submerge when disturbed, and for this reason we include them in our key to adults. Three families have species that are terrestrial as larvae and aquatic as adults (Curculionidae, Dryopidae, and Hydraenidae), a highly unusual combination among insects. Many other beetle families, which we do not include in our keys, have species that are riparian, semiaquatic, coastal, or marine (White et al. 1984). Because beetles exhibit such diverse adaptations, generalizations about life history patterns and feeding are difficult at the ordinal or even subordinal level. See More-Detailed Information for additional facts about natural history at the family level of aquatic or semiaquatic beetles.

Life History

Beetles are holometabolous. The aquatic species usually deposit eggs singly or in masses in or on submerged vegetation, stones, or other substrates. Most are univoltine, but some may undergo more than one generation per year. Some hydrophilid females have been reported to build a small case of vegetation into which they oviposit (Maillard 1970). Eggs of aquatic coleopterans hatch in one or two weeks, with diapause occurring rarely (White et al. 1984). Larvae undergo from three to eight molts. The pupal phase of all coleopterans is technically terrestrial, making this life stage of beetles the only one that has not successfully invaded the aquatic habitat. The mature larva either crawls out of the water and pupates in a pupal chamber constructed from mud or detritus within a few meters of shore or remains submerged in an air bubble formed within a silken cocoon (Pennak 1978). A few species have diapausing prepupae, but most complete transformation to adults in two to three weeks. Individuals of several families (Dytiscidae, Elmidae, Hydrophilidae) can be maintained in aquaria for years (White et al. 1984). Terrestrial adults of aquatic beetles are typically short-lived and sometimes nonfeeding, like those of the other orders of aquatic insects. Some of these short-lived adults emerge synchronously. Copulation in some aquatic species is preceded by acoustical signaling (stridulation), as in hemipterans, or by other courtship behaviors (White et al. 1984).

Habitat

Beetles are found in a very wide range of aquatic habitats. Two major groups can be characterized: those with most species occurring in ponds or slow-moving areas of streams (e.g., Dytiscidae, Haliplidae, Hydrophilidae), and those that live only in fast-flowing streams (e.g., Elmidae, Dryopidae, Psephenidae). The Gyrinidae, or whirligig beetles, occur on the surface of ponds in aggregations of up to thousands of individuals. Unlike the mating swarms of mayflies and hemipterans, these aggregations serve primarily to confuse predators (Heinrich and Vogt 1980). Whirligig beetles have other interesting defensive adaptations. For example, the Johnston's organ at the base of the antennae enables them to echolocate using surface wave signals; their compound eyes are divided into two pairs, one above and one below the water surface, enabling them to detect both aerial and aquatic predators; and they produce noxious chemicals that are highly effective at deterring predatory fish (Eisner and Meinwald 1966).

Some aquatic beetles live in specialized habitats such as marshes or acidic bogs (scirtid larvae), submerged macrophytes (chrysomelid larvae), emergent vegetation (Curculionidae; adults may sometimes submerge), mud or sand along the riparian zone (Heteroceridae, Limnichidae, Ptilodactylidae), and marine or brackish waters (Carabidae, Limnichidae, Staphylinidae). White et al. (1984) classify aquatic beetles as clingers, climbers, sprawlers, swimmers, divers, and burrowers.

Feeding

Like their habitats, the feeding habits of aquatic beetles are very diverse. Larvae can be herbivores (chewers or piercers), scavengers (gathering collectors), or voracious predators (engulfers or piercers). Dytiscid larvae, well known for their piercing mandibles, inject proteolytic enzymes into their prey or your hand, resulting in either subsequent ingestion of internal tissues or excruciating pain. Larval dytiscids prey even on small vertebrates (fish and tadpoles). Adult beetles also exhibit a wide range of feeding habits. Some species have been reported to scrape blue-green algae from substrates (White et al. 1984). Others are detritivores or predators. One adult dytiscid was observed consuming a small snapping turtle in an aquarium (C. O. Berg, pers. comm., 1980); however, the extent of these beetles' predation on vertebrates is not well known.

Respiration

Beetles exhibit four general types of respiration. Many larvae respire cutaneously, with additional oxygen taken in through tracheal gills. Adults of some families (Dytiscidae, Hydrophilidae) carry under their elytra an air store that is renewed by periodic trips to the water surface. Oxygen diffuses into the tracheae through open abdominal spiracles. A third group respires by means of a plastron, held in place by hydrofuge pubescence or fine sclerotized granules (Pennak 1978). Members of this group can remain submerged for longer periods without replenishing their oxygen supply than can beetles with an air store. Use of this mode of respiration saves riffle beetles (Dryopidae, Elmidae) from making many hazardous trips for air to the surface of fast-flowing streams. One highly specialized group (chrysomelid larvae) uses stylet-like caudal spines to pierce plant tissues and obtain oxygen from air spaces within the plants. However, these larvae have been observed to live for three weeks without penetrating plant tissues, implicating other (probably cutaneous) modes of respiration (Brigham et al. 1982).

Collection and Preservation

Aquatic beetles can be collected easily with a dip net or bottle trap (Hilsenhoff 1987) in open water, on substrates, or most often in association with pond vegetation. Extremely small beetles must be collected with fine mesh nets, and overwintering adults may be found burrowed into sediments or debris. Specimens can be preserved in vials of 80% ethanol with glycerine or can be pinned after a thorough rinse in 70% ethanol. Pennak (1978) recommends cleaning specimens that are covered with debris. Smaller beetle larvae should be cleared and mounted for observation under the compound microscope (Q. D. Wheeler, pers. comm., 1981).

Identification

Keys to both adults and larvae are included in this chapter. Although most adults can easily be identified to species, we include only generic keys here. Figures A and B illustrate the general morphological features of adult and larval beetles, respectively.

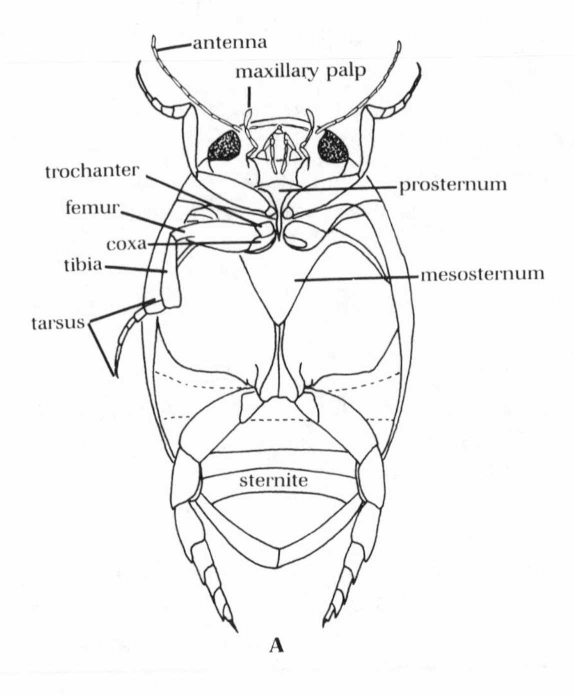

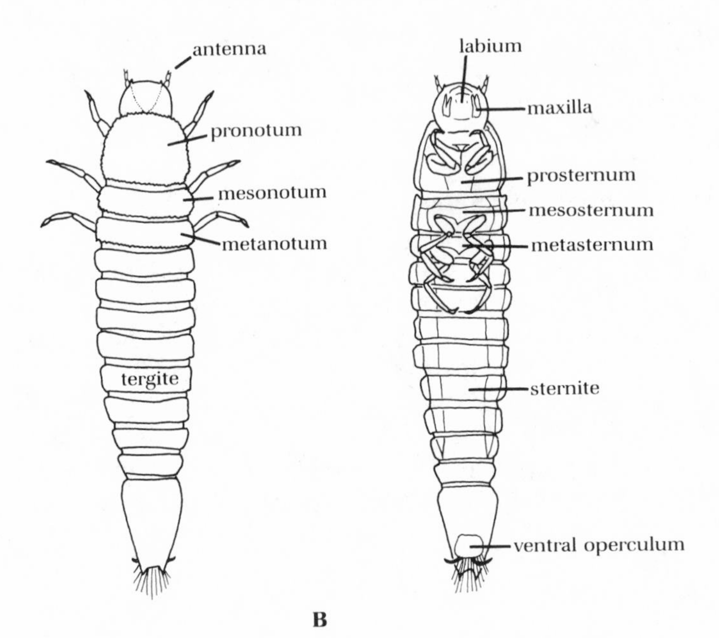

A. *Laccophilus* (Dytiscidae) adult, ventral view. **B.** Elmid larva; dorsal (left) and ventral (right) views. (A redrawn from Brown 1972; B modified from Pennak 1978.)

Features commonly used to identify adults are metacoxal plates, prosternal processes, antennae, elytral sculpturing, mouthparts, thoracic leg segmentation, eyes, scutellum, number and shape of claws, and body size. Larval keys are tentative, since the taxonomy of beetle larvae is poorly known. Many cannot be identified to species or even to genus (Hilsenhoff 1981). Characters used in our keys to separate genera of beetle larvae are tarsal claws, caudal appendages, leg segmentation, antennae, mouthparts, body shape, gills, and other abdominal appendages.

Although larval curculionids can be separated from larvae of other aquatic coleopteran families because they do not possess legs, the larvae of many genera and most species of aquatic curculionids have yet to be described, making construction of larval keys for this family impossible at present (White et al., 1984). Association of adults and larvae on host plants enables tentative identification of some taxa of immature curculionids.

We have not keyed riparian or beach-zone beetles, since few can be considered truly aquatic (White et al. 1984). White et al. (1984) provide a key to the adults of the staphylinid genera generally thought to be aquatic, but the larvae of this family are very incompletely described. Maritime species are treated by Moore and Legner (1976), and identification of semiaquatic and littoral genera are given by Arnett (1960) and Moore and Legner (1974). Adult and larval heterocerids are considered subaquatic, inhabiting littoral regions or burrows excavated beside streams; none of those taxa are considered in this key.

More-Detailed Information

Leech and Chandler 1956, Arnett 1960, Pennak 1978, Crowson 1981, Hilsenhoff 1981, Brigham et al. 1982, White et al. 1984, Brown 1987.

Checklist of Aquatic Coleoptera

Suborder Adephaga
- Dytiscidae
  - *Acilius*
  - *Agabetes*
  - *Agabus*
  - *Anodocheilus*[a]
  - *Bidessonotus*
  - *Carrhydrus*[a]
  - *Celina*
  - *Colymbetes*
  - *Copelatus*
  - *Coptotomus*
  - *Cybister*
  - *Deronectes*
  - *Desmopachria*
  - *Dytiscus*
  - *Graphoderus*
  - *Hydaticus*
  - *Hydroporus*
  - *Hydrovatus*
  - *Hygrotus*
  - *Ilybius*
  - *Laccophilus*
  - *Laccornis*
  - *Liodessus*
  - *Lioporius*[a]
  - *Matus*
  - *Neoscutopterus*[a]
  - *Oreodytes*
  - *Rhantus*
  - *Thermonectus*
  - *Uvarus*
- Gyrinidae
  - *Dineutus*
  - *Gyrinus*
- Haliplidae
  - *Haliplus*
  - *Peltodytes*
- Noteridae
  - *Hydrocanthus*
  - *Pronoterus*[a]
  - *Suphisellus*

Suborder Polyphaga
- Chrysomelidae[b]
  - *Donacia*
- Curculionidae[c]
- Dryopidae[c]
  - *Helichus*
- Elmidae
  - *Ancyronyx*
  - *Dubiraphia*
  - *Macronychus*
  - *Microcylloepus*
  - *Optioservus*
  - *Oulimnius*
  - *Promoresia*
  - *Stenelmis*

Hydraenidae[c]
Gymnochthebius
Hydraena
Limnebius
Ochthebius
Hydrophilidae
Anacaena
Berosus
Chaetarthria
Crenitis
Crenitulus[a]
Cymbiodyta
Derallus
Dibolocelus[a]
Enochrus
Helochares
Helocombus
Helophorus
Hydrobius
Hydrochara
Hydrochus
Hydrophilus
Laccobius
Paracymus
Sperchopsis
Tropisternus
Limnichidae[b]
Limnichus[a]
Lutrochus
Psephenidae[b]
Dicranopselaphus
Ectopria
Psephenus
Ptilodactylidae[b]
Anchytarsus
Scirtidae[b]
Cyphon
Elodes
Prionocyphon
Scirtes

[a]Larvae are unknown or undescribed.
[b]Adults are always, or usually, terrestrial.
[c]Larvae are terrestrial.

Key to Genera of Aquatic Coleoptera Adults

1a. Head formed into a snout or beak anteriorly (Fig. 1); antennae geniculate (Fig. 1) . . . . **Curculionidae** (semiaquatic; genera not keyed)
1b. Head not formed into a snout or beak; antennae not geniculate 2

2a (1b). Two pairs of eyes, a dorsal and ventral pair divided by sides of head; meso- and metathoracic legs extremely flattened; tarsi folded fanwise . . . . **Gyrinidae** 11
2b. One pair of eyes; meso- and metathoracic legs not extremely flattened; tarsi not folded fanwise 3

3a (2b). Metacoxae expanded into large plates that cover 2 or 3 abdominal sterna and bases of metafemora (Figs. 2, 3) . . . . **Haliplidae** 12
3b. Metacoxae not expanded into large plates 4

4a (3b). Prosternum with a postcoxal process (also referred to as the "prosternal process") that extends posteriorly to mesocoxae (Fig. 4); 1st visible abdominal sternite completely divided by metacoxal processes (Figs. 4–6) 5
4b. Prosternum with postcoxal process absent or short; 1st visible abdominal sternite extending for its entire breadth behind coxal processes (Fig. 7) 6

5a (4a). Scutellum concealed; enlarged median metasternal keel present (Fig. 5) . . . . **Noteridae** 13
5b. Scutellum concealed or exposed (as in Figs. 48, 49); enlarged median metasternal keel absent (Figs. 4, 6) . . . . **Dytiscidae** 15

6a (4b). Antennae terminating in an abrupt, globular, or elongate club (Fig. 8), with segments 4, 5, or 6 modified to form a cupule 7

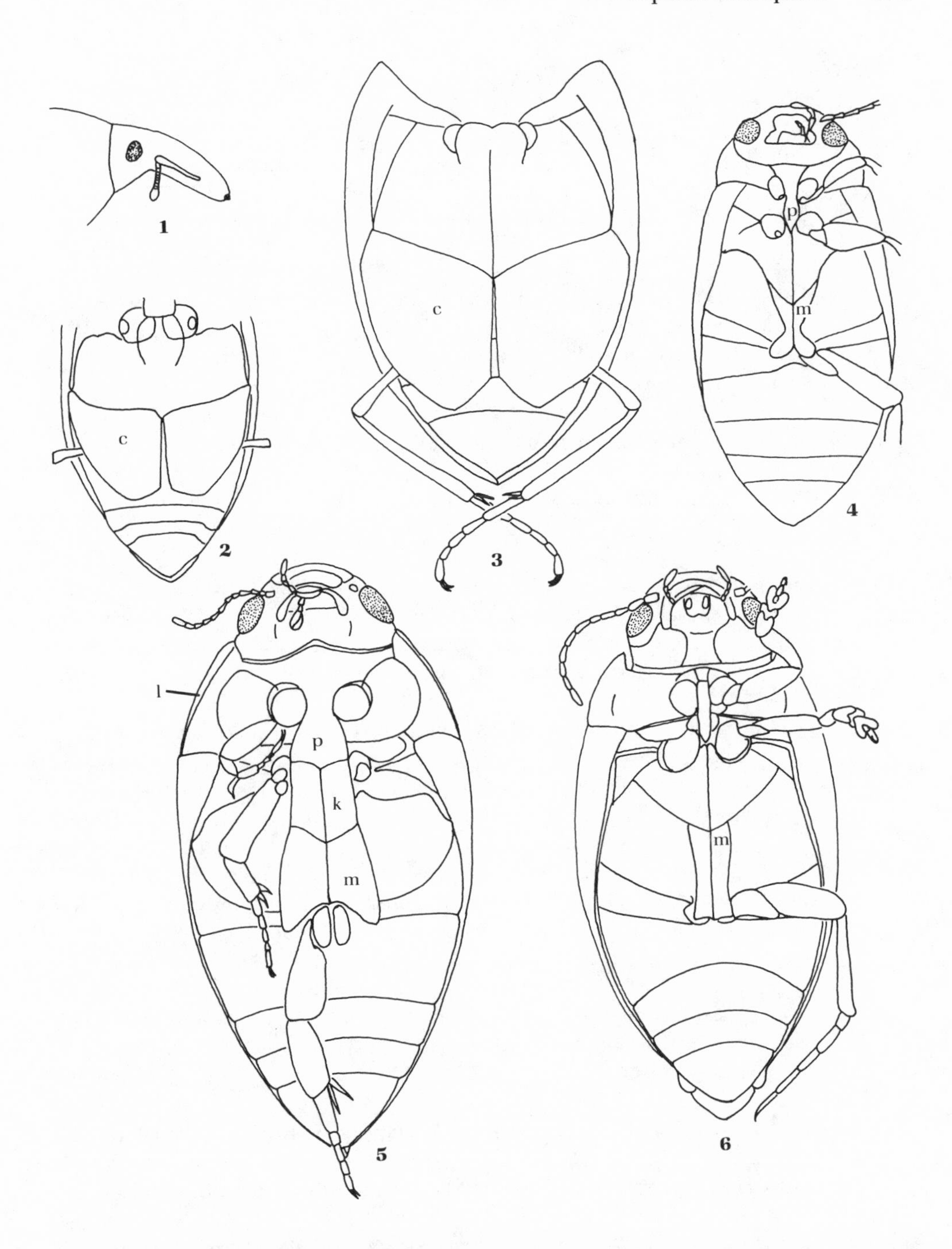

1. Head of Curculionidae, lv. **2.** Abdomen of *Haliplus* (Haliplidae), vv: c, coxal plate. **3.** Abdomen of *Peltodytes* (Haliplidae), vv: c, coxal plate. **4.** *Agabus* (Dytiscidae), vv: m, metacoxal process: p, postcoxal process. **5.** *Hydrocanthus* (Noteridae), vv: k, keel; l, lateral marginal line; m, metacoxal process; p, prosternal process. **6.** *Hydroporus* (Dytiscidae), vv: m, metacoxal process. lv, lateral view; vv, ventral view. (1, 2, 4 redrawn from Hilsenhoff 1981; 3, 5, 6 redrawn from Young 1954.)

6b. Antennae slender and elongate (Fig. 9), or very short and thick, with basal segment(s) enlarged (Fig. 10) 8

7a (6a). Antennae with 5 segments past cupule; body length < 2.5 mm . . . . **Hydraenidae** 46
7b. Antennae with 3 segments past cupule (Fig. 8); body length 1.5–40.0 mm . . . . **Hydrophilidae** 49

8a (6b). Antennae slender and elongate (Fig. 9), much longer than head; body length < 4.5 mm long . . . . **Elmidae** 69
8b. Antennae usually thick, with enlarged basal segments (Fig. 10), as long as or shorter than head 9

9a (8b). Hind coxae separate; antennae with 11 segments; body length 5.0–6.3 mm . . . . **Dryopidae,** *Helichus*
9b. Hind coxae contiguous; antennae with 10 or 11 segments; body length < 5 mm . . . . **Limnichidae** 10

10a (9b). **Limnichidae:** Small, body length < 2 mm; antennae with 10 segments; generally terrestrial . . . . *Limnichus*
10b. Larger, at least 2.5 mm long; antennae with 11 segments; generally cling to emergent objects, but may enter the water when disturbed . . . . *Lutrochus*

11a (2a). **Gyrinidae:** Scutellum exposed; elytron with 11 rows of sharp punctures; body length 3–8 mm . . . . *Gyrinus*
11b. Scutellum not visible; elytron smooth or with 9 vaguely impressed or scattered punctures; body length 9–16 mm . . . . *Dineutus*

12a (3a). **Haliplidae:** Last segment of maxillary palps as long or longer than penultimate segment (Fig. 11); pronotum with paired basal black impressions; only last abdominal sternite completely exposed (Fig. 3); body length 3.3–4.5 mm . . . . *Peltodytes*
12b. Last segment of maxillary palps shorter than penultimate segment (Fig. 12); pronotum sometimes with a basal black spot, but without paired black impressions; last 3 abdominal sternites usually exposed (Fig. 2); body length 2.5–4.9 mm . . . . *Haliplus*

13a (5a). **Noteridae:** Prosternal process round posteriorly; body length 2.5–3.0 mm . . . . *Pronoterus*
13b. Prosternal process truncate posteriorly (Fig. 5); body length 2.7–4.5 mm . . 14

14a (13b). Last segment of maxillary palps truncate; pronotum with lateral marginal lines running entire length; body length 3.7–4.5 mm (Fig. 5) . . . *Hydrocanthus*
14b. Last segment of maxillary palps emarginate; pronotum with lateral marginal lines from hind angle to about middle of segment; body length 2.7–3.0 mm . . . . *Suphisellus*

15a (5b). **Dytiscidae:** Prosternal process not in same plane as middle of prosternum (Fig. 14); pro- and mesotarsi either 4-segmented, or 5-segmented with 4th very small and 3rd bilobed (Fig. 13) 16

15b. Prosternal process in same plane as middle of prosternum (Fig. 15); pro- and mesotarsi 5-segmented with segments 3 and 4 subequal 30

16a (15a). Scutellum exposed; apices of elytra and last abdominal sternum strongly acuminate (Fig. 16); body length 4 mm .... *Celina*

16b. Scutellum covered by elytra; apices of elytra and abdomen not strongly acuminate 17

17a (16b). Broad apex of metacoxal processes divided into 3 parts, 2 widely separated narrow lateral lobes and a broader depressed middle region (Fig. 17) (for best viewing, orient the beetle with head toward you) .... *Hydrovatus*

17b. Metacoxal processes not in 3 parts; if lateral lobes present they cover base of trochanters 18

18a (17b). Metacoxal processes without lateral lobes, bases of metatrochanters entirely free (Fig. 18) 19

18b. Sides of metacoxal processes divergent, bases of metatrochanters covered by lateral lobes (Fig. 19) 23

19a (18a). Metatibiae straight, of almost uniform width from near base to apex (Fig. 20); metatarsal claws unequal; body length 1.8 mm .... *Desmopachria*

19b. Metatibiae slightly arcuate, narrow at base, widening gradually to apex (Fig. 21); metatarsal claws equal in length 20

20a (19b). Head without a transverse suture behind eyes; body length 1.6–2.0 mm .... *Uvarus*

20b. Head with a transverse suture behind eyes (Fig. 22) 21

21a (20b). Clypeal margin thickened anteriorly, upturned, tuberculate, or distinctly rimmed .... *Anodocheilus*

21b. Clypeal margin not thickened or only feebly so, not upturned, tuberculate, or rimmed 22

22a (21b). Pro- and mesotarsi clearly 5-segmented, and 4th segment short but not hidden by lobes of the 3rd; metacoxal lines strongly impressed, converging anteriorly across midmetasternum to nearly meet at mesocoxae (Fig. 23); body length 1.7–2.2 mm .... *Bidessonotus*

22b. Pro- and mesotarsi appearing 4-segmented, and 4th segment hidden by the lobes of the 3rd; metacoxal lines not continuing onto midmetasternum; body length 1.8–2.2 mm .... *Liodessus*

23a (18b). Bases of metafemora contacting metacoxal lobes (Fig. 24); body length 4.5–5.0 mm .... *Laccornis*

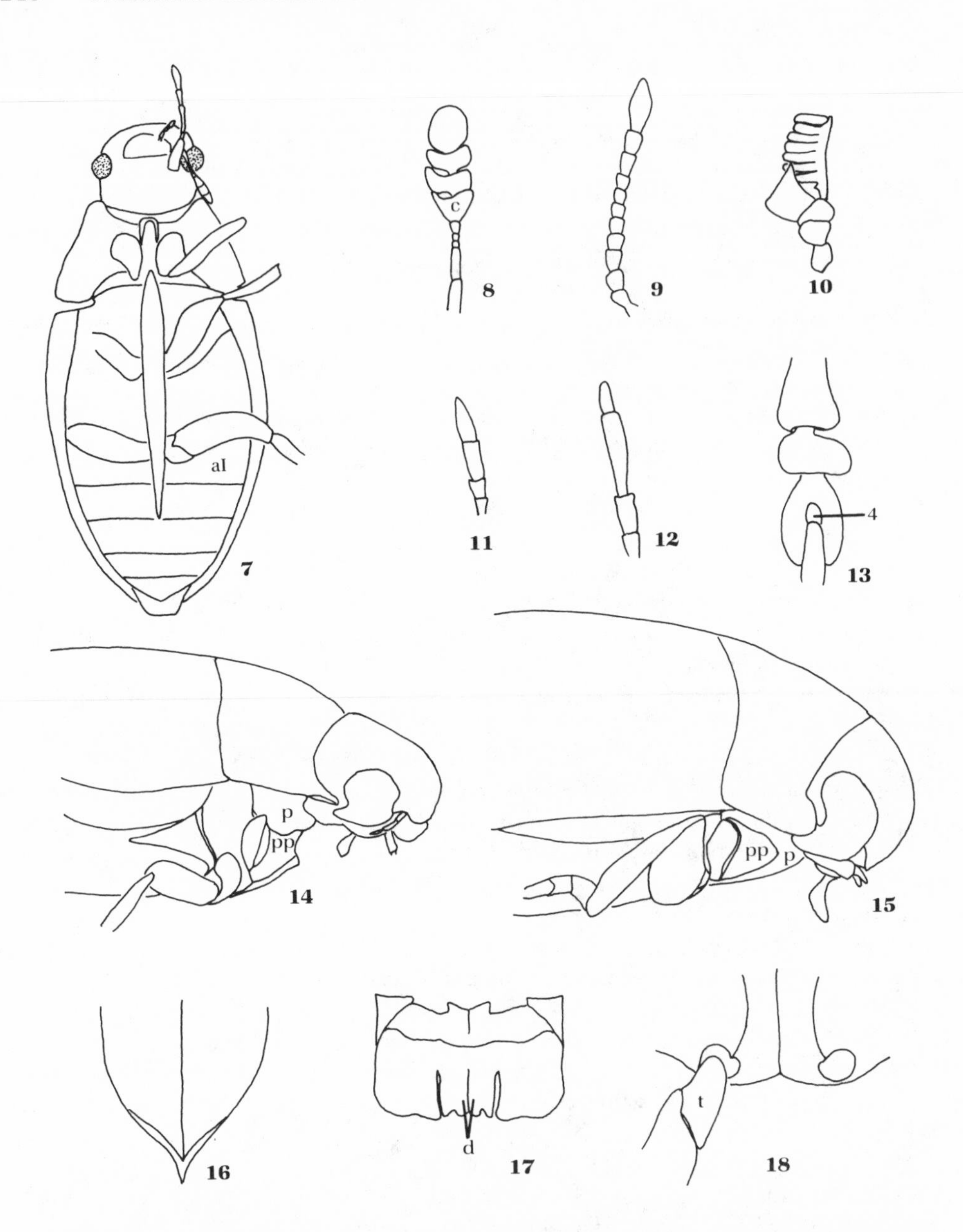

7. *Tropisternus* (Hydrophilidae), vv: aI, 1st abdominal sternite. **8.** Antenna of *Hydrochara* (Hydrophilidae), dv: c, cupule. **9.** Antenna of *Optioservus* (Elmidae), dv. **10.** Antenna of *Helichus* (Dryopidae). **11.** Maxillary palp of *Peltodytes* (Haliplidae). **12.** Maxillary palp of *Haliplus* (Haliplidae). **13.** Protarsus of *Hydroporus* (Dytiscidae), dv: 4, tarsal segment 4. **14.** Head and thorax of *Uvarus* (Dytiscidae) with forelegs removed, lv: p, prosternum; pp, prosternal process. **15.** Head and thorax of *Laccophilus* (Dytiscidae) with forelegs removed, lv: p, prosternum; pp, prosternal process. **16.** Apex of elytra and abdomen of *Celina* (Dytiscidae), dv. **17.** Metathorax of *Hydrovatus* (Dytiscidae), vv: d, depressed region. **18.** Apex of metacoxal process of *Liodessus* (Dytiscidae), vv: t, trochanter. dv, dorsal view; lv, lateral view; vv, ventral view. (7, 10–12, 16 redrawn from Hilsenhoff 1981; 8, 13, 17 modified from Arnett 1960; 9 redrawn from Brigham et al. 1982; 18 modified from Balfour-Browne 1940.)

23b. Metafemora separated from metacoxal lobes by basal part of trochanters 24

24a (23b). Diagonal carina crossing epipleural fold near base (Fig. 25); pro- and mesotarsi 4-segmented; body length 1.9–5.4 mm . . . . *Hygrotus*

24b. No carina crossing epipleural fold; pro- and mesotarsi 5-segmented, 4th usually small and hidden by bilobed 3rd 25

25a (24b). Posterior margin of metacoxal process truncate (Fig. 26), angularly prominent at middle (Fig. 27); or obtusely angulate (Fig. 28) so that median line of process is as long or longer than lateral coxal lines 26

25b. Posterior margin of metacoxal process incised or concave at middle (as in Fig. 19) so that median line is shorter than lateral coxal lines 29

26a (25a). Hind margin of metacoxal process either truncate (Fig. 26) or obtusely angulate (Fig. 28) . . . . *Hydroporus* (in part)

26b. Hind margin of metacoxal process sinuate and somewhat angularly prominent at middle (Fig. 27) 27

27a (26b). Prosternal process not protuberant; male with 4th and 5th antennal segments enlarged; male protarsi with cupule (Fig. 29); body length > 3.3 mm . . . . *Lioporius*

27b. Prosternal process usually protuberant (if not protuberant, then body length < 2.2 mm); male with 4th and 5th antennal segments not enlarged; male protarsi without cupule 28

28a (27b). Hind angles of pronotum rectangular or obtuse . . . . *Hydroporus* (in part)

28b. Hind angles of pronotum acute . . . . *Deronectes* (in part)

29a (25b). Metacoxal plates with many small punctures and scattered larger punctures; pronotum with distinct grooves laterally; body length 3.4–4.4 mm . . . . *Oreodytes*

29b. Metacoxal plates without larger punctures; pronotum without grooves; body length 4.3–5.0 mm . . . . *Deronectes* (in part)

30a (15b). Scutellum covered; metatarsi with single stout, straight claw; apex of prosternal process lanceolate; body length 4.0–6.0 mm . . . . *Laccophilus*

30b. Scutellum exposed; metatarsi with 2 claws, or if 1 claw, not as above; apex of prosternal process may or may not be lanceolate 31

31a (30b). Anterior margin of eyes emarginate above bases of antennae (Fig. 30) 32

31b. Anterior margin of eyes not emarginate above bases of antennae 41

32a (31a). Metafemora with linear group of short setae near posterolateral angle (Fig. 31) 33

32b. Metafemora without linear group of short setae near posterolateral angle 35

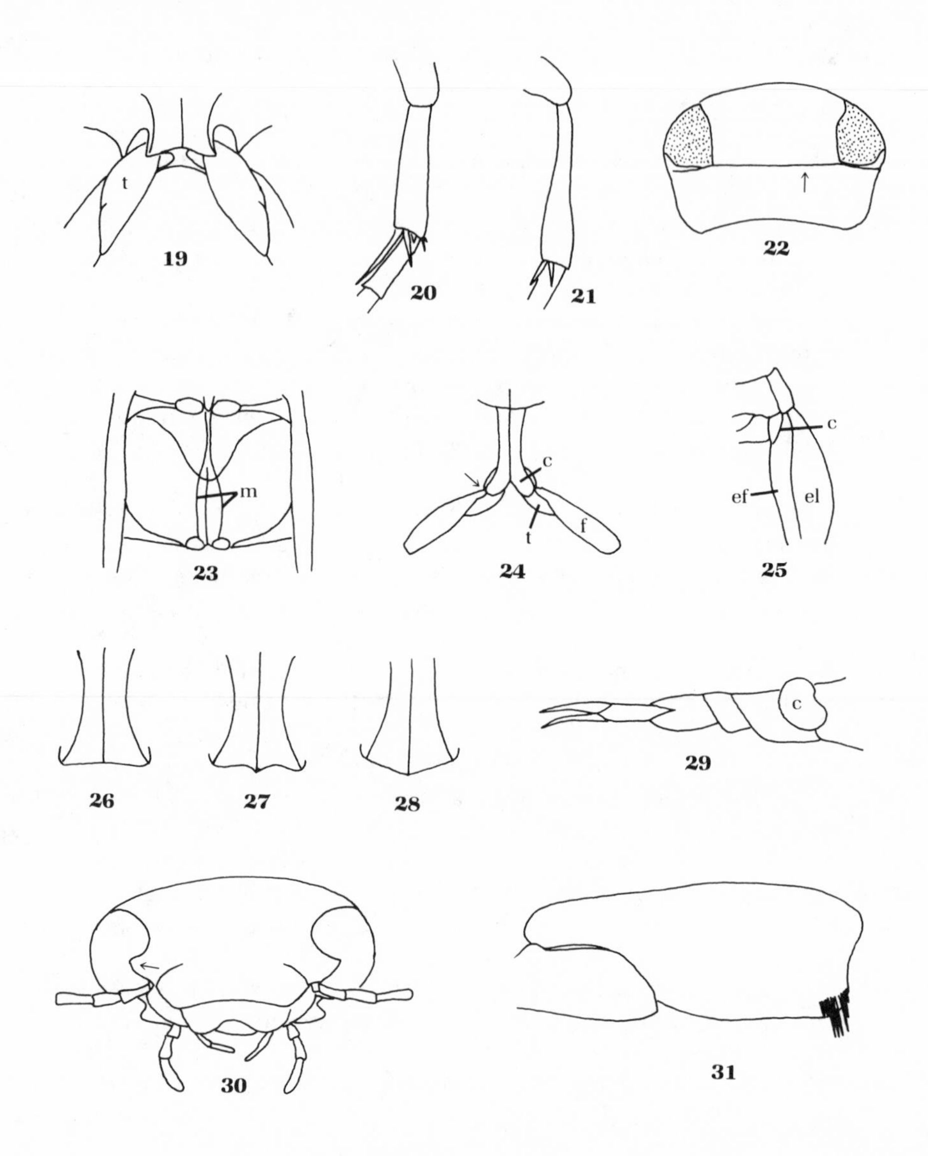

19. Apex of metacoxal process of *Hygrotus* (Dytiscidae), vv: t, trochanter. **20.** Metatibia of *Desmopachria* (Dytiscidae), lv. **21.** Metatibia of *Liodessus* (Dytiscidae), lv. **22.** Head of *Liodessus* (Dytiscidae), dv. **23.** Meso- and metasternum of *Bidessonotus* (Dytiscidae), vv: m, metacoxal lines. **24.** Base of hindlegs of *Laccornis* (Dytiscidae), vv: c, metacoxal lobe; f, metafemur; t, metatrochanter. **25.** Ventral margin of *Hygrotus* (Dytiscidae): c, carina; ef, epipleural fold; el, elytron. **26.** Posterior margin of metacoxal process of *Hydroporus* sp. a (Dytiscidae), vv. **27.** Posterior margin of metacoxal process of *Hydroporus* sp. b (Dytiscidae), vv. **28.** Posterior margin of metacoxal process of *Hydroporus* sp. c (Dytiscidae), vv. **29.** Protarsus of *Lioporius* (Dytiscidae), vv: c, cupule. **30.** Head of *Agabus* (Dytiscidae), av. **31.** Metafemur of *Agabus* (Dytiscidae), vv. av, anterior view; dv, dorsal view; lv, lateral view; vv, ventral view. (19 modified from Balfour-Browne 1940; 20–22, 24 redrawn from Hilsenhoff 1981; 23, 25 modified from Arnett 1960; 26–28 redrawn from Guignot 1931; 29 redrawn from Wolfe and Matta 1981; 30, 31 redrawn from Balfour-Browne 1950.)

33a (32a). Metatarsal claws equal; if slightly unequal, then both are very short, only ⅓ length of 5th tarsal segment; 6–11 mm long . . . . *Agabus*

33b. Metatarsal claws obviously unequal, outer one of each pair ⅔ or less the length of inner claw 34

34a (33b). Labial palps with penultimate segment triangular in cross section, the faces concave and unequal; genital valves of female dorsoventrally compressed, not armed with teeth; body length 10.5–13.3 mm . . . . *Carrhydrus*

34b. Labial palps with penultimate segment cylindrical; genital valves of female laterally compressed, sawlike, with a series of sharp teeth along dorsal edge; body length 8.0–11.5 mm . . . . *Ilybius*

35a (32b). Prosternal process with median longitudinal furrow; body length 8.5–9.0 mm . . . . *Matus*

35b. Prosternal process without median longitudinal furrow 36

36a (35b). Metacoxal lines coming close together posteriorly, almost touching midline, then turning laterally outward almost at a 90° angle to the midline (Fig. 32); metatarsal claws equal; body length 4.5–5.5 mm . . . . *Copelatus*

36b. Metacoxal lines turning laterally outward and posteriorly at greater than a 90° angle to the midline, and never almost touching midline (as in Fig. 4); metatarsal claws sometimes equal 37

37a (36b). Metatarsal claws equal or subequal 38

37b. Metatarsal claws obviously unequal 39

38a (37a). Apical segment of maxillary palps deeply emarginate (Fig. 33); body length 7.0–8.5 mm . . . . *Coptotomus*

38b. Apical segment of maxillary palps not emarginate (Fig. 34); body length 6.0–7.0 mm . . . . *Agabetes*

39a (37b). Anterior apex of metasternum, between mesocoxae, triangularly split to receive apex of prosternal process, triangular channel usually deep, with its apex about on a line with hind margins of mesocoxae; body length 9–11 mm . . . . *Rhantus*

39b. Anterior apex of metasternum depressed, with shallow pit or broad notch to receive apex of prosternal process, never with triangular excavation 40

40a (39b). Elytra sculptured with numerous parallel transverse grooves; body length 15–17 mm . . . . *Colymbetes*

40b. Elytra coarsely reticulate, without parallel transverse grooves; body length 14–16 mm . . . . *Neoscutopterus*

41a (31b). One large spur at apex of metatibiae twice as broad as other spur; beetle widest at posterior third; body length 28–33 mm . . . . *Cybister*

41b. Large spurs at apex of metatibiae subequal in width 42

42a (41b). Posterior margins of first 4 metatarsal segments without a fringe of golden setae; beetle widest near middle; body length 25–40 mm .... *Dytiscus*

42b. Posterior margins of first 4 metatarsal segments with a fringe of golden setae 43

43a (42b). Outer margin of metasternal "wings" straight (Fig. 35); outer spur at apex of metatibiae acute; body length 12–14 mm .... *Hydaticus*

43b. Outer margin of metasternal "wings" arcuate (Fig. 36); outer spur at apex of metatibiae blunt, more or less emarginate 44

44a (43b). Elytral punctation dense, usually fluted and hairy in females; body length 12–16 mm .... *Acilius*

44b. Elytral punctation extremely fine or absent, not fluted and hairy in females 45

45a (44b). Hind margin of mesofemora with stiff setae only ½ as long as width of femur (Fig. 37); body length 11–16 mm .... *Graphoderus*

45b. Hind margin of mesofemora with stiff setae as long as or longer than width of femur (Fig. 38); body length 9–13 mm .... *Thermonectus*

46a (7a). **Hydraenidae:** The 3rd segment of the maxillary palps longer and broader than the 4th (Fig. 39); pronotum with transparent or semitransparent borders at anterior and posterior margins; transparent lateral margins on pronotum present or absent 47

46b. Third and 4th segments of the maxillary palps subequal in length (Fig. 40), or with the 3rd much shorter than the 4th (Fig. 41); transparent lateral margins on pronotum absent 48

47a (46a). Pronotum with sides of sclerotized part deeply emarginate in both anterior and posterior ½, between these emarginations sides produced in sharp points; anterior angles of pronotum are lobate (Fig. 42) ..... *Gymnochthebius*

47b. Pronotum with sclerotized part of diverse forms, in some specimens very transverse, sides gradually rounded from anterior angles to behind the middle, then deeply notched (Fig. 43) or pronotum with sides sinuate and convergent to base; anterior angles of pronotum not markedly lobate .... *Ochthebius*

48a (46b). Maxillary palps with 2nd segment very elongate, much longer than the 3rd (Fig. 41) .... *Hydraena*

48b. Maxillary palps with 2nd and 3rd segments subequal in length (Fig. 40) .... *Limnebius*

49a (7b). **Hydrophilidae:** Pronotum with 5 longitudinal grooves (Fig. 44); body length 2.5–8.0 mm .... *Helophorus*

49b. Pronotum without such grooves 50

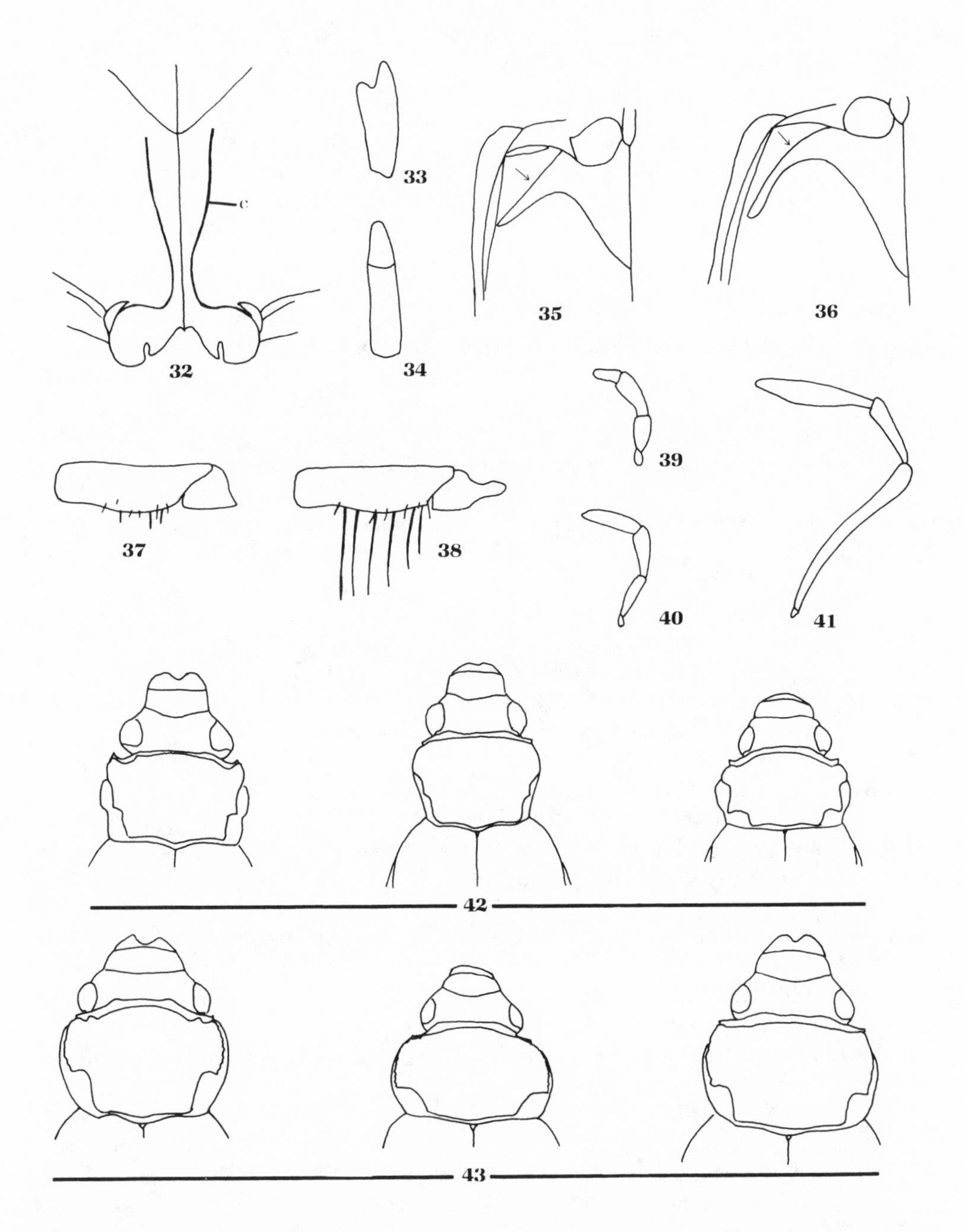

32. Metasternum of *Copelatus* (Dytiscidae), vv: c, coxal lines. **33.** Terminal segment of maxillary palp of *Coptotomus* (Dytiscidae). **34.** Terminal segment of maxillary palp of *Agabetes* (Dytiscidae). **35.** Right side of metasternum of *Hydaticus* (Dytiscidae), vv. **36.** Right side of metasternum of *Acilius* (Dytiscidae), vv. **37.** Mesofemur of *Graphoderus* (Dytiscidae). **38.** Mesofemur of *Thermonectus* (Dytiscidae). **39.** Maxillary palp of *Ochthebius* (Hydraenidae). **40.** Maxillary palp of *Limnebius* (Hydraenidae). **41.** Maxillary palp of *Hydraena* (Hydraenidae). **42.** Variation in pronotum shape in *Gymnochthebius* spp. (Hydraenidae), dv. **43.** Variation in pronotum shape in *Ochthebius* spp. (Hydraenidae), dv. dv, dorsal view; vv, ventral view. (32 modified from Balfour-Browne 1935; 33–38 redrawn or modified from Hilsenhoff 1981; 39–43 redrawn from Perkins 1980.)

50a (49b). Pronotum granular and conspicuously narrower than elytral base (Fig. 45); elytra may be narrowed anteriorly; scutellum very small; eyes protuberant; body length 2.0–4.0 mm . . . . *Hydrochus*
50b. Pronotum not appreciably narrower than base of elytra, but if it is, scutellum elongate 51

51a (50b). Metatarsi 5-segmented, with basal segment longer than 2nd; antennae usually longer than maxillary palps; segment 2 of maxillary palps much thicker than 3 or 4 . . . . Sphaeridiinae (terrestrial)
51b. Metatarsi 4-segmented, or 5-segmented with basal segment much shorter than 2nd; antennae subequal or shorter than maxillary palps; segment 2 of maxillary palps not or only slightly thicker than segments 3 or 4 52

52a (51b). Meso- and metasternum with a continuous median longitudinal keel produced into a posterior spine extending between metacoxae (Figs. 7, 46) . . . 53
52b. Sternal region without such a keel 56

53a (52a). Prosternum carinate (Fig. 47); metasternal keel not or hardly reaching beyond base of hind trochanters; body length 13–16 mm . . . . *Hydrochara*
53b. Prosternum sulcate to receive anterior part of mesosternal keel (Fig. 46); metasternal keel projecting beyond metatrochanters as a spine 54

54a (53b). Small species, body length 6–16 mm; last segment of maxillary palps equal to or longer than penultimate segment . . . . *Tropisternus*
54b. Large species, body length 30–45 mm; last segment of maxillary palps shorter than penultimate segment 55

55a (54b). Prosternum bifurcate, groove open anteriorly; body length 31–33 mm . . . . *Dibolocelus*
55b. Prosternum sulcate, closed anteriorly; body length 32–37 mm . . *Hydrophilus*

56a (52b). First 2 visible abdominal sternites on each side with common excavation covered by a bilobed plate . . . . *Chaetarthria*
56b. Abdomen with normal sternites 57

57a (56b). Meso- and metatibiae fringed with long hairs (Fig. 48); pronotum not continuous in outline with elytra (Fig. 48) 58
57b. Meso- and metatibiae without fringe of long hairs; pronotum and elytra continuous in outline (Fig. 49) 59

58a (57a). Body length 1.5–2.0 mm; black; very convex . . . . *Derallus*
58b. Body length 2.5–6.0 mm; brown to yellowish, usually with dark markings on pronotum and elytra; not very convex . . . . *Berosus*

59a (57b). Maxillary palps about same length as antennae, last segment as long as or longer than penultimate segment 60

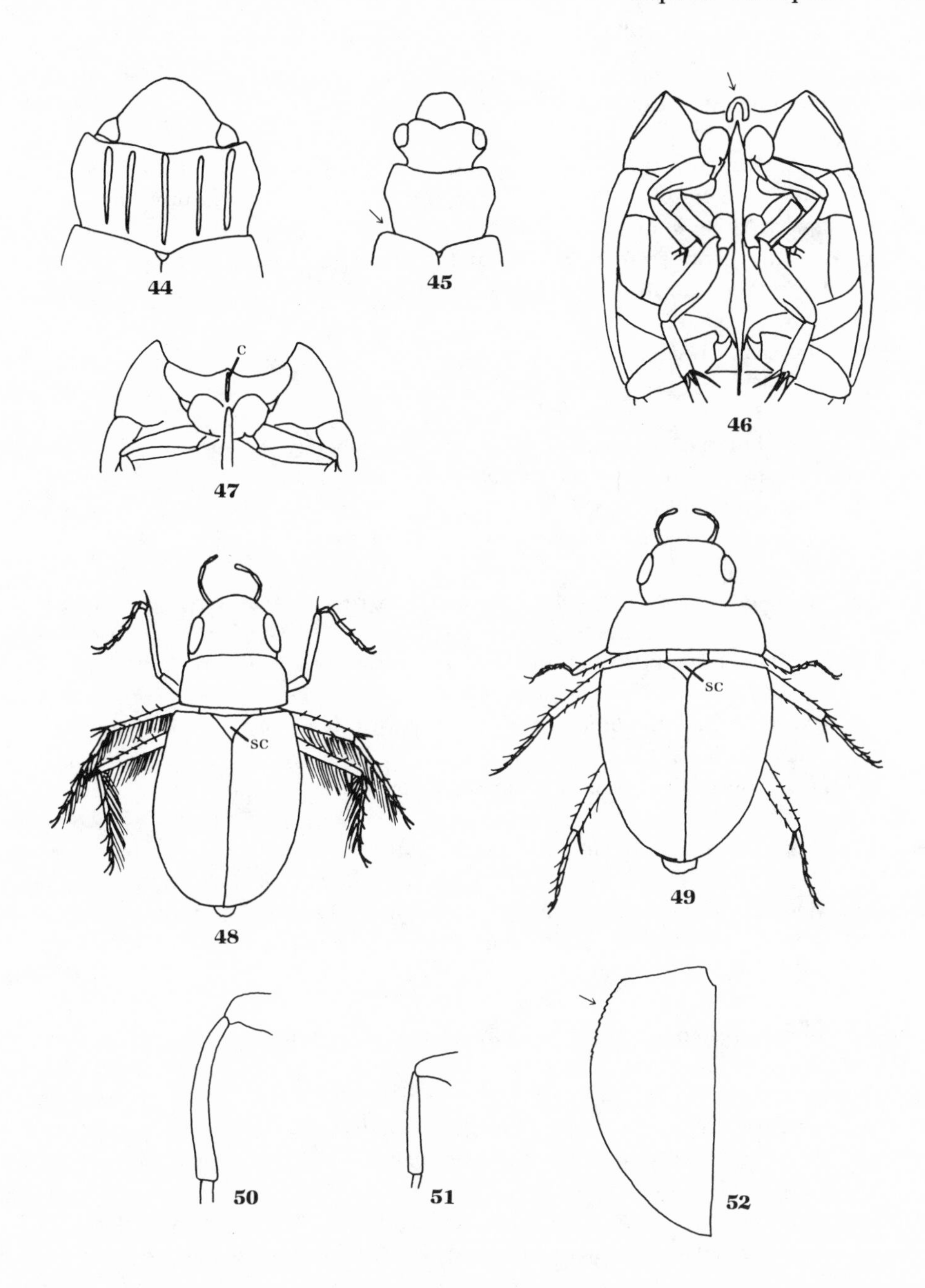

44. Head and pronotum of *Helophorus* (Hydrophilidae), dv. **45.** Head and pronotum of *Hydrochus* (Hydrophilidae), dv. **46.** Thorax of *Hydrophilus* (Hydrophilidae), vv. **47.** Prosternum of *Hydrochara* (Hydrophilidae), vv: c, carina. **48.** *Berosus* (Hydrophilidae), dv: sc, scutellum. **49.** *Hydrobius* (Hydrophilidae), dv: sc, scutellum. **50.** Metatibia of *Laccobius* (Hydrophilidae). **51.** Metatibia of *Anacaena* (Hydrophilidae). **52.** Left elytron of *Sperchopsis* (Hydrophilidae), dv. dv, dorsal view; vv, ventral view. (44–47 modified from Arnett 1960; 50–52 modified from Hilsenhoff 1981.)

59b. Maxillary palps longer than antennae, with last segment usually shorter than penultimate segment 66

60a (59a). Metatibiae arcuate (Fig. 50); metatrochanters large, about ⅓ as long as femora, their apices distinct from femora; elytra without sutural striae; body length 25–40 mm .... *Laccobius*

60b. Metatibiae not arcuate (Fig. 51); metatrochanters normal, closely applied to femora; elytra with sutural striae 61

61a (60b). Body length 4.5–10 mm 62

61b. Body length ≤ 3.3 mm 63

62a (61a). Lateral margins of elytra weakly serrate basally (Fig. 52); meso- and metatarsi with scattered fine hairs dorsally; body length 8–9 mm .... *Sperchopsis*

62b. Lateral margins of elytra without serrations; meso- and metatarsi with a fringe of fine hairs dorsally; body length 6–10 mm .... *Hydrobius*

63a (61b). Metatarsi longer than metatibiae; elytra narrowed posteriorly; body length 1.7–2.1 mm .... *Crenitulus*

63b. Metatarsi not longer than metatibiae; elytra not narrowed posteriorly 64

64a (63b). Prosternum with median longitudinal carina; body usually with a metallic sheen; body length 2.0–2.9 mm .... *Paracymus*

64b. Prosternum without median longitudinal carina 65

65a (64b). Mesosternum simple or with a small transverse protuberance anterior to mesocoxae; eyes protuberant (Fig. 53); body length 2.4–3.3 mm .... *Crenitis*

65b. Mesosternum with a prominent angularly elevated or dentiform protuberance anterior to mesocoxae (Fig. 54); eyes not protuberant (Fig. 55); body length 2.1–2.8 mm .... *Anacaena*

66a (59b). All tarsi 5-segmented, basal segment may be minute 67

66b. Meso- and metatarsi 4-segmented 68

67a (66a). Long pseudobasal segment of maxillary palps with concavity to posterior or side; mesosternum with ventrally projecting longitudinal plate in front of mesocoxae; body length 2.9–9.5 mm .... *Enochrus*

67b. Long pseudobasal segment of maxillary palps with concavity to anterior or center; mesosternum at most feebly protuberant; body length 5.0–6.0 mm .... *Helochares*

68a (66b). Mesosternum with a prominent conical process just anterior to mesocoxae; elytra with many impressed striae; maxillary palps long and slender; tarsal claws with broad basal tooth in male, less prominently toothed in female; body length 6.0–8.0 mm .... *Helocombus*

68b. Mesosternum with transverse carina just anterior to mesocoxae; only sutural striae of elytra impressed; maxillary palps shorter, stouter; tarsal claws simple in both sexes; body length 3.0–6.0 mm . . . . *Cymbiodyta*

69a (8a). **Elmidae:** legs very long (Fig. 56), mesofemora as long as or longer than basal width of elytra; coxae very widely separated and about equal in size, located more on side of body than on nearly flat venter; elytra never with longitudinal brownish yellow stripes 70

69b. Legs of normal size (Figs. 57–59), mesofemora less than ¾ basal width of elytra; coxae, especially the procoxae, not widely separated; elytra often with longitudinal brownish yellow stripes 71

70a (69a). Pronotum with 2 oblique, transverse depressions at anterior ⅓, without sublateral carinae; antennae filiform, with 11 segments, not clubbed; tibiae without patches of tomentum; elytra without sublateral carinae; body length 2.7–3.5 mm (Fig. 56) . . . . *Ancyronyx*

70b. Pronotum without transverse depressions, but with basal sublateral carinae; antennae 7-segmented, last segment may be enlarged; tibiae with 1 or 2 patches of tomentum; elytra with sublateral carinae; body length 2.7–3.7 mm . . . . *Macronychus*

71a (69b). Posterior lateral corner of 4th, or middle of lateral margin of 5th abdominal sternite not produced into a prominent upturned tooth; epipleura usually uniformly and narrowly tapered to a fine point (Fig. 60) 72

71b. Posterior lateral corner of 4th (Fig. 61), or middle of lateral margin of 5th (Fig. 62) abdominal sternite produced into a prominent upturned tooth; epipleura slightly (Fig. 61) to conspicuously (Fig. 62) widened to receive tooth, usually bluntly narrowing to apex posterior to tooth (Figs. 61, 62) 73

72a (71a). Protibiae with a thick patch of tomentum on anterior side; epipleura usually ending near base or middle of 5th abdominal segment; body length 2.0–3.5 mm (Fig. 57) . . . . *Dubiraphia*

72b. Protibiae without patch of tomentum; epipleura extending to posterior apex of elytra (Fig. 60); body length 2.7–4.2 mm (Fig. 59) . . . . *Stenelmis*

73a (71b). Tooth at the lateral margin of 5th abdominal segment fitting into a conspicuously widened area near the apex of the shortened epipleura (Fig. 62) . . . . *Microcylloepus*

73b. Posterior lateral angle of the 4th abdominal segment produced into a prominent tooth that is usually bent vertically to fit into a slightly widened area near the apex of the shortened epipleura (Fig. 61) 74

74a (73b). Pronotum with steplike sublateral carinae extending from posterior to anterior margin (Fig. 63); elytra with sublateral carinae . . . . *Oulimnius*

74b. Pronotum with sublateral carinae short or absent; elytra without sublateral carinae (Fig. 58) 75

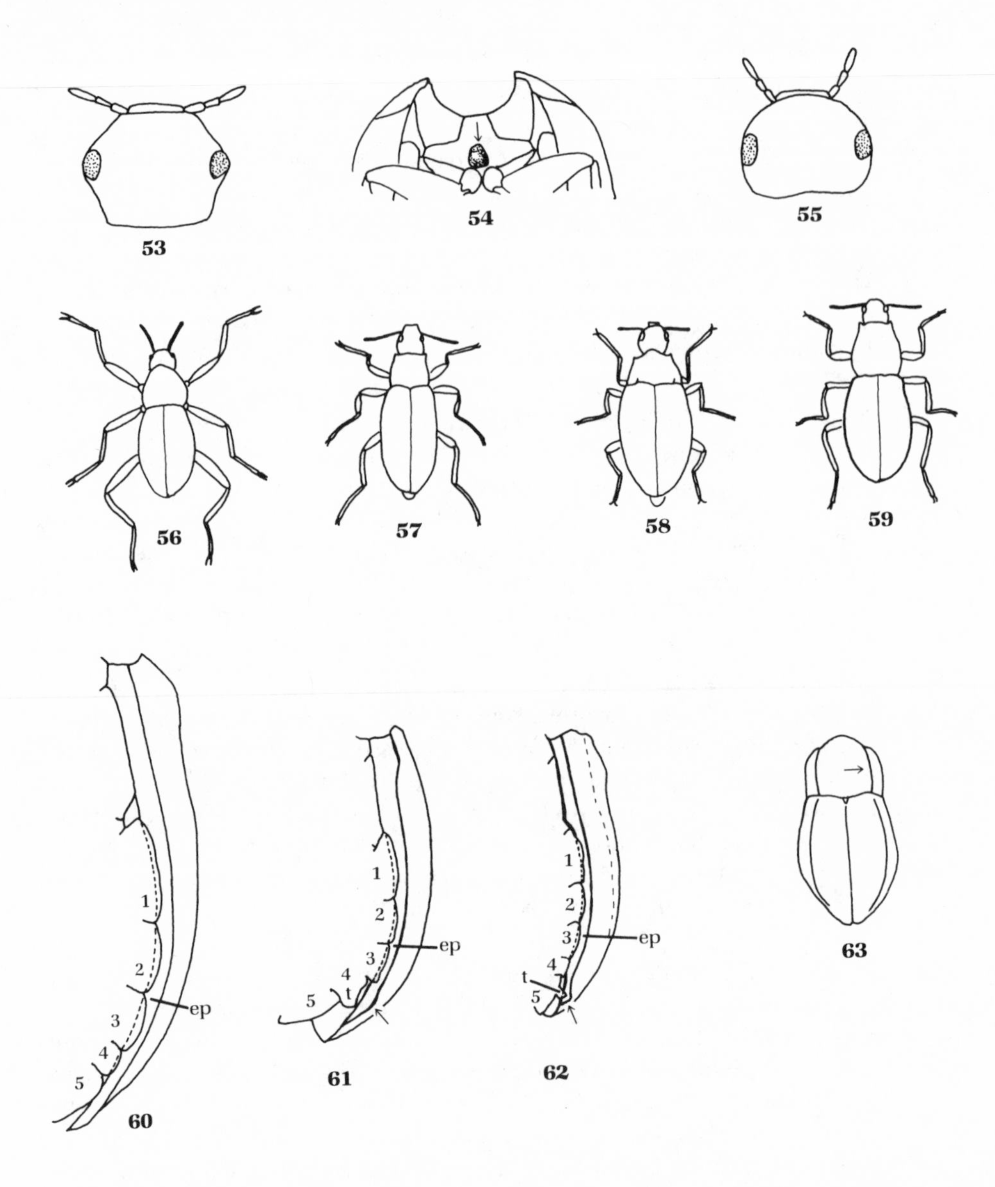

53. Head of *Crenitis* (Hydrophilidae), dv. **54.** Prosternum of *Anacaena* (Hydrophilidae), vv. **55.** Head of *Anacaena* (Hydrophilidae), dv. **56.** *Ancyronyx* (Elmidae), dv. **57.** *Dubiraphia* (Elmidae), dv. **58.** *Optioservus* (Elmidae), dv. **59.** *Stenelmis* (Elmidae), dv. **60.** Epipleuron and abdominal segments of *Stenelmis* (Elmidae), vv: ep, epipleuron. **61.** Epipleuron and abdominal segments of *Optioservus* (Elmidae), vv: ep, epipleuron; t, tooth. **62.** Epipleuron and abdominal segment of *Microcylloepus* (Elmidae), vv: ep, epipleuron; t, tooth. **63.** *Oulimnius* (Elmidae) thorax and abdomen, dv. dv, dorsal view; vv, ventral view. (53, 55–59 redrawn or modified from Hilsenhoff 1981; 54, 63 modified or redrawn from Arnett 1960; 60–62 redrawn from Leech and Chandler 1956.)

75a (74b). Body rather elongate; lateral and posterior margins of prothorax smooth; tarsi stout and nearly as long as tibiae, distal ½ of last segment as thick as tibiae, claws very large, curved more than 90° (i.e., 100°) . . . . *Promoresia*

75b. Body plump; lateral margins of prothorax slightly serrate, posterior margin with many small, loosely placed teeth; tarsi less developed, distal segment never as thick as tibiae, claws shorter and more slender, curved about 90°; body length 1.7–3.5 mm . . . . *Optioservus*

Key to Genera of Aquatic Coleoptera Larvae

1a. Each tarsus with 2 claws 2

1b. Each tarsus with 1 claw 4

2a (1a). Abdomen with 4 conspicuous hooks on last segment; at least 8 abdominal segments with pairs of lateral filaments (Fig. 1) . . . . **Gyrinidae** 11

2b. No hooks on last abdominal segment; if lateral abdominal filaments are present, there are only 6 pairs 3

3a (2b). Posterior ½ of abdomen little narrowed (Fig. 2); legs and cerci short . . . . **Noteridae** 13

3b. Posterior ½ of abdomen conspicuously narrowed (Fig. 3); legs long, cerci often elongate . . . . **Dytiscidae** 14

4a (1b). Legs distinctly 5-segmented (excluding claw); abdomen terminating in 1 or 2 long filaments (Fig. 4) . . . . **Haliplidae** 12

4b. Legs apparently 4-segmented (excluding claw); abdomen not terminating in long filaments 5

5a (4b). Mandibles large, readily visible from above (Fig. 5) . . . . **Hydrophilidae** 37

5b. Mandibles not readily visible from above 6

6a (5b). Antennae long, filiform, as long as head and thorax combined (Fig. 6) . . . . **Scirtidae** 54

6b. Antennae much shorter than head and thorax combined 7

7a (6b). Body oval and extremely flat (Fig. 7); head completely concealed from dorsal view . . . . **Psephenidae** 57

7b. Body elongate, round, or triangular in cross section; head exposed 8

8a (7b). All terga rounded and pale; grublike larva with 2 large, sclerotized dorsal hooks on spiracles of 8th abdominal segment (Fig. 8) . . . . **Chrysomelidae,** *Donacia*

8b. Body elongate and sclerotized; larva without sclerotized dorsal hooks on 8th abdominal segment 9

9a (8b). Ninth abdominal segment without a movable ventral operculum . . . . **Ptilodactylidae,** *Anchytarsus*
9b. Ninth abdominal segment with a movable ventral operculum closing a caudal chamber (Figs. 9, 10) 10

10a (9b). Abdomen with pleurites present on only the 1st 4 segments; each undivided thoracic pleurite with erect hairs along medial margin; thoracic sternites membranous or absent; terminal abdominal segment evenly rounded; head often retracted within body . . . . **Limnichidae,** *Lutrochus*
10b. Abdomen with pleurites on at least the 1st 6 segments (Fig. 10); thoracic pleurites without erect hairs; thoracic sternites sclerotized; terminal abdominal segment bifid or slightly emarginate posteriorly (Fig. 10) and with lateral ridges; head not retractile . . . . **Elmidae** 59

11a (2a). **Gyrinidae:** Head narrowed posteriorly to form a distinct collar (Fig. 11) . . . . *Dineutus*
11b. Elongate head not narrowed posteriorly to form a collar (Fig. 12) . . . . *Gyrinus*

12a (4a). **Haliplidae:** Each body segment with 2 or more long, spinelike filaments, each ½ as long as body (Fig. 13) . . . . *Peltodytes*
12b. Spines on body segments less than length of a segment . . . . *Haliplus*

13a (3a). **Noteridae:** Mandibles stout, bifid at tip; 3rd antennal segment no longer than 4th . . . . *Suphisellus*
13b. Mandibles slender, not bifid at tip; 3rd antennal segment at least twice as long as 4th . . . . *Hydrocanthus*

14a (3b). **Dytiscidae:** Lateral gills on abdominal segments 1–6 . . . . *Coptotomus*
14b. No lateral gills on abdominal segments 15

15a (14b). Head with a frontal projection (Figs. 14, 16) 16
15b. Head without a frontal projection (but may have teeth as in *Cybister*) (Figs. 15, 17) 23

16a (15a). Cerci shorter, with 1st segment less than 3 times the length of the 8th abdominal segment; cerci with only primary hairs (as in Fig. 18) 17
16b. Cerci very long, with 1st segment 3 or more times the length of the 8th abdominal segment; cerci with secondary hairs (Fig. 19) 18

17a (16a). Frontal projection notched laterally (Fig. 16) . . . . *Hydroporus, Hygrotus*
17b. Frontal projection without a notch on each side (Fig. 14) 19

18a (16b). Frontal projection notched laterally . . . . *Deronectes*
18b. Frontal projection without a notch on each side . . . . *Oreodytes*

19a (17b). Basal (widened) segment of cerci distinctly longer than last abdominal seg-

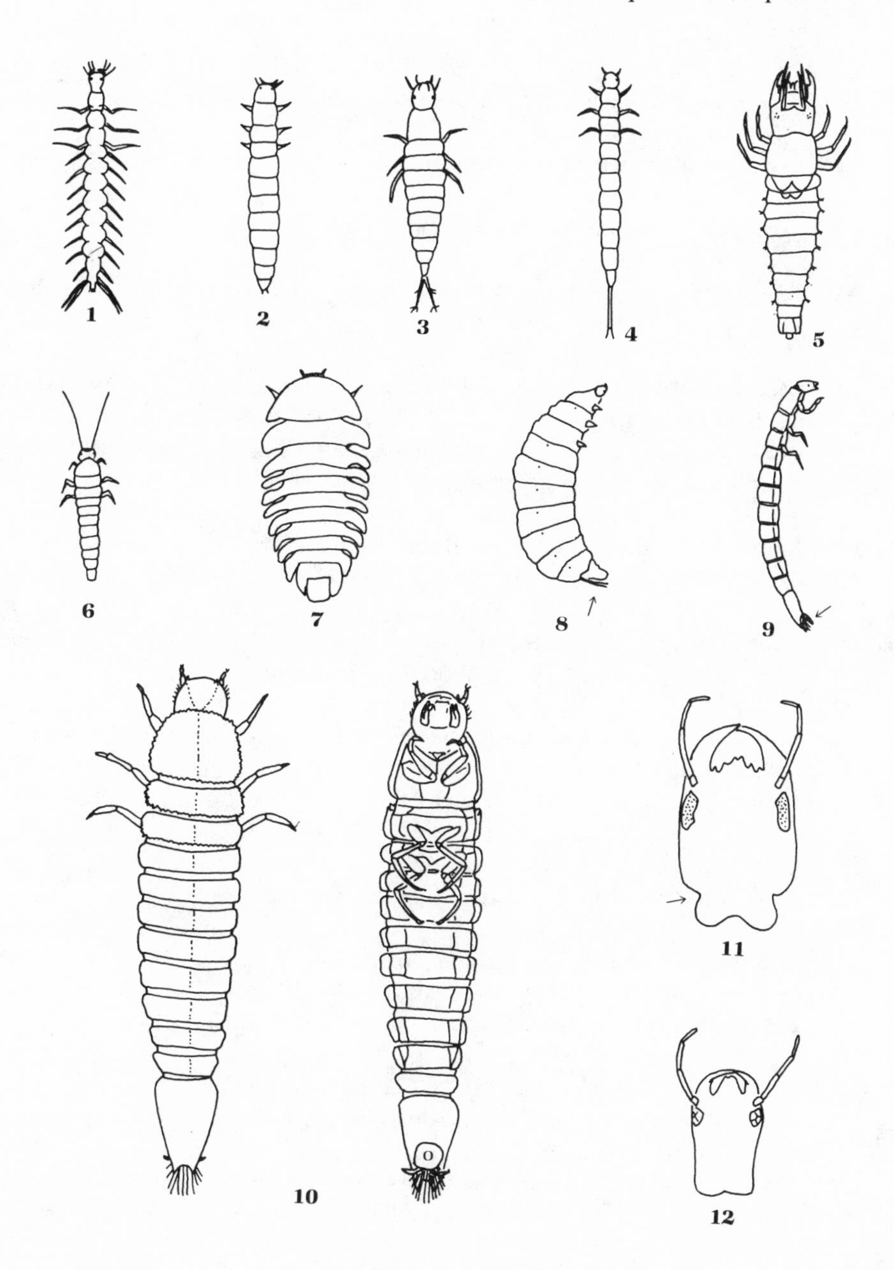

1. *Dineutus* (Gyrinidae), dv. **2.** *Pronoterus* (Noteridae), dv. **3.** *Agabus* (Dytiscidae), dv. **4.** *Haliplus* (Haliplidae), dv. **5.** *Tropisternus* (Hydrophilidae), dv. **6.** *Scirtes* (Scirtidae), dv. **7.** *Ectopria* (Psephenidae), dv. **8.** *Donacia* (Chrysomelidae), lv. **9.** *Stenelmis* (Elmidae), lv. **10.** *Neocylloepus* (Elmidae), dv (left), vv (right): o, ventral operculum. **11.** Head of *Dineutus* (Gyrinidae), dv. **12.** Head of *Gyrinus* (Gyrinidae), dv. dv, dorsal view; lv, lateral view; vv, ventral view. (1–9, 11, 12 modified or redrawn from Hilsenhoff 1981; 10 redrawn from Brown 1972.)

ment . . . . *Bidessonotus, Liodessus, Uvarus*
19b. Basal segment of cerci shorter than last abdominal segment 20

20a (19b). Larva greatly widened in middle, much wider than at pronotum (Fig. 14) 21
20b. Larva about as wide at middle as at pronotum 22

21a (20a). Tarsal claws ⅔ length of tarsus . . . . *Hydrovatus*
21b. Tarsal claws $<$ ⅓ length of tarsus . . . . *Desmopachria*

22a (20b). Cerci very short, about ¼ length of last abdominal segment; without recurved tracheal trunks projecting past last abdominal segment . . . . *Laccornis*
22b. Cerci nearly as long as last abdominal segment; recurved tracheal trunks projecting past last abdominal segment (Fig. 20) . . . . *Celina*

23a (15b). Abdominal segments 7 and 8 without a lateral fringe of long hairs 24
23b. Abdominal segments 7 and 8 with a lateral fringe of long hairs (Fig. 15) . . . 32

24a (23a). Cerci extremely short, ventral, difficult to see (Fig. 21) . . . . *Agabetes*
24b. Cerci at least ¼ length of last abdominal segment 25

25a (24b). Pro- and mesothoracic legs chelate, with inner apex of tibiae formed into a long serrated process parallel to and as long as tarsi (Fig. 22) . . . . *Matus*
25b. Legs not chelate 26

26a (25b). Cerci with only primary hairs, usually 7 in 2 whorls (Fig. 18) 27
26b. Cerci with numerous secondary hairs (Fig. 19) 29

27a (26a). Fourth antennal segment double, one half very short (Fig. 23); mandibles with an area of serrations on inner edge (Fig. 24) . . . . *Copelatus*
27b. Fourth antennal segment single; mandibles without serrations 28

28a (27b). Lateral margin of head more or less compressed or keeled; spines on posterolateral margins of head usually on a line that would intersect or pass just below ocelli (Fig. 25) . . . . *Ilybius*
28b. Lateral margin of head not keeled; spines on posterolateral margins of head usually on a line that would pass well below ocelli (Fig. 26) . . *Agabus* (in part)

29a (26b). Fourth antennal segment $>$ ⅔ length of 3rd 30
29b. Fourth antennal segment $<$ ½ length of 3rd 31

30a (29a). Cerci with numerous short, spinelike setae on outer edge; head not more than 2.5 mm wide in mature larvae . . . . *Rhantus*
30b. Cerci with at most 2 or 3 short setae; head often about 3 mm wide in mature larvae . . . . *Colymbetes*

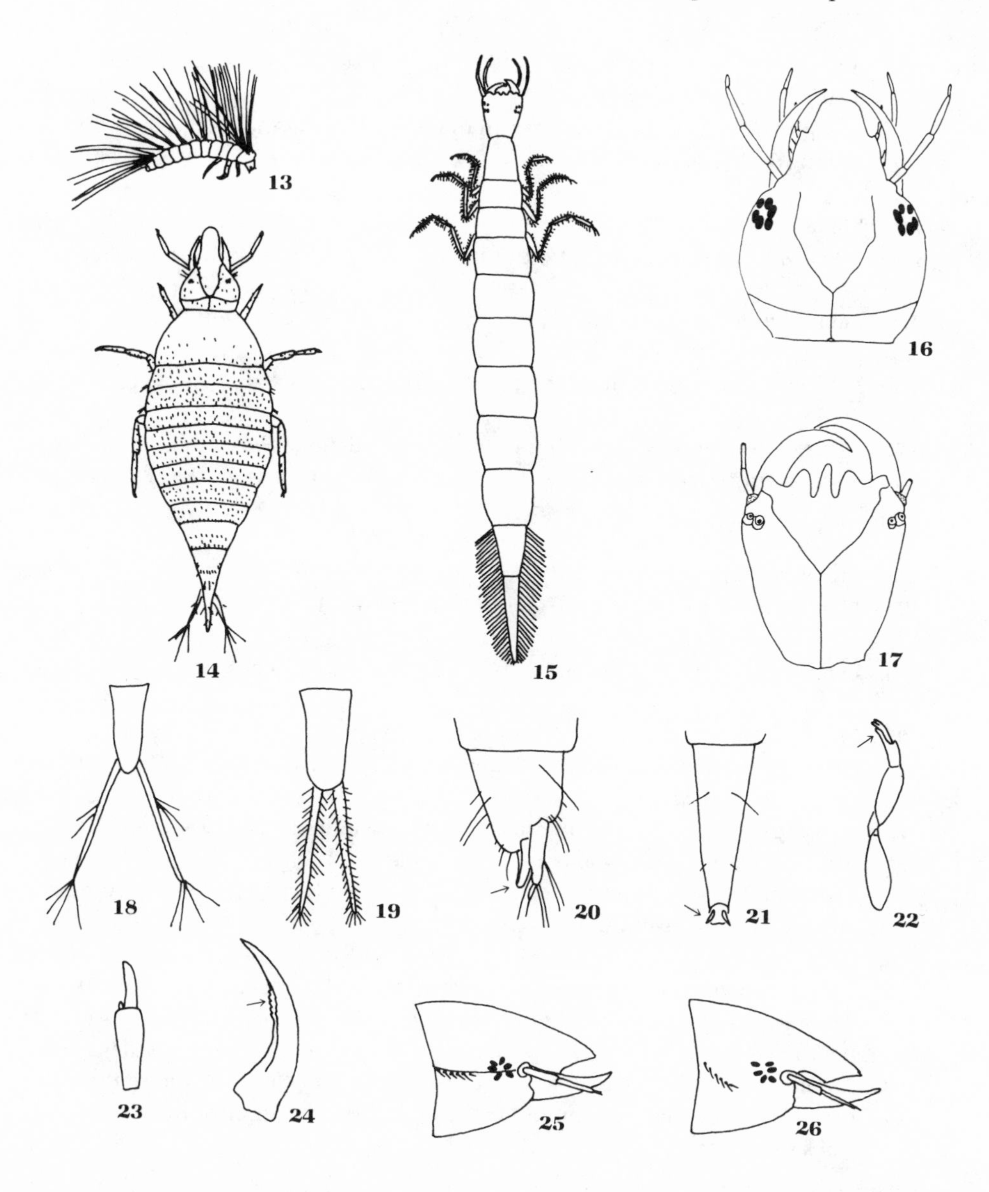

13. *Peltodytes* (Haliplidae), lv. **14.** *Desmopachria* (Dytiscidae), dv. **15.** *Cybister* (Dytiscidae), dv. **16.** Head of *Hydroporus* (Dytiscidae), dv. **17.** Head of *Cybister* (Dytiscidae), dv. **18.** Last abdominal tergum and cerci of *Agabus* (Dytiscidae), dv. **19.** Last abdominal tergum and cerci of *Laccophilus* (Dytiscidae), dv. **20.** Last abdominal segment and cerci of *Celina* (Dytiscidae), lv. **21.** Last abdominal segment and cerci of *Agabetes* (Dytiscidae), dv. **22.** Prothoracic leg of *Matus* (Dytiscidae), lv. **23.** Last two antennal segments of *Copelatus* (Dytiscidae). **24.** Mandible of *Copelatus* (Dytiscidae), dv. **25.** Head of *Ilybius* (Dytiscidae), lv. **26.** Head of *Agabus* (Dytiscidae), lv. dv, dorsal view; lv, lateral view. (13, 18–23 redrawn or modified from Hilsenhoff 1981; 14, 15 redrawn or modified from Barman 1972; 16 redrawn from Pennak 1978; 24 redrawn from Spangler 1962; 25, 26 redrawn from Balfour-Browne 1950.)

31a (29b). A row of spines on posterolateral margin of head; 4th antennal segment < ¼ as long as 3rd . . . . . *Laccophilus*
31b. No spines on posterolateral margin of head; 4th antennal segment about ⅓ as long as 3rd . . . . . *Agabus* (in part)

32a (23b). Maxillary stipes at least 4 times as long as wide (Fig. 27) 33
32b. Maxillary stipes broad, not more than 3 times as long as wide (Fig. 28) 35

33a (32a). Labroclypeus with long teeth anteriorly (Fig. 17); cerci absent (Fig. 15) . . . . . *Cybister*
33b. Labroclypeus without long teeth anteriorly; cerci present 34

34a (33b). Cerci with lateral fringes; labium without projecting lobes . . . . . *Dytiscus*
34b. Cerci without lateral fringes; labium with 2 projecting lobes (Fig. 29) . . . . . *Hydaticus*

35a (32b). Ligula apically bifid (Fig. 30) . . . . . *Acilius*
35b. Ligula simple or with 2 thin, elongate processes (Figs. 31, 32) 36

36a (35b). Ligula simple and nearly equal to or longer than 1st segment of labial palps (Fig. 31) . . . . . *Graphoderus*
36b. Ligula not as long as 1st segment of labial palps and with 2 elongate processes at tip (Fig. 32) . . . . . *Thermonectus*

37a (5a). **Hydrophilidae:** Nine complete abdominal segments, 10th reduced but distinct; integument noticeably chitinized (Fig. 33) . . . . . *Helophorus*
37b. Eight complete abdominal segments, 9th and 10th reduced and forming an atrium (atrium absent in *Berosus*) 38

38a (37b). Antennae with points of insertion nearer anterolateral angles of head than are insertion points of mandibles; labium and maxillae inserted in furrow beneath head . . . . . *Hydrochus*
38b. Antennae with points of insertion farther from anterolateral angles than are those of mandibles; labium and maxillae inserted at anterior margin of ventral side of head 39

39a (38b). Abdominal segments with broad (Fig. 34) or long (Fig. 35) lateral projections 40
39b. Lateral projections either absent or short and thin (Fig. 36) 41

40a (39a). First 7 abdominal segments with long, lateral tracheal gills (Fig. 35) . . *Berosus*
40b. Abdominal segments and posterior thoracic segments with broad lateral projections (Fig. 34) . . . . . *Crenitis*

41a (39b). Antennae biramal, terminal articles accompanied by a fingerlike antennal appendage (Fig. 37) 42

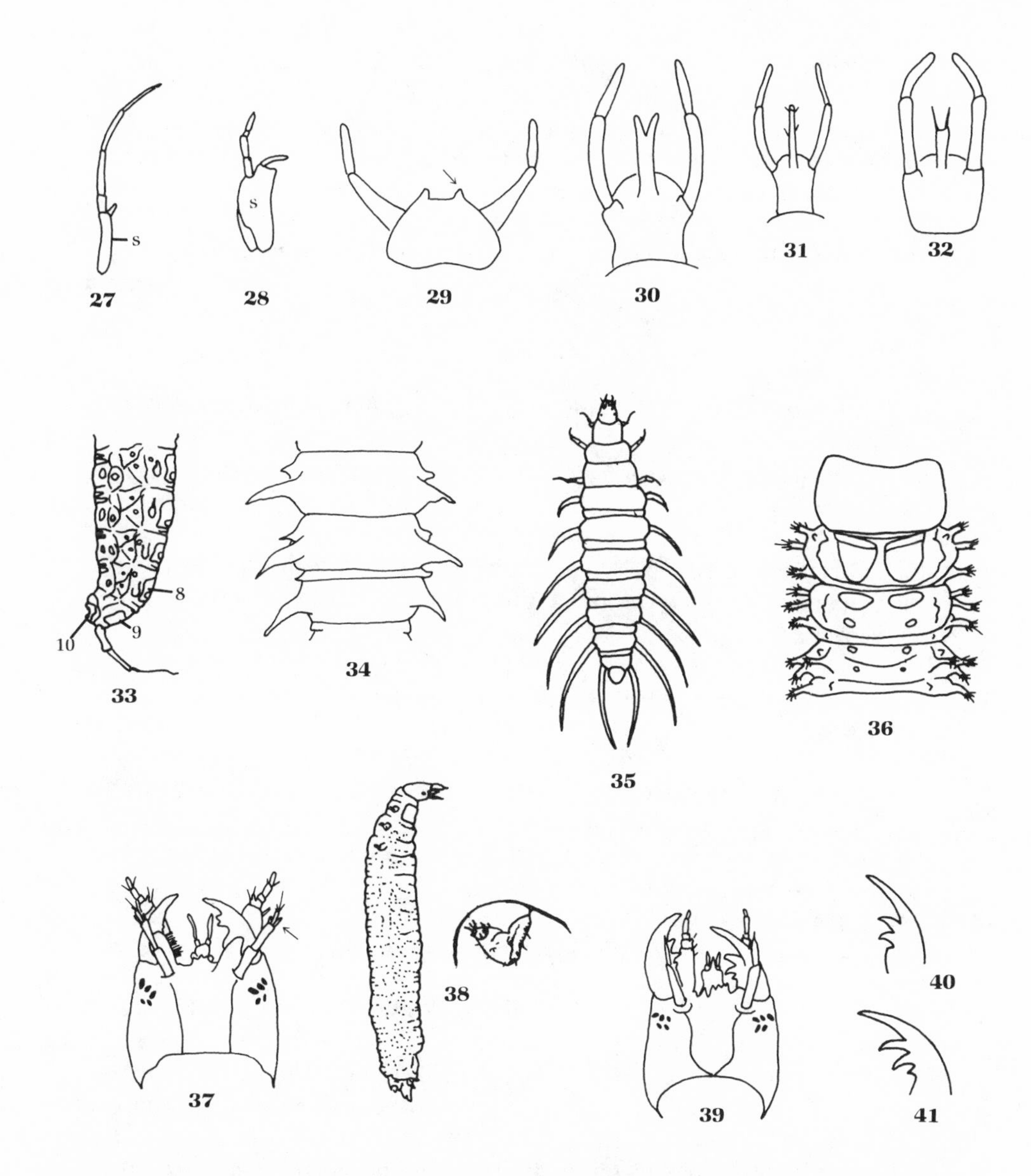

27. Right maxilla of *Dytiscus* (Dytiscidae), vv: s, stipe. **28.** Right maxilla of *Acilius* (Dytiscidae), vv: s, stipe. **29.** Labium of *Hydaticus* (Dytiscidae), vv. **30.** Labium of *Acilius* (Dytiscidae), vv. **31.** Labium of *Graphoderus* (Dytiscidae), vv. **32.** Labium of *Thermonectus* (Dytiscidae), vv. **33.** Terminal abdominal segments of *Helophorus* (Hydrophilidae), lv. **34.** Middle abdominal segments of *Crenitis* (Hydrophilidae), dv. **35.** *Berosus* (Hydrophilidae), dv. **36.** Thorax and 1st abdominal segment of *Derallus* (Hydrophilidae), dv. **37.** Head of *Laccobius* (Hydrophilidae), dv. **38.** *Chaetarthria* (Hydrophilidae), lv (inset shows vestigial leg, lv). **39.** Head of *Hydrobius* (Hydrophilidae), dv. **40.** Right mandible of *Paracymus* (Hydrophilidae), dv. **41.** Right mandible of *Anacaena* (Hydrophilidae), dv. dv, dorsal view; lv, lateral view; vv, ventral view. (27–32 modified or redrawn from Hilsenhoff 1981; 33, 37, 39 modified from Böving and Craighead 1930; 34 redrawn from Pennak 1978; 35 modified from Wilson 1923b; 36 modified from Bertrand 1972; 38 modified from Böving and Henriksen 1938.)

41b. Antennae not biramal, fingerlike antennal appendages absent 51

42a (41a). Legs vestigial, without claws (Fig. 38); ligula present .... *Chaetarthria*
42b. Legs sometimes very small but always complete, with claws; ligula present or absent 43

43a (42b). Frontal sutures parallel and not uniting to form an epicranial suture (Fig. 37); ligula absent .... *Laccobius*
43b. Frontal sutures not parallel (Fig. 39), may or may not unite to form an epicranial suture; ligula present 44

44a (43b). Epicranial suture absent; antennae short; legs reduced, barely visible from above 45
44b. Epicranial suture present; antennae longer; legs not reduced, easily visible from above 46

45a (44a). Frons truncate posteriorly; mandibles with 2 inner teeth (Fig. 40); anterior margin of pronotum without a fringe of stout setae; legs not visible from above .... *Paracymus*
45b. Frons rounded posteriorly; mandibles with 3 inner teeth (Fig. 41); anterior margin of pronotum with a fringe of stout setae; legs barely visible from above .... *Anacaena*

46a (44b). Mandibles asymmetrical, the left with 1 inner tooth, the right with 2; abdomen with prolegs on segments 3–7 .... *Enochrus*
46b. Mandibles symmetrical, each with 2–3 teeth; abdomen without prolegs .. 47

47a (46b). Clypeus with 4 or 5 distinct teeth (Figs. 42, 43) 48
47b. Clypeus with 6 or more distinct teeth 49

48a (47a). Middle tooth on clypeus smaller than the others (Fig. 43); prosternum entire .... *Sperchopsis*
48b. All teeth on clypeus subequal (Fig. 42); prosternum with a mesal fracture .... *Hydrobius*

49a (47b). Clypeus with 6 distinct teeth, grouped 2 on the left and 4 on the right (Fig. 44) .... *Helochares*
49b. Clypeus with more than 6 teeth, those on the right not clearly defined and with several smaller secondary teeth 50

50a (49b). Clypeus with teeth on the right side projecting farther than teeth on the left side (Fig. 45) .... *Cymbiodyta*
50b. Clypeus with left and right side teeth projecting equal distances .... *Helocombus*

51a (41b). Mesothorax, metathorax, and 1st abdominal segment each with 3 or 4 setiferous lateral gills on each side, abdominal segments 2–6 each with 4 setiferous lateral gills on each side (Fig. 36); femur without fringe of long hairs . . . . . *Derallus*

51b. Gills absent or, if present, with only 1 lateral gill on each side of abdominal segments; femur with fringe of long hairs 52

52a (51b). Head subspherical; antennae 4-segmented in 2nd and 3rd instars; each mandible with a single inner tooth, which is larger and bifid on right mandible (Fig. 46) . . . . . *Hydrophilus*

52b. Head subquandrangular, narrowed behind (Fig. 5); antennae 3-segmented; each mandible with more than 1, usually 2, inner teeth 53

53a (52b). Mentum with sides nearly straight (Fig. 47); lateral gills rudimentary tubular projections with several terminal setae . . . . . *Tropisternus*

53b. Mentum with sides convergent basally (Fig. 48); lateral gills fairly well developed and pubescent . . . . . *Hydrochara*

54a (6a). **Scirtidae:** Anterior margin of hypopharynx with a central cone bearing 1 pair of flat spines (Fig. 49); head with 3 ocelli on each side . . . . . *Elodes*

54b. Cone bearing 2 pairs of flat spines; head with 1 or 2 ocelli on each side . . . 55

55a (54b). Sides of abdominal segments with setae similar to those on dorsum, although usually more numerous . . . . . *Cyphon*

55b. Sides of at least abdominal segments 3–6 with a regular row of short, flattened setae that differ markedly from setae on dorsum 56

56a (55b). Anterior of labrum straight, with corners bent under to expose inner portion in dorsal view (Fig. 50) . . . . . *Prionocyphon*

56b. Anterior of labrum simply emarginate (Fig. 51) . . . . . *Scirtes*

57a (7a). **Psephenidae:** Abdominal pleura contiguous; gills on abdominal segments 2–6 . . . . . *Psephenus*

57b. Abdominal pleura separated from each other (Fig. 7); no gills on abdominal segments 2–6 58

58a (57b). Ninth abdominal segment almost rectangular (Fig. 52); lateral expansion of 8th segment forming part of margin of body outline (Fig. 52) . . . . . *Ectopria*

58b. Ninth abdominal segment not rectangular, the sides expanding from base toward broadly arcuate apex (Fig. 53); lateral expansion of 8th segment short, not forming part of lateral margin of body outline (Fig. 53) . . . . . *Dicranopselaphus*

59a (10b). **Elmidae:** Prothorax with sclerotized posterior sternum (Fig. 54) 60

59b. Prothorax without sclerotized posterior sternum (Fig. 55) 62

60a (59a). Posterolateral angles of anterior abdominal segments produced (Fig. 56) . . . . . *Ancyronyx*

60b. Posterolateral angles of abdominal segments not produced 61

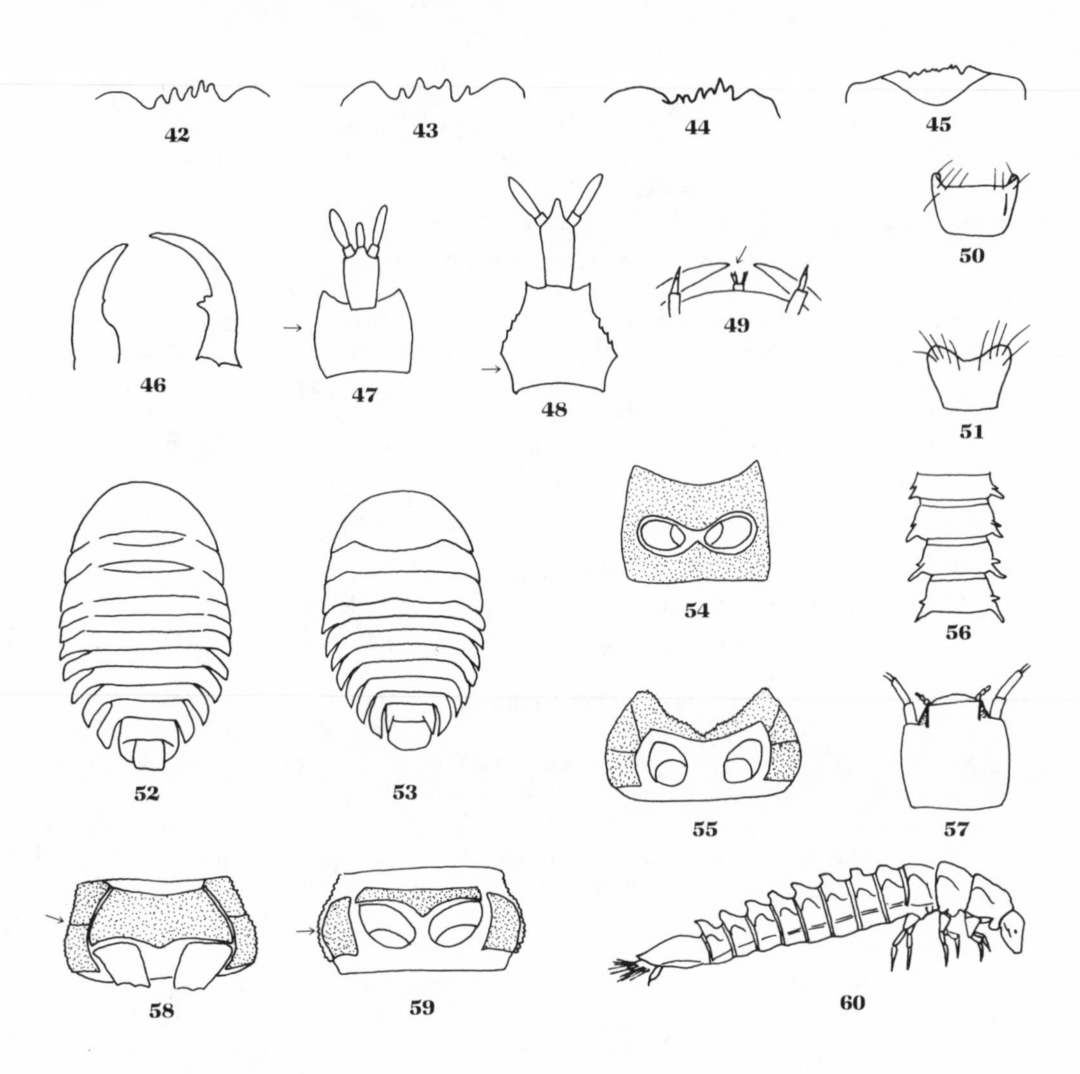

42. Anterior margin of labroclypeus of *Hydrobius* (Hydrophilidae), dv. **43.** Anterior margin of labroclypeus of *Sperchopsis* (Hydrophilidae), dv. **44.** Anterior margin of labroclypeus of *Helochares* (Hydrophilidae), dv. **45.** Anterior margin of labroclypeus of *Cymbiodyta* (Hydrophilidae), dv. **46.** Mandibles of *Hydrophilus* (Hydrophilidae), dv. **47.** Labium of *Tropisternus* (Hydrophilidae), vv. **48.** Labium of *Hydrochara* (Hydrophilidae), vv. **49.** Anterior margin of head of *Elodes* (Scirtidae), vv. **50.** Labrum of *Prionocyphon* (Scirtidae), dv. **51.** Labrum of *Scirtes* (Scirtidae), dv. **52.** *Ectopria* (Psephenidae), dv. **53.** *Dicranopselaphus* (Psephenidae), dv. **54.** Prosternum of *Stenelmis* (Elmidae), vv. **55.** Prosternum of *Optioservus* (Elmidae), vv. **56.** Abdominal segments 2–5 of *Ancyronyx* (Elmidae), dv. **57.** Head of *Stenelmis* (Elmidae), dv. **58.** Mesosternum and mesopleura of *Macronychus* (Elmidae), vv. **59.** Mesosternum and mesopleura of *Optioservus* (Elmidae), vv. **60.** *Promoresia* (Elmidae), lv. dv, dorsal view; lv, lateral view; vv, ventral view. (42, 43, 47–51, 54–59 redrawn from Hilsenhoff 1981; 44 modified from Panzera 1932; 45 modified from Richmond 1920; 46 modified from Wilson 1923a; 52, 53 modified from Brown 1972; 60 modified from West 1929.)

61a (60b). Anterior margin of head with a distinct tooth on each side (Fig. 57) .... *Stenelmis*
61b. Anterior margin of head without a distinct tooth on each side .... *Microcylloepus*

62a (59b). Last abdominal segment 5 times as long as wide .... *Dubiraphia*
62b. Last abdominal segment less than 3 times as long as wide 63

63a (62b). Mesopleuron divided (Fig. 58) 64
63b. Mesopleuron undivided (Fig. 59) 65

64a (63a). Six abdominal pleura .... *Macronychus*
64b. Seven abdominal pleura (as in Fig. 10) .... *Oulimnius*

65a (63b). Abdominal tergites with 1 mesal and 2 mesolateral humps, 1 on each side (Fig. 60) .... *Promoresia*
65b. Abdominal tergites not humped .... *Optioservus*

References on Aquatic Coleoptera Systematics

(*Used in construction of key.)

Anderson, R.D. 1983. Revision of the Nearctic species of *Hygrotus* groups IV, V, VII (Coleoptera: Dytiscidae). Ann. Entomol. Soc. Amer. 76:173–196.

*Arnett, R.H., Jr. 1960. The beetles of the United States. A manual for identification. Catholic Univ. of America Press, Washington, D.C. 1,112 pp.

*Arnett, R.H., Jr., N.M. Downie, and H.E. Jacques. 1980. How to know the beetles, 2nd ed. Pictured Key Nature Series. W.C. Brown Co. Publ., Dubuque, Iowa. 416 pp.

*Balfour-Browne, F. 1935. Systematic notes upon British aquatic Coleoptera. Part VII. Entomol. Mo. Mag. 71:195–201.

*——. 1940. British water beetles, vol. 1. Ray Soc., London. 375 pp.

*——. 1950. British water beetles, vol. 2. Ray Soc., London. 394 pp.

Barman, E.H., Jr. 1972. The biology and immature stages of selected species of Dytiscidae (Coleoptera) of central New York State. Ph.D. dissertation. Cornell Univ., Ithaca, N.Y. 207 pp.

Berthélemy, C. 1986. Remarks on the genus *Hydraena* and revision of the subgenus *Phothydraena* (Coleoptera: Hydraenidae). Ann. Limnol. 22:181–193.

*Bertrand, H. 1972. Larves et nymphes des coléoptères aquatiques du globe. Imprimerie F. Paillart. Paris. 804 pp.

Beutel, R.G., and R.E. Roughley. 1988. On the systematic position of the family Gyrinidae (Coleoptera: Adephaga). Z. Zool. Syst. Evol. 26:380–400.

Böving, A.G. 1910. Natural history of larvae of Donaciinae. Int. Rev. ges. Hydrobiol. Biol. Suppl. 1:1–108.

——. 1929. On the classification of beetles according to larval characters. Brooklyn Entomol. Soc. Bull. 24:55–97.

*Böving, A.G., and F.C. Craighead. 1930. An illustrated synopsis of the principal larval forms of the order Coleoptera. Entomol. Amer. 11:1–351.

*Böving, A.G., and K.L. Henriksen. 1938. The developmental stages of the Danish Hydrophilidae (Ins., Coleoptera). Dansk Naturhist. Foren. København Vidensk. Meddel. 102:27–162.

*Brigham, W.U., M.W. Sanderson, and D.S. White. 1982. Aquatic Coleoptera. *In* A.R. Brigham, W.U. Brigham, and A. Gnilka (eds.). Aquatic insects and oligochaetes of North and South Carolina, pp. 10.1–10.136. Midwest Aquatic Enterprises, Mahomet, Ill.

Brown, H.P. 1970. A key to the dryopoid genera of the New World (Coleoptera: Dryopoidea). Entomol. News 81:115.

*——. 1972. Aquatic dryopoid beetles (Coleoptera) of the United States. Biota Freshwat. Ecosyst. Ident. Man. 6:1–83.

——. 1975. A distributional checklist of North American genera of aquatic dryopoid and dascilloid beetles. Coleopt. Bull. 29:149–160.

——. 1981. A distributional survey of the world genera of aquatic dryopoid beetles (Coleoptera: Dryopoidae, Elmidae, and Psephenidae *sens lat*). Pan-Pac. Entomol. 57:133–148.

——. 1987. Biology of riffle beetles. Annu. Rev. Entomol. 32:253–274.

Brown, H.P., and C.M. Murvosh. 1974. A revision of the genus *Psephenus* (water-penny beetles) of the United States and Canada (Coleoptera: Psephenidae). Trans. Amer. Entomol. Soc. 100:289–340.

Brown, H.P., and D.S. White. 1978. Notes on separation and identification of North American riffle beetles (Coleoptera: Dryopidae: Elmidae). Entomol. News 89:1–13.

Campbell, J.M. 1980. Distribution patterns of Coleoptera in eastern Canada. Can. Entomol. 112:1161–1176.

Crowson, R.A. 1981. The biology of the Coleoptera. Academic Press, New York. 802 pp.

Eisner, T., and J. Meinwald. 1966. Defensive secretions in arthropods. Science 158:1341–1350.

Evans, M.E.G. 1983. Early evolution of the Adephaga—some locomotor speculations. Coleopt. Bull. 36:597–607.

Fall, H.C. 1922. The North American species of *Gyrinus* (Coleoptera). Trans. Amer. Entomol. Soc. 47:269–306.

——. 1923. A revision of the North American species of *Hydroporous* and *Agaporous*. John D. Sherman, Jr., Mt. Vernon, N.Y. 129 pp.

Finni, G.R., E.C. Masteller, and R.L. Hasse. 1978. An annotated list of elmids, dryopids, and psephenids from Pennsylvania (Coleoptera). Entomol. News 89:17–26.

*Guignot, F. 1931. Les Hydrocantheres de France. Les Freres Douladoure, Toulouse. 1,019 pp.

Gunderson, R.W. 1978. Nearctic *Enochrus:* biology, keys, descriptions, and distribution (Coleoptera: Hydrophilidae). R.W. Gunderson, St. Cloud, Minn. 54 pp.

Hatch, H.M. 1925a. Phylogeny and phylogenetic tendencies of Gyrinidae. Pap. Mich. Acad. Sci. Arts Lett. 5:429–476.

——. 1925b. An outline of the ecology of the Gyrinidae. Bull. Brooklyn Entomol. Soc. Bull. 20:101–114.

——. 1926. The morphology of Gyrinidae. Pap. Mich. Acad. Sci. Arts Lett. 7:311–350.

——. 1928. Studies on Dytiscidae. Bull. Brooklyn Entomol. Soc. 23:217–229.

Heinrich, B., and F.D. Vogt. 1980. Aggregation and foraging behavior of whirligig beetles (Gyrinidae). Behav. Ecol. Sociobiol. 7:179–186.

Hickman, J.R. 1930. Life histories of Michigan Haliplidae (Coleoptera). Pap. Mich. Acad. Sci. Arts Lett. 11:399–424.

——. 1931. Contributions to the biology of the Haliplidae (Coleoptera). Ann. Entomol. Soc. Amer. 24:129–142.

Hilsenhoff, W.L. 1980. *Coptotomus* (Coleoptera: Dytiscidae) in eastern North America with descriptions of two new species. Trans. Amer. Entomol. Soc. 105:461–471.

*——. 1981. Aquatic insects of Wisconsin. Nat. Hist. Counc. Publ. no. 2. Univ. Wisc., Madison. 60 pp.

——. 1987. Effectiveness of bottle traps for collecting Dytiscidae. Coleopt. Bull. 41:377–380.

Hilsenhoff, W.L., and W.U. Brigham. 1978. Crawling water beetles of Wisconsin (Coleoptera: Haliplidae). Great Lakes Entomol. 11:11–22.

Hilsenhoff, W.L., and B.H. Tracy. 1985. Techniques for collecting water beetles from lentic habitats. Proc. Acad. Nat. Sci. Phila. 137:8–11.

Hinton, H.E. 1939. An inquiry into the natural classification of the Dryopoidea, based partly on a study of their internal anatomy (Coleoptera). Trans. R. Entomol. Soc. Lond. 89:133–184.

——. 1955. On the respiratory adaptations, biology, and the taxonomy of the Psephenidae, with notes on some related families. Proc. Zool. Soc. Lond. 125:543–568.

Hodgson, E.S. 1953. Collection and laboratory maintenance of Dytiscidae (Coleoptera). Entomol. News 64:36–37.

Hoebeke, E.R. 1978. Catalogue of the Coleoptera types in the Cornell University insect collection. Search: Agric. 8:1–31.

Hoffman, C.E. 1940. Morphology of the immature stages of some northern Michigan Donaciini (Chrysomelidae, Coleoptera). Pap. Mich. Acad. Sci. Arts Lett. 25:243–292.

Holeski, P.M. 1978. A method for sampling shore beetles. Entomol. News 89:191–192.

Hosseinie, S.O. 1966. Studies on the biology and life histories of aquatic beetles of the genus *Tropisternus* (Coleoptera: Hydrophilidae). Ph.D. dissertation. Indiana Univ. 109 pp.

James, H.G. 1969. Immature stages of five diving beetles (Coleoptera: Dytiscidae), notes on their habits and life history, and a key to aquatic beetles of vernal woodland pools in southern Ontario. Proc. Entomol. Soc. Ont. 100:52–97.

Larson, D.A. 1975. The predaceous water beetles (Coleoptera: Dytiscidae) of Alberta: systematics, natural history and distribution. Quaest. Entomol. 11:245–498.

Larson, D.J. 1987. Revision of North American species of *Ilybius* Erickson (Coleoptera: Dytiscidae), with systematic notes on Palearctic species. J. N.Y. Entomol. Soc. 95:341–413.

Lawrence, J.F., and A.F. Newton, Jr. 1982. Evolution and classification of beetles. Annu. Rev. Ecol. Syst. 13:261–290.

*Leech, H.B., and H.P. Chandler. 1956. Aquatic Coleoptera. *In* R.L. Usinger (ed.). Aquatic insects of California, pp. 293–371. Univ. Calif. Press, Berkeley.

Leech, H.B., and M.W. Sanderson. 1959. Coleoptera. *In* W.T. Edmondson (ed.). Fresh-water biology, 2nd ed., pp. 981–1023. John Wiley and Sons, New York.

Maillard, V.P. 1970. Etude comparee de la construction du cocon de ponte chez *Hydrophilus piceus* L. et *Hydrochara caraboides* L. (Insecte Coleopt. Hydrophilidae). Bull. Soc. Zool. Fr. 95:71–84.

Malcolm, S.E. 1971. The water beetles of Maine: including the families Gyrinidae, Haliplidae, Dytiscidae, Noteridae, and Hydrophilidae. Tech. Bull. Univ. Maine Life Sci. Agric. Exp. Stn. 48:1–49.

Matheson, R. 1912. The Haliplidae of North America, north of Mexico. J. N.Y. Entomol. Soc. 20:156–193.

Matta, J.F. 1974. The insects of Virginia, no. 8. The aquatic Hydrophilidae of Virginia (Coleoptera: Polyphaga). Va. Polytech. Inst. Res. Div. Bull. 94:1–44.

——. 1976. The insects of Virginia, no. 10. The Haliplidae of Virginia (Coleoptera, Adephaga). Va. Polytech. Inst. Res. Div. Bull. 109:1–26.

——. 1983. Description of the larva of *Uvarus granarius* (Aubé) with a key to the Nearctic Hydroporinae larvae. Coleopt. Bull. 37:203–207.

——. 1986. *Agabus* (Coleoptera: Dytiscidae) larvae of southeastern United States. Proc. Entomol. Soc. Wash. 88:515–520.

Matta, J.F., and G.W. Wolfe. 1981. A revision of the subgenus *Heterosternuta* Strand of *Hydroporus* Clairville (Coleoptera: Dytiscidae). Pan-Pac. Entomol. 57:176–219.

Michael, A.G., and J.F. Matta. 1977. The Dytiscidae of Virginia (Coleoptera: Adephaga). Va. Polytech. Inst. Res. Div. Bull. 124:1–53.

Mingo, T.M. 1979. Distribution of aquatic Dryopoidea (Coleoptera) in Maine. Entomol. News 90:177–186.

Moore, I., and E.F. Legner. 1974. Keys to the genera of Staphylinidae of America north of Mexico exclusive of the Aleocharinae (Coleoptera: Staphylinidae). Hilgardia 42:548–563.

——. 1976. Intertidal rove beetles (Coleoptera: Staphylinidae). *In* L. Cheng (ed.). Marine insects, pp. 521–551. North Holland Publ. Co., Amsterdam.

Musgrave, P.N. 1935. Notes on collecting Dryopidae. Can. Entomol. 67:61–63.

Needham, J.G., and H.V. Williamson. 1907. Observations on the natural history of diving beetles. Amer. Nat. 41:477–494.

Nelson, R.E. 1988. Some notes on winter collecting of Coleoptera. Coleopt. Bull. 42:55–56.

Nilsson, A.N. 1988. A review of primary setae and pores on legs of larval Dytiscidae (Coleoptera). Can. J. Zool. 66:2283–2294.

*Panzera, O. 1932. Descrizione delle larve di *Helochares griseus* Fabr. e *H. lividus* Forst. (Coleoptera, Hydrophilidae). Mem. Soc. Entomol. Ital. 11:52–63.

*Pennak, R.L. 1978. Freshwater invertebrates of the United States, 2nd ed. John Wiley and Sons, New York. 803 pp.

*Perkins, D.P. 1980. Aquatic beetles of the family Hydraenidae in the western hemisphere. Classification, biogeography, and inferred phylogeny (Insecta: Coleoptera). Quaest. Entomol. 16:5–554.

*Richmond, E.A. 1920. Studies on the biology of the aquatic Hydrophilidae. Bull. Amer. Mus. Nat. Hist. 42:1–94.

Roughley, R.E., and D.H. Pengelly. 1981. Classification, phylogeny, and zoogeography of *Hydaticus* Leach (Coleoptera: Dytiscidae) of North America. Quaest. Entomol. 17:249–309.

Sanderson, M.W. 1938. A monographic revision of the North American species of *Stenelmis* (Dryopidae: Coleoptera). Univ. Kans. Sci. Bull. 39:635–717.

Sinclair, R.M. 1964. Water quality requirements of the family Elmidae (Coleoptera) with keys to the larvae and adults of the eastern genera. Tenn. Stream Poll. Control Bd., Nashville. 14 pp.

Smetana, A. 1974. Revision of the genus *Cymbiodyta* Bed. (Coleoptera: Hydrophilidae). Mem. Entomol. Soc. Can. 93:1–113.

——. 1978. Revision of the subfamily Sphaeridiinae of America north of Mexico (Coleoptera, Hydrophilidae). Mem. Entomol. Soc. Can. 105:1–292.

——. 1980. Revision of the genus *Hydrochara* Berth. (Coleoptera: Hydrophilidae). Mem. Entomol. Soc. Can. 111:1–100.

——. 1985a. Revision of the subfamily Helophorinae of the Nearctic region (Coleoptera: Hydrophilidae). Mem. Entomol. Soc. Can. 131:1–154.

——. 1985b. Synonymical notes on Hydrophilidae. Coleopt. Bull. 39:328–353.

——. 1988. Review of the family Hydrophilidae of Canada and Alaska (Coleoptera). Mem. Entomol. Soc. Can. 142:1–316.

*Spangler, P.J. 1962. Natural history of Plummers Island, Maryland. XIV. Biological notes and description of the larva and pupa of *Copelatus glyphicus* (Say) (Coleoptera: Dytiscidae). Proc. Biol. Soc. Wash. 75:19–23.

Swenson, G. 1977. Water beetle records from central New York. Coleopt. Bull. 31:116.

——. 1979. Water beetle records from the vicinity of Ithaca, N.Y. (Coleoptera: Dytiscidae). Coleopt. Bull. 33:477–479.

——. 1982. Water beetle records from the vicinity of Ithaca, N.Y. II. (Coleoptera: Dytiscidae and Hydrophilidae). Coleopt. Bull. 36:350–351.

Van der Eijk, R.H. 1986a. Population dynamics of gyrinid beetles. II. Reproduction. Oecologia 69:31–40.

——. 1986b. Population dynamics of gyrinid beetles. III. Survival of adults. Oecologia 69:41–46.

Wallace, F.L., and R.C. Fox. 1980. A comparative morphological study of the hind-wing venation of the order Coleoptera. Part II. Proc. Entomol. Soc. Wash. 82:609–654.

Wallis, J.B. 1933. Revision of the North American species (north of Mexico) of the genus *Haliplus* Latreille. Trans. R. Can. Inst. 19:1–76.

——. 1939. The genus *Ilybius* Er. in North America (Coleoptera, Dytiscidae). Can. Entomol. 71:192–199.

*West, L.S. 1929. A preliminary study of larval structure in the Dryopidae. Ann. Entomol. Soc. Amer. 22:691–727.

White, D.S. 1978. A revision of the Nearctic *Optioservus* (Coleoptera: Elmidae) with descriptions of new species. Syst. Entomol. 3:59–74.

*White, D.S., W.U. Brigham, and J.T. Doyen. 1984. Aquatic Coleoptera. *In* R.W. Merritt and K.W. Cummins (eds.). An introduction to the aquatic insects of North America, 2nd ed., pp. 361–437. Kendall/Hunt Publ. Co., Dubuque, Iowa.

White, R.E. 1983. A field guide to the beetles of North America. Houghton Mifflin, Boston. 368 pp.

*Wilson, C.B. 1923a. Life history of the scavenger water-beetle *Hydrous (Hydrophilus) triangularis,* and its economic relation to fish breeding. Bull. U.S. Bur. Fish. 39:9–38.

*——. 1923b. Water beetles in relation to pondfish culture, with life histories of those found in fishponds at Fairport, Iowa. Bull. U.S. Bur. Fish. 39:231–345.

Wolfe, G.W. 1983. Nomenclatural changes in Hydroporini (Coleoptera: Dytiscidae). Can. Entomol. 115:1547–1548.

*Wolfe, G.W., and J.F. Matta. 1981. Notes on nomenclature and classification of *Hydroporus* subgenera with the description of a new genus of Hydroporini (Coleoptera: Dytiscidae). Pan-Pac. Entomol. 57:149–175.

Wooldridge, D.P. 1967. The aquatic Hydrophilidae of Illinois. Trans. Ill. State Acad. Sci. 60:422–431.

*Young, F.N. 1954. The water beetles of Florida. Univ. Fla. Stud., Biol. Sci. Ser. 5:1–238.

——. 1958. Notes on the care and rearing of *Tropisternus* in the laboratory (Coleoptera: Hydrophilidae). Ecology 39:166–167.

Zimmerman, J.R. 1970. A taxonomic revision of the aquatic beetle genus *Laccophilus* (Dytiscidae) of North America. Mem. Entomol. Soc. Amer. 26:1–275.

——. 1981. A revision of the *Colymbetes* of North America (Dytiscidae). Coleopt. Bull. 35:1–52.

Zimmerman, J.R., and A.H. Smith. 1975a. The genus *Rhantus* (Coleoptera: Dytiscidae) in North America. Part I. General account of the species. Trans. Amer. Entomol. Soc. 101:33–123.

——. 1975b. A survey of the *Deronectes* (Coleoptera: Dytiscidae) of Canada, United States, and northern Mexico. Trans. Amer. Entomol. Soc. 101:651–722.

11 Megaloptera

Classification

The order Megaloptera is a small order of insects in the infraclass Neoptera, division Endopterygota. The Megaloptera are closely related to the Neuroptera; in fact, some authors consider them a suborder of Neuroptera (Brigham 1982). However, we treat them, as do most North American workers, as a separate order. The Megaloptera comprise only two families, the Corydalidae (fishflies and dobsonflies) and the Sialidae (alderflies). Larvae of all species of Megaloptera are aquatic and attain the largest size of all aquatic insects. Larval Corydalidae are sometimes called hellgrammites or toe biters.

Life History

Females of this holometabolous order lay elongate eggs in masses on vegetation overhanging the aquatic habitat, on large rocks projecting from the water, or on bridge abutments (Evans and Neunzig 1984). After about a week at cool temperatures, eggs hatch at night and first-instar larvae fall into the water. As young larvae swallow air, gas bubbles form in their guts, possibly providing the buoyancy necessary to transport to riffles first instars that land in pools (Brigham 1982). The metabolic consequences of this air bubble are unknown for most species. First-instar sialids are setaceous and apparently photopositive, swimming toward light for a period after hatching. First-instar corydalids are not comparably setaceous and burrow quickly into the substrate, especially those that hatch over dry stream beds (E. D. Evans, pers. comm., 1987). Megalopteran larvae go through 10–12 instars before crawling out of the water onto shore to pupate. Some have been reported to pupate as far as 50 m from the shore. Larvae of certain species construct a pupal chamber under stones or detritus; others pupate in dry stream beds (Pennak 1978). Both sialids and corydalids have exarate pupae with functional mandibles. Pupae of some species (*Neohermes*) are active and are capable of biting and of pushing themselves along with their cerci (Brigham 1982). Those species move away from the pupal chamber before molting the pupal exuviae. After about two weeks, adults emerge during the spring and summer months. Adults, depending on the species, are active during daylight or at night. Although sialid adults do not readily fly, corydalid adults may fly some distance from water and are attracted to lights (Evans and Neunzig 1984). Adults of both families are short-lived (a few days), and because they have atrophied guts, they do not feed on solid food, instead ingesting liquids. Adult sialids use acoustical signals (abdominal vibrations or tapping), behavior similar to that of the stoneflies, for mate recognition and courtship (Rupprecht 1975). Mating corydalids use scents that can be detected easily when adults are preserved in ethanol (E. D. Evans, pers. comm., 1987). Most sialids have one- or two-year life cycles, whereas corydalids in cold mountain streams and in intermittent streams may live for up to five years (E. D. Evans, pers. comm., 1987).

Habitat

Corydalids are found in well-oxygenated streams and lakes, as well as in productive ponds or swamps where dissolved oxygen may be very low. Sialids occur in the same broad habitat categories, but usually require muddy or silty deposits and accumulated detritus. Some species of Megaloptera can exist in intermittent streams and ponds by burrowing into the substrate during dry periods. Sialids are classified by Evans and Neunzig (1984) as burrowers, while corydalids are generally clingers or climbers.

Feeding

The larvae of both families of Megaloptera are active predators, feeding on aquatic insects, annelids, crustaceans, and mollusks. Little is known about the predatory behavior of these groups, despite their conspicuous presence and apparently voracious appetites. Evans and Neunzig (1984) classify them as engulfers.

Respiration

Megalopteran larvae are easily distinguished from most coleopteran and trichopteran larvae by their paired lateral abdominal gill filaments. These may be simple or segmented and, in certain corydalids, are supplemented with tufts of tracheal gills (Pennak 1978). Some corydalids also have functional posterior spiracles with paired dorsal respiratory tubes, which enable them to respire when they leave the water to pupate or when the dissolved-oxygen levels are very low.

Collection and Preservation

Megalopteran larvae can be collected with dip nets or by sorting through detritus. Some workers collect them with pitfall traps that capture the larvae as they leave the water to pupate. Final storage of specimens in ethanol best maintains their color and shape, but experts recommend first killing specimens by boiling them or fixing them in a formalin-based fluid (Brigham 1982). E. D. Evans (pers. comm., 1987) suggests injecting formalin preservative into the mouths of larger corydalid larvae to obtain specimens that are well preserved and distended.

Identification

Genera of larval megalopterans are easily separated by examining the structure of terminal appendages, dorsal respiratory tubes, and lateral processes.

More-Detailed Information

Chandler 1956, Pennak 1978, Brigham 1982, Evans and Neunzig 1984.

Checklist of Megaloptera Larvae

Corydalidae
Chauliodes
Corydalus
Neohermes
Nigronia

Sialidae
Sialis

Key to Genera of Megaloptera Larvae

1a. With 7 pairs of lateral segmented abdominal appendages; a long terminal anal filament; anal prolegs absent (Fig. 1); larvae rarely exceed 25 mm in length; alderflies . . . . **Sialidae,** *Sialis*

1b. With 8 pairs of lateral abdominal appendages, unsegmented or imperfectly segmented; no terminal anal filament; anal prolegs present (Figs. 2, 3); mature larvae 30–65 mm long . . . . **Corydalidae** 2

2a (1b). **Corydalidae:** With a tuft of filamentous gills at the base of each lateral abdominal appendage (Fig. 2); dobsonflies (hellgrammites) . . . . *Corydalus*

2b. Without a tuft of gills at bases of appendages; fishflies 3

3a (2b). Last pair of abdominal spiracles (segment 8) at the apex of 2 long respiratory tubes extending beyond the prolegs (Fig. 3) . . . . *Chauliodes*

3b. Last pair of abdominal spiracles not at apex of long respiratory tubes extending beyond the prolegs (Figs. 4, 5) 4

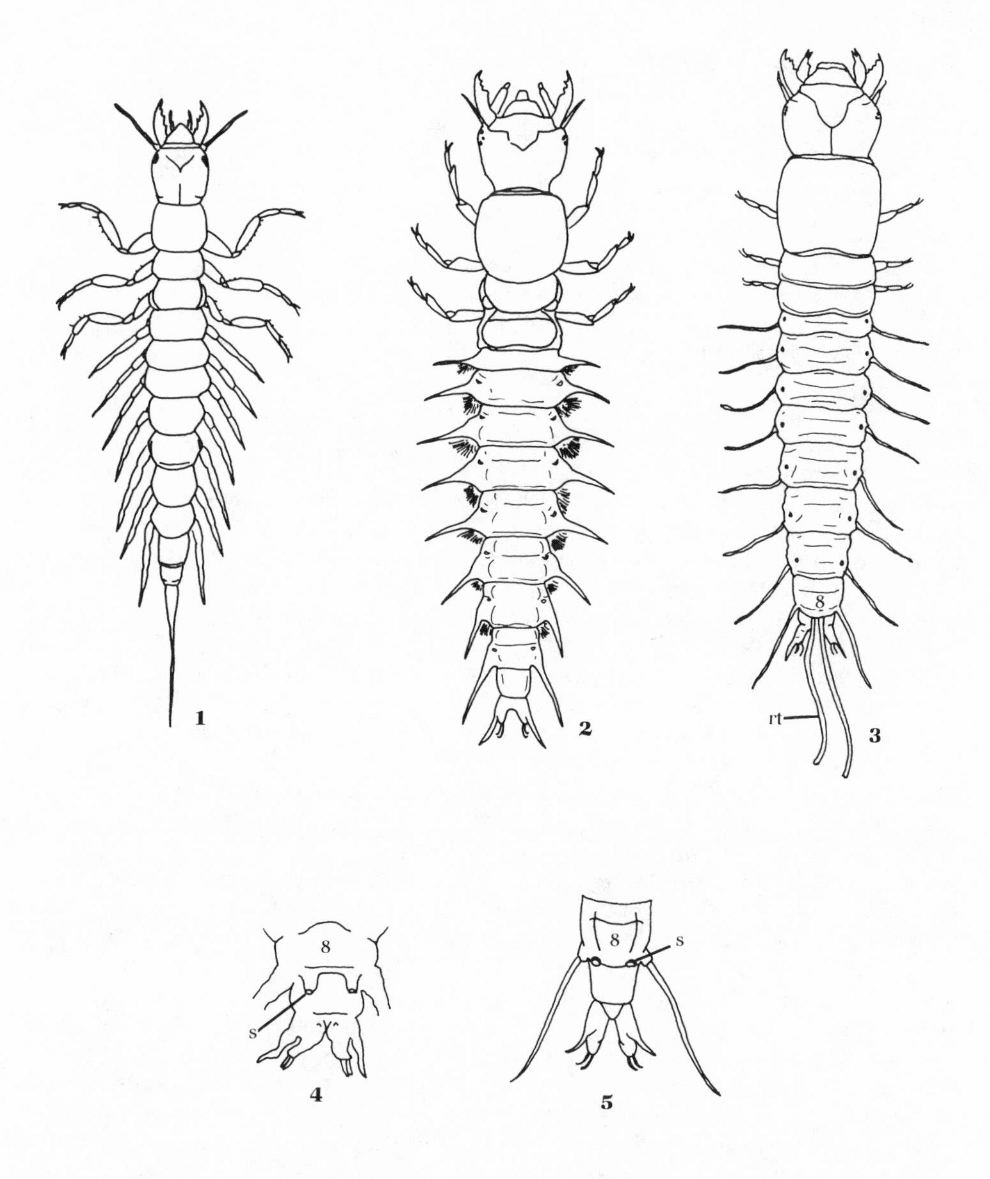

1. *Sialis* (Sialidae), dv. **2.** *Corydalus* (Corydalidae), dv. **3.** *Chauliodes* (Corydalidae), dv: rt, respiratory tube. **4.** Last abdominal segments of *Nigronia* (Corydalidae), dv: s, spiracle. **5.** Last abdomimal segments of *Neohermes* (Corydalidae), dv: s, spiracle. dv, dorsal view. (1–3 modified from Gurney and Parfin 1959; 4 redrawn from Neunzig 1966; 5 redrawn from Brigham 1982.)

4a (3b). Each spiracle of 8th abdominal segment at end of short, tapered dorsal respiratory tube, about 1½ times as long as wide (Fig. 4); larvae without obvious dorsal pattern on head and thorax . . . . *Nigronia*

4b. Each spiracle of 8th abdominal segment on posterior edge of segment, not on tube (Fig. 5); larvae with distinct dorsal pattern on head and thorax . . . . *Neohermes*

References on Megaloptera Systematics

(*Used in construction of key.)

*Brigham, W.U. 1982. Megaloptera. *In* A.R. Brigham, W.U. Brigham, and A. Gnilka (eds.). Aquatic insects and oligochaetes of North and South Carolina, pp. 7.1–7.12. Midwest Aquatic Enterprises, Mahomet, Ill.

Canterbury, L.E., and S.E. Neff. 1980. Eggs of *Sialis* (Sialidae: Megaloptera) in eastern North America. Can. Entomol. 112:409–420.

Chandler, H.P. 1956. Megaloptera. *In* R.L. Usinger (ed.). Aquatic insects of California, pp. 229–233. Univ. Calif. Press, Berkeley.

Cuyler, R.D. 1958. The larvae of *Chauliodes* Latreille (Megaloptera, Corydalidae). Ann. Entomol. Soc. Amer. 51:582–586.

Davis, K.C. 1903. Sialidae of North and South America. Bull. N.Y. State Mus. 68:442–486.

*Evans, E.D. 1978. Megaloptera and aquatic Neuroptera. *In* R.W. Merritt and K.W. Cummins (eds.). An introduction to the aquatic insects of North America, pp. 133–146. Kendall/Hunt Publ. Co., Dubuque, Iowa.

*Evans, E.D., and H.H. Neunzig. 1984. Megaloptera and aquatic Neuroptera. *In* R.W. Merritt and K.W. Cummins (eds.). An introduction to the aquatic insects of North America, 2nd ed., pp. 261–270. Kendall/Hunt Publ. Co., Dubuque, Iowa.

Flint, O.S., Jr. 1965. The genus *Neohermes* (Megaloptera: Corydalidae). Psyche 72:255–263.

Glorioso, M.J. 1981. Systematics of the dobsonfly subfamily Corydalinae (Megaloptera: Corydalidae). Syst. Entomol. 6:253–290.

*Gurney, A.B., and S. Parfin. 1959. Neuroptera. *In* W.T. Edmondson (ed.). Fresh-water biology, pp. 973–980. John Wiley and Sons, New York.

*Hilsenhoff, W.L. 1981. Aquatic insects of Wisconsin. Nat. Hist. Counc. Publ. no. 2. Univ. Wisc., Madison. 60 pp.

*Neunzig, H.H. 1966. Larvae of the genus *Nigronia* Banks (Neuroptera: Corydalidae). Proc. Entomol. Soc. Wash. 68:11–16.

*Pennak, R.W. 1978. Freshwater invertebrates of the United States, 2nd ed. John Wiley and Sons, New York. 803 pp.

Ross, H.H. 1937. Studies of Nearctic aquatic insects. I. Nearctic alderflies of the genus *Sialis* (Megaloptera: Sialidae). Bull. Ill. Nat. Hist. Surv. 21:57–78.

Rupprecht, R. 1975. Die Kommunikation von *Sialis* (Megaloptera) durch vibrationssignale. J. Ins. Physiol. 21:305–320.

12 Aquatic Neuroptera

Classification

The order Neuroptera is closely related to the order Megaloptera in the infraclass Neoptera, division Endopterygota. Only 1 of 40 families of this order—Sisyridae—has aquatic larvae. Two genera of the Sisyridae occur in North America, both having terrestrial eggs, pupae, and adults. Larvae are highly specialized predators on freshwater sponges (see Feeding); thus, they are called spongillaflies.

Life History

As is true of the Megaloptera, eggs of this holometabolous group are laid above the surface of the water on overhanging objects. Females lay groups of 5–10 eggs and cover them with a protective silken tent. Larvae hatch in about a week, drop into the water, and drift or swim to find a suitable sponge host. Sisyrids go through three larval instars, then climb out of the water to pupate. Some individuals have been reported to crawl for several hours as far as 20 m inland to find a suitable site for spinning a cocoon (Evans and Neunzig 1984). Pupation lasts about a week and is usually initiated at dusk in late spring and summer. Adults live about two weeks. Sisyrids may go through two to five generations each year depending on ambient temperature and food conditions. They overwinter as third-instar larvae or prepupae (Brigham 1982).

Habitat

Sisyrid larvae live exclusively in association with freshwater sponges, either on the surface or in the body cavities of their hosts. Evans and Neunzig (1984) classify them as climbers, clingers, or burrowers. Larvae are found on most of the common species of freshwater sponges, and they do not seem to show host specificity. Pennak (1978) reports that sisyrids have been observed on bryozoans and algae but states that these occurrences are probably accidental. Brigham (1982) suggests that larval spongillaflies may feed on bryozoans or algae but that they never do so far from sponges. While the habitat of freshwater sponges and, thus, of sisyrids, ranges from cool, clean lakes and streams to relatively polluted ponds, the former is more typical (Pennak 1978).

Feeding

Larval sisyrid mouthparts are highly modified for piercing the tissues and sucking the contents of sponges. The alimentary canal is closed between the mid- and hindgut and retains undigestible residue until the adult stage is reached (Pennak 1978). Once settled on a sponge, sisyrid larvae usually do not leave unless the sponge dies. The typical mode of feeding is in bouts of one or two minutes, followed by one- to five-minute search bouts (Brigham 1982). Although many texts label sisyrids "parasites," more properly they should be classified as predators. Evans and Neunzig (1984) include them in the piercing predator functional feeding group. Adult spongillaflies feed on nectar and pollen.

Respiration

Sisyrid larvae have paired, segmented tracheal gills on the ventral side of the first seven abdominal segments. Pennak (1978) reports that gills are periodically fluttered, presumably to behaviorally regulate oxygen intake.

Collection and Preservation

Sisyrids can be collected along with their sponge hosts. Sponges grow on any stable submerged object, such as stones, logs or branches, and piers, in relatively unsilted habitats. Specimens can be preserved and stored in 70% ethanol.

Identification

Sisyrid larvae are very small and somewhat difficult to identify. The two genera in our key, *Climacia* and *Sisyra*, are separated by the length of the dorsal tubercles and presence or absence of spines at the bases of their setae.

More-Detailed Information

Chandler 1956, Pennak 1978, Brigham 1982, Evans and Neunzig 1984.

Checklist of Aquatic Neuroptera Larvae

Sisyridae
Climacia
Sisyra

Key to Genera of Aquatic Neuroptera Larvae

1a. Pair of dorsal setae present on abdominal segment 8 (Fig. 1); in the 3rd instar, the median tubercle on abdominal segment 6 is elongate, and its seta is longer and thinner than those on the 2 adjacent tubercles (Fig. 2) . . *Climacia*

1b. Pair of dorsal setae absent on abdominal segment 8; in the 3rd instar, the lateralmost tubercle on abdominal segment 6 is elongate and its seta is longer and thinner than those on the 2 adjacent tubercles (Fig. 3) . . . . *Sisyra*

References on Aquatic Neuroptera Systematics

(*Used in construction of key.)

Brigham, W.R. 1982. Aquatic Neuroptera. *In* A.R. Brigham, W.U. Brigham, and A. Gnilka (eds.). Aquatic insects and oligochaetes of North and South Carolina, pp. 8.1–8.4. Midwest Aquatic Enterprises, Mahomet, Ill.

*Chandler, H.P. 1956. Aquatic Neuroptera. *In* R.L. Usinger (ed.). Aquatic insects of California, pp. 234–236. Univ. Calif. Press, Berkeley.

*Evans, E.D. 1978. Megaloptera and aquatic Neuroptera. *In* R.W. Merritt and K.W. Cummins (eds.). An introduction to the aquatic insects of North America, pp. 133–146. Kendall/Hunt Publ. Co., Dubuque, Iowa.

*Evans, E.D., and H.H. Neunzig. 1984. Megaloptera and aquatic Neuroptera. *In* R.W. Merritt and K.W. Cummins (eds.). An introduction to the aquatic insects of North America, 2nd ed., pp. 261–270. Kendall/Hunt Publ. Co., Dubuque, Iowa.

Gurney, A.B., and S. Parfin. 1959. Neuroptera. *In* W.T. Edmondson (ed.). Fresh-water biology, 2nd ed., pp. 873–880. John Wiley and Sons, New York.

*Hilsenhoff, W.L. 1981. Aquatic insects of Wisconsin. Nat. Hist. Counc. Publ. no. 2. Univ. Wisc., Madison. 60 pp.

Parfin, S.I., and A.B. Gurney. 1956. The spongilla flies, with special reference to those of the Western Hemisphere (Sisyridae, Neuroptera). Proc. U.S. Natl. Mus. 105:421–529.

Pennak, R.W. 1978. Freshwater invertebrates of the United States, 2nd ed. John Wiley and Sons, New York. 803 pp.

*Pupedis, R.J. 1980. Generic differences among the new world spongilla-fly larvae and a description of the female of *Climacia striata* (Neuroptera: Sisyridae). Psyche 83:305–314.

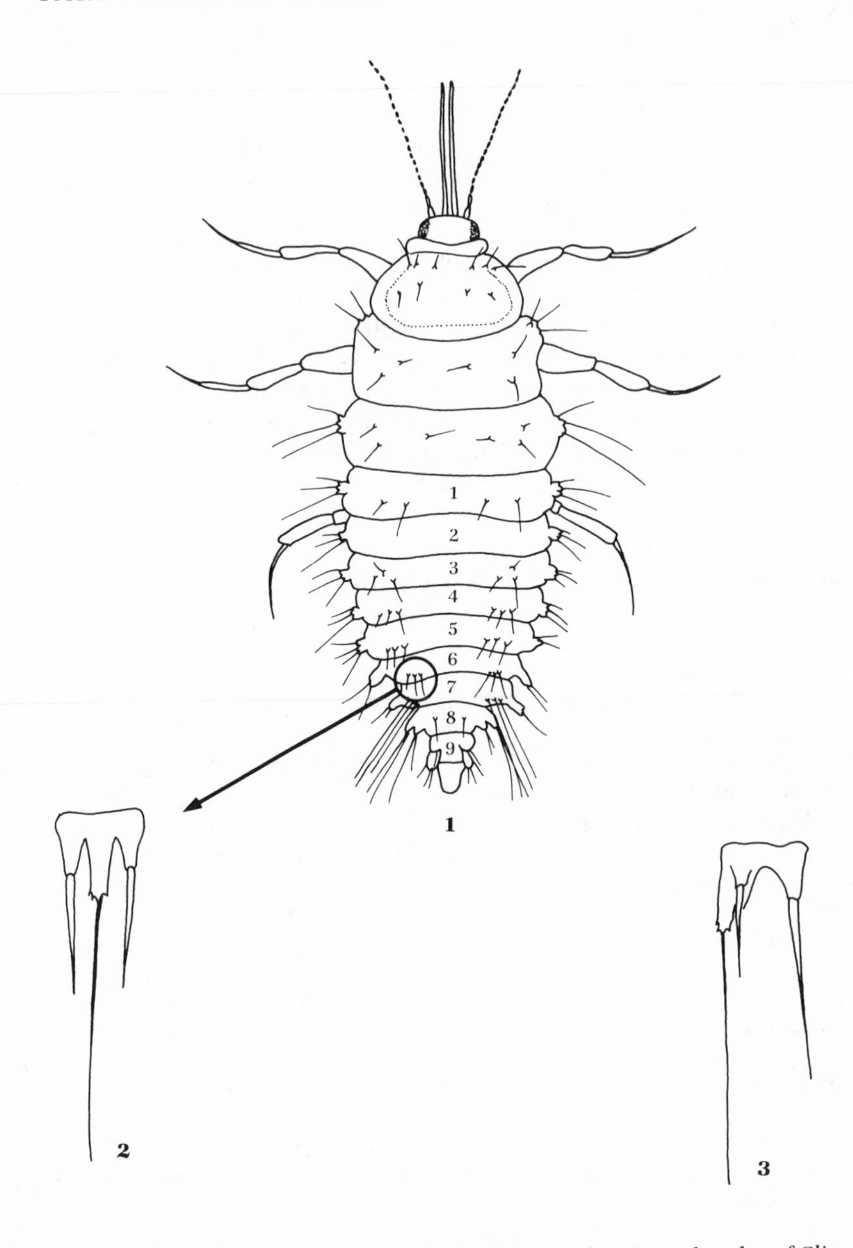

1. Third instar of *Climacia*, dv. **2.** Left 6th abdominal tergite showing tubercles of *Climacia*, dv. **3.** Left 6th abdominal tergite showing tubercles of *Sisyra*, dv. dv, dorsal view. (1 redrawn from Chandler 1956; 2, 3 modified from Pupedis 1980.)

13 | Aquatic Diptera

Classification

The order Diptera (true flies), one of the largest orders of insects, belongs to the infraclass Neoptera, division Endopterygota. McAlpine et al. (1981) divide the Diptera into two suborders: the Nematocera and the Brachycera, which is further divided into two infraorders, the Orthorrhapha and the Cyclorrhapha. Thirteen families of Nematocera, and six families each of the Orthorrhapha and Cyclorrhapha, have aquatic species. Approximately 10% of all dipteran species are aquatic in their larval stage. Eggs and pupae of these species are also aquatic, whereas adults are always terrestrial. See More-Detailed Information for references that include the natural history of specific families of aquatic Diptera.

Life History

Egg-laying behavior is diverse in the aquatic members of this holometabolous order. Some scatter eggs just below the surface on vegetation or on mineral substrates; others deposit eggs in gelatinous masses at, below, or above the water surface on emergent objects. One family (the Athericidae) oviposits on overhanging vegetation as in the orders Megaloptera and Neuroptera. Females of this family die after laying their eggs, and their bodies provide nourishment to the egg mass. Successive females oviposit in the same place, and a bolus of eggs and dead bodies forms that drops into the water when eggs hatch. Aldrich (1912) reported that a tribe of California Indians considered these egg masses to be culinary treats. In species of the families Culicidae and Simuliidae, larger females are more fecund, as in the order Ephemeroptera. Hatching may occur after a short period of development, or it may be delayed by egg diapause as in the culicids, enabling eggs to withstand periods of drought.

The number of larval instars varies among dipterans; nematocerans have four, cyclorrhaphans have three, and orthorrhaphans may have eight or nine. Larval stages as short as several weeks and as long as two years have been reported, depending on the species, water temperature, and food conditions. Nematocerans and orthorrhaphans (except the stratiomyids) cast the last larval skin before pupation, which occurs under water. Pupal appendages are somewhat free from the body, yet are more or less fused to each other. Culicid and some chironomid pupae are actually active, wriggling at the water surface. Stratiomyids retain their last larval exuvia, which loosely encloses the pupa. In cyclorrhaphans, the last larval exuvia hardens to form a puparium that encloses and protects the pupa. The pupal stage lasts approximately two weeks, except in species that overwinter as pupae. Most species overwinter as hibernating eggs, as larvae, or, rarely, adults. Some arctic chironomid larvae coil up in cocoons and overwinter under freezing conditions. They produce cryoprotectants that enable them either to supercool or to survive intra- and extracellular ice formation (Baust and Rojas 1985).

Both nematocerans and brachycerans emerge from their pupal cases at the anterior end; the puparia of some cyclorrhaphans have a T-shaped dorsal split. Females of many species are anautogenous, requiring a blood meal to acquire enough protein to produce eggs (Ceratopogonidae, Culicidae, Simuliidae, Tabanidae), and some are vectors of animal or human diseases or both. Mating of many nematocerans and some brachycerans (e.g., Tabanidae) occurs in swarms. Some species swarm over sand or cow dung or many objects that serve as swarm markers; others swarm around their hosts for both mating and feeding. After mating occurs, females leave the swarm to oviposit. Males of some empidids offer nuptial gifts (usually food) to their mates and copulate while the females are eating (Cloudsley-Thompson 1976).

Most aquatic dipterans are univoltine, but under favorable conditions some may complete more than one generation a year. Some species may take two or three years to complete development in colder climates. One arctic chironomid is reported to have a seven-year life cycle (Butler 1984).

Habitat

Dipteran larvae occur in almost every conceivable aquatic habitat, from the bracts of pitcher plants (Culicidae: *Wyeomyia*), tree holes (e.g., Chironomidae and Culicidae), saturated soil, and mud puddles, to streams, ponds, large lakes, rivers, and even the marine rocky intertidal zone. Stratiomyids have been recorded from geyser-fed thermal pools that reach temperatures up to 49°C, and ephydrids from natural seeps of crude petroleum (Merritt and Cummins 1984). Dipterans are found in rushing streams (Blephariceridae, Deuterophlebiidae), stagnant pools, hot springs, and frozen sediments. The only aquatic habitat where dipterans have not been recorded is the open ocean. Merritt and Cummins (1984) classify them as clingers, sprawlers, planktonic swimmers, burrowers, climbers, and miners.

Feeding

The dietary diversity of aquatic dipteran larvae parallels the habitat diversity. Some families are shredders of coarse particulate detritus (Tipulidae); others are specialized for filter feeding (Simuliidae). These black fly larvae have unique labral fan structures that enable them to filter very fine particles suspended in the water column. They attach the hooks of the posterior end of their abdomens to the substrate by spinning a silken pad, and can thus withstand torrential flows in areas that are fairly safe from predators; fast-flowing streams also bring food particles rapidly past their labral fans. Some families are specialized scrapers (Blephariceridae), consuming periphyton and associated materials from the surface of the substrate of fast-flowing streams. Other families are predaceous. For example, larval chaoborids use prehensile antennae to capture zooplankton in the epilimnion at night, and during the day, in lakes with fish, they migrate to the hypolimnion to feed on benthic oligochaetes and chironomids. The larvae of the highly specialized family Sciomyzidae are predaceous on aquatic snails (Berg and Knutson 1978). Merritt and Cummins (1984) report representatives of every functional feeding group among the aquatic dipterans.

Respiration

Aquatic dipteran larvae show a variety of interesting adaptations for acquiring oxygen. Culicids and dixids respire while suspended at the water surface through posterior respiratory tubes or directly through caudal spiracles (i.e., they are metapneustic). Other families with caudal spiracles burrow into the substrate (Dolichopodidae, Tabanidae, Tipulidae, and Stratiomyidae) and thus are restricted to shallow water so that they can migrate easily to the air-water interface. Some metapneustic families (Ptychopteridae, Syrphidae) have extensile respiratory tubes attached to their caudal spiracles; these tubes enable them to remain submerged in oxygen-poor waters for long periods. The free-swimming chironomids, ceratopogonids, and chaoborids, which are apneustic, rely on cutaneous respiration that is sometimes supplemented by gills. The haemolymph-containing "blood gills" of culicids are not respiratory structures but are osmoregulatory structures. A few families of aquatic Diptera are amphipneustic, having open spiracles on prothoracic and last abdominal segments (Empididae, Psychodidae). Tipulids have both open spiracles and anal gills, a combination that enables them to respire in air and in water (similar to the corydalid megalopterans). Thus, crane fly larvae can survive in damp leaf packs that may become exposed to air when water levels drop.

Collection and Preservation

Most dipteran larvae can be collected by sorting through detritus or samples of the benthos from mineral substrates. Some groups that adhere tightly to the substrate (e.g., Blephariceridae) can be collected from individual stones. Most dipteran larvae can be preserved in 70% ethanol, after which chironomid larvae and some larval chaoborids, culicids, and simuliids should be slide-mounted for observation under a compound microscope (see Ch. 14 for details). Pennak (1978) suggests killing dipteran larvae in boiling water and preserving them in 75–85% ethanol or 4% formalin. However, H. J. Teskey (pers. comm., 1987) cautions against the use of formalin for preserving dipterans because specimens so preserved are difficult to slide-mount. Boiling and soaking specimens in 10% KOH enables more effective study of detailed structure of larval mouthparts.

Identification

The identities of the larvae of many families of aquatic Diptera are poorly known. Species identification is possible for only a few families. Even generic determinations cannot be made for aquatic representatives of some families of Brachycera. The lack of associations between larval and adult forms is a critical problem for these families. We do not even know, for example, whether the larvae of some species known as adults are aquatic or terrestrial. We have not included generic keys to those families (Dolichopodidae, Ephydridae, Muscidae, Scathophagidae). We have also omitted the genera of Tipulidae that live in damp soil or marginal detritus and the riparian, subaquatic, marine, and estuarine genera in this and other families. Students who wish to identify those groups should refer to McAlpine et al. 1981, Byers 1984, and Teskey 1984.

The characters used in our keys to separate families and genera of aquatic dipterans include head capsule morphology, caudal spiracular plate structure or other respiratory structures, proleg structure, antennal segmentation, body segmentation and setation, mouthpart morphology, and body pigmentation. Because of its complexity, the family Chironomidae is treated separately, in Chapter 14.

More-Detailed Information

Wirth and Stone 1956; Pennak 1978; Hilsenhoff 1981; McAlpine et al. 1981, 1987; Teskey 1981a, 1984; Webb and Brigham 1982; Byers 1984; Merritt and Schlinger 1984; Newson 1984; Peterson 1984.

Checklist of Aquatic Diptera Larvae

Suborder Brachycera
- Infraorder Cyclorrhapha
  - Ephydridae[a]
  - Muscidae[a]
  - Phoridae
    - *Dohrniphora*
  - Scathophagidae[a]
  - Sciomyzidae
    - *Antichaeta*
    - *Atrichomelina*
    - *Colobaea*
    - *Dictya*
    - *Elgiva*
    - *Hedria*
    - *Pherbellia*
    - *Pteromicra*
    - *Renocera*
    - *Sciomyza*
    - *Sepedon*
    - *Tetanocera*
  - Syrphidae
    - *Chrysogaster*
    - *Eristalis*
    - *Helophilus*
- Infraorder Orthorrhapha
  - Athericidae
    - *Atherix*
  - Dolichopodidae[a]
  - Empididae
    - *Chelifera*
    - *Clinocera*
    - *Dolichocephala*
    - *Hemerodromia*
    - *Oreogeton*
    - *Roederiodes*
    - *Wiedemannia*

Pelecorhynchidae
Glutops
Stratiomyidae
Caloparyphus
Euparyphus
Hedriodiscus
Nemotelus
Odontomyia
Oxycera
Stratiomys
Tabanidae
Atylotus
Chlorotabanus
Chrysops
Diachlorus
Haematopota
Hybomitra
Merycomyia
Tabanus
Suborder Nematocera
Blephariceridae
Blepharicera
Ceratopogonidae
Alluaudomyia
Atrichopogon
Bezzia
Ceratopogon
Culicoides
Dasyhelea
Forcipomyia
Leptoconops
Mallochohelea
Monohelea
Palpomyia
Probezzia
Serromyia
Sphaeromias
Stilobezzia
Chaoboridae
Chaoborus
Eucorethra
Mochlonyx
Chironomidae (see Ch. 14)
Culicidae
Aedes
Anopheles
Culex
Culiseta
Mansonia
Orthopodomyia
Psorophora
Toxorhynchites
Uranotaenia
Wyeomyia
Dixidae
Dixa
Dixella
Psychodidae
Pericoma
Psychoda
Telmatoscopus
Threticus
Ptychopteridae
Bittacomorpha
Bittacomorphella
Ptychoptera
Simuliidae
Cnephia
Ectemnia
Greniera
Metacnephia
Prosimulium
Simulium
Stegopterna
Twinnia
Tanyderidae
Protoplasa
Thaumaleidae
Thaumalea
Tipulidae
Antocha
Brachypremna
Cryptolabis
Dactylolabis
Dicranota
Erioptera
Helius
Hexatoma
Leptotarsus
Limnophila
Limonia
Lipsothrix
Molophilus
Pedicia
Phalacrocera
Prionocera
Pseudolimnophila
Rhabdomastix
Tipula

[a]Generic keys not included, for reasons given in Identification.

Key to Genera of Aquatic Diptera Larvae

1a. Mandibles moving against one another in a horizontal or oblique plane (Figs. 1, 2); head capsule usually complete and fully exposed (may be sclerotized or unsclerotized), except retracted in Tipulidae ..... Suborder NEMATOCERA 2

1b. Mandibles or mouth hooks moving parallel to one another in a vertical plane (Figs. 3–5); head capsule variously reduced posteriorly, partially or almost completely retracted within thorax even if such retracted portions comprise only a few slender rods .... Suborder BRACHYCERA 15

2a (1a). NEMATOCERA: Head capsule with longitudinal incisions of varying depths dorsolaterally (Figs. 1, 2, 6), in extreme cases head reduced to several slender rods (Fig. 1); head capsule capable of partial or complete retraction within thorax; metapneustic or apneustic (Fig. 7); if metapneustic, posterior spiracles bordered by 1–3 pairs of short lobes often fringed with hairs (Figs. 54–64, 66–69, 76–81) .... **Tipulidae** 26

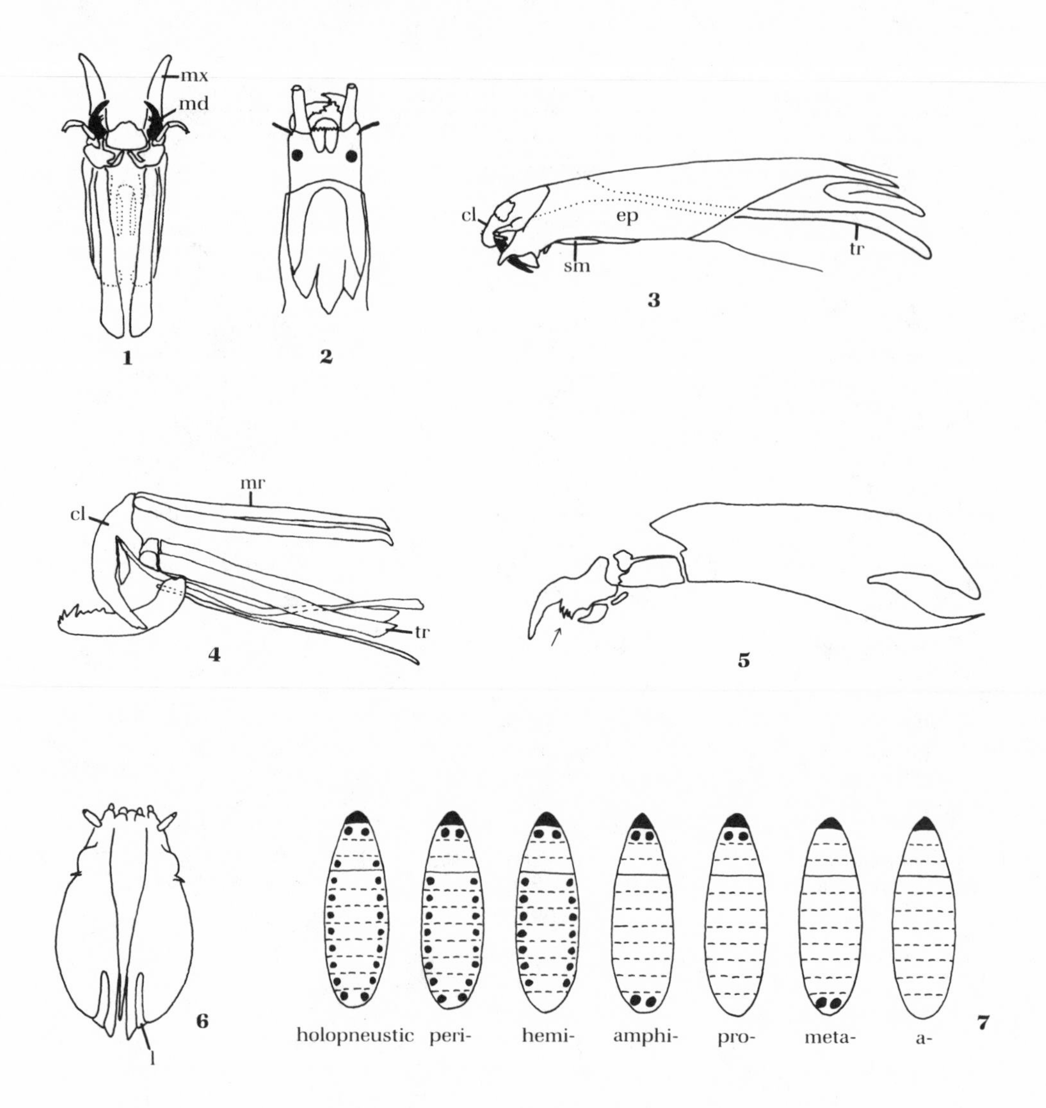

1. Head capsule of *Limnophila* (Tipulidae), dv: md, mandible; mx, maxilla. **2.** Head capsule of *Dicranota* (Tipulidae), dv. **3.** Head capsule of *Tabanus* (Tabanidae), lv: cl, clypeus; ep, epicranium; sm, submentum; tr, tentorial rod. **4.** Head skeleton of Empidinae (Empididae), lv: cl, clypeus; mr, metacephalic rod; tr, tentorial rod. **5.** Cephalopharyngeal skeleton of *Sepedon* (Sciomyzidae), lv. **6.** Head capsule of *Prionocera* (Tipulidae), lv: l, longitudinal incision. **7.** Diagrammatic representation of spiracular systems, dv. dv, dorsal view; lv, lateral view. (1 and 6 modified from Alexander and Byers 1981; 2 redrawn from Johannsen 1934; 3 modified from Pechuman and Teskey 1981; 4 modified from Steyskal and Knutson 1981; 5 modified from Knutson 1987; 7 redrawn from Teskey 1981a.)

2b. Head capsule complete, without longitudinal incisions and incapable of retraction within thorax; amphipneustic, metapneustic, or apneustic; if metapneustic, posterior spiracles usually without bordering fringed lobes 3

3a (2b). Body without distinct constriction between head and thorax; body divided into 6 major divisions, the 1st comprising the fused head, thorax, and 1st abdominal segment (Fig. 8); each division has a median suctorial disk ventrally (Fig. 8) .... **Blephariceridae,** *Blepharicera*

3b. Body with distinct constriction between head and thorax (Figs. 9–14); suctorial disks absent 4

4a (3b). A pair of elongate prolegs present on each of 8 abdominal segments (Fig. 14) .... **Nymphomyiidae,** *Palaeodipteron*

4b. Prolegs usually absent, but if present, on no more than 3 abdominal segments 5

5a (4b). Abdomen terminating in a long, slender, retractile respiratory tube (Fig. 15); abdominal and thoracic segments with multiple transverse ridges or rows of small setae or setiferous papillae; 1st 3 abdominal segments with a pair of ventral prolegs, sometimes very small, bearing a single, slender, curved claw (Fig. 15) .... **Ptychopteridae** 48

5b. Having none of the above characteristics 6

6a (5b). Thoracic segments fused and indistinctly differentiated, forming a single segment that is wider than abdominal segments (Fig. 9); thoracic and abdominal segments with prominent lateral fanlike tufts of long setae, and/or terminal segment with an anal setal fan (Figs. 9–12) 7

6b. Thoracic segments usually individually distinguishable, thorax and abdomen about equal in diameter or abdomen wider (Fig. 16) (except in some Dixidae); setae on thoracic and abdominal segments not tufted, and anal fan of terminal segment absent (Figs. 13, 17, 20) 8

7a (6a). Mouth brushes lacking or, if present, not visible in dorsal view (Figs. 10–12); antennae prehensile, with long apical setae .... **Chaoboridae** 50

7b. Prominent mouth brushes present on either side of labrum and visible in dorsal view (Fig. 9); antennae not prehensile and with only short apical setae (Fig. 9) .... **Culicidae** 52

8a (6b). Paired crochet-bearing prolegs ventrally on 1st and usually 2nd abdominal segments (Fig. 13); abdomen with 2 flattened dorsolateral postspiracular lobes having setose margins projecting above a conical, dorsally sclerotized segment bearing the terminal anus and anal papillae (Fig. 13) .... **Dixidae** 61

8b. Abdominal segments without prolegs; abdomen not as above 9

9a (8b). Prothorax with 1 or a pair of prolegs ventrally (Figs. 16, 17, 20–23) 10

9b. Prothorax lacking prolegs 13

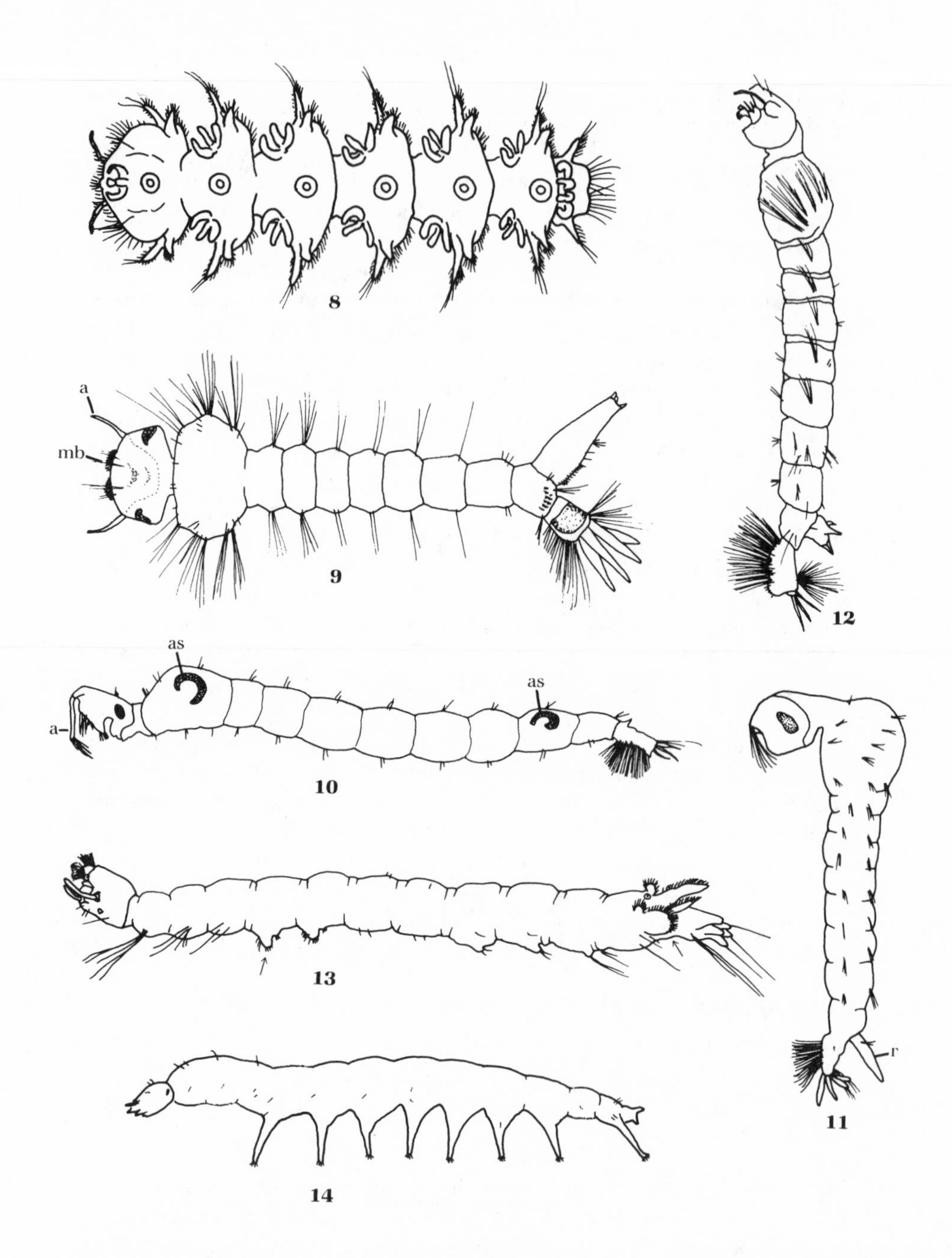

8. *Philorus* (Blephariceridae), vv. **9.** *Aedes* (Culicidae), dv: a, antenna; mb, mouth brush. **10.** *Chaoborus* (Chaoboridae), lv: a, antenna; as, air sac. **11.** *Mochlonyx* (Chaoboridae), lv: r, respiratory tube. **12.** *Eucorethra* (Chaoboridae), lv. **13.** *Dixa* (Dixidae), dlv. **14.** *Palaeodipteron* (Nymphomyiidae), lv. dlv, dorsolateral view; dv, dorsal view; lv, lateral view. (8 redrawn from Hogue 1981; 9, 10, 12 redrawn from Johannsen 1934; 11 redrawn from Webb and Brigham 1982; 13 redrawn from Peters 1981; 14 redrawn from Kevan and Cutten 1981.)

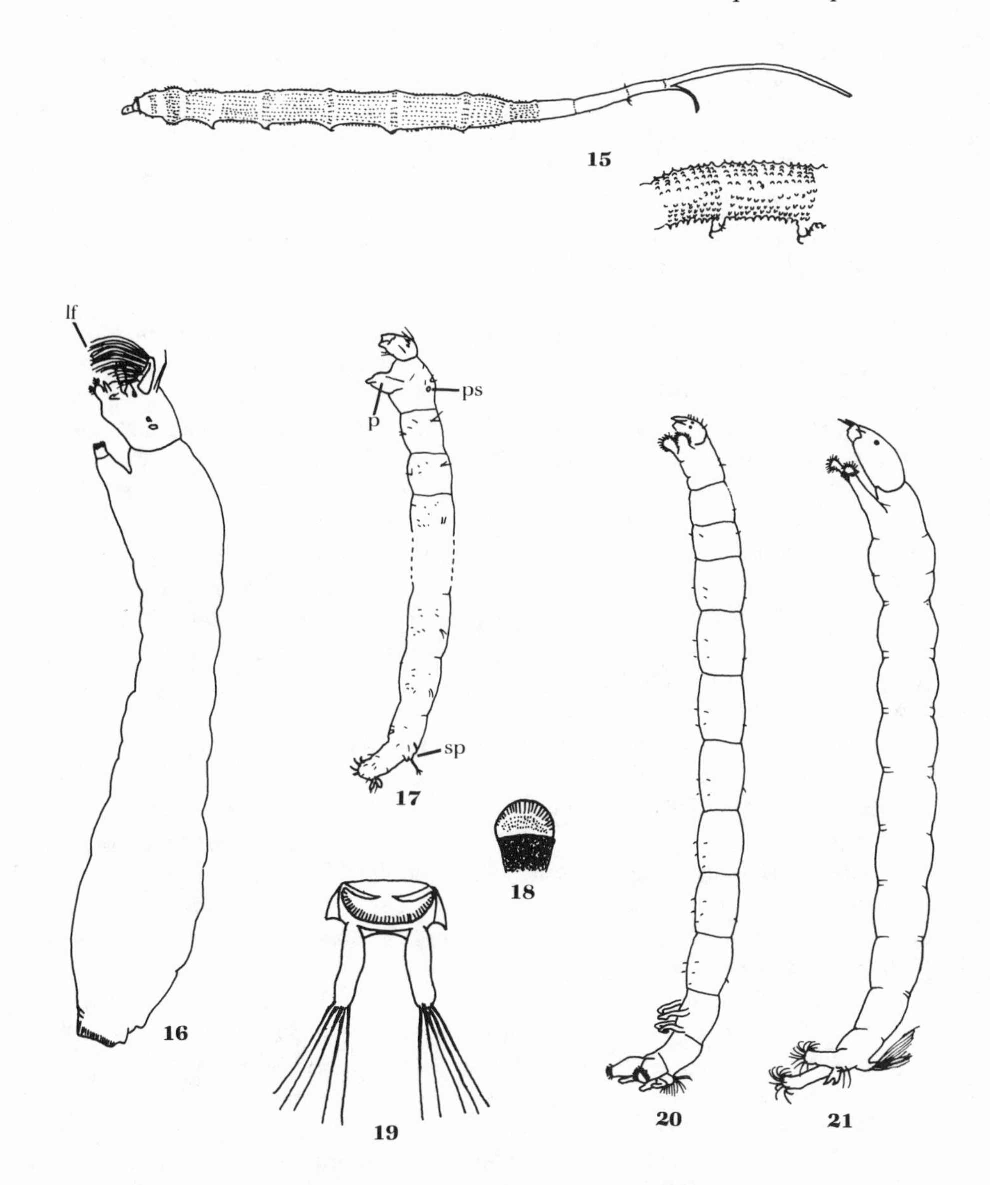

15. *Ptychoptera* (Ptychopteridae) with enlargement of 1st 2 abdominal segments of *Bittacomorpha* (Ptychopteridae), lv. **16.** *Simulium* (Simuliidae), lv: lf, labral fan. **17.** *Thaumalea* (Thaumaleidae), lv: p, proleg; ps, prothoracic spiracle; sp, spiracular plate. **18.** Prothoracic spiracle of *Thaumalea* (Thaumaleidae). **19.** Dorsal spiracular plate and associated lobes of *Thaumalea* (Thaumaleidae), dv. **20.** *Chironomus* (Chironomidae), lv. **21.** *Ablabesymyia* (Chironomidae), lv. dv, dorsal view; lv, lateral view. (15 redrawn from Alexander 1981a; 16 redrawn from Peterson 1981; 17 redrawn from Stone and Peterson 1981; 18, 19 redrawn from Hennig 1950; 20, 21 redrawn from Oliver 1981.)

10a (9a). Head capsule usually with a pair of conspicuous, folding, labral fans dorsolaterally (Fig. 16); abdominal segments 5–8 swollen, posterior segment terminating in a ring or circlet of numerous radiating rows of minute hooks (Fig. 16) .... **Simuliidae** 62

10b. Head capsule lacking labral fans; abdominal segments 5–8 not swollen, posterior abdominal segment not as above, although anal proleg(s) bearing crochets may be present 11

11a (10b). Amphipneustic (Figs. 7, 17); prothoracic spiracles on short stalks (Fig. 18), and posterior spiracles opening into a transverse cleft between fingerlike processes on 8th abdominal segment (Fig. 19); prothoracic and anal prolegs unpaired .... **Thaumaleidae,** *Thaumalea*

11b. Apneustic (Figs. 7, 22–26); prothoracic or anal prolegs usually paired even if distinction is only a slight separation of the apical spines (Fig. 21) 12

12a (11b). All body segments with prominent dorsal tubercles and/or setae (Fig. 23) .... **Ceratopogonidae** (in part) 69

12b. All body segments lacking prominent dorsal tubercles and setae (Figs. 20–22) .... **Chironomidae** (see Ch. 14 for key to genera)

13a (9b). Posterior 2 abdominal segments with long filamentous processes (Fig. 27), pairs of such processes arising laterally on the penultimate segment, dorsolaterally on the terminal segment, and from near the apex of 2 elongate cylindrical prolegs that project posteroventrally from the terminal segment (Fig. 27) .... **Tanyderidae,** *Protoplasa*

13b. Posterior 2 abdominal segments without long filamentous processes; at most, only a single anal proleg present 14

14a (13b). All body segments secondarily divided into 2 or 3 subdivisions with some or all of these subdivisions bearing dorsal sclerotized plates (Fig. 28); remainder of integument with numerous dark spots, which, together with the dorsal plates, impart a grayish brown coloration; amphipneustic (Fig. 7), posterior spiracles at apex of a relatively short, conical respiratory tube (Fig. 28) .... **Psychodidae** 81

14b. Body segments usually not secondarily divided; integument smooth, shiny, and creamy white (when live), lacking all surface features except a few setae (Figs. 25, 26) and sometimes a retractile anal proleg bearing a few crochets (Fig. 24); apneustic (Fig. 7) .... **Ceratopogonidae** (in part) 70

15a (1b). BRACHYCERA: Sclerotized portions of head capsule exposed externally, although sometimes greatly reduced, in which case slender tentorial and metacephalic rods prominent internally (Figs. 3–4, 29–35) .... Infraorder ORTHORRHAPHA 16

15b. External sclerotized portions of head capsule absent; head reduced to an internal cephalopharyngeal skeleton (Figs. 5, 36–44) .... Infraorder CYCLORRHAPHA 21

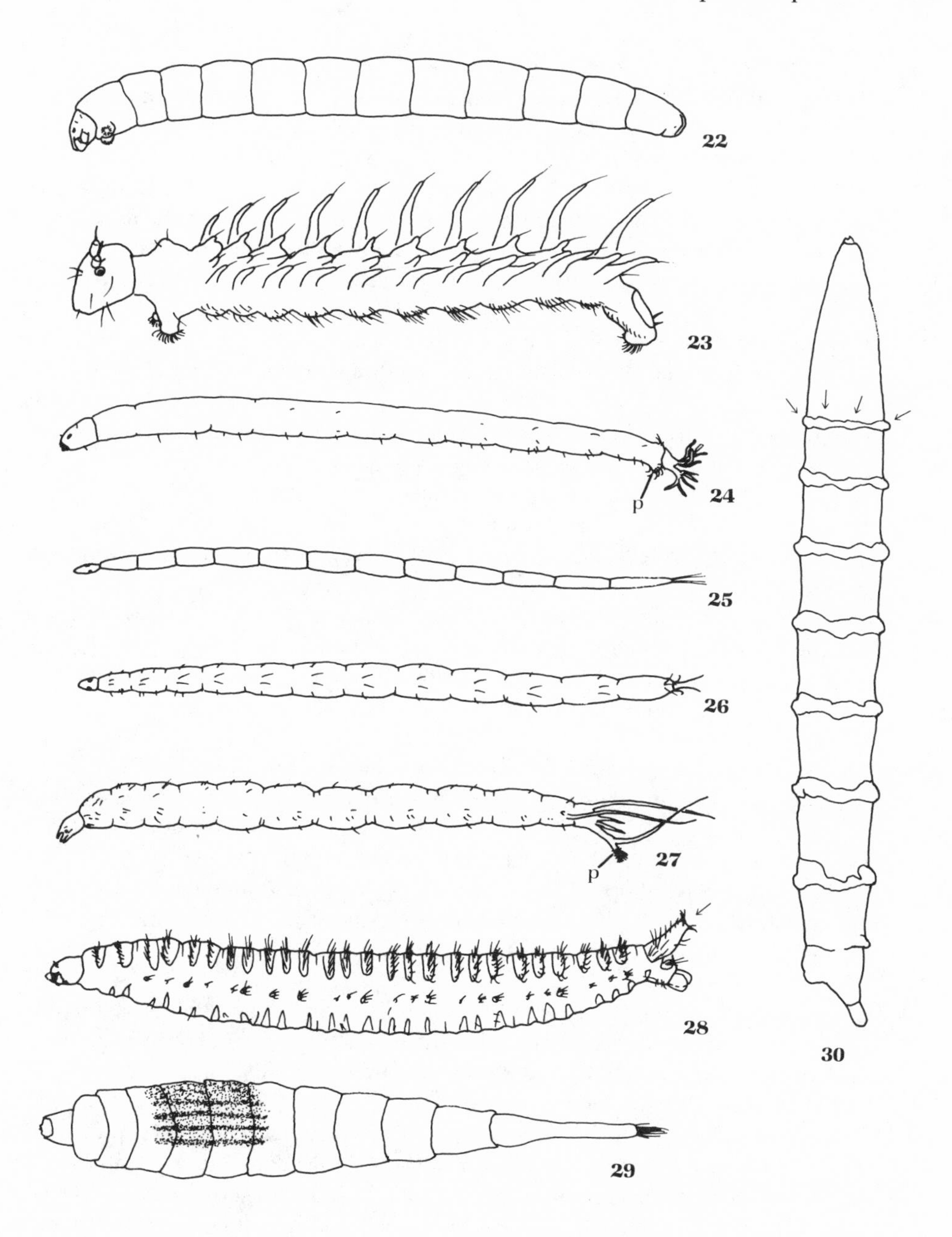

22. *Pseudosmittia* (Chironomidae), lv. **23.** *Atrichopogon* (Ceratopogonidae), dlv. **24.** *Dasyhelea* (Ceratopogonidae), lv: p, proleg. **25.** *Bezzia* (Ceratopogonidae), dv. **26.** *Culicoides* (Ceratopogonidae), dv. **27.** *Protoplasa* (Tanyderidae), lv: p, proleg. **28.** *Pericoma* (Psychodidae), lv. **29.** *Stratiomys* (Stratiomyidae), dv. **30.** *Tabanus* (Tabanidae), lv. dlv, dorsolateral view; dv, dorsal view; lv, lateral view. (22 redrawn from Oliver 1981; 23–26 redrawn from Downes and Wirth 1981; 27 redrawn from Alexander 1981b; 28 redrawn from Quate and Vockeroth 1981; 29 redrawn from James 1981; 30 redrawn from Pechuman and Teskey 1981.)

16a (15a). ORTHORRHAPHA: Body somewhat depressed dorsoventrally; integument toughened and leathery from calcium deposits that are evident as numerous small reticulately arranged facets (Fig. 29); head capsule capable of only slight independent movement; usually with distinct lateral eye prominences . . . . **Stratiomyidae** 82

16b. Body not depressed; integument not toughened and leathery; head capsule capable of extensive independent movement; without distinct eye prominences 17

17a (16b). Integument smooth and shiny with beadlike segmentation, lacking tubercles and prolegs (Fig. 35); without terminal abdominal lobes or appendages . . . . **Pelecorhynchidae,** *Glutops*

17b. Body with either prolegs, fleshy swellings, or creeping welts; with or without terminal abdominal lobes or appendages 18

18a (17b). Body cylindrical; 1st 7 abdominal segments girdled by fleshy swellings with 3–4 pairs of prolegs that may bear hooks (Figs. 30, 31); terminal segments of abdomen tapered or rounded, with spiracles present, but never with terminal lobes or appendages . . . . **Tabanidae** 88

18b. Body variable; abdomen with at most 1 pair of prolegs per segment (Figs. 32, 34), or with creeping welts (Fig. 33); spiracles present or absent; terminal abdominal segment with 1 or 2 pairs of appendages or lobes 19

19a (18b). Apex of abdomen with a pair of caudal processes that are longer than prolegs; each abdominal segment with a pair of prolegs (Fig. 32); spiracles absent . . . . **Athericidae,** *Atherix*

19b. Apex of abdomen with caudal processes reduced to fleshy lobes (Fig. 33) or with processes shorter than prolegs (Figs. 45, 46); abdominal segments with creeping welts (Fig. 33) or with prolegs lacking from 1st abdominal segment (Fig. 34); spiracles present or absent 20

20a (19b). Metapneustic (Fig. 7); 4 smooth primary lobes on last abdominal segment, with posterior spiracles situated at the base of the upper 2 lobes; transverse ventral creeping welts present on abdominal segments (Fig. 33); metacephalic rods expanded posteriorly (Fig. 47) . . . . **Dolichopodidae** (not keyed to genus)

20b. Usually apneustic (Fig. 7), with last abdominal segment with 1–4 rounded lobes bearing apical setae, and abdominal segments bearing paired prolegs with apical crochets (Figs. 34, 45, 46); if metapneustic, then posterior segment with only a single lobe below spiracles, and abdominal segments with ventral creeping welts (Fig. 48); metacephalic rods slender posteriorly (Fig. 4) . . . . **Empididae** 94

21a (15b). CYCLORRHAPHA: Pharyngeal skeleton with the "hypopharynx" produced as a median tooth (Fig. 36); known from pitcher plants and from waste-treatment facilities, where it is a predator on psychodid larvae . . . . **Phoridae,** *Dohrniphora* (Fig. 37)

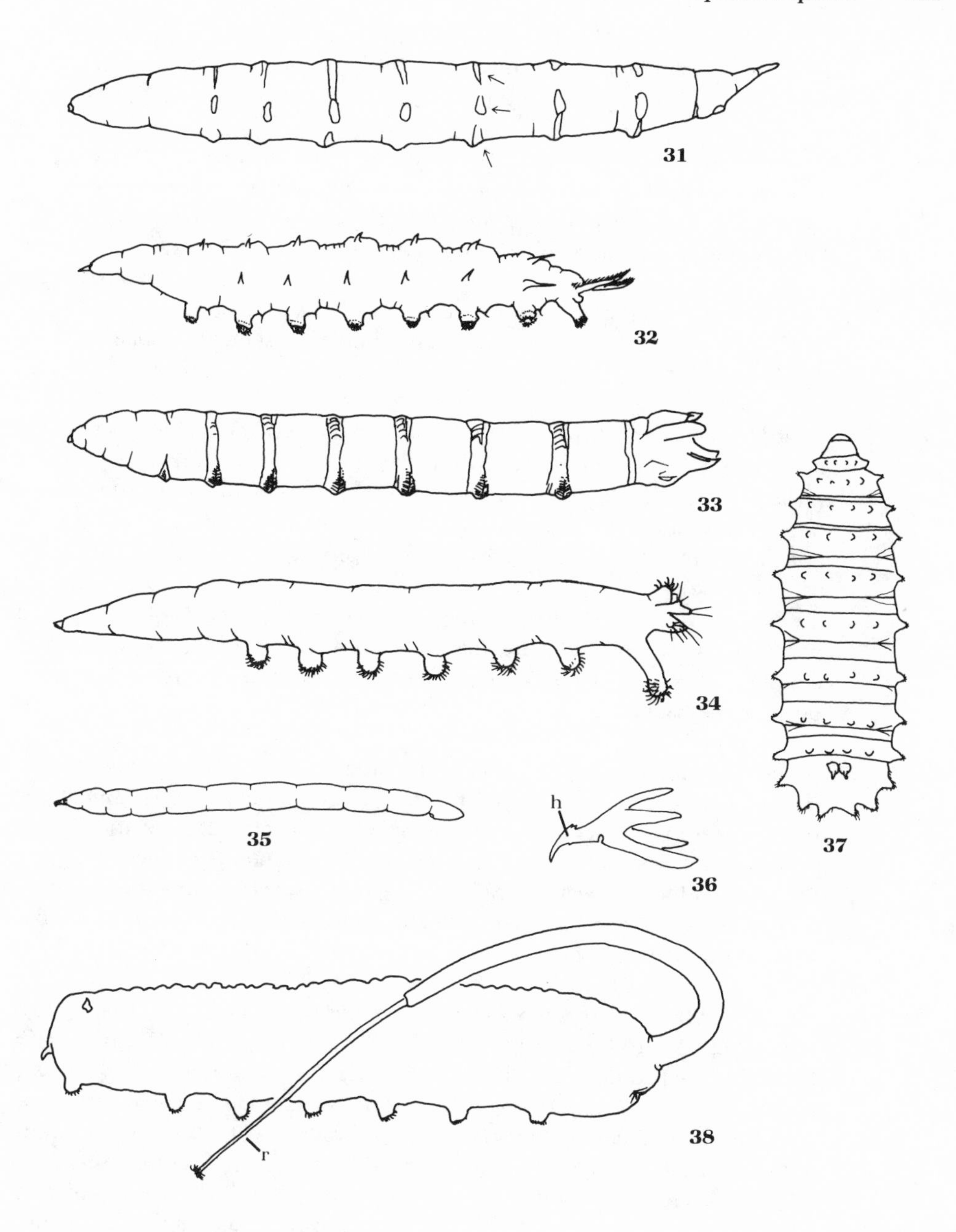

31. *Chrysops* (Tabanidae), lv. **32.** *Atherix* (Athericidae), lv. **33.** *Rhaphium* (Dolichopodidae), lv. **34.** *Hemerodromia* (Empididae), dlv. **35.** *Glutops* (Pelecorhynchidae), lv. **36.** Pharyngeal skeleton of *Dohrniphora* (Phoridae), lv: h, hypopharynx. **37.** *Dohrniphora* (Phoridae), dv. **38.** *Eristalis* (Syrphidae), lv: r, respiratory tube. dlv, dorsolateral view; dv, dorsal view; lv, lateral view. (31 redrawn from Pechuman and Teskey 1981; 32 redrawn from Webb 1981; 33 redrawn from Robinson and Vockeroth 1981; 34 redrawn from Steyskal and Knutson 1981; 35 redrawn from Teskey 1981c; 36, 37 modified from Jones 1918; 38 redrawn from Teskey 1981b.)

21b. Pharyngeal skeleton not so produced; rarely associated with pitcher plants 22

22a (21b). Posterior spiracular plates fused or very closely approximated, usually on apex of a retractile respiratory tube (Figs. 38, 39), cephalopharyngeal skeleton lacking mouth hooks or similar structure, a (possibly homologous) ribbed filter chamber in area normally occupied by mouth hooks (Fig. 40) .... **Syrphidae** 102

22b. Posterior spiracular plates always distinctly separated whether mounted on a retractile respiratory tube or not; mouth hooks present, filter chamber absent 23

23a (22b). Cephalopharyngeal skeleton with a sclerotized ventral arch below base of mouth hooks, its anterior margin usually toothed (Fig. 5); spiracular plate at posterior end surrounded by several lobes; body segments often extensively covered with short, fine hairs; posterior segment often somewhat tapered, its apex with tubercles surrounding posterior spiracles that are only slightly elevated (Fig. 41); parasitic/predaceous on snails .... **Sciomyzidae** 104

23b. Cephalopharyngeal skeleton lacking a ventral arch; spiracular plate absent; if body extensively covered with short, fine hairs, then a respiratory tube present (Figs. 42, 43), or each spiracle situated on a short tubular projection on posterior segment 24

24a (23b). Posterior abdominal segment somewhat tapered, sometimes ending in a retractile respiratory tube (Figs. 42, 43); integument of posterior abdominal segments covered with setae (Fig. 42, enlargement) or spinules, or with setaceous tubercles on some segments ... **Ephydridae** (not keyed to genus)

24b. Posterior abdominal segment rather truncate and/or integument with setae only on intersegmental areas (Fig. 49); tubercles, if present, restricted to posterior abdominal segment 25

25a (24b). Posterior abdominal segment lacking tubercles other than those bearing spiracles (Fig. 49); prothoracic spiracles, if present, fan-shaped and usually with fewer than 10 papillae (Fig. 50); accessory oral sclerite present below mouth hooks (Fig. 44) .... **Muscidae** (not keyed to genus)

25b. Posterior abdominal segment often with several pairs of tubercles surrounding spiracles (Fig. 51); prothoracic spiracle a sievelike plate, or with many papillae arranged in a 2-horned fan (Fig. 52); accessory oral sclerite absent .... **Scathophagidae** (not keyed to genus)

26a (2a). **Tipulidae:** Both dorsal and lateral longitudinal rows of conspicuous, usually elongate, fleshy projections on thoracic and abdominal segments; those of thoracic segments simple, posterior ones on most abdominal segments either deeply forked or, if simple, approximately 10 times as long as their basal diameter (Fig. 53); in aquatic or semiaquatic mosses .... *Phalacrocera*

26b. Thoracic and abdominal segments not as above (Figs. 54, 55); lateral abdominal projections, if present, blunt, shorter than their basal diameter 27

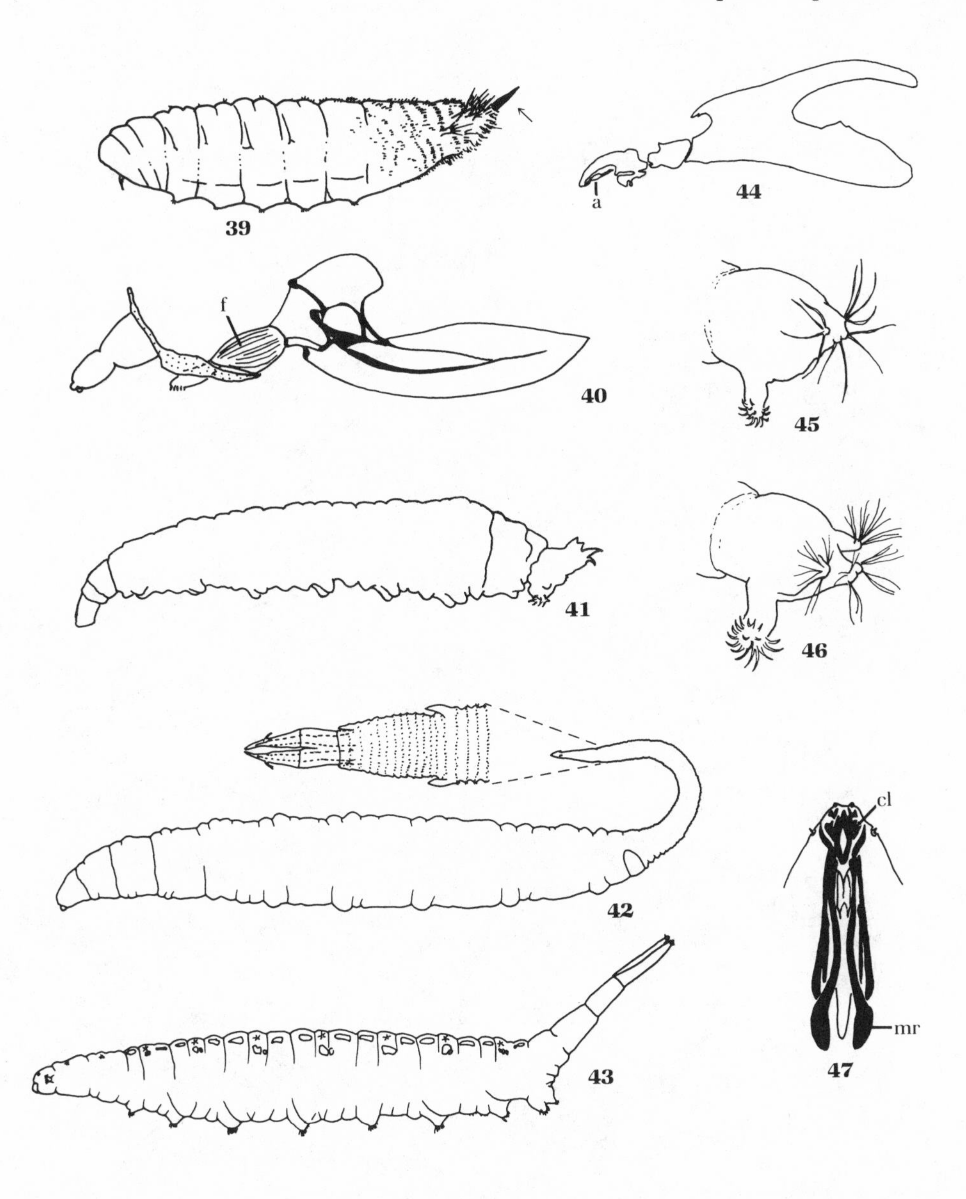

39. *Chrysogaster* (Syrphidae), lv. **40.** Cephalopharyngeal skeleton of *Eristalis* (Syrphidae), lv: f, filter chamber. **41.** *Sepedon* (Sciomyzidae), lv. **42.** *Notiphila* (Ephydridae) with enlargement of apex of respiratory tube, lv. **43.** *Ephydra* (Ephydridae), lv. **44.** Cephalopharyngeal skeleton of *Potamia* (Muscidae), lv: a, accessory oral sclerite. **45.** Caudal segment of *Chelifera* (Empididae), dlv. **46.** Caudal segment of *Clinocera* (Empididae), dlv. **47.** Head skeleton of *Rhaphium* (Dolichopodidae), dv: cl, clypeus; mr, metacephalic rod. dlv, dorsolateral view; dv, dorsal view; lv, lateral view. (39 redrawn from Varley 1937; 41 redrawn from Knutson 1987; 42, 43 redrawn from Wirth et al. 1987; 44 redrawn from Huckett and Vockeroth 1987; 45, 46 redrawn from Steyskal and Knutson 1981; 47 redrawn from Robinson and Vockeroth 1981.)

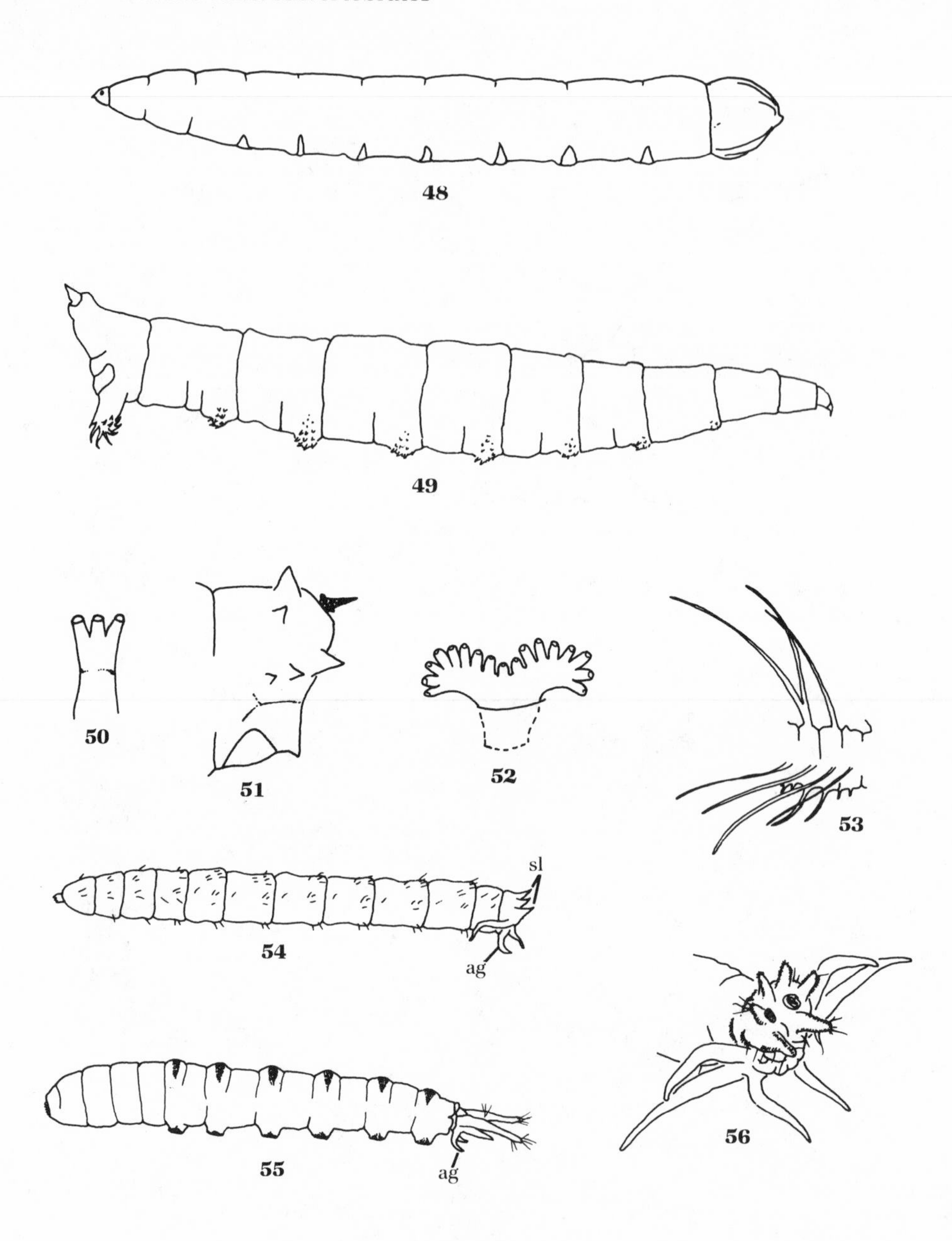

48. *Rhamphomyia* (Empididae), lv. **49.** *Limnophora* (Muscidae), lv. **50.** Prothoracic spiracle of *Phaonia* (Muscidae). **51.** Tip of abdomen of *Cordilura* (Scathophagidae), lv. **52.** Typical 2-horned prothoracic spiracle of Scathophagidae. **53.** Abdominal segments of *Phalacrocera* (Tipulidae), lv. **54.** *Tipula* (Tipulidae), lv: ag, anal gill; sl, spiracular lobes. **55.** *Antocha* (Tipulidae), lv: ag, anal gill. **56.** Caudal segments of *Tipula* (Tipulidae), oplv. lv, lateral view; oplv, oblique posterolateral view. (48 redrawn from Steyskal and Knutson 1981; 49 redrawn from Johannsen 1935; 50 redrawn from Wallace 1971; 51 redrawn from Wallace and Neff 1971; 52 redrawn from Teskey 1981b; 53 redrawn from Alexander 1920; 54 redrawn from Pennak 1978; 55, 56 redrawn from Alexander and Byers 1981.)

27a (26b). Spiracular plate bordered by 6 (rarely, 8) usually subconical lobes, ordinarily 2 dorsal, 2 dorsolateral, 2 below spiracles; lobes sometimes short and blunt (Figs. 56, 57) 28

27b. Spiracular plate bordered by 5 (rarely, 7) or fewer lobes, often 1 dorsomedial, 2 lateral, 2 below spiracles; shape of lobes variable (Fig. 58) 31

28a (27a). Anal gills pinnately branched (Fig. 59); dorsal lobes of spiracular plate short, bluntly rounded, lower lobes more than 2 times as long as their basal width; aquatic or semiaquatic in small streams . . . . *Leptotarsus*

28b. Anal gills not pinnately branched; lobes of spiracular plate variable 29

29a (28b). All lobes of spiracular plate elongate; lateral and ventral lobes 3–4 times as long as their basal width; lobes bordered by numerous long hairs, outer hairs 2–3 times as long as width of lobe at point of attachment of hair (Fig. 60) . . . . *Prionocera*

29b. Lobes of spiracular plate not all elongate, longest ones rarely more than 2 times as long as their basal width; bordering hairs usually sparse, or, if numerous, not long (Figs. 57, 61) 30

30a (29b). Abdominal segments and posterior of metathorax densely pilose, giving larva a wooly appearance; thoracic segments otherwise with only short pubescence, nearly bare by contrast; spiracular plate relatively small with dorsal lobes low, inconspicuous; lobes of spiracular plate with darkened faces . . . . *Brachypremna*

30b. Pilosity of abdomen not in contrast with that on thorax; lobes of spiracular plate relatively large and highly variable, from short and round to elongate and subconical (Figs. 54, 57, 62); faces of lobes of plate not darkened, or with only 1 or 2 thin darkened lines (Figs. 57, 62) . . . . *Tipula*

31a (27b). Spiracles absent, tracheal system closed; dorsal and lateral lobes of 9th abdominal segment absent or extremely reduced, ventral lobes elongate, deeply separated, slightly divergent, with a few tufts of hairs (Fig. 55); anal papillae elongate; dorsal and ventral creeping welts conspicuous on abdominal segments 2–7 (Fig. 55) . . . . *Antocha*

31b. Spiracles present, usually conspicuous (may be concealed if lobes of plate are infolded); dorsal and lateral lobes usually present, but absent in some species (Fig. 63) 32

32a (31b). Dorsal and lateral lobes of spiracular plate absent or extremely reduced and ventral lobes elongate (Fig. 63) 33

32b. Median dorsal and/or lateral lobes of spiracular plate well developed if ventral lobes are elongate (Fig. 64); ventral lobes usually short, less often absent 35

33a (32a). Paired prolegs with sclerotized apical crochets on venter of abdominal segments 3–7 (Fig. 65); found in wet to saturated soil along streams . . . . *Dicranota* (most)

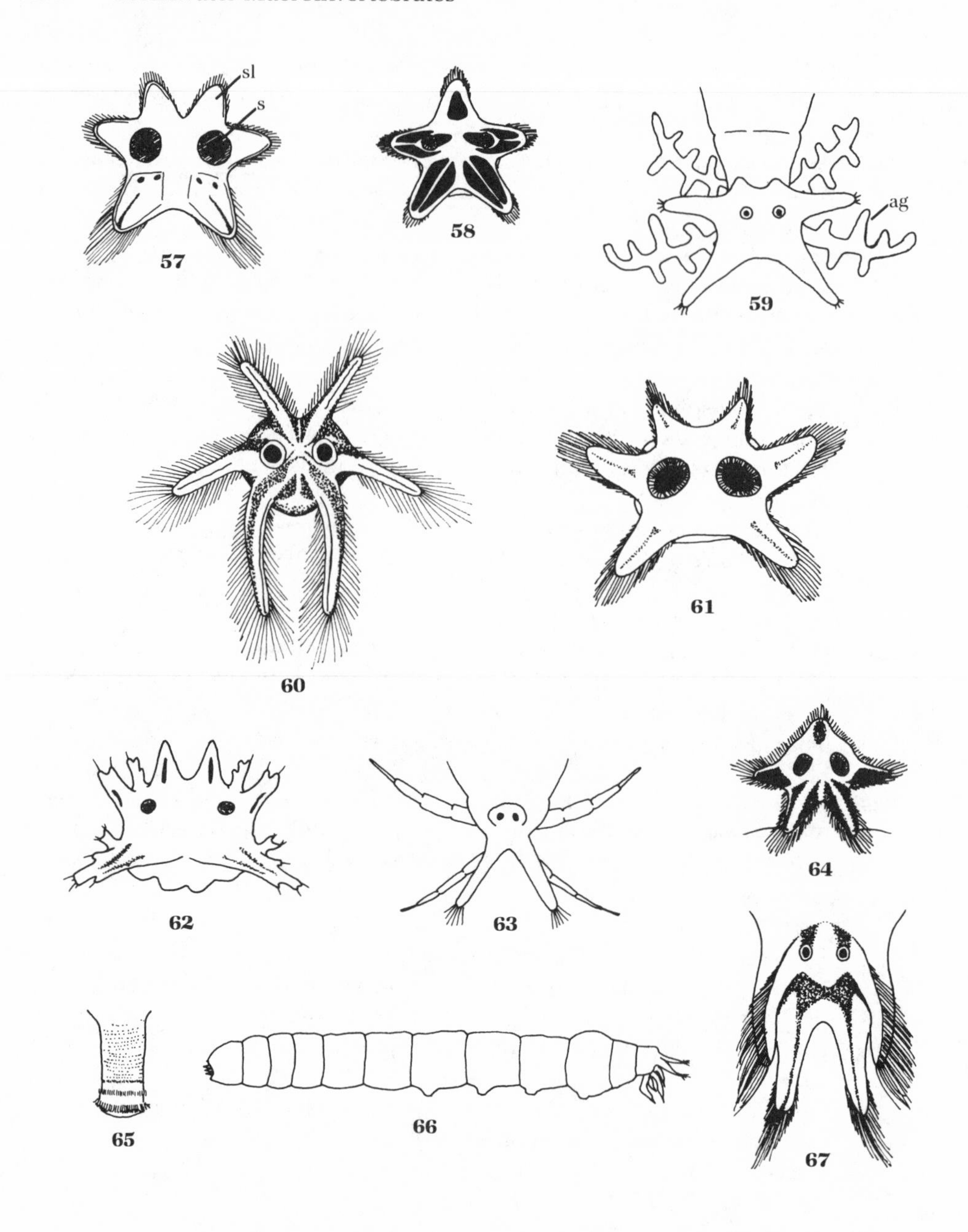

57. Spiracular plate of *Tipula* (Tipulidae), pv: s, spiracle; sl, spiracular lobe. **58.** Spiracular plate of *Molophilus* (Tipulidae), pv. **59.** Spiracular plate of *Leptotarsus* (Tipulidae), pv: ag, anal gill. **60.** Spiracular plate of *Prionocera* (Tipulidae), pv. **61.** Spiracular plate of *Holorusia* (Tipulidae), pv. **62.** Spiracular plate of *Tipula* (Tipulidae), pv. **63.** Spiracular plate of *Pedicia* (Tipulidae), pv. **64.** Spiracular plate of *Helius* (Tipulidae), pv. **65.** Abdominal proleg of *Dicranota* (Tipulidae). **66.** *Pedicia* (Tipulidae), dlv. **67.** Spiracular plate of *Hexatoma* (Tipulidae), pv. dlv, dorsolateral view; pv, posterior view. (57, 61–65, 67 redrawn from Johannsen 1934; 58–60 redrawn from Alexander 1920; 66 redrawn from Alexander and Byers 1981.)

33b. Abdomen without prolegs; roughened creeping welts or broad tubercles on basal annulus of segments 4–7 (Fig. 66) 34

34a (33b). Creeping welts on both dorsum and venter, bearing microscopic spicules . . . . *Dicranota* (subgenus *Rhaphidolabina*)

34b. Creeping welts or broad tubercles on venter only (Fig. 66), without spicules, but with microscopic roughened surface; found in wet soil of swampy woods, springs, and seepage areas . . . . *Pedicia*

35a (32b). Spiracular plate with 4 or 5 peripheral lobes (Figs. 67, 68) 36

35b. Spiracular plate with only 3 lobes, or without distinct lobes (Fig. 69) 47

36a (35a). Blades of maxillae do not project from retracted head; internal portion of head extensively sclerotized dorsally and laterally, with shallow posterior incisions (can be determined by cutting prothoracic skin at one side, or often can be seen through skin) (Fig. 70) 37

36b. Blades of maxillae project from retracted head (Fig. 1); internal portion of head divided by deep posterior incisions into elongate, slender, rodlike-to-spatulate sclerites (Fig. 71), or if sclerites are platelike, they are darkly sclerotized only along margins, giving appearance of separate rods (Fig. 72) 45

37a (36a). Abdominal segments without creeping welts; hypostomal bridge (called *mentum* or *maxillary plates* by some authors) almost always interrupted medially by membranous area (Fig. 73) (hypostomal plates in contact but not fused in *Pseudolimnophila;* Fig. 74) 38

37b. Creeping welts present on basal ring of abdominal segments, or abdominal segments with transverse bands or patches of dense pilosity of both basal and apical rings; hypostomal bridge complete (hypostomal plates fused), although may be deeply incised posteriorly (Fig. 75) 41

38a (37a). Plane of spiracular plate roughly perpendicular to long axis of body (Fig. 76); plate surrounded by 5 lobes 39

38b. Plane of spiracular plate diagonal to long axis of body (Fig. 79); plate surrounded by 4 lobes (Figs. 78, 79) 40

39a (38a). Hypostomal plates with anterior margins toothed (Fig. 73); spiracular plate with black spots on ventral and dorsolateral lobes divided medially by pale line (Fig. 58) . . . . *Molophilus*

39b. Hypostomal plates with anterior margins untoothed; spiracular plate with solid black spots undivided on all lobes (Fig. 77) . . . . *Erioptera* (subgenus *Trimicra*)

40a (38b). Ventral lobes of spiracular plate darkly pigmented on upper surface, fringed with hairs that are longer than the lobes (Fig. 79); spiracles dark; hypostoma forming a pair of toothed plates (Fig. 74); hypostomal plates each bearing 7 or 8 anterior teeth (Fig. 74); found in thin organic mud in swampy woods, pond margins, marshy areas . . . . *Pseudolimnophila*

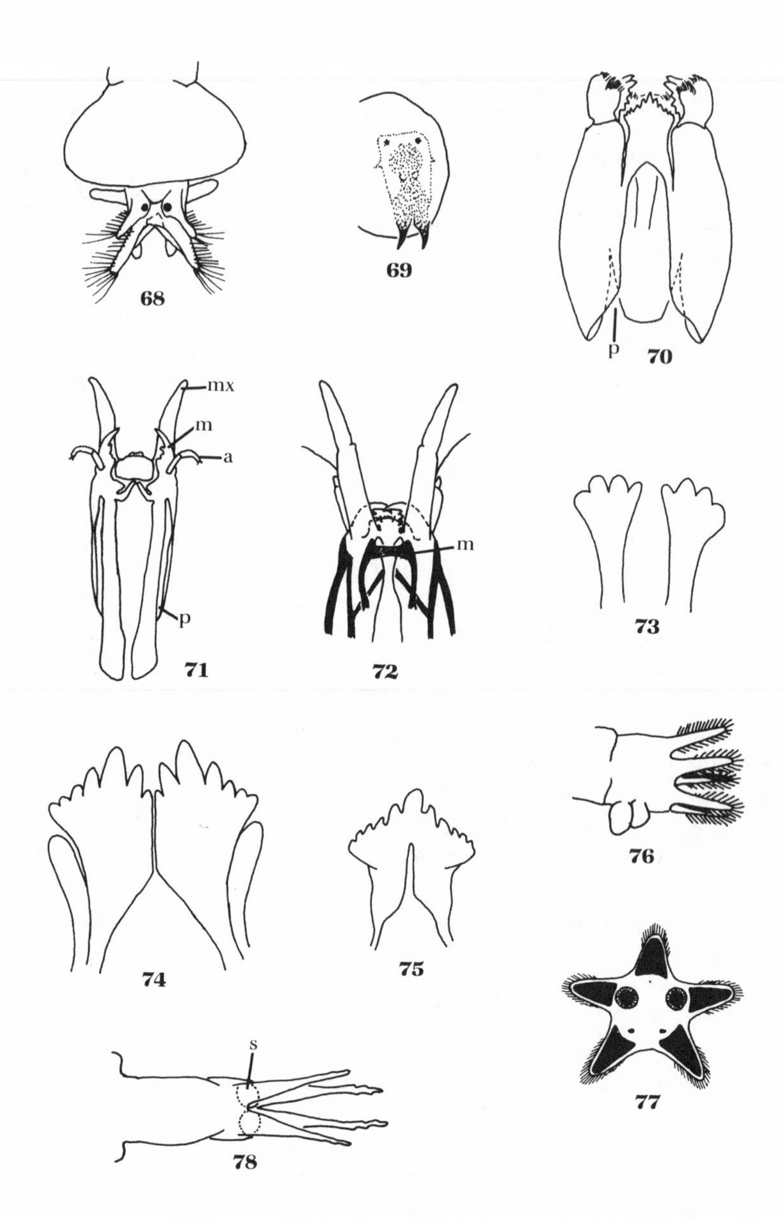

68. Posterior end of abdomen of *Hexatoma* (Tipulidae), dv. **69.** Spiracular plate of *Rhabdomastix* (Tipulidae), plv. **70.** Head capsule of *Limonia* (Tipulidae), dv: p, posterior incisions. **71.** Head capsule of *Limnophila* (Tipulidae), dv: a, antenna; m, mandible; mx, maxillary palp; p, posterior incision. **72.** Head capsule of *Limnophila* (Tipulidae), vv: m, mentum. **73.** Hypostomal plates of *Molophilus* (Tipulidae), vv. **74.** Hypostomal plates of *Pseudolimnophila* (Tipulidae), vv. **75.** Hypostomal bridge of *Limonia* (Tipulidae), vv. **76.** Caudal end of *Erioptera* (Tipulidae), lv. **77.** Spiracular plate of *Erioptera* (Tipulidae), pv. **78.** Caudal end of *Cryptolabis* (Tipulidae), dv: s, spiracle. dv, dorsal view; lv, lateral view; plv, posterolateral view; pv, posterior view; vv, ventral view. (68, 72, 74, 75, 77 redrawn from Johannsen 1934; 69 redrawn from Hynes 1969; 70, 71 redrawn from Alexander and Byers 1981; 73 redrawn from Alexander 1920; 76 modified from Gerbig 1913; 78 redrawn from Hynes 1963.)

40b. Ventral lobes of spiracular plate not darkly pigmented on upper surface, not fringed with long hairs (Fig. 78); spiracles pale; hypostoma reduced to small, longitudinal rods below maxillae on each side; found in sandy bottoms of clear, cold streams . . . . *Cryptolabis*

41a (37b). Spiracular plate with 5 peripheral lobes (Fig. 64) 42

41b. Spiracular plate with 4 peripheral lobes (Fig. 67) (if vestigial median dorsal lobe is present, it is unpigmented) 43

42a (41a). Abdominal segments 2–7 with both dorsal and ventral creeping welts on basal ring; lobes of spiracular plate wider at base than they are long, broadly rounded, unpigmented or with only limited darkened spots; large genus, in numerous aquatic and terrestrial habitats . . . . *Limonia* (in part)

42b. Abdominal segments 2–7 with only ventral creeping welts on basal ring; ventral lobes of spiracular plate longer than their width at base, darkened at margins and pale in the middle (Fig. 64); hypostomal bridge with 5 teeth; body brownish, with long, appressed (closely applied to body) pubescence; occurs in marsh borders in decomposing aquatic vegetation, or in marsh areas in woods . . . . *Helius*

43a (41b). Abdominal segments 2–7 without distinct creeping welts, all segments with transverse bands or patches of dense pilosity; occurs in moss and algae on wet rocky cliffs . . . . *Dactylolabis*

43b. Abdominal segments 2–7 with distinct creeping welts, without transverse bands or patches of dense pilosity 44

44a (43b). Ventral lobes of spiracular plate longer than their width at base, tapering to subacute apex, fringed with long hairs (as in Fig. 80); lobes of spiracular plate narrowly darkened at margins; found in sodden, decayed wood, at or just below water level . . . . *Lipsothrix*

44b. Ventral lobes of spiracular plate shorter than their width at base, broadly rounded and without long marginal hairs (Fig. 81); lobes of spiracular plate with isolated spots of dark pigmentation, generally pale (Fig. 81); large genus in numerous aquatic and terrestrial habitats . . . . *Limonia* (in part)

45a (36b). Lobes of spiracular plate (5) short, bluntly rounded; ventral lobes not fringed with long hairs; in some species, a densely sclerotized, hornlike projection near apex of median dorsal lobe and lateral lobes; found in sandy bottoms and margins of clear streams . . . . *Rhabdomastix* (in part)

45b. Lobes of spiracular plate (usually 4) not all short and bluntly rounded, ventral ones usually elongate; ventral lobes fringed with long hairs; no sclerotized, hornlike projections from upper lobes (Fig. 68) 46

46a (45b). Midventral region of head before line of attachment of skin entirely membranous, without darkened transverse bar just beneath surface; found in sand or gravel near margins of clear, cool brooks and streams . . . . *Hexatoma*

In *Hexatoma* especially, but also in some other genera in similar habitats, larvae may be found with the 7th abdominal segment much swollen (Fig. 68), possibly as an aid in locomotion or anchorage; this swelling may persist in preserved specimens.

46b. Midventral region of head before line of attachment of skin entirely membranous, with darkened, narrow transverse bar (part of hypopharynx) visible just beneath surface (Fig. 72); large genus with carnivorous, aquatic larvae found usually in organic mud in swampy woods, pond margins, less often in bottom mud or sand of small streams . . . . *Limnophila*

47a (35b). Internal portion of head divided by deep posterior incisions into elongate, slender, or spatulate sclerites (can be determined by cutting prothoracic skin at one side, or often can be seen through skin); abdominal segments 2–7 without dorsal and ventral creeping welts, spiracular plate lightly pigmented, vertically subrectangular, with 2 clawlike projections at ventral margin (Fig. 69); spiracles minute, pale, separated by about 3 times the diameter of a spiracle (Fig. 69); yellowish, aquatic, found in sandy bottoms and margins of clear streams . . . . *Rhabdomastix* (in part)

47b. Internal portion of head extensively sclerotized dorsally and laterally; sclerites platelike, with shallow posterior incisions (Fig. 70); abdominal segments 2–7 with both dorsal and ventral creeping welts (of differing structure in some species) on basal ring; spiracular plate may be darkly pigmented and is roughly circular or broadly oval to transversely subrectangular; spiracles often large, oval, inclined together dorsally (Fig. 81); large genus in numerous aquatic and terrestrial habitats . . . . *Limonia* (in part)

48a (5a). **Ptychopteridae:** Body pale; prolegs weakly developed (Fig. 82); mandibles with 3 outer teeth . . . . *Ptychoptera*

48b. Body rusty red or black; prolegs well developed (Fig. 83); mandibles with a single outer tooth 49

49a (48b). Body black; integument covered with long projections that are encased in a black, horny substance . . . . *Bittacomorphella*

49b. Body mostly red; integument covered with transverse rows of short tubercles (Fig. 83) . . . . *Bittacomorpha*

50a (7a). **Chaoboridae:** Abdominal segment 8 with dorsal respiratory tube (Fig. 11) . . . . *Mochlonyx*

50b. Abdominal segment 8 without respiratory tube (Figs. 10, 12) 51

51a (50b). Dark or black-spotted air sacs under integument (Fig. 10) and fat bodies in thorax and abdominal segment 7 . . . . *Chaoborus*

51b. Air sacs lacking (Fig. 12) . . . . *Eucorethra*

52a (7b). **Culicidae:** Abdominal segment 8 without a respiratory tube (Fig. 84), or if present, very short . . . . *Anopheles*

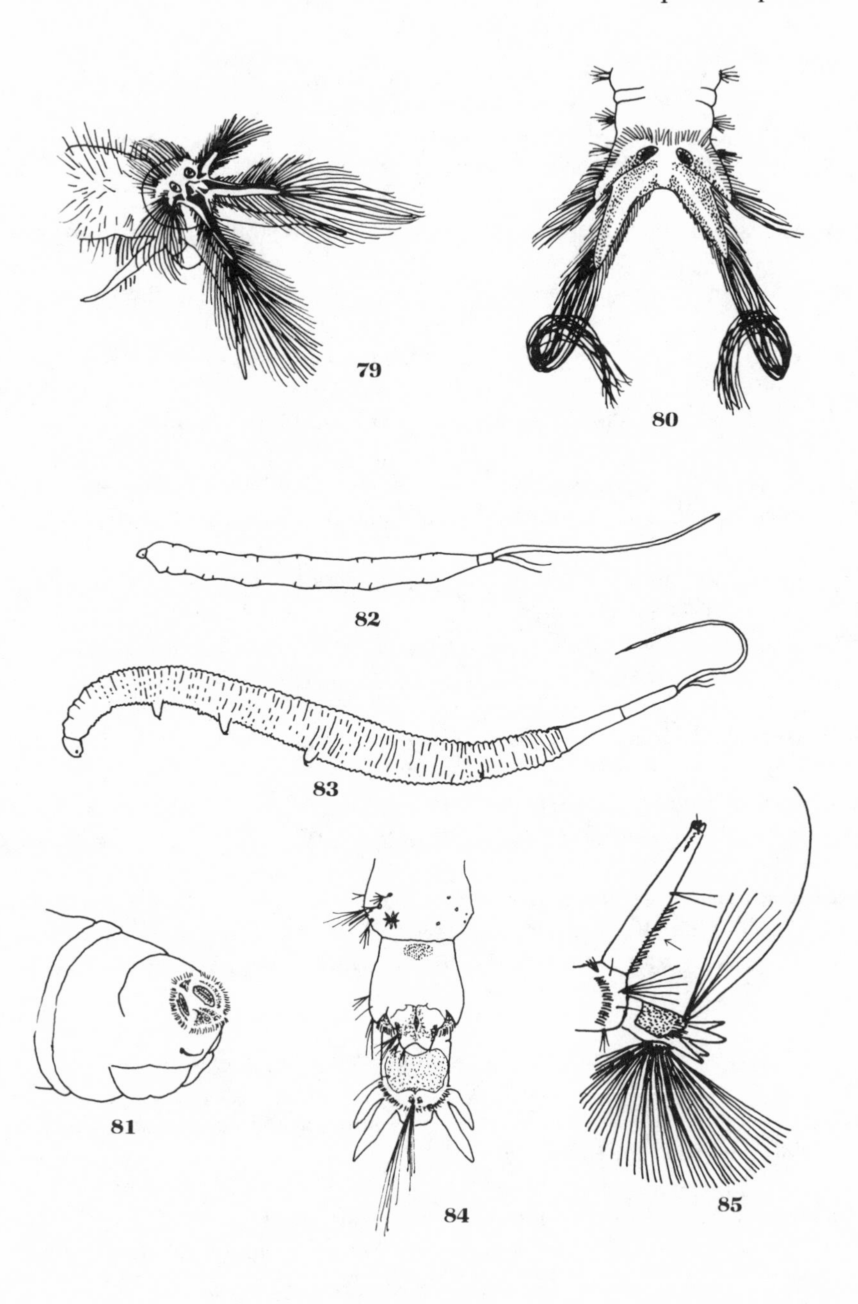

79. Caudal segments of *Pseudolimnophila* (Tipulidae), oplv. **80.** Spiracular plate of *Limnophila* (Tipulidae), dv. **81.** Caudal segment of *Limonia* (Tipulidae), opv. **82.** *Ptychoptera* (Ptychopteridae), lv. **83.** *Bittacomorpha* (Ptychopteridae), lv. **84.** Abdomen of *Anopheles* (Culicidae), dv. **85.** *Aedes* (Culicidae) respiratory tube and anal segment, lv. dv, dorsal view; lv, lateral view; oplv, oblique posterolateral view; opv, oblique posterior view. (79, 81 redrawn from Alexander and Byers 1981; 80 redrawn from Alexander 1920; 82, 83 modified from Webb and Brigham 1982; 84 redrawn from Stone 1981; 85 redrawn from Howard et al. 1912.)

52b. Abdominal segment 8 with a long respiratory tube (Figs. 85–87, 90–92) 53

53a (52b). Respiratory tube with a pecten (Fig. 85) 54
53b. Respiratory tube without a pecten (Figs. 86, 87) 59

54a (53a). Labral brush prehensile, composed of about 10 stout curved rods (Fig. 88) . . . . *Toxorhynchites*
54b. Labral brush not prehensile 55

55a (54b). Head longer than wide; upper and lower head hairs on dorsal side of head spinelike (Fig. 89) . . . . *Uranotaenia*
55b. Head wider than long; upper and lower head hairs not spinelike 56

56a (55b). Respiratory tube with 1 pair of large basoventral hair tufts (Fig. 90) . . *Culiseta*
56b. Respiratory tube without basoventral hair tufts (Figs. 85, 91) 57

57a (56b). Respiratory tube with more than 1 pair of ventral tufts (Fig. 91) or tufts interspersed with single long hairs . . . . *Culex*
57b. Respiratory tube with only a single pair of ventral tufts (Figs. 85, 92) or none (Fig. 86) 58

58a (57b). Ventral brush of anal segment with several tufts arising out of a sclerotized ring (Fig. 92) . . . . *Psorophora*
58b. Ventral brush of anal segment with all tufts posterior to sclerotized ring (Fig. 93), or sclerotized ring incomplete ventrally (Fig. 85) . . . . *Aedes*

59a (53b). Respiratory tube triangular and very short, may have teeth on margin near apex (Fig. 94); head wider than long . . . . *Mansonia*
59b. Respiratory tube conical and elongate, without teeth near apex; head as long as wide 60

60a (59b). Respiratory tube with many single hairs; anal segment without median ventral brush (Fig. 86) . . . . *Wyeomyia*
60b. Respiratory tube with a single pair of highly branched tufts; anal segment with median ventral brush (Fig. 87) . . . . *Orthopodomyia*

61a (8a). **Dixidae:** Dorsum of abdomen bare or nearly so (Fig. 95) . . . . *Dixella*
61b. Dorsum of abdomen with rosettes of hairs on segments 2–7 (Fig. 96) . . . *Dixa*

62a (10a). **Simuliidae:** Lateral margins of head capsule strongly convex (Fig. 97); labral fans absent (Fig. 97); anal sclerite Y-shaped (Fig. 98) . . . . *Twinnia*
62b. Lateral margins of head capsule only slightly convex, almost straight (Figs. 102, 105); labral fans present (Fig. 99); anal sclerite X-shaped (Fig. 100) or nearly rectangular (Fig. 101), or absent 63

63a (62b). Postocciput nearly complete dorsally and enclosing cervical sclerites (Fig. 102); basal 2 segments of antennae pale, strongly contrasting with darkly pig-

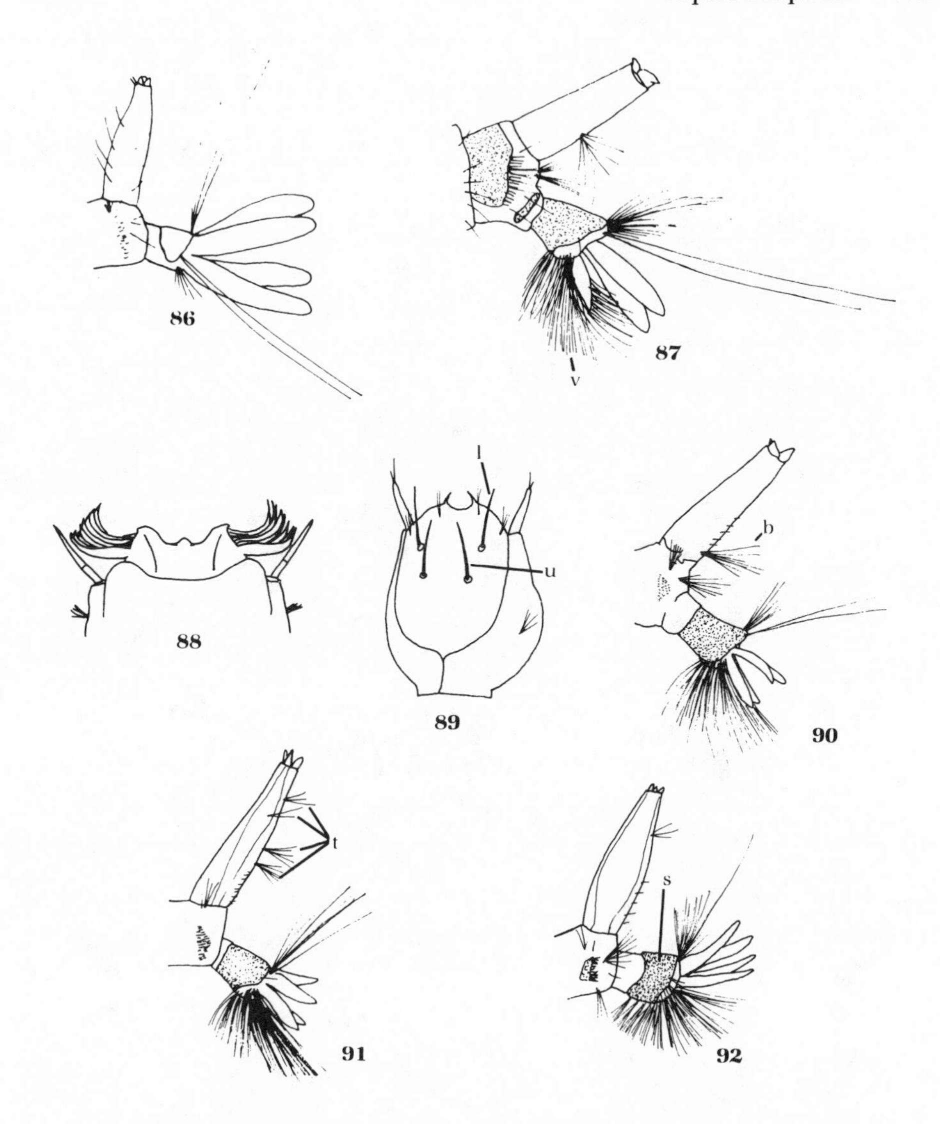

86. Caudal abdominal segments of *Wyeomyia* (Culicidae), lv. **87.** Caudal abdominal segments of *Orthopodomyia* (Culicidae), lv: v, ventral brush. **88.** Anterior part of head of *Toxorhynchites* (Culicidae), dv. **89.** Head of *Uranotaenia* (Culicidae), dv: l, lower head hairs; u, upper head hairs. **90.** Respiratory tube and anal segment of *Culiseta* (Culicidae), lv: b, basal tuft. **91.** Respiratory tube and anal segment of *Culex* (Culicidae), lv: t, tufts. **92.** Respiratory tube and anal segment of *Psorophora* (Culicidae), lv: s, sclerotized ring. dv, dorsal view; lv, lateral view. (86, 91, 92 redrawn from Howard et al. 1912; 87, 89, 90 redrawn from Ross and Horsfall 1965; 88 redrawn from Stone 1981.)

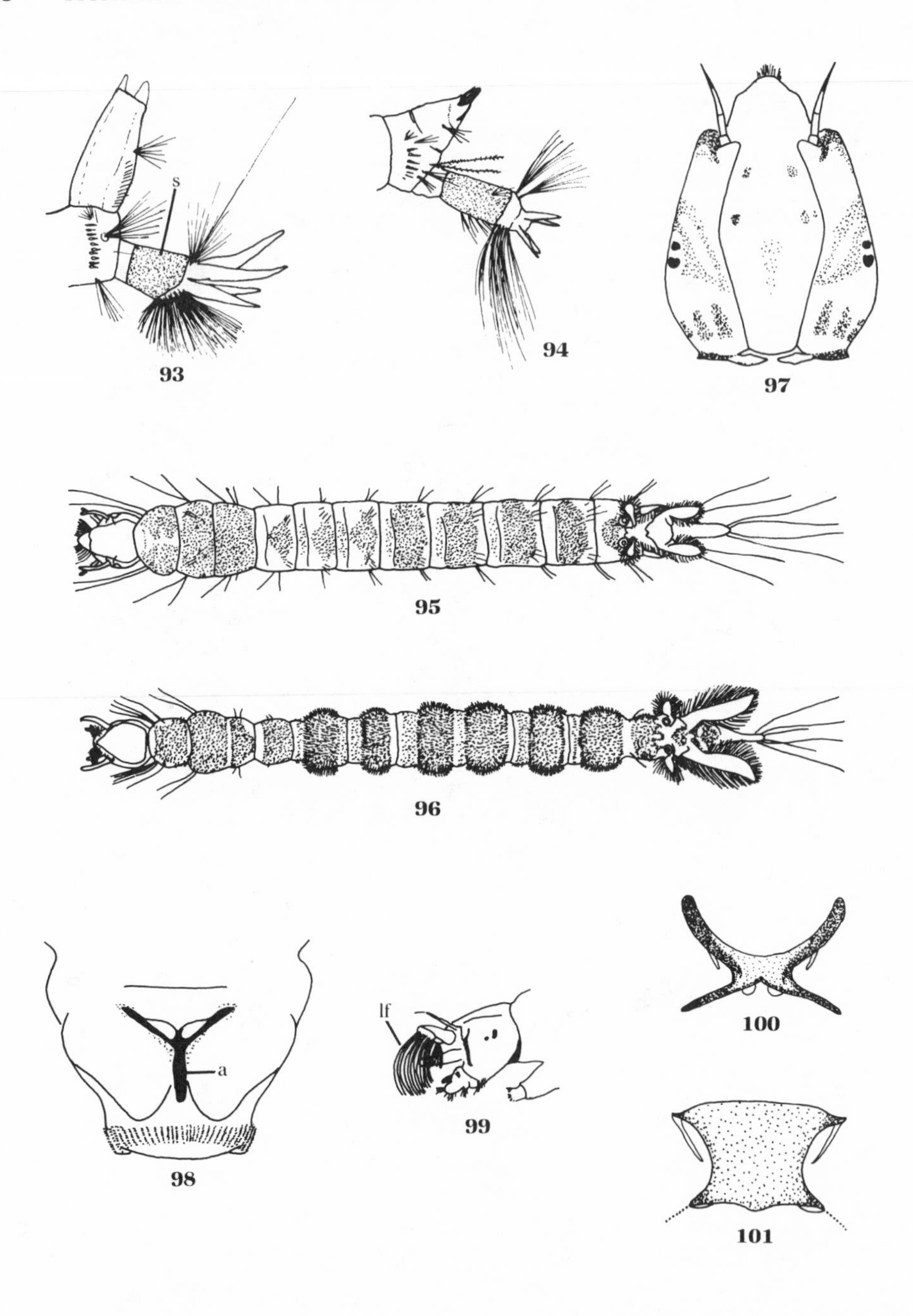

93. Respiratory tube and anal segment of *Aedes*, lv: s, sclerotized ring. **94.** Caudal segment of *Mansonia* (Culicidae), lv. **95.** *Dixella* (Dixidae), dv. **96.** *Dixa* (Dixidae), dv. **97.** Head of *Twinnia* (Simuliidae), dv. **98.** Posterior end of abdomen of *Gymnopais* (Simuliidae), dv: a, anal sclerite. **99.** Head of *Simulium* (Simuliidae), lv: lf, labral fan. **100.** Anal sclerite of *Prosimulium* (Simuliidae), dv. **101.** Anal sclerite of *Prosimulium* (Simuliidae), dv. dv, dorsal view; lv, lateral view. (93 redrawn from Howard et al. 1912; 94 redrawn from Stone 1981; 95, 96 modified from Nowell 1951; 97–99 redrawn from Peterson 1981; 100, 101 redrawn from Wood et al. 1963.)

mented distal segments (Fig. 102); median tooth of hypostomium (Fig. 103) distinctly trifid (Fig. 104) . . . . *Prosimulium*

63b. Postocciput with a distinct and usually wide gap dorsally, not enclosing cervical sclerites (Fig. 105); basal 2 segments of antennae at least partially pigmented yellow to brown, not strongly contrasting with distal segments; median tooth of hypostomium single (Fig. 106) 64

64a (63b). Hypostomium with median tooth and outer lateral (corner) teeth moderately large and subequal in height, with 3 smaller but nearly equal sublateral (intermediate) teeth between (Fig. 106); anal gills simple or compound lobes . . . . *Simulium*

64b. Hypostomium either with uniformly small teeth (Figs. 107, 108) or with teeth clustered in 3 prominent groups (Fig. 109); anal gills simple lobes 65

65a (64b). Postgenal cleft anteriorly reaching or extending beyond posterior margin of hypostomium (Fig. 110), cleft broad throughout, apex rather truncate; hypostomial teeth all very small . . . . *Metacnephia*

65b. Postgenal cleft not extending much more than half distance to posterior margin of hypostomium, often much less, usually an inverted U or V shape (Figs. 103, 111); hypostomial teeth variable but often large and distinct (Fig. 109) 66

66a (65b). Abdominal segment 8 with 2 ventral, cone-shaped tubercules (Fig. 112) . . . 67

66b. Abdominal segment 8 without 2 ventral, cone-shaped tubercles, but a single transverse bulge may be present (Fig. 113) 68

67a (66a). Abdomen abruptly and greatly expanded at segment 5, with its lateral margins projecting ventrally beyond central portion of segment (Fig. 112); anal sclerite absent; hypostomal teeth small and indistinct . . . . *Ectemnia*

67b. Abdomen of normal shape, not abruptly or greatly expanded at segment 5; anal sclerite present; lateral hypostomal teeth very large, longer than or subequal in length to median tooth (Fig. 114) . . . . *Greniera*

68a (66b). Postgenal cleft moderately deep, its anterior margin an inverted U shape (Fig. 111); hypostomial teeth uniformly small (Fig. 108); abdominal segment 8 simple, without a transverse, midventral bulge . . . . *Cnephia*

68b. Postgenal cleft narrow and shallow, its anterior margin an inverted V shape (Fig. 103); hypostomial teeth situated in 3 distinct groups (Fig. 109); abdominal segment 8 with a single, transverse, midventral bulge (Fig. 113) . . . . *Stegopterna*

69a (12a). **Ceratopogonidae** (in part): Body oval in cross section; lateral processes of body at least as long as segments (Fig. 23) . . . . *Atrichopogon*

69b. Body circular in cross section; lateral processes absent or less than half as long as segments (Fig. 115) . . . . *Forcipomyia*

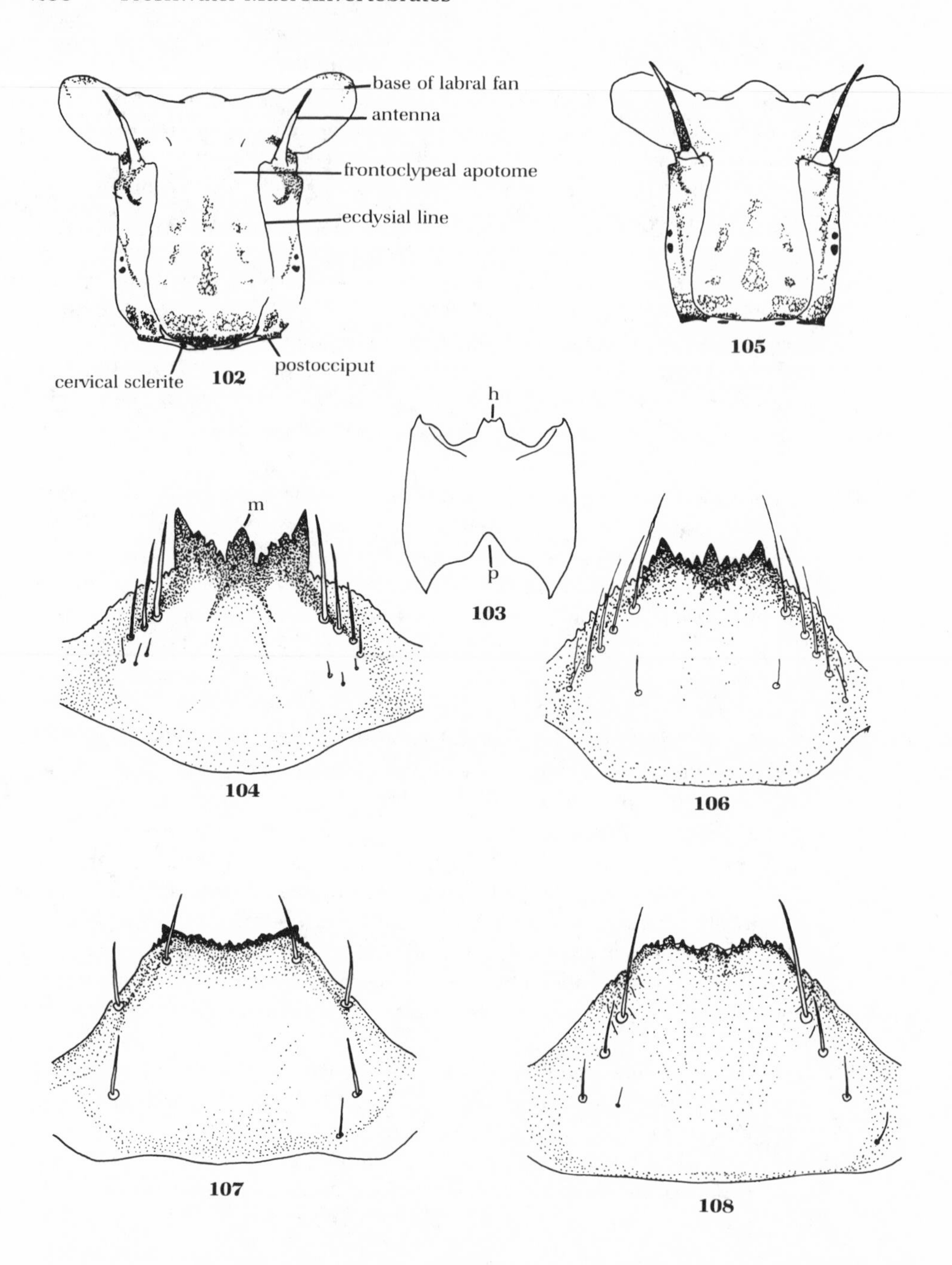

102. Head of *Prosimulium* (Simuliidae), dv. **103.** Head capsule of *Stegopterna* (Simuliidae), vv: h, hypostomium; p, postgenal cleft. **104.** Anterior portion of head capsule of *Prosimulium* (Simuliidae), vv: m, median tooth. **105.** Head of *Simulium* (Simuliidae), dv. **106.** Anterior of head capsule of *Simulium* (Simuliidae), vv. **107.** Anterior of head capsule of *Ectemnia* (Simuliidae), vv. **108.** Anterior of head capsule of *Cnephia* (Simuliidae), vv. dv, dorsal view; vv, ventral view. (102–108 redrawn from Peterson 1981.)

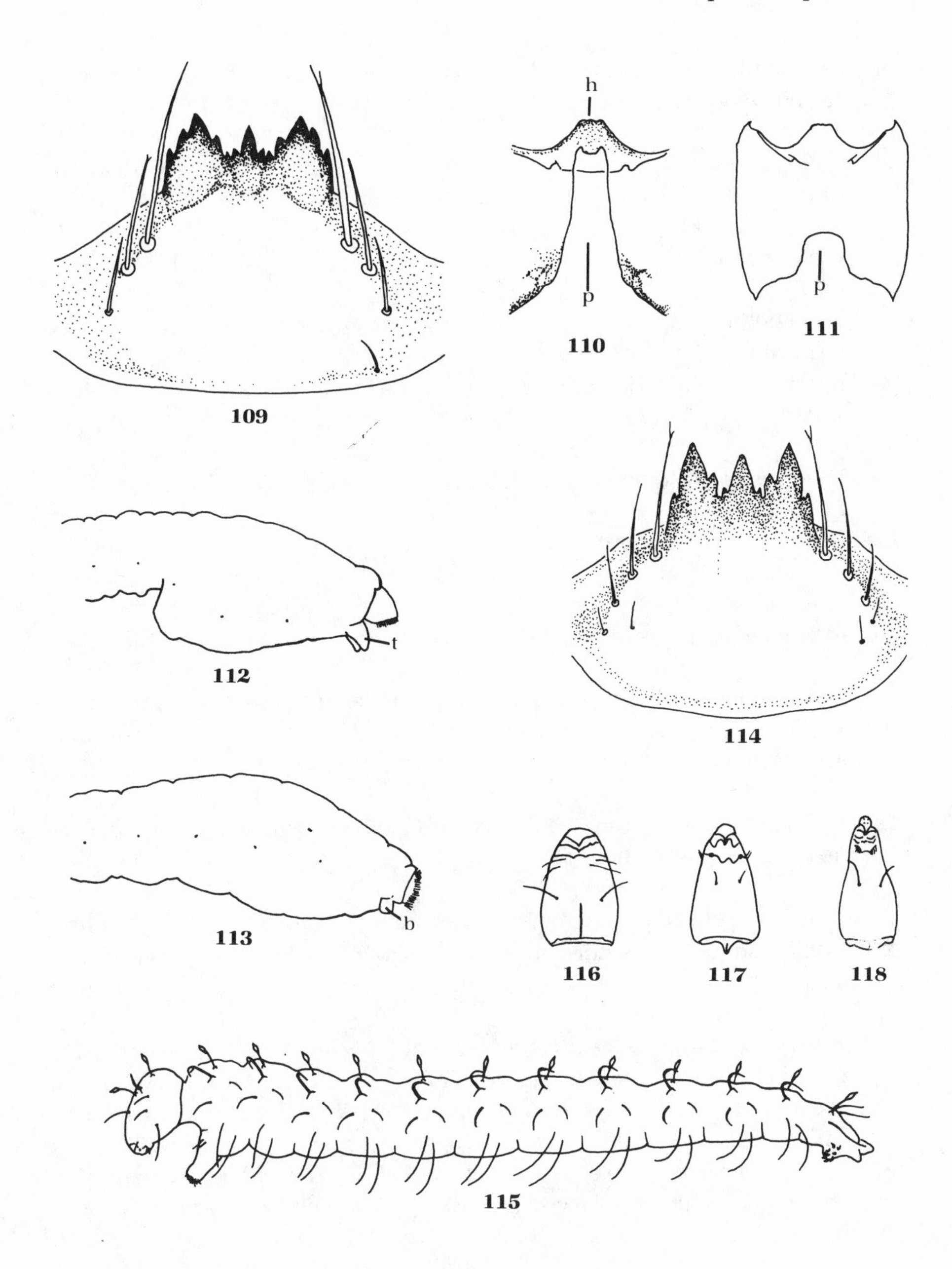

109. Anterior of head capsule of *Stegopterna* (Simuliidae), vv. **110.** Head capsule of *Metacnephia* (Simuliidae), vv: h, hypostomium; p, postgenal cleft. **111.** Head capsule of *Cnephia* (Simuliidae), vv: p, postgenal cleft. **112.** Abdomen of *Ectemnia* (Simuliidae), lv: t, tubercle. **113.** Abdomen of *Stegopterna* (Simuliidae), lv: b, transverse bulge. **114.** Anterior of head capsule of *Greniera* (Simuliidae), vv. **115.** *Forcipomyia* (Ceratopogonidae), lv. **116.** Head of *Ceratopogon* (Ceratopogonidae), vv. **117.** Head of *Sphaeromais* (Ceratopogonidae), vv. **118.** Head of *Probezzia* (Ceratopogonidae), vv. lv, lateral view; vv, ventral view. (109–114 redrawn from Peterson 1981; 115 modified from Downes and Wirth 1981; 116–118 redrawn from Hilsenhoff 1981.)

70a (14b). **Ceratopogonidae** (in part): Head capsule not sclerotized . . . . . . *Leptoconops*
70b. Head capsule sclerotized 71

71a (70b). Proleg present on last abdominal segment, may be partially or wholly withdrawn and evident only as terminal hooks on last segment (Fig. 24) . . . . *Dasyhelea*
71b. Proleg absent 72

72a (71b). Head not more than 2 times as long as wide and with posterior collar of uniform width ventrally (Fig. 116) 73
72b. Head more than 2 times as long as wide (Fig. 118), *or* posterior collar of head with a triangular or hemispherical expansion ventrally (Figs. 117, 118) 78

73a (72a). Head very small compared with body, and ⅓ length of anal segment (Fig. 119) . . . . *Serromyia*
73b. Head larger than above 74

74a (73b). Antennae obviously protruding . . . . *Stilobezzia*
74b. Antennae not protruding 75

75a (74b). Anal segment with long dark setae, equal to length of segment (Fig. 120) . . . . *Alluaudomyia*
75b. Anal segment with setae less than ½ length of segment 76

76a (75b). Anal segment with stout black setae, ⅓ length of segment . . . . *Monohelea*
76b. Anal segment with short pale setae 77

77a (76b). A long sclerotized suture on ventral side of head (Fig. 116) . . . . . *Ceratopogon*
77b. No long sclerotized suture on ventral side of head (Fig. 26 illustrates body shape) . . . . *Culicoides*

78a (72b). Four pairs of thick setae on anal segment ½–¾ length of segment (Fig. 25); posterior collar of head of uniform width ventrally, or, if ventral, triangular expansion present (*Palpomyia tibialis*), head less than 1.7 times as long as wide . . . . *Bezzia, Palpomyia*
78b. Four pairs of thick setae on anal segment about ⅓ length of segment; posterior collar of head with a ventral triangular or hemispherical expansion (Figs. 117, 118) 79

79a (78b). Head less than 1.8 times as long as wide; ventral collar of head with triangular expansion (Fig. 117) . . . . *Sphaeromias*
79b. Head more than 1.8 times as long as wide; ventral collar of head with hemispherical expansion (Fig. 118) 80

80a (79b). Head more than 2 times as long as wide, with anterior strongly narrowed (Fig. 118) . . . . *Probezzia*

80b. Head less than 2 times as long as wide, anterior not strongly narrowed . . . . . . *Mallochohelea*

81a (14a). **Psychodidae:** Twenty-six dorsal plates (Fig. 28); paired adanal plates and a single preanal plate (Fig. 121) . . . . *Pericoma, Telmatoscopus*

81b. Dorsal plates absent or fewer than 26; adanal plate single, transverse, preanal plates absent . . . . *Psychoda, Threticus*

82a (16a). **Stratiomyidae:** Last abdominal segment with a coronet of plumose or pinnate setae on dorsum . . . . *Nemotelus*

82b. Last abdominal segment with a coronet of plumose or pinnate setae at apex 83

83a (82b). Antennae dorsal, not at apex of ocular lobe (Fig. 122) 84

83b. Antennae at apex of ocular lobe (Fig. 123) 85

84a (83a) Ventral curved spines on posterior margin of abdominal segment 7 (may be concealed) . . . . *Caloparyphus, Euparyphus*

84b. Ventral curved spines absent from posterior margin of abdominal segment 7 . . . . *Oxycera*

85a (83b). Ventral curved spines on posterior margin of abdominal segment 7 (may be concealed) 86

85b. Ventral curved spines absent from posterior margin of abdominal segment 7 87

86a (85a). Ventral curved spines on posterior margins of abdominal segments 6 and 7 . . . . *Hedriodiscus* or *Odontomyia* (in part)

86b. Ventral curved spines absent from abdominal segment 6, may be present on 7 . . . . *Odontomyia* (in part)

87a (85b). Last abdominal segment less than 2 times as long as basal width (Fig. 124); integument covered with shield-shaped spines (Fig. 124) . . . . *Odontomyia* (in part)

87b. Last abdominal segment more than 2 times as long as basal width (Fig. 29); integument without shield-shaped spines . . . . *Stratiomys*

88a (18a). **Tabanidae:** Three pairs of prolegs (dorsal, lateral, and ventral) on each of 1st 7 abdominal segments (Fig. 31) 89

88b. Four pairs of prolegs (dorsal, lateral, ventral, and ventrolateral) on each of 1st 7 abdominal segments (Fig. 30) 91

89a (88a). Pubescence restricted to anterior or posterior margin or posterior border of prolegs of 1 or more segments . . . . *Chrysops*

89b. Body surface, except respiratory tube, completely clothed with a dense covering of short pubescence 90

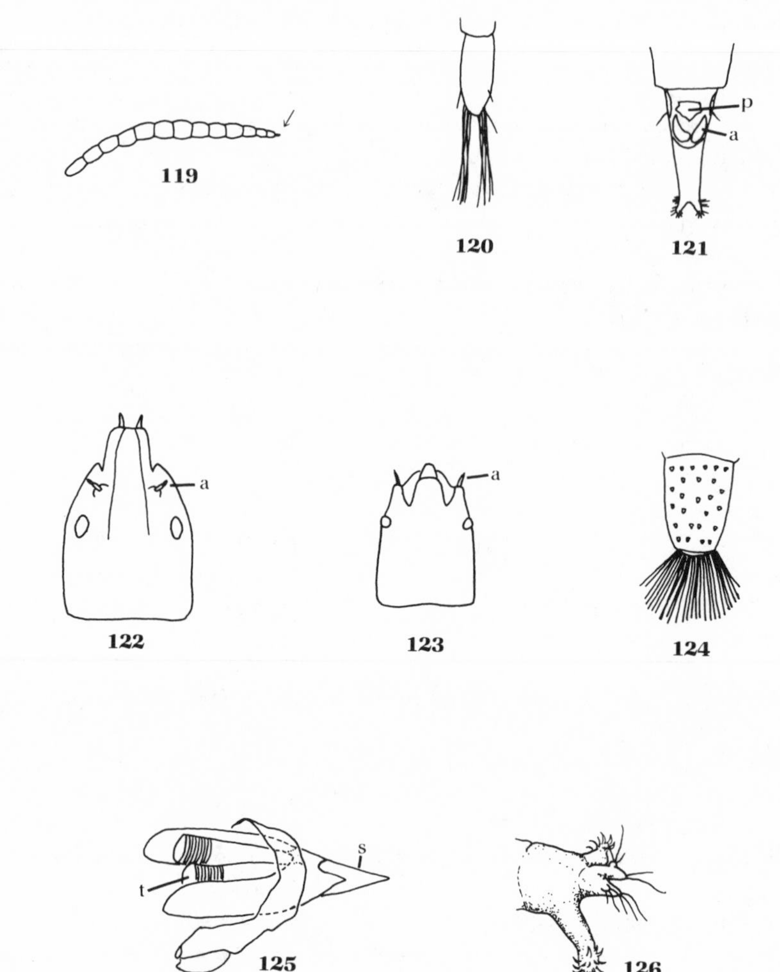

119. *Serromyia* (Ceratopogonidae), lv. **120.** Caudal segment of *Alluaudomyia* (Ceratopogonidae). **121.** Caudal abdominal segments of *Pericoma* (Psychodidae), vv: a, adanal plate; p, preanal plate. **122.** Head of *Euparyphus* (Stratiomyidae), dv: a, antenna. **123.** Head of *Odontomyia* (Stratiomyidae), dv: a, antenna. **124.** Caudal segment of *Odontomyia* (Stratiomyidae), dv. **125.** Dissected caudal segment of *Merycomyia* (Tabanidae), dlv: s, spine; t, tracheal trunk. **126.** Caudal segment of *Hemerodromia* (Empididae), plv. dlv, dorsolateral view; dv, dorsal view; plv, posterolateral view; vv, ventral view. (119, 121–124 modified or redrawn from Hilsenhoff 1981; 120 redrawn from Webb and Brigham 1982; 125 redrawn from Pechuman and Teskey 1981; 126 redrawn from Steyskal and Knutson 1981.)

90a (89b). Pubescence conspicuously mottled with dark and paler areas; 3rd antennal segment shorter than 2nd; respiratory tube equal to or only slightly longer than its basal width . . . . *Chlorotabanus*

90b. Pubescence not conspicuously mottled; 3rd antennal segment longer than 2nd; respiratory tube about 2 times as long as its basal width . . . . *Diachlorus*

91a (88b). Respiratory tube comprising distal ends of 2 opposed sclerotized plates between which tracheal trunks terminate in a spiracular spine (Fig. 125); incomplete, inconspicuous lateral striations present on abdominal segments . . . . *Merycomyia*

91b. Respiratory tube always membranous and lacking sclerotized plates, although may have tracheal trunks that sometimes terminate in a spiracular spine; striations present or absent on abdominal segments 92

92a (91b). Respiratory tube very short, projecting no more than ½ the length of its basal width, with extremely fine integumental striations on all aspects of body that are usually visible only under high magnification (striations spaced at ca. 5 μm) . . . . *Haematopota*

92b. If respiratory tube shorter than its basal diameter, then integumental striations spaced at usually > 20 μm 93

93a (92b). Median lateral surfaces of anal segment lacking pubescent markings; striations present on dorsal and ventral surfaces of all abdominal segments, or, if absent, pubescence restricted, at most, to a prothoracic annulus and vestiture on anal ridge . . . . *Hybomitra*

93b. Either median lateral surfaces of anal segment with pubescent markings, or striations absent or poorly developed on dorsal or ventral surface or both surfaces of abdominal segments . . . . *Atylotus, Tabanus*

94a (20b). **Empididae:** Usually apneustic (except posterior spiracles present in *Oreogeton* and *Roederiodes*); 7 or 8 pairs of abdominal prolegs present; caudal segment with elongate lobes (Fig. 46, 126) or rounded, with short tubercles (Fig. 45) ending in long setae, rarely ending in only a tuft of apical setae; aquatic or semiaquatic 95

94b. Amphipneustic; without abdominal prolegs; caudal segment rounded, with a single midventral bare lobe; usually terrestrial, found in dry or moist soil, leaf litter, rotting wood . . . . (not keyed)

95a (94a). Caudal segment rounded posteriorly, at most with small dorsal and apical tubercles; each tubercle with 1–3 pairs of long setae (Fig. 45); 7 pairs of abdominal prolegs . . . . *Chelifera*

95b. Caudal segment with prominent lobes (Figs. 46, 126); 7 or 8 pairs of abdominal prolegs 96

96a (95b). Caudal segment with a single more or less medially divided setose lobe (Fig. 126); 7 pairs of abdominal prolegs . . . . *Hemerodromia*

96b. Caudal segment with dorsal and apical lobes; 8 pairs of abdominal prolegs 97

97a (96b). Caudal segment with 2 dorsolateral lobes and with 1 more or less divided apical lobe (Fig. 46) .... *Clinocera* (subgenus *Hydrodromia*)
97b. Caudal segment with 2 pairs of lobes 98

98a (97b). Prolegs and caudal lobes rounded, very short .... *Dolichocephala*
98b. Prolegs and caudal lobes elongate, longer 99

99a (98b). Caudal segment with apical lobe short and with dorsolateral lobes long 100
99b. Caudal segment with lobes of equal length 101

100a (99a). Posterior spiracles present .... *Roederiodes*
100b. Posterior spiracles absent .... *Clinocera* (subgenus *Clinocera*)

101a (99b). Posterior spiracles present, on dorsolateral lobes; abdominal segments 1–7 each with a group of 6 setae on either side; last thoracic segment with a row of brushy protuberances dorsally .... *Oreogeton*
101b. Posterior spiracles absent; abdominal segments without setae laterally; last thoracic segment without brushy protuberances .... *Wiedemannia*

102a (22a). **Syrphidae:** Respiratory tube, when extended, about ½ length of body .... *Chrysogaster*
102b. Respiratory tube, when extended, much longer than body (Fig. 38) 103

103a (102b). Anterior half of longitudinal tracheal trunks straight .... *Eristalis*
103b. Anterior half of longitudinal tracheal trunks undulating .... *Helophilus*
Dissection may be necessary if integument is not transparent. See Gäbler 1932 for further details.

104a (23a). **Sciomyzidae:** Found in snail egg masses; integument with a thick coat of transparent spinules; palmate hairs surrounding posterior spiracles .... *Antichaeta*
104b. Never in snail egg masses; integument without a thick coat of transparent spinules 105

105a (104b). Body smooth; unpigmented or very lightly pigmented; with patches of spinules on creeping welts ventrally; posterior spiracles without palmate "float hairs"; usually found inside snail shells; never in the water and rarely seen free on the ground .... *Atrichomelina, Colobaea, Pherbellia, Pteromicra, Sciomyza*
105b. Body warty; lightly to very darkly pigmented; without a series of creeping welts ventrally (Fig. 41); posterior spiracles surrounded by palmate "float hairs" (Fig. 127); commonly found free and on the ground 106

106a (105b). Attenuate posteriorly; postanal portion of segment 12 (terminal segment) 2 or more times as long as wide; lightly pigmented .... *Elgiva*
106b. Less attenuate to blunt posteriorly; postanal portion of terminal segment (segment 12) about as long as wide; pigmentation variable 107

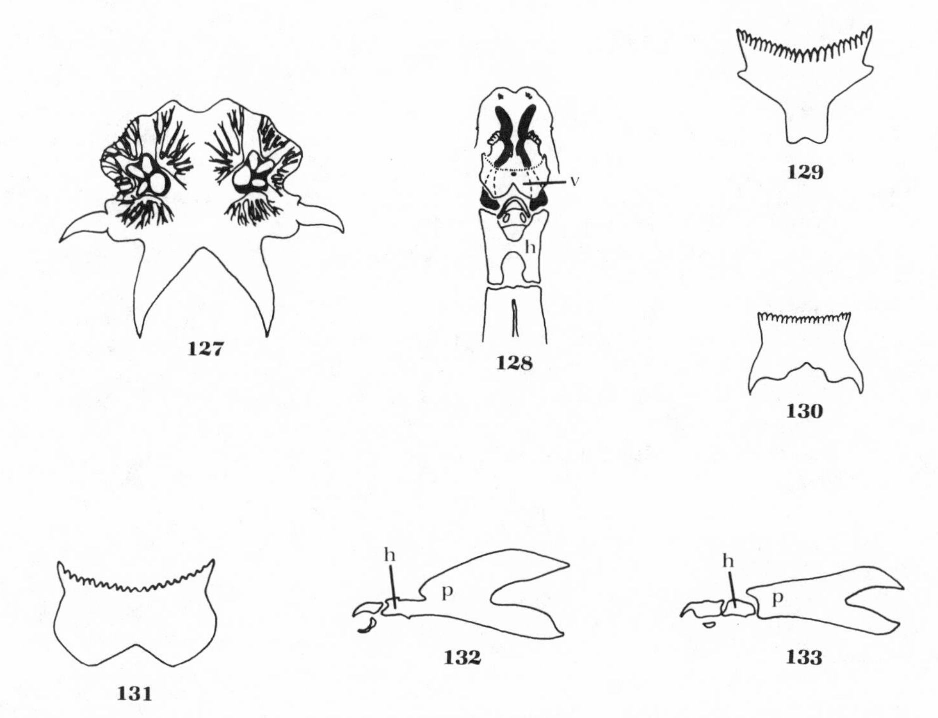

127. Spiracular plate of *Sepedon* (Sciomyzidae), pv. **128.** Cephalopharyngeal skeleton of *Sepedon* (Sciomyzidae), vv: h, hypopharyngeal sclerite; v, ventral arch. **129.** Ventral arch of *Dictya* (Sciomyzidae), vv. **130.** Ventral arch of *Sepedon* (Sciomyzidae), vv. **131.** Ventral arch of *Elgiva* (Sciomyzidae), vv. **132.** Sclerites of head of *Renocera* (Sciomyzidae), lv: h, hypostomal sclerite; p, pharyngeal sclerite. **133.** Sclerites of head of *Elgiva* (Sciomyzidae), lv: h, hypostomal sclerite; p, pharyngeal sclerite. lv, lateral view; pv, posterior view; vv, ventral view. (127 modified from Neff and Berg 1966; 128 modified from Knutson 1987; 129–133 modified or redrawn from Hilsenhoff 1981.)

107a (106b). Ventral arch (Fig. 128) subtriangular (Fig. 129); integument black, often iridescent; 8 lobes on spiracular plate . . . . *Dictya*

107b. Ventral arch bilobed (Figs. 128, 130, 131); pigmentation variable; 8 or 10 lobes on spiracular plate 108

108a (107b). Spiracular plate with 10 lobes (Fig. 127); lightly pigmented; ventral arch with posterolateral projections (Figs. 128, 130) . . . . *Sepedon*

108b. Spiracular plate with 8 lobes; darkly pigmented; ventral arch without posterolateral projections (Fig. 131) 109

109a (108b). Hypostomal and pharyngeal sclerites fused to form hypopharyngeal sclerite (Fig. 132) . . . . *Renocera*

109b. Hypostomal and pharyngeal sclerites separate (Fig. 133) 110

110a (109b). Lateral, ventrolateral, and ventral lobes of spiracular plate elongate, subequal . . . . *Hedria*

110b. Lateral lobes of spiracular plate much shorter than ventrolateral and ventral lobes, which are elongate or short . . . . *Tetanocera*

References on Aquatic Diptera Systematics

(*Used in construction of key.)

Aldrich, J.M. 1912. Flies of the leptid genus *Atherix* used as food by California Indians (Diptera). Entomol. News 23:159–163.

Alexander, C.P. 1919. The crane-flies of New York. Part I. Distribution and taxonomy of the adult flies. Mem. Cornell Univ. Agric. Exp. Stn. 25:767–993.

*——. 1920. The crane-flies of New York. Part II. Biology and phylogeny. Mem. Cornell Univ. Agric. Exp. Stn. 38:699–1133.

——. 1942. Guide to the insects of Connecticut. Part VI. The Diptera or true flies of Connecticut. Fasc. 1. Family Tanyderidae. Family Ptychopteridae. Family Trichoceridae. Family Anisopodidae. Family Tipulidae. Bull. Conn. State Geol. Nat. Hist. Surv. 64:183–486.

——. 1962. The crane flies of Maine. Tech. Ser. Bull. Univ. Maine Agric. Exp. Stn. T4:1–24.

——. 1963. Guide to the insects of Connecticut. Part VI. The Diptera or true flies of Connecticut. Fasc. 8. Family Blephariceridae. Family Deuterophlebiidae. Bull. Conn. State Geol. Nat. Hist. Surv. 93:39–83.

*——. 1981a. Tanyderidae. *In* J.F. McAlpine, B.V. Peterson, G.E. Shewell, H.J. Teskey, J.R. Vockeroth, and D.M. Wood (eds.). Manual of Nearctic Diptera, vol. 1, pp. 149–152. Res. Branch Agric. Can. Monogr. no. 27. Ottawa.

*——. 1981b. Ptychopteridae. *In* J.F. McAlpine, B.V. Peterson, G.E. Shewell, H.J. Teskey, J.R. Vockeroth, and D.M. Wood (eds.). Manual of Nearctic Diptera, vol. 1, pp. 325–328. Res. Branch Agric. Can. Monogr. no. 27. Ottawa.

*Alexander, C.P., and G.W. Byers. 1981. Tipulidae. *In* J.F. McAlpine, B.V. Peterson, G.E. Shewell, H.J. Teskey, J.R. Vockeroth, and D.M. Wood (eds.). Manual of Nearctic Diptera, vol. 1, pp. 153–190. Res. Branch Agric. Can. Monogr. no. 27. Ottawa.

Andreyeva, R.V. 1982. On ecologo-morphological typing of tabanid larvae (Diptera, Tabanidae). Entomol. Rev. 61:48–54.

Atchley, W.R., W.W. Wirth, C.T. Gaskins, and S.L. Strauss. 1981. A bibliography and keyword index of the biting midges (Diptera: Ceratopogonidae). U.S. Dep. Agric. Sci. Educ. Admin., Bibliogr. Lit. Agric. no. 13. Washington, D.C. 544 pp.

Barr, A.R. 1986. Bases for mosquito systematics. J. Amer. Mosq. Control Assoc. 2:261–266.

Bates, M. 1949. The natural history of mosquitoes. Macmillan, New York. 379 pp.

Bauer, L.S., and J. Granett. 1979. The black flies of Maine. Tech. Bull. Maine Life Sci. Agric. Exp. Stn. 95:1–18.

Baust, J.G., and R.R. Rojas. 1985. Review—insect cold hardiness: facts and fancy. J. Insect Physiol. 31:755–760.

Beck, W., Jr. 1977. Environmental requirements and pollution tolerance of common freshwater Chironomidae. US EPA–600/4–77–024. 270 pp.

Belton, P. 1983. The mosquitoes of British Columbia. B.C. Prov. Mus. Handb. no. 41. Victoria. 189 pp.

Berg, C.O., and L. Knutson. 1978. Biology and systematics of the Sciomyzidae. Annu. Rev. Entomol. 23:239–258.

Boreham, M.M. 1981. A new method for sampling mangrove mud for *Culicoides* larvae and pupae, with notes on factors affecting its use. Mosq. News 41:1–6.

Brenner, R.J., and E.W. Cupp. 1980. Rearing black flies (Diptera: Simuliidae) in a closed system of water circulation. Tropenmed. Parasitol. 31:247–258.

Burger, J.F., D.J. Lake, and M.L. McKay. 1981. The larval habitats and rearing of some common *Chrysops* species (Diptera: Tabanidae) in New Hampshire. Proc. Entomol. Soc. Wash. 83:373–389.

Butler, M.G. 1984. Life histories of aquatic insects. *In* V.H. Resh and D.M. Rosenberg (eds.). The ecology of aquatic insects, pp. 24–55. Praeger, New York.

*Byers, G.W. 1984. Tipulidae. *In* R.W. Merritt and K.W. Cummins (eds.). An introduction to the aquatic insects of North America, 2nd ed., pp. 491–514. Kendall/Hunt Publ. Co., Dubuque, Iowa.

Carpenter, S.J. 1968. Review of recent literature on mosquitoes of North America. Calif. Vector Views 15:71–98.

Carpenter, S.J., and W.J. LaCasse. 1955. Mosquitoes of North America (north of Mexico). Univ. Calif. Press, Berkeley. 360 pp.

Carter, J.C., M.J. Dadswell, J.C. Roff, and W.G. Sprules. 1980. Distribution and zoogeography of planktonic crustaceans and dipterans in glaciated eastern North America. Can. J. Zool. 58:1355–1387.

Catalan, Z.B., and D.S. White. 1982. Creating and maintaining cultures of *Chironomus tentans* (Diptera: Chironomidae). Entomol. News 93:54–58.

Cloudsley-Thompson, J.L. 1976. Evolutionary trends in the mating of Arthropoda. Meadowfield Press, Durham, England. 86 pp.

Cook, E.F. 1956. The Nearctic Chaoborinae (Diptera: Culicidae). Tech. Bull. Univ. Minn. Agric. Exp. Stn. 218:1–102.

*——. 1981. Chaoboridae. *In* J.F. McAlpine, B.V. Peterson, G.E. Shewell, H.J. Teskey, J.R. Vockeroth, and D.M. Wood (eds.). Manual of Nearctic Diptera, vol. 1, pp. 335–340. Res. Branch Agric. Can. Monogr. no. 27. Ottawa.

Craig, D.A., and A. Borkent. 1980. Intra- and inter-familial homologies of maxillary palpal sensilla of larval Simuliidae (Diptera: Culicomorpha). Can. J. Zool. 58:2264–2279.

Crampton, G.C. 1942. Guide to the insects of Connecticut. Part VI. The Diptera or true flies of Connecticut. Fasc. 1. The external morphology of the Diptera. Bull. Conn. State Geol. Nat. Hist. Surv. 64:10–165.

Cresson, E.T., Jr. 1920. A revision of the Nearctic Sciomyzidae (Diptera, Acalyptratae). Trans. Amer. Entomol. Soc. 46:27–89.

——. 1942. Synopses of North American Ephydridae. I. Psilopinae. Trans. Amer. Entomol. Soc. 68:101–128.

——. 1944. Synopses of North American Ephydridae. II. Hydrellini, Hydrinini, and Ilytheini. Trans. Amer. Entomol. Soc. 70:159–180.

——. 1946. Synopses of North American Ephydridae. III. Notiphilini. Trans. Amer. Entomol. Soc. 72:227–240.

——. 1949. Synopses of North American Ephydridae. IV. Napaeinae. Trans. Amer. Entomol. Soc. 74:225–260.

Crosskey, R.W. 1985. The authorship, dating and application of suprageneric names in Simuliidae (Diptera). Entomol. Month. Mag. 121:167–178.

Cupp, E.W., and A.E. Gordon (eds.). 1983. Notes on the systematics, distribution, and bionomics of black flies (Diptera: Simuliidae) in the northeastern United States. Search: Agric. 25:1–75.

Curran, C.H. 1942. Guide to the insects of Connecticut. Part VI. The Diptera or true flies of Connecticut. Fasc. 1. Key to families. Bull. Conn. State Geol. Nat. Hist. Surv. 64:175–182.

——. 1965. The families and genera of North American Diptera, 2nd ed. Henry Tripp, Woodhaven, N.Y. 515 pp.

Currie, D.C. 1986. An annotated list of and keys to the immature black flies of Alberta (Diptera: Simuliidae). Mem. Entomol. Soc. Can. 134:1–90.

Darsie, F.R., and R.A. Ward. 1981. Identification and geographical distribution of the mosquitoes of North America north of Mexico. Mosq. Syst. Suppl. 1:1–313.

Darsie, R.F., Jr. 1951. Pupae of the culicine mosquitoes of the northeastern United States (Diptera, Culicidae, Culicini). Mem. Cornell Univ. Agric. Exp. Stn. 304:1–67.

Deonier, D.L. (ed.). 1979. First symposium on the systematics and ecology of Ephydridae (Diptera), April 20, 1979, Erie, Penn. N. Amer. Benthol. Soc. 147 pp.

Dow, M.I., and E.C. Turner, Jr. 1976. A revision of the Nearctic species of the genus *Bezzia* (Diptera: Ceratopogonidae). Va. Polytech. Inst. Res. Div. Bull. 103:1–162.

*Downes, J.A., and W.W. Wirth. 1981. Ceratopogonidae. *In* J.F. McAlpine, B.V. Peterson, G.E. Shewell, H.J. Teskey, J.R. Vockeroth, and D.M. Wood (eds.). Manual of Nearctic Diptera, vol. 1, pp. 393–422. Res. Branch Agric. Can. Monogr. no. 27. Ottawa.

Drees, B.M., L. Butler, and L.L. Pechuman. 1980. Horse flies and deer flies of West Virginia: an illustrated key (Diptera: Tabanidae). W. Va. Agric. For. Exp. Stn. 674:1–67.

Dyar, H.G. 1906. Key to the known larvae of the mosquitoes of the United States. U.S. Dep. Agric. Entomol. Circ. 72:1–6.

Endris, R.G., P.V. Perkins, D.G. Young, and R.N. Johnson. 1982. Techniques for laboratory rearing of sand flies (Diptera: Psychodidae). Mosq. News 42:400–407.

Exner, K., and O.A. Craig. 1976. Larvae of Alberta Tanyderidae (Diptera: Nematocera). Quaest. Entomol. 12:219–237.

Fairchild, G.B. 1950. Guide to the insects of Connecticut. Part VI. The Diptera or true flies of Connecticut. Fasc. 4. Family Tabanidae. Bull. Conn. State Geol. Nat. Hist. Surv. 75:3–31.

Ferrar, P. 1987. A guide to the breeding habits and immature stages of Diptera Cyclorrhapha. Parts 1 and 2. Entomonograph 8:1–907.

Fisher, T.W., and R.E. Orth. 1983. The marsh flies of California (Diptera: Sciomyzidae). Bull. Calif. Insect Surv. 24:1–117.

Foot, R.H., and H.D. Pratt. 1954. The *Culicoides* of the eastern United States. Publ. Health Monogr. 18:1–53.

Freeman, J.V. 1987. Immature stages of *Tabanus conterminus* and keys to larvae and pupae of common Tabanidae from United States east coast salt marshes. Ann. Entomol. Soc. Amer. 80:613–623.

Gäbler, H. 1932. Beitrag zur Kenntnis dur *Eristalis*-Larven. Stett. Entomol. Z. 93:143–147.

*Gerbig, F. 1913. Über Tipuliden-Larven mit besonderer Berücksichtigung der Respirationsorgane. Zool. Jahrb. Abt. Syst. 35:127–184.

Golini, V.I. 1981. A simple technique for rearing pupae of Simuliidae and other Diptera. Entomol. Scand. 12:426–428.

Grogan, W.L., Jr., and W.W. Wirth. 1975. A revision of the genus *Palpomyia* Meigen of northeastern North America (Diptera: Ceratopogonidae). Md. Agric. Exp. Stn. Contrib. 5076:1–49.

Harbach, R.E., and K.L. Knight. 1981. Corrections and additions to "The taxonomist's glossary of mosquito anatomy." Mosq. Syst. 13:201–217.

Harper, P.P. 1980. Phenology and distribution of aquatic dance flies (Diptera: Empididae) in a Laurentian watershed. Amer. Midl. Nat. 104:110–117.

Harrison, B.A., and E.L. Peyton. 1984. The value of the pupal stage to anopheline taxonomy, with notes on anamalous setae (Diptera: Culicidae). Mosq. Syst. 16:201–210.

Harrison, J.R., R. Loiselle, and D.J. Leprince. 1980. Inventaire des moustiques (Diptera: Culicidae) de sud du Québec 1973–1978. Ann. Entomol. Soc. Québ. 25:195–206.

Hatfield, L.D., J.L. Riner, and B.R. Norment. 1985. A dredge sampler for mosquito larvae. J. Amer. Mosq. Control Assoc. 1:372–373.

Hays, K.L. 1956. A synopsis of the Tabanidae (Diptera) of Michigan. Misc. Publ. Univ. Mich. Mus. Zool. 98:1–71.

Heath, B.L., and W.P. McCafferty. 1975. Aquatic and semiaquatic Diptera of Indiana. Purdue Univ. Agric. Exp. Stn. Res. Bull. 930:1–17.

*Hennig, W. 1950. Die Larvenformen der Dipteren, 2 Teil. Akademie-Verlag. Berlin. 458 pp.

*Hilsenhoff, W.L. 1975. Aquatic insects of Wisconsin, with generic keys and notes on biology, ecology and distribution. Tech. Bull. Wisc. Dep. Nat. Resour. 89:1–53.

*——. 1981. Aquatic insects of Wisconsin. Nat. Hist. Counc. Publ. no. 2. Univ. Wisc., Madison. 60 pp.

Hogue, C.L. 1978. The net-winged midges of eastern North America, with notes on new taxonomic characters in the family Blephariceridae (Diptera). Contrib. Sci. 219:1–41.

*——. 1981. Blephariceridae. *In* J.F. McAlpine, B.V. Peterson, G.E. Shewell, H.J. Teskey, J.R. Vockeroth, and D.M. Wood (eds.). Manual of Nearctic Diptera, vol. 1, pp. 191–198. Res. Branch Agric. Can. Monogr. no. 27. Ottawa.

*Howard, L.O., H.G. Dyar, and F. Knab. 1912. The mosquitoes of North and Central America and the West Indies. Publ. Carnegie Inst. no. 159(II):1–150.

*Huckett, H.C., and J.R. Vockeroth. 1987. Muscidae. *In* J.F. McAlpine, B.V. Peterson, G.E. Shewell, H.J. Teskey, J.R. Vockeroth, and D.M. Wood (eds.). Manual of Nearctic Diptera, vol. 2, pp. 1115–1132. Res. Branch Agric. Can. Monogr. no. 28. Ottawa.

*Hynes, C.D. 1963. Description of the immature stages of *Cryptolabis magnistyla* Alexander (Diptera: Tipulidae). Pan-Pac. Entomol. 39:255–260.

*——. 1969. The immature stages of the genus *Rhabdomastix* (Diptera: Tipulidae). Pan-Pac. Entomol. 45:229–237.

James, M. 1959. Diptera. *In* W.T. Edmondson (ed.). Fresh-water biology, 2nd ed., pp. 1057–1079. John Wiley and Sons, New York.

*——. 1981. Stratiomyidae. *In* J.F. McAlpine, B.V. Peterson, G.E. Shewell, H.J. Teskey, J.R. Vockeroth, and D.M. Wood (eds.). Manual of Nearctic Diptera, vol. 1, pp. 497–512. Res. Branch Agric. Can. Monogr. no. 27. Ottawa.

James, M.T., and G.C. Steyskal. 1952. A review of the Nearctic Stratiomyini (Diptera, Stratiomyidae). Ann. Entomol. Soc. Amer. 45:385–412.

Jamnback, H. 1965. The *Culicoides* of New York State (Diptera: Ceratopogonidae). Bull. N.Y. State Mus. 399:1–154.

Johannsen, O.A. 1923. North American Dixidae. Psyche 30:52–58.

*——. 1934. Aquatic Diptera. Part I. Nematocera, exclusive of Chironomidae and Ceratopogonidae. Mem. Cornell Univ. Agric. Exp. Stn. 164:1–71.

*——. 1935. Aquatic Diptera. Part II. Orthorrhapha-Brachycera and Cyclorrhapha. Mem. Cornell Univ. Agric. Exp. Stn. 177:1–62.

*Jones, F.M. 1918. *Dohrniphora venusta* Coquillett (Diptera) in *Serracenia flava*. Entomol. News 29:299–303.

Kellogg, V.L. 1903. The net-winged midges (Blephariceridae) of North America. Proc. Calif. Acad. Sci. Zool. Ser. 3:187–232.

*Kevan, D.K.M., and F.E.A. Cutten. 1981. Nymphomyiidae. *In* J.F. McAlpine, B.V. Peterson, G.E. Shewell, H.J. Teskey, J.R. Vockeroth, and D.M. Wood (eds.). Manual of Nearctic Diptera, vol. 1, pp. 203–208. Res. Branch Agric. Can. Monogr. no. 27. Ottawa.

Knausenberger, W.I., and E.C. Turner. 1977. Techniques for associating developmental stages of Ceratopogonidae and other Diptera. J. N.Y. Entomol. Soc. 85:183.

*Knutson, L.V. 1987. Sciomyzidae. *In* J.F. McAlpine, B.V. Peterson, G.E. Shewell, H.J. Teskey, J.R. Vockeroth, and D.M. Wood (eds.). Manual of Nearctic Diptera, vol. 2, pp. 927–940. Res. Branch Agric. Can. Monogr. no. 28. Ottawa.

Leonard, M.D. 1930. A revision of the dipterous family Rhagionidae (Leptidae) in the United States. Mem. Amer. Entomol. Soc. 7:1–181.

Loew, H., and R. Osten-Sacken. 1862. Monographs of the Diptera of North America. Part 1. Smithson. Misc. Coll. 6:1–221.

Mackey, A.P., and H.M. Brown. 1980. The pupae of *Atherix ibis* (F.) and *A. marginata* (F.) (Diptera: Rhagionidae). Entomol. Gaz. 31:157–161.

Malloch, J.R. 1917. A preliminary classification of Diptera, exclusive of Pupipara, based upon larval and pupal characters, with keys to imagines in certain families. Part I. Ill. State Lab. Nat. Hist. 12:161–409.

Matheson, R. 1944. Handbook of the mosquitoes of North America, 2nd ed. Comstock Publ. Co., Ithaca, N.Y. 314 pp.

——. 1945. Guide to the insects of Connecticut. Part VI. The Diptera or true flies of Connecticut. Fasc. 2. Family Culicidae. Bull. Conn. State Geol. Nat. Hist. Surv. 68:2–48.

Mathis, W.H. 1982a. Studies of Ephydrinae (Diptera: Ephydridae), VI. Review of the tribe Dagini. Smithson. Contrib. Zool. 345:1–30.

——. 1982b. Studies of the Ephydrinae (Diptera: Ephydridae), VII. Revision of the genus *Setacera* Cresson. Smithson. Contrib. Zool. 350:1–57.

——. 1986. Studies of Psilopinae (Diptera: Ephydridae), I. A revision of the shore fly genus *Placopsidella* Kertész. Smithson. Contrib. Zool. 430:1–30.

Mathis, W.N., and F.C. Thompson (eds.). 1982. Recent advances in Diptera systematics. Mem. Entomol. Soc. Wash. 10:1–227.

*McAlpine, J.F., B.V. Peterson, G.E. Shewell, H.J. Teskey, J.R. Vockeroth, and D.M. Wood (eds.). 1981. Manual of Nearctic Diptera, vol. 1. Res. Branch Agric. Can. Monogr. no. 27. Ottawa. 674 pp.

*——. 1987. Manual of Nearctic Diptera, vol. 2. Res. Branch Agric. Can. Monogr. no. 28. Ottawa. 658 pp.

McFadden, M.W. 1967. Soldier fly larvae in America north of Mexico. Proc. U.S. Natl. Mus. 121:1–72.

Means, R.G. 1979. Mosquitoes of New York. Part I. The genus *Aedes* Meigen with identification keys to the genera of Culicidae. Bull. N.Y. State Mus. 430:1–221.

*Merritt, R.W., and K.W. Cummins (eds.). 1984. An introduction to the aquatic insects of North America, 2nd ed. Kendall/Hunt Publ. Co., Dubuque, Iowa. 722 pp.

Merritt, R.W., D.H. Ross, and B.V. Peterson. 1978. Larval ecology of some lower Michigan black flies (Diptera: Simuliidae) with keys to the immature stages. Great Lakes Entomol. 11:177–208.

Merritt, R.W., and E.I. Schlinger. 1984. Aquatic Diptera. Part II. Adults of aquatic Diptera. *In* R.W. Merritt and K.W. Cummins (eds.). An introduction to the aquatic insects of North America, 2nd ed. pp. 467–490. Kendall/Hunt Publ. Co., Dubuque, Iowa.

Metcalf, C.L. 1932. Black flies and other biting flies of the Adirondacks. Bull. N.Y. State Mus. 289:1–78.

Molloy, D.P. 1988. Guide to recent literature on black flies (Diptera: Simuliidae). Bull. Soc. Vector Ecol. 13:126–220.

Nagatoni, A. 1982. Geographical distribution of the lower Brachycera (Diptera). Pac. Insects 24:139–150.

*Neff, S.E., and C.O. Berg. 1966. Biology and immature states of malacophagous Diptera of the genus *Sepedon* (Sciomyzidae). Va. Agric. Exp. Stn. Bull. 566:1–113.

*Newson, H.D. 1984. Culicidae. *In* R.W. Merritt and K.W. Cummins (eds.). An introduction to the aquatic insects of North America, 2nd ed., pp. 515–533. Kendall/Hunt Publ. Co., Dubuque, Iowa.

Nielsen, L.T. 1980. The current status of mosquito systematics. Mosq. Syst. 12:1–6.

*Nowell, W.R. 1951. The dipterous family Dixidae in western North America (Insecta: Diptera). Microentomology 16:187–270.

——. 1963. Guide to the insects of Connecticut. Part VI. The Diptera or true flies of Connecticut. Fasc. 8. Family Dixidae. Bull. Conn. State Geol. Nat. Hist. Surv. 93:85–111.

Oldroyd, H. 1964. The natural history of flies. Weidenfeld and Nicolson, London. 324 pp.

*Oliver, D.R. 1981. Chironomidae. *In* J.F. McAlpine, B.V. Peterson, G.E. Shewell, H.J. Teskey, J.R. Vockeroth, and D.M. Wood (eds.). Manual of Nearctic Diptera, vol. 1, pp. 423–458. Res. Branch Agric. Can. Monogr. no. 27. Ottawa.

Orth, R.E., and L. Knutson. 1987. Systematics of snail-killing flies of the genus *Elgiva* in North America and biology of *E. divisa* (Diptera: Sciomyzidae). Ann. Entomol. Soc. Amer. 80:829–840.

Pechuman, L.L. 1972. The horse flies and deer flies of New York (Diptera: Tabanidae). Search: Agric. Entomol. (Ithaca) 2(5):1–72.

——. 1981. The horse flies and deer flies of New York (Diptera: Tabanidae), 2nd ed. Search: Agric. 18:1–68.

*Pechuman, L.L., and H.J. Teskey. 1981. Tabanidae. *In* J.F. McAlpine, B.V. Peterson, G.E. Shewell, H.J. Teskey, J.R. Vockeroth, and D.M. Wood (eds.). Manual of Nearctic Diptera, vol. 1, pp. 463–478. Res. Branch Agric. Can. Monogr. no. 27. Ottawa.

*Pennak, R.W. 1978. Freshwater invertebrates of the United States, 2nd ed. John Wiley and Sons, New York. 803 pp.

*Peters, T.M. 1981. Dixidae. *In* J.F. McAlpine, B.V. Peterson, G.E. Shewell, H.J. Teskey, J.R. Vockeroth, and D.M. Wood (eds.). Manual of Nearctic Diptera, vol. 1, pp. 329–334. Res. Branch Agric. Can. Monogr. no. 27. Ottawa.

Peters, T.M., and E.F. Cook. 1966. The Nearctic Dixidae (Diptera). Misc. Publ. Entomol. Soc. Amer. 5:233–278.

Peterson, B.V. 1970. The *Prosimulium* of Canada and Alaska (Diptera: Simuliidae). Mem. Entomol. Soc. Can. 69:1–216.

*——. 1981. Simuliidae. *In* J.F. McAlpine, B.V. Peterson, G.E. Shewell, H.J. Teskey, J.R. Vockeroth, and D.M. Wood (eds.). Manual of Nearctic Diptera, vol. 1, pp. 355–392. Res. Branch Agric. Can. Monogr. no. 27. Ottawa.

——. 1984. Simuliidae. *In* R.W. Merritt and K.W. Cummins (eds.). An introduction to the aquatic insects of North America, 2nd ed., pp. 534–550. Kendall/Hunt Publ. Co., Dubuque, Iowa.

——. 1987. Phoridae. *In* J.F. McAlpine, B.V. Peterson, G.E. Shewell, H.J. Teskey, J.R. Vockeroth, and D.M. Wood (eds.). Manual of Nearctic Diptera, vol. 2, pp. 689–712. Res. Branch Agric. Can. Monogr. no. 28. Ottawa.

Pritchard, G. 1983. Biology of Tipulidae. Annu. Rev. Entomol. 28:1–22.

Quate, L.W. 1955. A revision of the Psychodidae (Diptera) in America north of Mexico. Univ. Calif. Publ. Entomol. 10:103–273.

——. 1960. Guide to the insects of Connecticut. Part VI. The Diptera or true flies of Connecticut. Fasc. 7. Family Psychodidae. Bull. Conn. State Geol. Nat. Hist. Surv. 92:3–47.

*Quate, L.W., and J.R. Vockeroth. 1981. Psychodidae. *In* J.F. McAlpine, B.V. Peterson, G.E. Shewell, H.J. Teskey, J.R. Vockeroth, and D.M. Wood (eds.). Manual of Nearctic Diptera, vol. 1, pp. 293–300. Res. Branch Agric. Can. Monogr. no. 27. Ottawa.

Restifo, R.A. 1982. Illustrated key to the mosquitoes of Ohio. Ohio Biol. Surv. Biol. Notes 17:1–56.

*Robinson, H., and J.R. Vockeroth. 1981. Dolichopodidae. *In* J.F. McAlpine, B.V. Peterson, G.E. Shewell, H.J. Teskey, J.R. Vockeroth, and D.M. Wood (eds.). Manual of Nearctic Diptera, vol. 1, pp. 625–640. Res. Branch Agric. Can. Monogr. no. 27. Ottawa.

*Ross, H.H., and W.R. Horsfall. 1965. A synopsis of the mosquitoes of Illinois (Diptera, Culicidae). Ill. Nat. Hist. Surv. Biol. Notes 52:1–50.

Sabrosky, C.W. 1967. Corrections to "A catalog of the Diptera of America north of Mexico." Bull. Entomol. Soc. Amer. 13:115–125.

Saether, O.A. 1970. Nearctic and Palearctic *Chaoborus* (Diptera: Chaoboridae). Bull. Fish. Res. Bd. Can. 174:1–57.

Savignac, R., and A. Maire. 1981. A simple characteristic for recognizing second and third instar larvae of five Canadian mosquito genera (Diptera: Culicidae). Can. Entomol. 113:13–20.

Skarlato, O.A. (ed.). 1979 (trans. 1985). Systematics of Diptera (Insecta). Ecological and morphological principles. U.S. Dep. Agric., Natl. Sci. Found. Amerind Publ. Co., New Delhi, India. (Zool. Inst. AN SSR Publ., Leningrad, USSR). 185 pp.

Smith, M.E. 1969. The *Aedes* mosquitoes of New England (Diptera: Culicidae). II. Larvae: keys to instars, and to species inclusive of first instar. Can. Entomol. 101:41–51.

Sofield, R.K., and E.J. Hansens. 1980. Rearing of *Tabanus nigrovittatus* Macquart (Diptera: Tabanidae) from egg to adult. J. N.Y. Entomol. Soc. 88:75.

Sommerman, K.M., and R.P. Simmet. 1967. Versatile mosquito trap. Mosq. News 27:412–417.

Steffan, W.A., and N.L. Evenhuis. 1981. Biology of *Toxorhynchites*. Annu. Rev. Entomol. 26:159–181.

Steffan, W.A., N.L. Evenhuis, and D.L. Manning. 1980. Annotated bibliography of *Toxorhynchites* (Diptera: Culicidae). J. Med. Entomol. Suppl. 3:1–140.

Steyskal, G.C. 1950. The genus *Sepedon* Latrielle in the Americas (Diptera: Sciomyzidae). Wassman J. Biol. 8:271–297.

——. 1954. The American species of the genus *Dictya* Meigen (Diptera: Sciomyzidae). Ann. Entomol. Soc. Amer. 47:511–539.

*Steyskal, G.C., and L.V. Knutson. 1981. Empididae. *In* J.F. McAlpine, B.V. Peterson, G.E. Shewell, H.J. Teskey, J.R. Vockeroth, and D.M. Wood (eds.). Manual of Nearctic Diptera, vol. 1, pp. 607–624. Res. Branch Agric. Can. Monogr. no. 27. Ottawa.

Stone, A. 1938. The horseflies of the subfamily Tabaninae of the Nearctic region. U.S. Dep. Agric. Misc. Publ. 305:1–171.

——. 1963. An annotated list of the genus-group names in the family Simuliidae (Diptera). U.S. Dep. Agric. Tech. Bull. 1284:1–28.

——. 1964. Guide to the insects of Connecticut. Part VI. The Diptera or true flies of Connecticut. Fasc. 9. Family Simuliidae and family Thaumaleidae. Bull. Conn. State Geol. Nat. Hist. Surv. 97:1–126.

——. 1980. History of Nearctic dipterology. Vol. 1, part 1: Flies of the Nearctic regions. G.D.C. Griffiths (ed.). E. Schweizerbartische Verlagbuchhandlung Stuttgart, Germany. 62 pp.

*——. 1981. Culicidae. *In* J.F. McAlpine, B.V. Peterson, G.E. Shewell, H.J. Teskey, J.R. Vockeroth, and D.M. Wood (eds.). Manual of Nearctic Diptera, vol. 1, pp. 341–350. Res. Branch Agric. Can. Monogr. no. 27. Ottawa.

Stone, A., and H.A. Jamnback. 1955. The black flies of New York State (Diptera: Simuliidae). Bull. N.Y. State Mus. 349:1–144.

*Stone, A., and B.V. Peterson. 1981. Thaumaleidae. *In* J.F. McAlpine, B.V. Peterson, G.E. Shewell, H.J. Teskey, J.R. Vockeroth, and D.M. Wood (eds.). Manual of Nearctic Diptera, vol. 1, pp. 351–354. Res. Branch Agric. Can. Monogr. no. 27. Ottawa.

Stone, A., and C.W. Sabrosky. 1965. A catalog of the Diptera of America north of Mexico. U.S. Dep. Agric. Res. Ser. Agric. Handb. no. 276, Washington, D.C. 1696 pp.

Stubbs, A., and P. Chandler. 1978. A dipterist's handbook. Amat. Entomol. 15:1–225.

Sturtevant, A.H., and M.R. Wheeler. 1954. Synopses of Nearctic Ephydridae (Diptera). Trans. Amer. Entomol. Soc. 79:151–261.

Teskey, H.J. 1969. Larvae and pupae of some eastern North American Tabanidae (Diptera). Mem. Entomol. Soc. Can. 63:1–147.

*——. 1981a. Morphology and terminology—larvae. *In* J.F. McAlpine, B.V. Peterson, G.E. Shewell, H.J. Teskey, J.R. Vockeroth, and D.M. Wood (eds.). Manual of Nearctic Diptera, vol. 1, pp. 65–88. Res. Branch Agric. Can. Monogr. no. 27. Ottawa.

*——. 1981b. Key to families—larvae. *In* J.F. McAlpine, B.V. Peterson, G.E. Shewell, H.J. Teskey, J.R. Vockeroth, and D.M. Wood (eds.). Manual of Nearctic Diptera, vol. 1, pp. 125–148. Res. Branch Agric. Can. Monogr. no. 27. Ottawa.

*——. 1981c. Pelecorhynchidae. *In* J.F. McAlpine, B.V. Peterson, G.E. Shewell, H.J. Teskey, J.R. Vockeroth, and D.M. Wood (eds.). Manual of Nearctic Diptera, vol. 1, pp. 459–462. Res. Branch Agric. Can. Monogr. no. 27. Ottawa.

——. 1983. A revision of the eastern North American species of *Atylotus* (Diptera: Tabanidae) with keys to adult and immature stages. Proc. Entomol. Soc. Can. 114:21–43.

*——. 1984. Larvae of aquatic Diptera. *In* R.W. Merritt and K.W. Cummins (eds.). An introduction to the aquatic insects of North America, 2nd ed., pp. 448–466. Kendall/Hunt Publ. Co., Dubuque, Iowa.

Teskey, H.J., and J.F. Burger. 1976. Further larvae and pupae of eastern North American Tabanidae (Diptera). Can. Entomol. 108:1085–1096.

Thier, R.W., and B.A. Foote. 1980. Biology of mud-shore Ephydridae (Diptera). Proc. Entomol. Soc. Wash. 82:517–535.

Thomsen, L.C. 1937. Aquatic Diptera. Part V. Ceratopogonidae. Mem. Cornell Univ. Agric. Exp. Stn. 210:57–80.

*Varley, G.C. 1937. Aquatic insect larvae, which obtain oxygen from the roots of plants. Proc. R. Entomol. Soc. (A)12:55–60.

Vockeroth, J.R. 1987. Scathophagidae. *In* J.F. McAlpine, B.V. Peterson, G.E. Shewell, H.J. Teskey, J.R. Vockeroth, and D.M. Wood (eds.). Manual of Nearctic Diptera, vol. 2, pp. 1085–1098. Res. Branch Agric. Can. Monogr. no. 28. Ottawa.

Vockeroth, J.R., and F.C. Thompson. 1987. Syrphidae. *In* J.F. McAlpine, B.V. Peterson, G.E. Shewell, H.J. Teskey, J.R. Vockeroth, and D.M. Wood (eds.). Manual of Nearctic Diptera, vol. 2, pp. 713–744. Res. Branch Agric. Can. Monogr. no. 28. Ottawa.

Wagner, R. 1980. The Nearctic Trichomyiinae (Diptera: Psychodidae). Pan-Pac. Entomol. 56:273–276.

——. 1982. Some new synonymies in Psychodidae (Diptera). Aquat. Insects 4:28.

——. 1984. Contributions to Nearctic Psychodidae (Diptera, Nematocera). Pan-Pac. Entomol. 60:238–244.

*Wallace, J.B. 1971. Immature stages of some muscoid calyptrate Diptera inhabiting mushrooms in the southern Appalachians. J. Georgia Entomol. Soc. 6:218–229.

*Wallace, J.B., and S.E. Neff. 1971. Biology and immature stages of the genus *Cordilura* (Diptera: Scatophagidae) in the eastern United States. Ann. Entomol. Soc. Amer. 64:1310–1330.

Walsh, D.J., D. Yeboah, and M.H. Colbo. 1981. A spherical sampling device for black fly larvae. Mosq. News 41:18–21.

Waugh, W.T., and W.W. Wirth. 1976. A revision of the genus *Dasyhelea* Kieffer of the eastern United States north of Florida (Diptera: Ceratopogonidae). Ann. Entomol. Soc. Amer. 69:219–247.

Webb, D.W. 1977. The Nearctic Athericidae (Insecta: Diptera). J. Kans. Entomol. Soc. 50:473–495.

*——. 1981. Athericidae. *In* J.F. McAlpine, B.V. Peterson, G.E. Shewell, H.J. Teskey, J.R. Vockeroth, and D.M. Wood (eds.). Manual of Nearctic Diptera, vol. 1, pp. 479–482. Res. Branch Agric. Can. Monogr. no. 27. Ottawa.

*Webb, D.W., and W.U. Brigham. 1982. Aquatic Diptera. *In* A.R. Brigham, W.U. Brigham, and A. Gnilka (eds.). Aquatic insects and oligochaetes of North and South Carolina, pp. 11.1–11.111. Midwest Aquatic Enterprises, Mahomet, Ill.

Wighton, D.C. 1980. New species of Tipulidae from the Paleocene of central Alberta, Canada. Can. Entomol. 112:621–628.

Wilder, D.D. 1981. A review of the genus *Roederiodes* Coquillett with the description of a new species (Diptera: Empididae). Pan-Pac. Entomol. 57:415–421.

Wirth, W.W. 1952. The genus *Alluaudomyia* Kieffer in North America (Diptera: Heleidae). Ann. Entomol. Soc. Amer. 45:423–434.

*Wirth, W.W., N.N. Mathis, and J.R. Vockeroth. 1987. Ephydridae. *In* J.F. McAlpine, B.V. Peterson, G.E. Shewell, H.J. Teskey, J.R. Vockeroth, and D.M. Wood (eds.). Manual of Nearctic Diptera, vol. 2, pp. 1027–1048. Res. Branch Agric. Can. Monogr. no. 28. Ottawa.

Wirth, W.W., and A. Stone. 1956. Aquatic Diptera. *In* R.L. Usinger, ed. Aquatic insects of California, with keys to North American genera and California species, pp. 372–482. Univ. Calif. Press, Berkeley.

Wood, D.M., P.T. Dang, and R.A. Ellis. 1979. The insects and arachnids of Canada. Part 6. The mosquitoes of Canada (Diptera: Culicidae). Agric. Can. Publ. 1686:1–390.

*Wood, D.M., B.V. Peterson, D.M. Davies, and H. Gyorkos. 1963. The black flies (Diptera: Simuliidae) of Ontario. Part II. Larval identification, with descriptions and illustrations. Proc. Entomol. Soc. Ont. 93(1962):99–129.

14 Chironomidae

Robert W. Bode

The midges (order Diptera, family Chironomidae; see Ch. 13) account for most of the macroinvertebrates in freshwater environments. In many aquatic habitats this group constitutes more than half of the total number of macroinvertebrate species present. The family is also the most widely distributed group of insects, having adapted to nearly every type of aquatic or semiaquatic environment. The subfamilies Chironominae, Orthocladiinae, and Tanypodinae contain the great majority of the species in the family in North America. Of these, the Tanypodinae and Chironominae are generally most common in lentic warm-water habitats, while the Orthocladiinae are found mostly in lotic and cold-water habitats. The less common subfamilies Diamesinae, Prodiamesinae, and Podonominae have habitat preferences similar to those of the Orthocladiinae.

Life History

Like other dipterans, chironomids have four life stages: egg, larva, pupa, and adult. Eggs are laid in a gelatinous mass in the water and may hatch within a few days. The larval stage has four instars; the first instar is often planktonic and differs greatly from the fourth instar morphologically. Instars can be separated most easily by the relative size of the head capsule. The duration of the larval stage may be from two weeks to several years; it seems to depend mostly on temperature. Mature fourth-instar larvae, or prepupae, are distinguished from other instars by a swollen thorax that contains the developing pupal and adult structures. During the pupal stage, the larval skin, or exuvia, is shed. Most pupae are sedentary, remaining in a loose case, although those of the subfamily Tanypodinae are free-living swimmers. The pupal stage lasts no more than a few days, after which the mature pupa moves to the water surface, where emergence takes place. The adult stage usually lasts only a few days. Mating takes place primarily in swarms, and most adults do not feed.

Habitat

Within aquatic ecosystems, chironomid larvae occur in nearly every type of habitat. The Orthocladiinae are most commonly found on rock or gravel substrates, whereas the Tanypodinae and Chironominae are more common in sand, silt, and clay. Species of Orthocladiinae and Chironominae have also been found on submerged wood and aquatic plants. Some genera of Orthocladiinae have larvae that are partially or entirely terrestrial, occurring in moist soils, tree holes, bird nests, and dung. Many larvae dwell in tubes or cases that they construct from substrate particles and salivary secretions.

Water quality also determines chironomid distribution, and within the family a wide range of tolerance is displayed. Some Tanypodinae and Chironominae are very tolerant of low levels of dissolved oxygen. *Chironomus plumosus* larvae are able to withstand a pH value of 2.3 (Harp and Campbell 1967). *Cricotopus bicinctus* is known for its tolerance for many substances, including electroplating wastes (Surber 1959) and crude oil (Rosenberg and Wiens 1976). Other members of the family are known for their intolerance for poor water quality. Assemblages of species are used as indicators of different types of water quality (Saether 1979c).

Feeding

The majority appear to be opportunistic omnivores, feeding on diatoms, detritus, and other small plants and animals (Pinder 1986). However, some general feeding preferences and tendencies exist, including predation by the Tanypodinae and algal feeding by the Orthocladiinae. There have been a few reports of more-specialized feeding modes, including filter feeding (Walshe 1951), plant mining (Berg 1950), and parasitizing other macroinvertebrates (Steffan 1968). Chironomid larvae exhibit a variety of feeding habits. Much can be learned about the diet of a chironomid larva by viewing the gut contents of a cleared and slide-mounted specimen.

Collection and Identification

The collection of chironomid larvae is very straightforward, since they are so ubiquitous and appear in nearly every aquatic sampling. Kick-net sampling is probably the least effective method of collecting the larvae because many attached species are not adequately sampled, and many others are small enough to pass through the net. Dredge or grab sampling is best for soft sediments; rock surfaces should be sampled by scraping all attached materials into a sieve. A method of collecting chironomids that has recently received wide attention is surface collection of floating pupal exuviae using sieves. Such collections have been used for studies of phenology (Coffman 1973) and faunal diversity, and for assessments of water quality (Wilson and Bright 1973, Ferrington 1987).

Almost all identifications of chironomid larvae require that the specimens be slide-mounted and examined through a compound microscope. The larval head capsule must be cleared of muscle tissue to allow proper viewing of the mouthparts. For routine identifications, clearing can be achieved with the use of a self-clearing mounting medium such as polyvinyl lectophenol or CMCP. For more-permanent mounts, however, Euparal, Canada Balsam, or Diaphane are recommended as mounting media, and clearing must be done prior to mounting. This is accomplished by placing the specimen in a 10% solution of KOH, either heated to a sub-boiling temperature for 5–10 minutes or left overnight at room temperature. Some workers recommend separating the head and body, clearing only the head and mounting the two parts under separate coverslips. Since the clearing procedure may result in loss of such structures as eyespots, lauterborn organs, and ventral tubules, we recommend that students observe these characters under a dissecting microscope before clearing the specimen. The primary concern in mounting the head capsule is to orient it dorsoventrally and flatten it completely so that all structures are in the same plane of view; moderate pressure must be applied to the coverslip to accomplish this. Chironomid larvae should be identified using a compound microscope with at least 400× magnification; oil immersion and 1,000× magnification are occasionally required, but usually not for identifications to genus. An ocular micrometer or grid is needed for determining measurements such as antennal ratios and body lengths.

The identification of chironomid larvae has often been difficult for several reasons. First has been the lack of larval keys; the majority of species have been described only in the adult stage. This problem is gradually being corrected as taxonomists rear species and describe the larvae and pupae. A second problem has been confusion in terminology, with several names being used by different authors for the same structures. Since the publication of Saether's (1980) glossary, this problem has mostly been alleviated, and we recommend this reference for all work on the family. A third difficulty with chironomid identification has been the changing nomenclature, mostly at the generic level. However, Wiederholm's Holarctic keys and diagnoses (1983, 1986) and Ashe's catalog of genera (1983) have defined most genera, and the nomenclature in the family should soon stabilize.

A few problems continue to make identification of some chironomid larvae difficult. Early instars are always a problem; keys and diagnoses are based on fourth-instar larvae, and earlier instars may differ so much in structure and size ratios that they do not fit the description for their own taxon. Other specimens may have abraded mouthparts, or may be poorly mounted, making identification impossible. In such instances one can only attempt to correlate unidentifiable specimens with mature, undamaged specimens from the same collection and to infer the determination from this correlation. This practice is made much easier if one compiles and maintains a reference collection of identified specimens.

The key presented here is intended to facilitate identification to genus of the majority of chironomid larvae from northeastern North America, although some genera in the northern extremes of northeastern Canada are omitted. The key has been constructed so as to minimize the use of structures, such as the pecten epipharyngis (PE), that are difficult to discern on routinely mounted specimens. Many of the structures used in separating genera within the family are illustrated in Figures 1–3. Couplets have been simplified as much as possible to include only features needed to separate the genera. Additional references may be needed for identifying some specimens; in such instances the user is referred to the keys and diagnoses in Wiederholm 1983. For the majority of specimens, however, we believe that the following key will enable easy and accurate determination to genus.

More-Detailed Information

Murray 1980, Oliver 1971, Roback 1983, Coffman and Ferrington 1984, Pinder 1986.

Checklist of Chironomid Larvae

Subfamily Chironominae
Asheum
Axarus
Beckidia
Chernovskiia
Chironomus
Cladopelma
Cladotanytarsus
Constempellina
Cryptochironomus
Cryptotendipes
Cyphomella
Demeijerea
Demicryptochironomus
Dicrotendipes
Einfeldia
Endochironomus
Gillotia
Glyptotendipes
Goeldichironomus
Harnischia
Hyporhygma
Kiefferulus
Lauterborniella
Lipiniella
Microchironomus
Micropsectra
Microtendipes
Nilothauma
Omisus
Pagastiella
Parachironomus
Paracladopelma
Paralauterborniella
Paratanytarsus
Paratendipes
Phaenopsectra
Polypedilum
Pseudochironomus
Rheotanytarsus
Robackia
Saetheria
Stelechomyia
Stempellina
Stempellinella
Stenochironomus
Stictochironomus
Sublettea
Tanytarsus
Tribelos
Xenochironomus
Zavrelia
Zavreliella

Subfamily Diamesinae
Boreoheptagyia
Diamesa
Pagastia
Potthastia
Protanypus
Pseudodiamesa
Pseudokiefferiella
Sympotthastia

Subfamily Orthocladiinae
Acamptocladius
Acricotopus
Brillia
Bryophaenocladius
Camptocladius
Cardiocladius
Chaetocladius
Corynoneura
Cricotopus
Diplocladius
Doithrix
Doncricotopus
Epoicocladius
Eukiefferiella
Euryhapsis
Georthocladius
Gymnometriocnemus
Heleniella
Heterotanytarsus
Heterotrissocladius
Hydrobaenus
Krenosmittia
Limnophyes
Lopescladius
Metriocnemus
Nanocladius
Orthocladius
Parachaetocladius
Paracricotopus
Parakiefferiella
Paralimnophyes
Parametriocnemus
Paraphaenocladius
Paratrissocladius
Parorthocladius
Psectrocladius
Pseudorthocladius
Pseudosmittia
Psilometriocnemus
Rheocricotopus
Rheosmittia
Smittia
Symbiocladius
Symposiocladius
Synorthocladius
Thienemanniella
Tokunagaia
Trissocladius

Tvetenia
Unniella
Xylotopus
Zalutschia
Subfamily Podonominae
Boreochlus
Lasiodiamesa
Paraboreochlus
Parochlus
Trichotanypus
Subfamily Prodiamesinae
Monodiamesa
Odontomesa
Prodiamesa
Subfamily Tanypodinae
Ablabesmyia
Alotanypus
Apsectrotanypus
Brundiniella
Clinotanypus
Coelotanypus
Conchapelopia
Derotanypus
Djalmabatista
Guttipelopia
Hayesomyia
Helopelopia
Hudsonimyia
Krenopelopia
Labrundinia
Larsia
Macropelopia
Meropelopia
Monopelopia
Natarsia
Nilotanypus
Paramerina
Pentaneura
Procladius
Psectrotanypus
Rheopelopia
Tanypus
Telopelopia
Thienemannimyia
Trissopelopia
Zavrelimyia

Key to Genera of Chironomid Larvae

1a. Head capsule with a ligula bearing 4–8 teeth (Fig. 2); mentum usually poorly developed, at least partially membranous; antennae retractile into head capsule .... **Tanypodinae** 6

1b. Head capsule without ligula; mentum usually well developed and sclerotized (Fig. 3); antenna not retractile 2

2a (1b). Premandibles absent; procercus long, more than 5 times as long as wide (Figs. 4, 5); antennal segment 3 annulate (Fig. 6) .... **Podonominae** 36

2b. Premandibles present (Fig. 3); procercus short, less than 4 times as long as wide; antennal segment 3 may or may not be annulate 3

3a (2b). Ventromental plates usually well developed and striate (except *Stenochironomus*) and without cardinal beard; plates positioned at right angle to body axis (Fig. 3) .... **Chironominae** 113

3b. Ventromental plates usually not well developed, or, if so, not striate and usually with cardinal beard (Fig. 3) beneath; plates usually positioned obliquely to body axis (except *Odontomesa*) 4

4a (3b). Antennal segment 3 usually annulate (Fig. 6), or, if not annulate, antenna 4-segmented and mandible with 5–6 inner teeth; postoccipital margin usually prominent and black .... **Diamesinae** 40

4b. Antennal segment 3 not annulate; antenna, mandible and postoccipital margin variable 5

5a (4b). Antenna 4-segmented; ventromental plates usually fairly well developed and with cardinal beard beneath .... **Prodiamesinae** 48

5b. Antenna usually with more than 4 segments, or, if 4-segmented, either ventromental plates small or cardinal beard absent .... **Orthocladiinae** 50

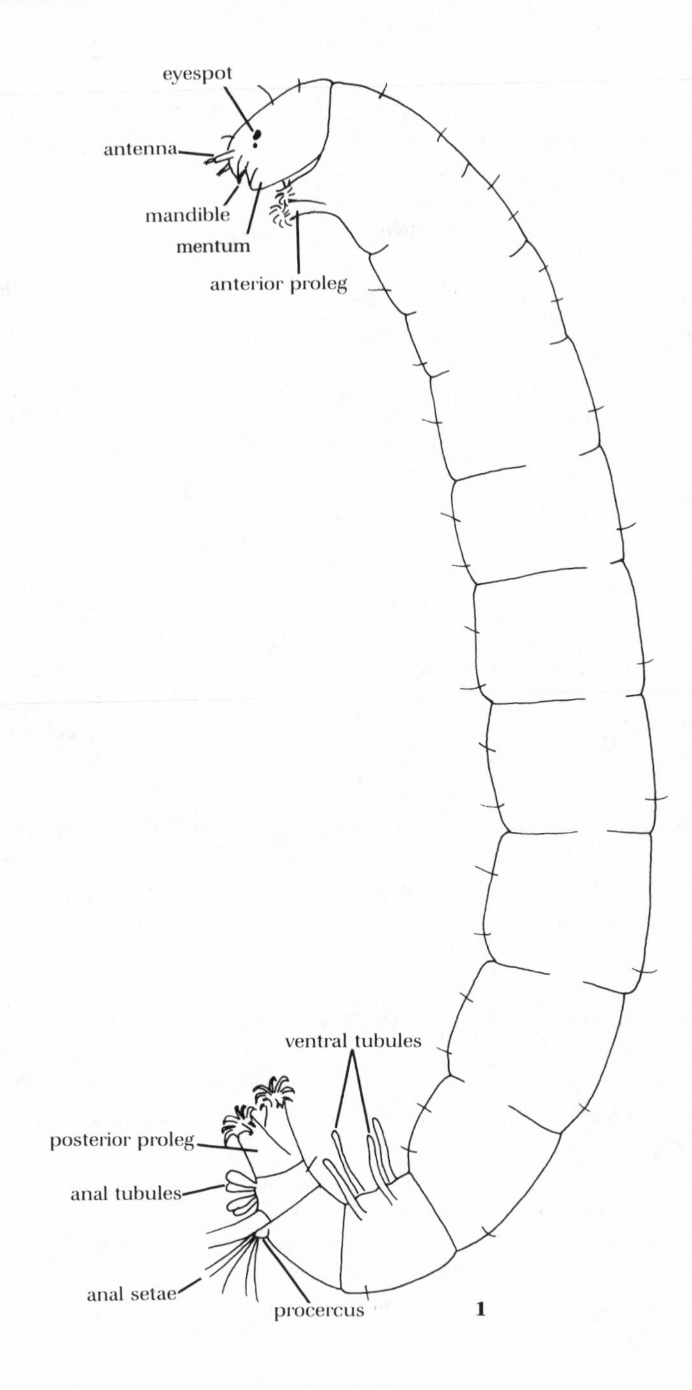

1. Chironomid larva, lateral view.

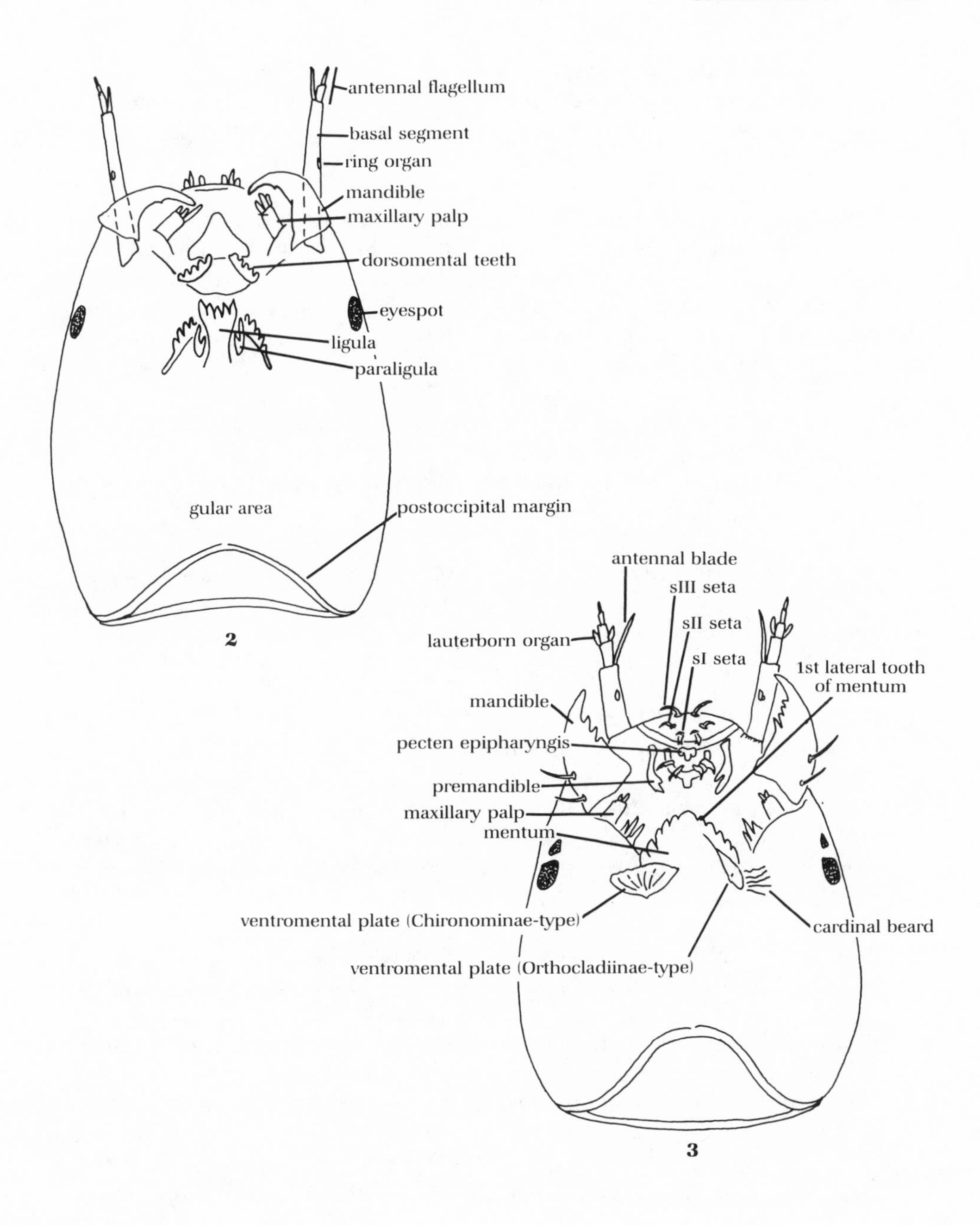

2. Tanypodinae head capsule, vv. **3.** Head capsule characters of Orthocladiinae and Chironominae, vv. vv, ventral view.

6a (1a). **Tanypodinae:** Ligula with 6 or 7 (rarely 5) pale teeth (Figs. 7, 8); dorsomental teeth present, but often difficult to discern, and not situated on plate 7

6b. Ligula with 4 or 5 pale or dark teeth; dorsomental teeth either absent or present on plates 8

7a (6a). Ligula with 7 teeth (Fig. 7); mandible moderately hooked and without large inner tooth (Fig. 9); antennal ratio (AR) < 8; thorax with 2 spurlike processes ..... *Coelotanypus*

7b. Ligula with 6 teeth (Fig. 8); mandible strongly hooked and with large inner tooth (Fig. 10); antennal ratio (AR) > 10; thorax lacking spurlike processes ..... *Clinotanypus*

8a (6b). Dorsomental plates present with more than 3 conspicuous teeth (Fig. 11); body segments with fringe of setae 9

8b. Dorsomental plates absent or with 3 or less conspicuous teeth; body segments without fringe of setae 17

9a (8a). Ligula convex with 5 pale teeth (Fig. 12); mandible with enlarged base (Fig. 13) ..... *Tanypus*

9b. Ligula usually concave, or, if convex, only 4 pale teeth present; mandible with base not enlarged 10

10a (9b). Ligula with 4 or 5 black teeth (Figs. 14, 15) 11

10b. Ligula with 4 or 5 pale teeth (Figs. 16, 17) 12

11a (10a). Ligula with 4 black teeth (Fig. 14); antennal blade more than twice as long as flagellum ..... *Djalmabatista*

11b. Ligula with 5 black teeth (Fig. 15); antennal blade about as long as flagellum ..... *Procladius*

12a (10b). Ligula convex with 4 pale teeth; mandible with 4 or more inner teeth 13

12b. Ligula concave with 5 pale teeth; mandible with less than 4 inner teeth ... 14

13a (12a). Ligula with outer tooth notched apically (Fig. 16); mandible with dorsal teeth (Fig. 18); dorsomental plates strongly concave ..... *Derotanypus*

13b. Ligula with outer tooth unnotched (Fig. 19); mandible without dorsal teeth (Fig. 20); dorsomental plates weakly concave ..... *Psectrotanypus*

14a (12b). Dorsomental plates each with 4–5 teeth; antennal segment 2 short, 2–2.5 times as long as wide 15

14b. Dorsomental plates each with 6–9 teeth; antennal segment 2 about 3.5–4 times as long as wide 16

15a (14a). Dorsomental plates each with 5 teeth; lobe on inner apex pointed (Fig. 21) ..... *Brundiniella*

15b. Dorsomental plates each with 4–5 teeth; lobe on inner apex not pointed (Fig. 22) ..... *Apsectrotanypus*

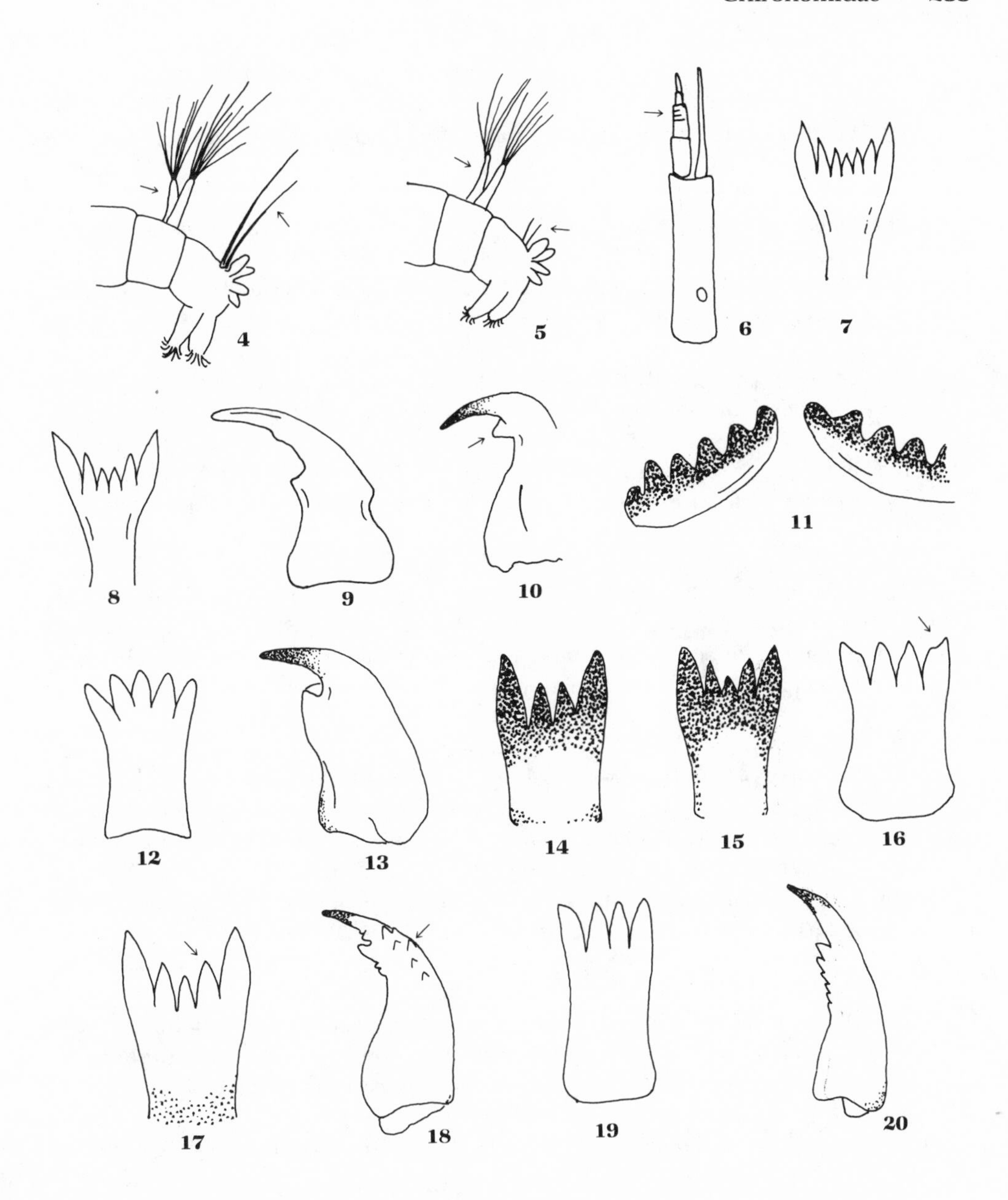

4. Posterior abdominal segments of *Paraboreochlus*, lv. **5.** Posterior abdominal segments of *Boreochlus*, lv. **6.** Antenna of Diamesinae. **7.** Ligula of *Coelotanypus*, vv. **8.** Ligula of *Clinotanypus*, vv. **9.** Mandible of *Coelotanypus*, vv. **10.** Mandible of *Clinotanypus*, vv. **11.** Dorsomental plates of *Procladius*, vv. **12.** Ligula of *Tanypus*, vv. **13.** Mandible of *Tanypus*, vv. **14.** Ligula of *Djalmabatista*, vv. **15.** Ligula of *Procladius*, vv. **16.** Ligula of *Derotanypus*, vv. **17.** Ligula of *Macropelopia*, vv. **18.** Mandible of *Derotanypus*, vv. **19.** Ligula of *Psectrotanypus*, vv. **20.** Mandible of *Psectrotanypus*, vv. lv, lateral view; vv, ventral view. (5, 14, 16–18 redrawn from Wiederholm 1983.)

16a (14b). Ligula deeply concave, inner tooth only slightly outcurved (Fig. 17) .... *Macropelopia*

16b. Ligula shallowly concave, inner tooth strongly outcurved (Fig. 23) .... *Alotanypus*

17a (8b). Ligula with median tooth distinctly longer than inner tooth (Figs. 24, 25); head capsule slender, about 1.5 times as long as wide 18

17b. Ligula with median tooth shorter than or equal to inner tooth (Figs. 26, 27); head capsule usually not slender, less than 1.5 times as long as wide 19

18a (17a). Posterior proleg with 1 bifid claw (Fig. 28); seta of posterior proleg toothed near base; median tooth of ligula nearly twice as wide as inner tooth (Fig. 24) .... *Labrundinia*

18b. Posterior proleg with 1 pectinate claw (Fig. 29); seta of posterior proleg not toothed near base; median tooth of ligula only slightly wider than inner tooth (Fig. 25) .... *Nilotanypus*

19a (17b). Antenna short, less than 0.4 times as long as head capsule (Fig. 30); antennal ratio (AR) < 3.5; mandible with large inner tooth (Fig. 31) .... *Natarsia*

19b. Antenna moderately long, more than 0.4 times as long as head capsule; antennal ratio (AR) variable; mandible with or without inner tooth 20

20a (19b). Maxillary palp with 2 or more basal segments (Figs. 32–34) 21

20b. Maxillary palp with only 1 basal segment (Figs. 35, 36) 22

21a (20a). Maxillary palp with 2 basal segments, 1st segment less tha ⅓ as long as 2nd (Fig. 32); ligula nearly straight (Fig. 26) .... *Paramerina*

21b. Maxillary palp either with more than 2 basal segments or with 2 basal segments subequal in length (Figs. 33, 34); ligula usually concave (Fig. 37) .... *Ablabesmyia*

22a (20b). Head capsule granular (Fig. 38); posterior prolegs with 3 dark claws and some claws with inner teeth (Fig. 39); AR ca. 6.0 .... *Guttipelopia*

22b. Head capsule not granular; posterior prolegs variable; AR variable 23

23a (22b). Ligula strongly concave (Fig. 27); mandible with large inner tooth (Fig. 40) 24

23b. Ligula variable; if strongly concave, never with large inner tooth on mandible 25

24a (23a). Posterior prolegs with some claws pectinate and with 1 claw dark; antennal segment 2 darker than basal segment .... *Monopelopia*

24b. Posterior prolegs with all claws simple and yellow; antennal segment 2 not darker than basal segment .... *Krenopelopia*

25a (23b). Mandible with at least 1 large inner tooth; ligula weakly concave to straight 26

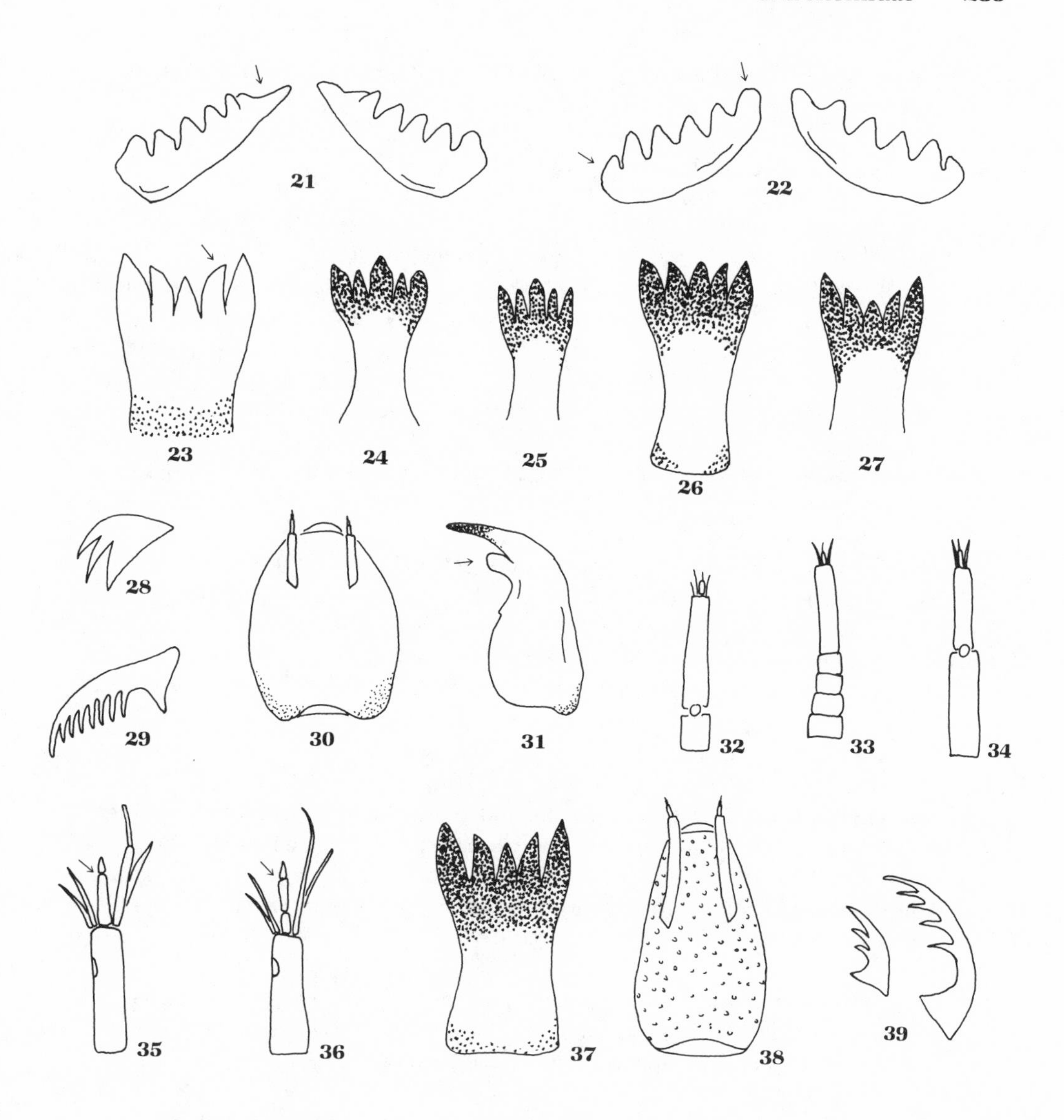

21. Dorsomental plates of *Brundiniella* vv. **22.** Dorsomental plates of *Apsectrotanypus*, vv. **23.** Ligula of *Alotanypus*, vv. **24.** Ligula of *Labrundinia*, vv. **25.** Ligula of *Nilotanypus*, vv. **26.** Ligula of *Paramerina*, vv. **27.** Ligula of *Monopelopia*, vv. **28.** Claw of posterior proleg of *Labrundinia*, dv. **29.** Claw of posterior proleg of *Nilotanypus*, dv. **30.** Head capsule of *Natarsia*, dv. **31.** Mandible of *Natarsia*, vv. **32.** Maxillary palp of *Paramerina*. **33.** Maxillary palp of *Ablabesmyia* (in part). **34.** Maxillary palp of *Ablabesmyia* (in part). **35.** Maxillary palp of *Conchapelopia*. **36.** Maxillary palp of *Thienemannimyia*. **37.** Ligula of *Ablabesmyia*, vv. **38.** Head capsule of *Guttipelopia*, dv. **39.** Claws of posterior prolegs of *Guttipelopia*, dv. dv, dorsal view; vv, ventral view. (21–23, 27 redrawn from Wiederholm 1983.)

25b. Mandible with at most minute inner teeth; ligula weakly to strongly concave 29

26a (25a). Ligula straight, all teeth equal in height (Figs. 41, 42) 27
26b. Ligula weakly concave, middle teeth shorter than lateral teeth (Fig. 43) 28

27a (26a). Supraanal seta enlarged, situated on papilla (Fig. 46); ligula narrowed basally (Fig. 41) . . . . *Pentaneura*
27b. Supraanal seta not enlarged (Fig. 47); ligula widened basally (Fig. 42) . . . . *Zavrelimyia*

28a (26b). Mandible only about ⅓ as long as antenna; procercus 4–6 times as long as wide . . . . *Larsia*
28b. Mandible about ½ as long as antenna; procercus about 3 times as long as wide . . . . *Telopelopia*

29a (25b). Maxillary palp with ring organ in middle ⅓ of basal segment (Fig. 44) 30
29b. Maxillary palp with ring organ in distal ⅓ of basal segment (Fig. 45) 31

30a (29a). Maxillary palp with basal segment about 7 times as long as wide; antennal segment 1 at least 420 μm long . . . . *Trissopelopia*
30b. Maxillary palp with basal segment about 4 times as long as wide; antennal segment 1 < 300 μm long . . . . *Hudsonimyia*

31a (29b). Ligula usually weakly concave (Fig. 48); mandible lacking even minute inner teeth (Fig. 49); AR < 3.8; basal seta of posterior proleg sometimes forked . . . . *Rheopelopia*
31b. Ligula usually strongly concave (Fig. 50); mandible with at least minute inner teeth (Fig. 51); AR usually > 3.8; basal seta of posterior proleg simple 32

32a (31b). Maxillary palp with apical appendage b (visible only under 400×) 2-segmented (Fig. 35) 33
32b. Maxillary palp with apical appendage b 3-segmented (Fig. 36) 35

33a (32a). AR usually > 4.9; antennae greater than 2 times length of mandible . . . . *Meropelopia*
33b. AR < 4.9; antennae less than 2 times length of mandible 34

34a (33b). AR 4.1–4.5; ligula 90–100 μm long . . . . *Hayesomyia senata*
34b. AR 4.6–4.9; ligula 110–120 μm long . . . . *Thienemannimyia*

35a (32b). Antennal segment 1 > 365 μm long; ligula > 125 μm long . . . . *Helopelopia*
35b. Antennal segment 1 < 365 μm long; ligula < 125 μm long . . . . *Conchapelopia*

36a (2a). **Podonominae:** Procercus monochromatic; antenna shorter than mandible . . . . *Parochlus*

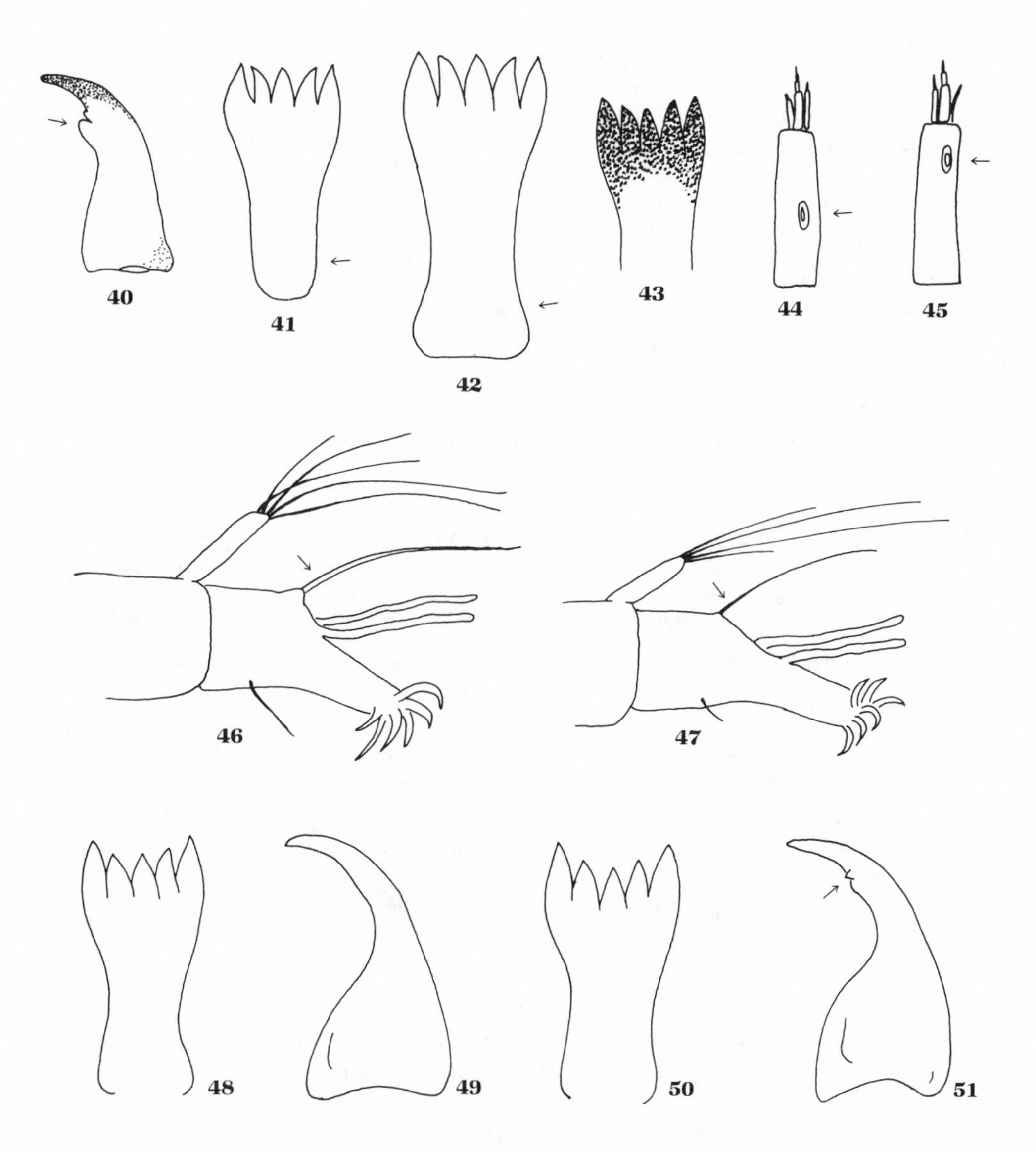

40. Mandible of *Zavrelimyia*, vv. **41.** Ligula of *Pentaneura*, dv. **42.** Ligula of *Zavrelimyia*, vv. **43.** Weakly concave ligula, vv. **44.** Maxillary palp showing ring organ. **45.** Maxillary palp showing ring organ. **46.** Posterior abdominal segments of *Pentaneura*, lv. **47.** Posterior abdominal segments of *Zavrelimyia*, lv. **48.** Ligula of *Rheopelopia*, vv. **49.** Mandible of *Rheopelopia*, vv. **50.** Strongly concave ligula, vv. **51.** Mandible with inner teeth, vv. dv, dorsal view; lv, lateral view; vv, ventral view.

36b. Procercus light anteriorly and darker posteriorly; antenna longer than mandible 37

37a (36b). Median tooth of mentum trifid and deeply recessed (Fig. 52) . . *Trichotanypus*

37b. Median tooth of mentum not trifid and recessed 38

38a (37b). Mentum with 10 or more pairs of lateral teeth (Fig. 53); procercus with 11–15 anal setae . . . . *Lasiodiamesa*

38b. Mentum with 7 pairs of lateral teeth; procercus with 5–8 anal setae 39

39a (38b). Procercus with 8 anal setae; 2 long, black supraanal setae present (Fig. 4) . . . . *Paraboreochlus*

39b. Procercus with 5 anal setae; supraanal setae reduced (Fig. 5) or absent . . . . *Boreochlus*

40a (4a). **Diamesinae:** Dorsal surface of head with 4–5 tubercles (Fig. 54) . . . . *Boreoheptagyia*

40b. Head lacking tubercles 41

41a (40b). Mentum lacking distinct teeth on central ⅔, with only 2 pairs of lateral teeth (Fig. 55); antenna 4-segmented; mandible with 5–6 inner teeth (Fig. 56) . . . . *Protanypus*

41b. Mentum either with teeth on central ⅔, or entirely lacking teeth; antenna 5-segmented; mandible with less than 5 inner teeth 42

42a (41b). Mentum lacking teeth; premandible with about 15 small teeth; mandible with no dark inner teeth . . . . *Potthastia longimana* group

42b. Mentum with some teeth present; premandible with 13 or fewer teeth; mandible usually with 4 dark inner teeth 43

43a (42b). Mentum with 3 large central teeth, median slightly recessed between 1st pair of lateral teeth, remaining lateral teeth indistinct (Fig. 57) . . . . *Pseudodiamesa*

43b. Mentum without the combination of characters described above 44

44a (43b). Body with some dark setae that are longer than corresponding body segment . . . . *Pseudokiefferiella*

44b. Body lacking dark setae that are longer than corresponding body segment 45

45a (44b). Median portion of mentum truncate with 4–6 minute teeth (Fig. 58); lateral teeth covered by large ventromental plates; procercus longer than wide . . . . *Pagastia*

45b. Mentum not as described above; procercus not longer than wide 46

46a (45b). Premandible simple (Fig. 59); median area of mentum with broad light tooth (Fig. 60) . . . . *Potthastia gaedii* group

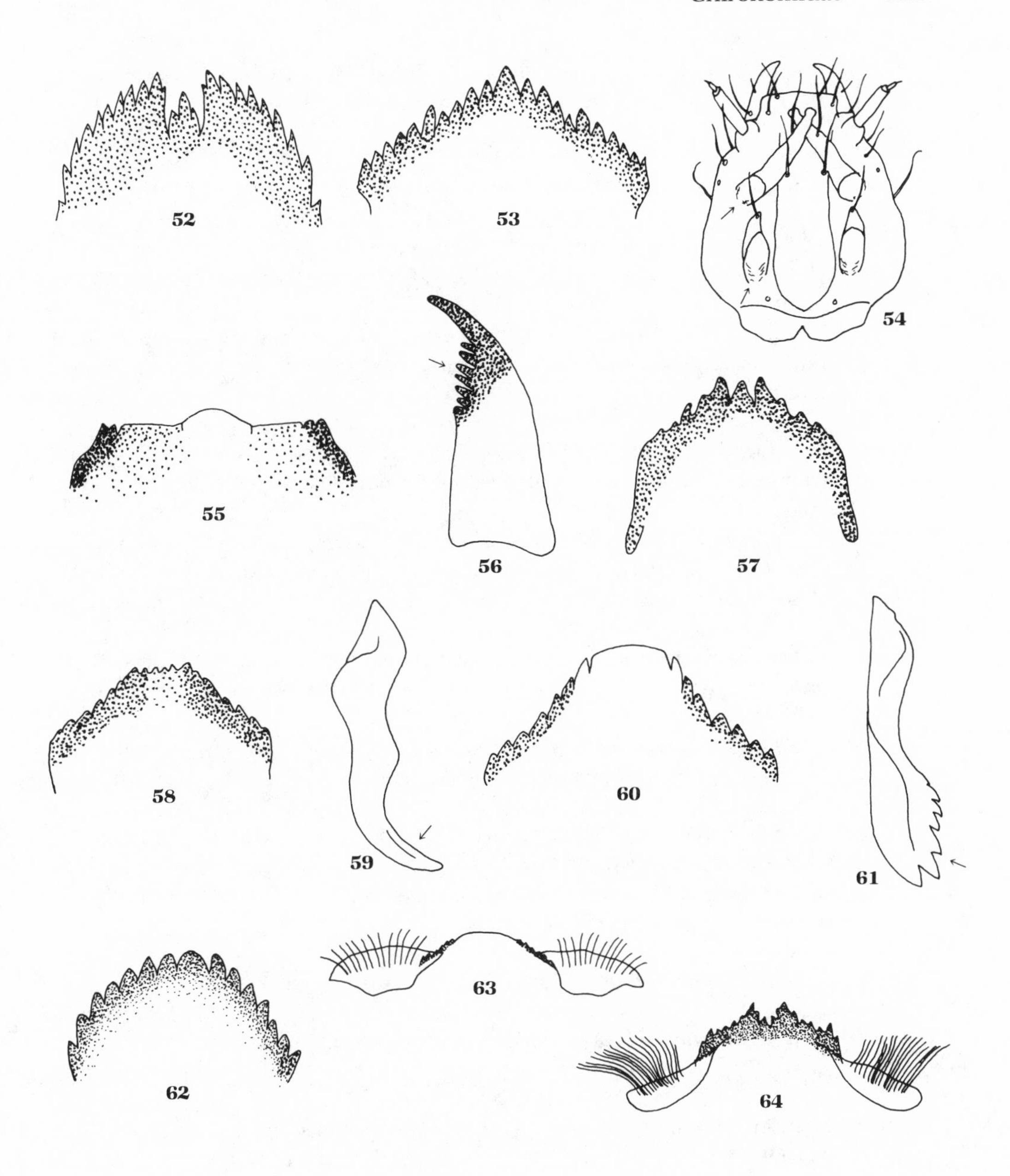

52. Mentum of *Trichotanypus*, vv. **53.** Mentum of *Lasiodiamesa*, vv. **54.** Head of *Boreoheptagyia*, dv. **55.** Mentum of *Protanypus*, vv. **56.** Mandible of *Protanypus*, vv. **57.** Mentum of *Pseudodiamesa*, vv. **58.** Mentum of *Pagastia*, vv. **59.** Premandible of *Potthastia*, vv. **60.** Mentum of *Potthastia gaedii* group, vv. **61.** Premandible of *Diamesa*, vv. **62.** Mentum of *Diamesa*, vv. **63.** Mentum and ventromental plates of *Odontomesa*, vv. **64.** Mentum and ventromental plates of *Prodiamesa*, vv. dv, dorsal view; vv, ventral view. (52–54, 57 redrawn from Wiederholm 1983.)

46b. Premandible with at least one inner tooth (Fig. 61); median teeth of mentum not broad and light (Fig. 62) 47

47a (46b). Premandible with 1–3 inner teeth .... *Sympotthastia*
47b. Premandible with 5 or more inner teeth .... *Diamesa*

48a (5a). **Prodiamesinae:** Mentum with broad convex median tooth (Fig. 63); mandible expanded basally, with several lateral setae and 30–40 seta interna filaments .... *Odontomesa*
48b. Mentum with either paired median teeth or single concave tooth; mandible not so expanded basally, lacking groups of lateral setae, and with fewer seta interna filaments 49

49a (48b). Mentum with paired median teeth deeply recessed between 1st lateral teeth; ventromental plates with well-developed beard (Fig. 64); mandible with short apical tooth .... *Prodiamesa*
49b. Mentum with single median tooth that is broad and concave, apparently double (Fig. 65); ventromental plates with only inconspicuous beard; mandible with long apical tooth .... *Monodiamesa*

50a (5b). **Orthocladiinae:** Anal end usually lacking procerci (Fig. 66), or, if present, procerci lacking anal setae and posterior prolegs; anterior prolegs often partially fused 51
50b. Anal end with procerci bearing anal setae (Fig. 67) and posterior prolegs; anterior prolegs almost always separated 57

51a (50a). Preanal segment directed ventrally; anal segment and posterior prolegs at right angle to body axis (Fig. 66) 52
51b. Preanal segment not directed ventrally; anal segment and posterior prolegs on same axis as body (Fig. 68) 3

52a (51a). Anal tubules absent; posterior prolegs each apically divided .... *Gymnometriocnemus*
52b. Anal tubules present; posterior prolegs not apically divided .... *Bryophaenocladius*

53a (51b). Mentum with toothless median area and 4–5 small spinelike lateral teeth (Fig. 69); mandible with 4 spinelike inner teeth; ectoparasitic on mayflies .... *Symbiocladius*
53b. Mentum usually fully toothed; mandibular teeth not spinelike; not parasitic on mayflies 54

54a (53b). Antenna very reduced, less than ½ as long as long as mandible; mandible lacking seta interna; SI and SII bifid 55
54b. Antenna at least ½ as long as mandible; mandible with seta interna; SI plumose, SII simple 56

55a (54a). Posterior prolegs and anal claws absent; premandible with 3 teeth . . . . *Camptocladius*

55b. Posterior prolegs and anal claws usually present; premandible with 4 teeth . . . . *Pseudosmittia*

56a (54b). Antennal segment 2 shorter than segment 1; anal tubules long, with many constrictions; premandible simple . . . . *Georthocladius*

56b. Antennal segment 2 longer than segment 1; anal tubules short, without constrictions; premandible bifid . . . . *Smittia*

57a (50b). Antenna elongate, more than ½ as long as head capsule (Figs. 70, 71) 58

57b. Antenna not so elongate, less than ½ as long as head capsule 62

58a (57a). Mentum with central portion concave (Fig. 72); anterior prolegs with group of preapical claws . . . . *Heterotanytarsus*

58b. Mentum not with central portion concave; anterior prolegs lacking preapical claws 59

59a (58b). Antenna 4-segmented, longer than head capsule (Fig. 70) 60

59b. Antenna 5-segmented, shorter than head capsule (Fig. 71) 61

60a (59a). Antenna with 2nd segment longer than 1st and with terminal segment elongate; procercus with 1 long anal seta . . . . *Lopescladius*

60b. Antenna with 2nd segment shorter than 1st and with terminal segment minute; procercus with 4 short anal setae . . . . *Corynoneura*

61a (59b). Antenna with 2nd segment as long as 1st and bearing alternate Lauterborn organs; premandible with several small teeth . . . . *Rheosmittia*

61b. Antenna with 2nd segment shorter than 1st and bearing terminal lauterborn organs; premandible simple . . . . *Thienemanniella*

62a (57b). Procercus with one anal seta elongate, at least ¼ as long as body (Fig. 73) 63

62b. No anal seta ¼ as long as body 65

63a (62a). Mentum with pointed, dome-shaped median tooth and 6 pairs of narrow lateral teeth (Fig. 74); premandible bifid . . . . *Krenosmittia*

63b. Mentum with 1 pair of median teeth and 4 pairs of lateral teeth; premandible simple 64

64a (63b). Mandible with 1–2 inner teeth; antenna at least ⅘ as long as mandible . . . . *Parachaetocladius*

64b. Mandible with 3 inner teeth; antenna only ⅔ as long as mandible . . . . *Pseudorthocladius*

65a (62b). Mentum with 3 minute recessed central teeth and 13–16 pairs of lateral

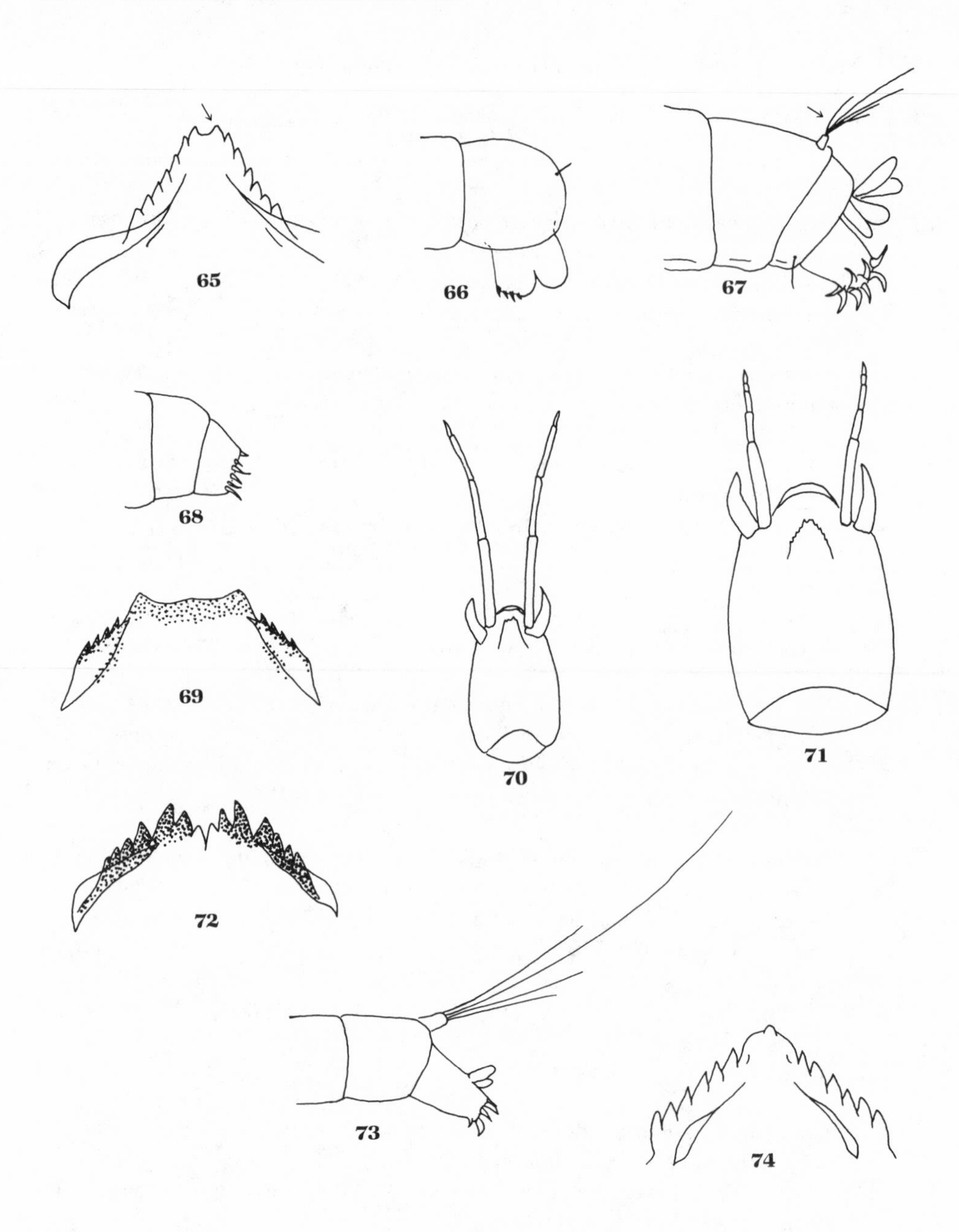

65. Mentum and ventromental plates of *Monodiamesa*, vv. **66.** Posterior abdominal segments of *Gymnometriocnemus*, lv. **67.** Posterior abdominal segments of *Eukiefferiella*, lv. **68.** Posterior abdominal segments of *Smittia*, lv. **69.** Mentum of *Symbiocladius*, vv. **70.** Head capsule with elongate antennae, vv. **71.** Head capsule with elongate antennae, vv. **72.** Mentum of *Heterotanytarsus*, vv. **73.** Posterior abdominal segments of *Krenosmittia*, lv. **74.** Mentum of *Krenosmittia*, vv. lv, lateral view; vv, ventral view. (66, 69, 72 redrawn from Wiederholm 1983.)

teeth on dorsomental plates (Fig. 75); premandible with 4 teeth . . . . *Acamptocladius*

65b. Not with the combination of characters described above 66

66a (65b). Mentum arched and truncate, with about 6 pale teeth across apex and 5 pairs of darker sloping lateral teeth (Fig. 76); antenna 4-segmented; abdomen with prominent setae; phoretic on mayflies . . . . *Epoicocladius*

66b. Not with the combination of characters described above 67

67a (66b). Mentum with single median tooth and 8–9 pairs of lateral teeth (Fig. 77); premandible simple . . . . *Orthocladius (Euorthocladius)* (in part)

67b. Mentum with less than 8 pairs of lateral teeth 68

68a (67b). Ventromental plates with cardinal beard underneath (Figs. 78, 79) 69

68b. Ventromental plates without cardinal beard underneath 76

69a (68a). SI simple 70

69b. SI bifid, palmate, or plumose 72

70a (69a). Mandible with apical tooth longer than combined width of inner teeth . . . . *Doncricotopus*

70b. Mandible with apical tooth shorter than combined width of inner teeth . . 71

71a (70b). Mentum with 2 median teeth forming elevated apex (Fig. 78); mandible with spine on inner margin . . . . *Synorthocladius*

71b. Mentum with 3 median teeth forming elevated apex (Fig. 80); mandible lacking spine on inner margin . . . . *Parorthocladius*

72a (69b). Premandible apically simple (Fig. 81) 73

72b. Premandible apically bifid (Fig. 82) 75

73a (72a). Mandible lacking seta interna; mentum with lateral teeth dark, median tooth paler, broad and weakly divided by notches into 4 parts (Fig. 83); head capsule with heavily sclerotized mentum, mandible, and postoccipital margin . . . . *Acricotopus*

73b. Mandible usually with seta interna present; mentum not dark, with single or paired median teeth; head capsule not heavily sclerotized 74

74a (73b). SI distinctively palmate (Fig. 84); mandible with apical tooth usually pale and longer than combined width of inner teeth (Fig. 85); ventromental plates generally not large (Fig. 86) . . . . *Psectrocladius*

74b. SI usually bifid (Fig. 87); mandible with apical tooth not pale and usually not longer than combined width of inner teeth; ventromental plates usually quite large (Fig. 88) . . . . *Rheocricotopus*

75a (72b). Ventromental plates with strong beard beneath (Fig. 79); mandible with 4 in-

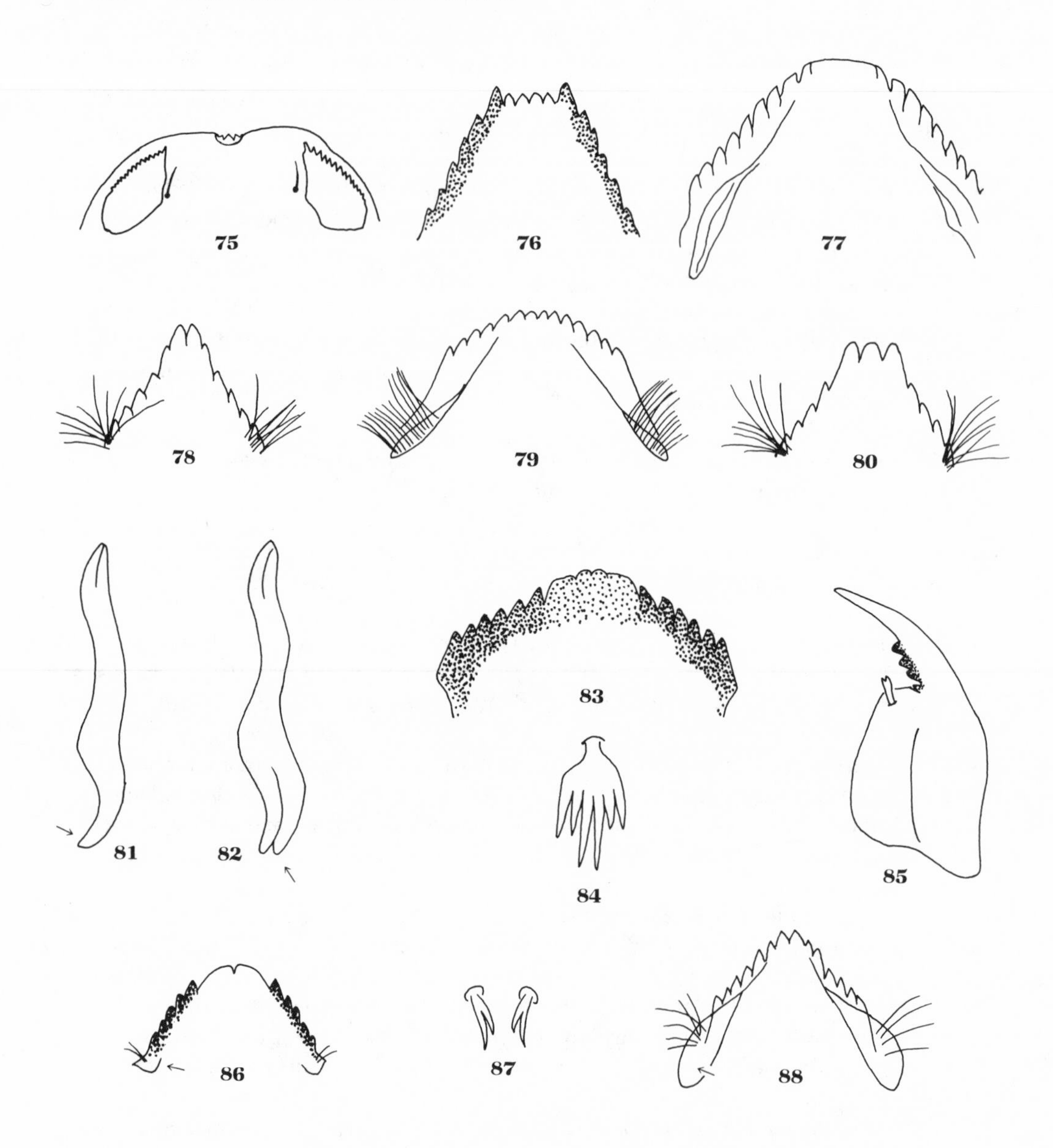

75. Mentum of *Acamptocladius*, vv. **76.** Mentum of *Epoicocladius*, vv. **77.** Mentum of *Orthocladius (Euorthocladius)* (in part), vv. **78.** Mentum of *Synorthocladius*, vv. **79.** Mentum and ventromental plate of *Diplocladius*, vv. **80.** Mentum of *Parorthocladius*, vv. **81.** Premandible of *Psectrocladius*, vv. **82.** Premandible of *Diplocladius*, vv. **83.** Mentum of *Acricotopus*, vv. **84.** Palmate SI seta of *Psectrocladius*. **85.** Mandible of *Psectrocladius*, vv. **86.** Mentum of *Psectrocladius*, dv. **87.** SI setae of *Rheocricotopus*. **88.** Mentum of *Rheocricotopus*, dv. dv, dorsal view; vv, ventral view. (75, 76, 80, 81 redrawn from Wiederholm 1983.)

ner teeth; mentum with 14 teeth, mostly subequal in size, central 4 elevated .... *Diplocladius*

75b. Ventromental plates with weak beard beneath (Fig. 89); mandible with 3 inner teeth; mentum with 1st pair of lateral teeth reduced .... *Zalutschia*

76a (68b). Mentum with broad, pale median tooth bearing pair of small points medially (Fig. 90); ventromental plates wide, well developed, and often wrinkled; mandible pale with long apical tooth and 3 small dark inner teeth .. *Nanocladius*

76b. Mentum variable but not bearing pair of points medially; ventromental plates usually not so well developed and not wrinkled; mandible not usually pale, and with short apical tooth, number of inner teeth variable 77

77a (76b). Mentum with 5 or fewer pairs of lateral teeth 78

77b. Mentum with 6 or more pairs of lateral teeth 101

78a (77a). Mentum with 11 dark teeth, 1st and 2nd pairs of lateral teeth small and appressed to median tooth, together appearing as single, wide convex tooth; 3 additional pairs of larger lateral teeth present (Fig. 91) .... *Cricotopus trifascia* group

78b. Median 5 teeth of mentum not forming a convex area separated from other teeth 79

79a (78b). Anal tubules very long; posterior prolegs greatly reduced (Fig. 92); mentum with 4 pairs of lateral teeth, last pair bifid .... *Doithrix*

79b. Anal tubules not long; posterior prolegs not reduced; mentum variable ... 80

80a (79b). Mentum with greatly elongated triangular median tooth and 2 pairs of basal lateral teeth (Fig. 93) .... *Symposiocladius*

80b. Mentum not as described above 81

81a (80b). Antenna 4-segmented; mentum with paired long median teeth separated by notch, sometimes with additional small tooth in notch, 5 pairs of lateral teeth (Fig. 95), 5th pair sometimes obscuring 4th pair 82

81b. Antenna 4–7 segmented mentum variable 83

82a (81a). Antennal segment 2 divided at basal ⅓ (Fig. 94); submental setae located on posterior ½ of submentum .... *Brillia*

82b. Antennal segment 2 not divided; submental setae located midway between mentum and postoccipital margin .... *Euryhapsis*

83a (81b). Mentum with 4 pairs of lateral teeth, median teeth paired and long (Fig. 96); abdomen with fringe of setal tufts; large, up to 16 mm long .... *Xylotopus*

83b. Mentum not as above; abdomen lacking fringe of setae; usually < 16 mm long 84

84a (83b). Premandible with 1 apical tooth 85

84b. Premandible with 2 or more apical teeth 92

85a (84a). Mentum with median tooth and 1st pair of lateral teeth elevated and somewhat separated from remaining teeth (Fig. 97); antennal segment 3 longer than segment 4; procercus with strong preapical spur .... *Paracricotopus*

85b. Mentum usually not with median and 1st pair of lateral teeth elevated; if elevated, antennal segment 3 shorter than segment 4 and procercus lacking spur 86

86a (85b). Mentum with only 4 pairs of lateral teeth or apparently only 4 pairs of lateral teeth, if 5th pair present much narrower than other 4 pairs (Figs. 98–100) 87

86b. Mentum with 5 pairs of lateral teeth present, all of about the same width (Fig. 101) 89

87a (86a). Mentum with paired pointed median teeth (Fig. 98); antenna 7-segmented .... *Paratrissocladius*

87b. Mentum with median tooth single, or low and bifid; antenna 5-segmented 88

88a (87b). Median tooth of mentum wide with median point (Fig. 99); mandible with 3 long serrations on inner margin; abdomen with some setae nearly as long as segment .... *Eukiefferiella devonica* group

88b. Median tooth of mentum wide but not pointed (Fig. 100), sometimes apparently bifid; mandible lacking serrations on inner margin; abdomen lacking strong setation .... *Tokunagaia*

89a (86b). Head capsule usually yellow; abdominal setae at least ½ as long as body segment; procercus 1½–2 times as long as wide .... *Tvetenia*

89b. Head capsule usually brown; abdominal setae usually less than ½ as long as body segment, or if longer, procercus less than 1½ times as long as wide 90

90a (89b). Mentum with median tooth pointed and outermost lateral teeth oriented anteriorly, preceding teeth oriented diagonally (Fig. 101); mandible with 2 inner teeth, reduced; antennal segment 3 greatly reduced .... *Cardiocladius albiplumus*

90b. Mentum with median tooth either single and rounded, or paired; mandible with 3 inner teeth, not reduced; antennal segment 3 variable 91

91a (90b). Procercus with 2 anal setae longer and thicker than others; anal segment greatly reduced (Fig. 102) .... *Cardiocladius* (in part)

91b. Procercus with anal setae of similar length and thickness; anal segment not reduced (Fig. 67) .... *Eukiefferiella* (in part)

92a (84b). Mentum with laterally projecting lobe at base of outer lateral tooth, median

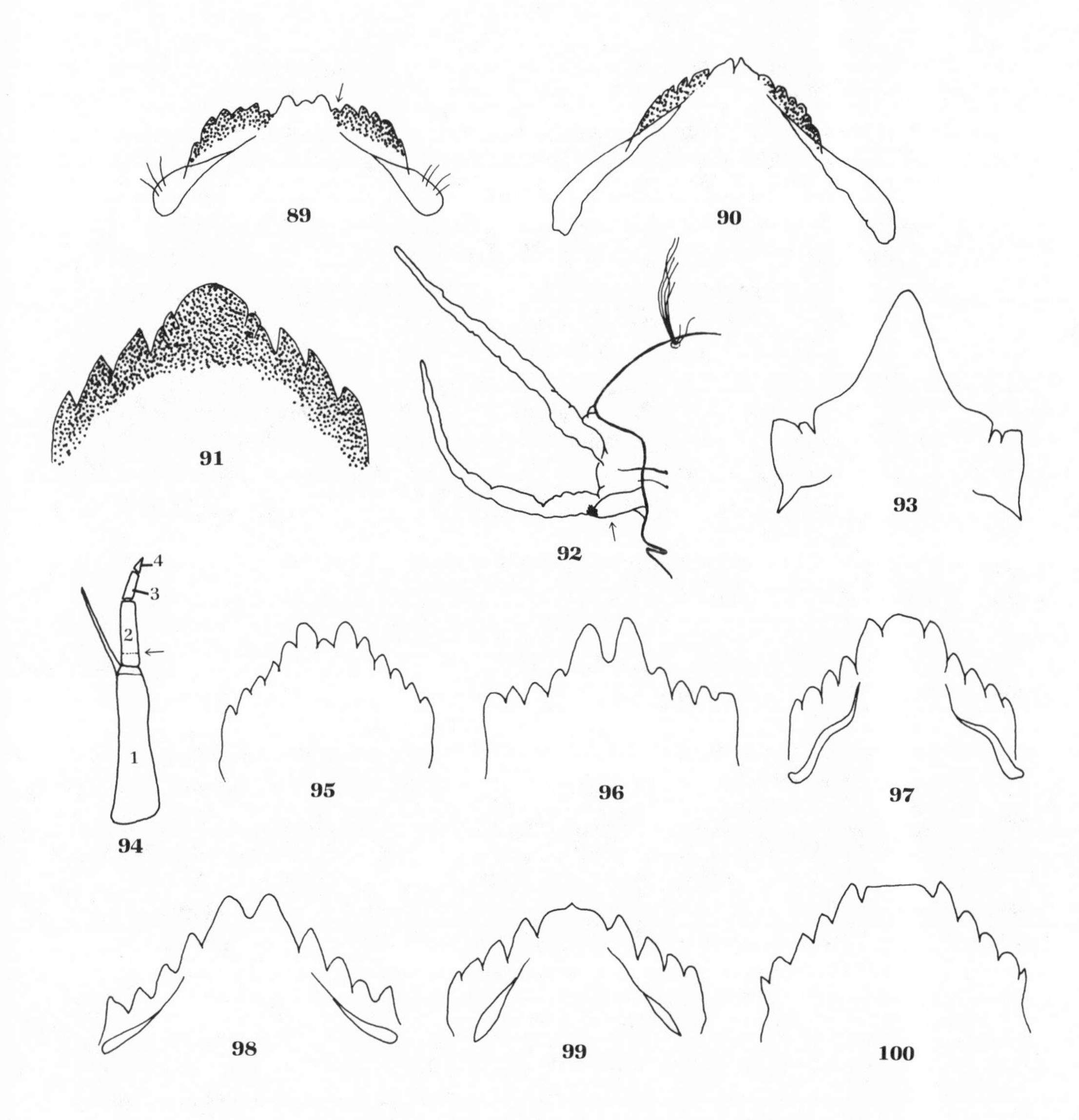

89. Mentum and ventromental plate of *Zalutschia*, vv. **90.** Mentum and ventromental plate of *Nanocladius*, vv. **91.** Mentum of *Cricotopus trifascia* group, vv. **92.** Posterior of *Doithrix*, lv. **93.** Mentum of *Symposiocladius*, vv. **94.** Antenna of *Brillia*. **95.** Mentum of *Brillia*, vv. **96.** Mentum of *Xylotopus*, vv. **97.** Mentum of *Paracricotopus*, vv. **98.** Mentum of *Paratrissocladius*, vv. **99.** Mentum of *Eukiefferiella devonica* group, vv. **100.** Mentum of *Tokunagaia*, vv. lv, lateral view; vv, ventral view. (92, 96–98, 100 redrawn from Wiederholm 1983.)

teeth paired and usually projecting (Figs. 103–105); abdomen usually with long setae 93

92b. Mentum lacking lobe at base of outer lateral tooth (Figs. 106, 107); median tooth or teeth variable; abdomen usually lacking strong setation 95

93a (92a). Antenna with segment 2 nearly as long as segment 1, divided near basal ⅓, segments 3–5 reduced, shorter than blade (Fig. 108) . . . . *Heleniella*

93b. Antenna with segment 2 shorter than segment 1, blade subequal to or longer than flagellum (Figs. 109, 110) 94

94a (93b). Median teeth of mentum substantially longer than 1st lateral teeth (Fig. 104); antennal blade longer than flagellum (Fig. 109); SI simple . . . *Paralimnophyes*

94b. Median teeth of mentum usually not much longer than 1st lateral teeth (Fig. 105); antennal blade subequal to flagellum (Fig. 110); SI plumose or toothed . . . . *Limnophyes*

95a (92b). Antenna at least 0.8 times as long as mandible 96

95b. Antenna less than 0.8 times as long as mandible 99

96a (95a). Antenna 5-segmented 97

96b. Antennae 6- or 7-segmented 98

97a (96a). Antennal ratio (AR) < 1.0 . . . . *Paraphaenocladius* (in part)

97b. Antennal ratio (AR) > 1.0 . . . . *Parametriocnemus*

98a (96b). Antenna 6-segmented, with ring organ near middle of segment 1 (Fig. 111); mentum with large paired median teeth (Fig. 112); SI apically toothed, but not plumose . . . . *Psilometriocnemus*

98b. Antenna 7-segmented, with ring organ in basal quarter of segment 1 (Fig. 113); mentum with single or paired median teeth (Fig. 114); SI plumose . . . . *Heterotrissocladius*

99a (95b). Mentum with median tooth single or paired, not more than 2 times as wide as 1st lateral teeth and often narrower than 1st lateral teeth; median teeth often recessed; ventromental plate poorly developed . . . . *Metriocnemus* (in part)

99b. Mentum with median tooth single or paired, more than 2 times as wide as 1st lateral teeth; ventromental plate usually well developed 100

100a (99b). Mentum with median tooth or teeth shorter than 1st lateral teeth, often having concave anterior margins (Fig. 115); medium to large body . . . . *Chaetocladius*

100b. Mentum with median tooth or teeth longer than 1st lateral teeth, with anterior margins entire (Fig. 116); small to medium body . . . . *Paraphaenocladius* (in part)

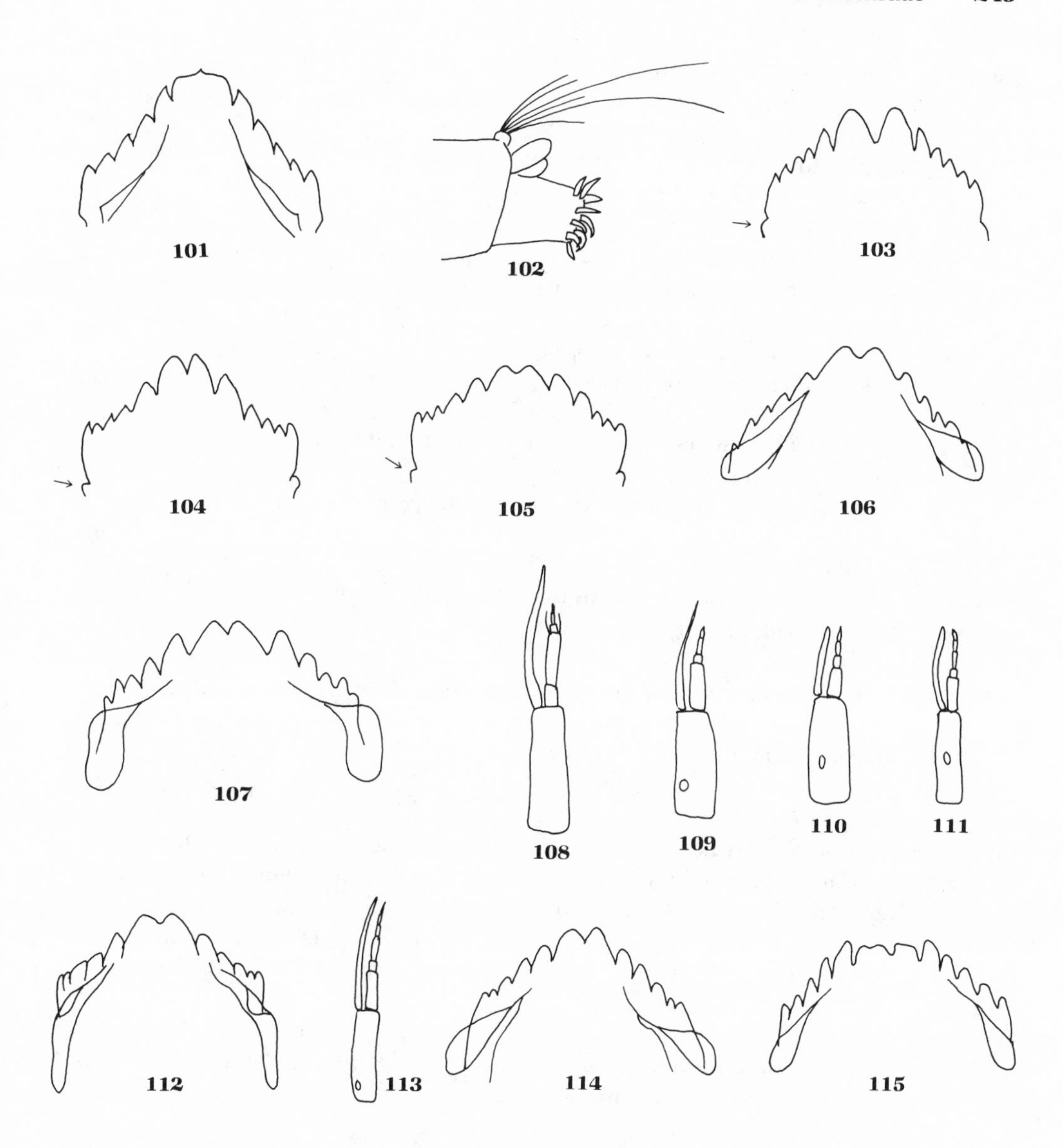

101. Mentum of *Cardiocladius albiplumus*, vv. **102.** Posterior abdominal segments of *Cardiocladius*, lv. **103.** Mentum of *Heleniella*, vv. **104.** Mentum of *Paralimnophyes*, vv. **105.** Mentum of *Limnophyes*, vv. **106.** Mentum of *Paraphaenocladius* (in part), vv. **107.** Mentum of *Parametriocnemus*, vv. **108.** Antenna of *Heleniella*. **109.** Antenna of *Paralimnophyes*. **110.** Antenna of *Limnophyes*. **111.** Antenna of *Psilometriocnemus*. **112.** Mentum of *Psilometriocnemus*, vv. **113.** Antenna of *Heterotrissocladius*. **114.** Mentum of *Heterotrissocladius*, vv. **115.** Mentum of *Chaetocladius*, vv. lv, lateral view; vv, ventral view. (111, 112 redrawn from Wiederholm 1983.)

101a (77b). Median tooth of mentum well recessed, 1st lateral tooth small and appressed to large 2nd lateral tooth (Fig. 118); ventromental plate well developed; antennal ratio (AR) ca. 2.0 (Fig. 117) . . . . *Unniella*

101b. Median tooth of mentum not markedly recessed; ventromental plates variable; antennal ratio (AR) usually < 2.0 102

102a (101b). Mandible with 4 inner teeth; premandible with 2–4 apical teeth; mentum with median teeth sometimes slightly recessed; anal setae often short . . . . *Metriocnemus* (in part)

102b. Mandible with 3 inner teeth; premandible with 1–2 apical teeth; mentum variable; anal setae not reduced 103

103a (102b). Antenna with 6 or 7 segments (although often difficult to discern); ventromental plate generally well developed 104

103b. Antenna with 4 or 5 segments; ventromental plate generally poorly developed 106

104a (103a). Mentum with single median tooth (Fig. 119) . . . . *Parakiefferiella*

104b. Mentum with paired median teeth 105

105a (104b). Median teeth of mentum shorter than 1st pair of lateral teeth (Fig. 120) . . . . *Trissocladius*

105b. Median teeth of mentum not shorter than 1st pair of lateral teeth (Fig. 121) . . . . *Hydrobaenus*

106a (103b). Mentum with 1st pair of lateral teeth appressed to elevated median tooth, central 5 teeth often paler than rest (Figs. 122, 123); mandible crenulate on outer margin (Fig. 124); lauterborn organs large and prominent . . . . *Orthocladius annectens*

106b. May possess some of these characters, but not with above combination 107

107a (106b). Some abdominal segments with setal tufts . . . . *Cricotopus* (in part)

107b. Abdominal segments lacking setal tufts 108

108a (107b). Inner margin of mandible serrate (Fig. 125) . . . . *Cricotopus bicinctus*

108b. Inner margin of mandible smooth 109

109a (108b). Mentum with more than 6 pairs of lateral teeth . . . . *Orthocladius (Euorthocladius)* (in part)

109b. Mentum with 6 pairs of lateral teeth 110

110a (109b). Antenna 4-segmented; mentum with 2nd lateral tooth reduced . . . . *Orthocladius (Pogonocladius)* (in part)

110b. Antenna 5-segmented; 2nd lateral tooth of mentum not reduced 111

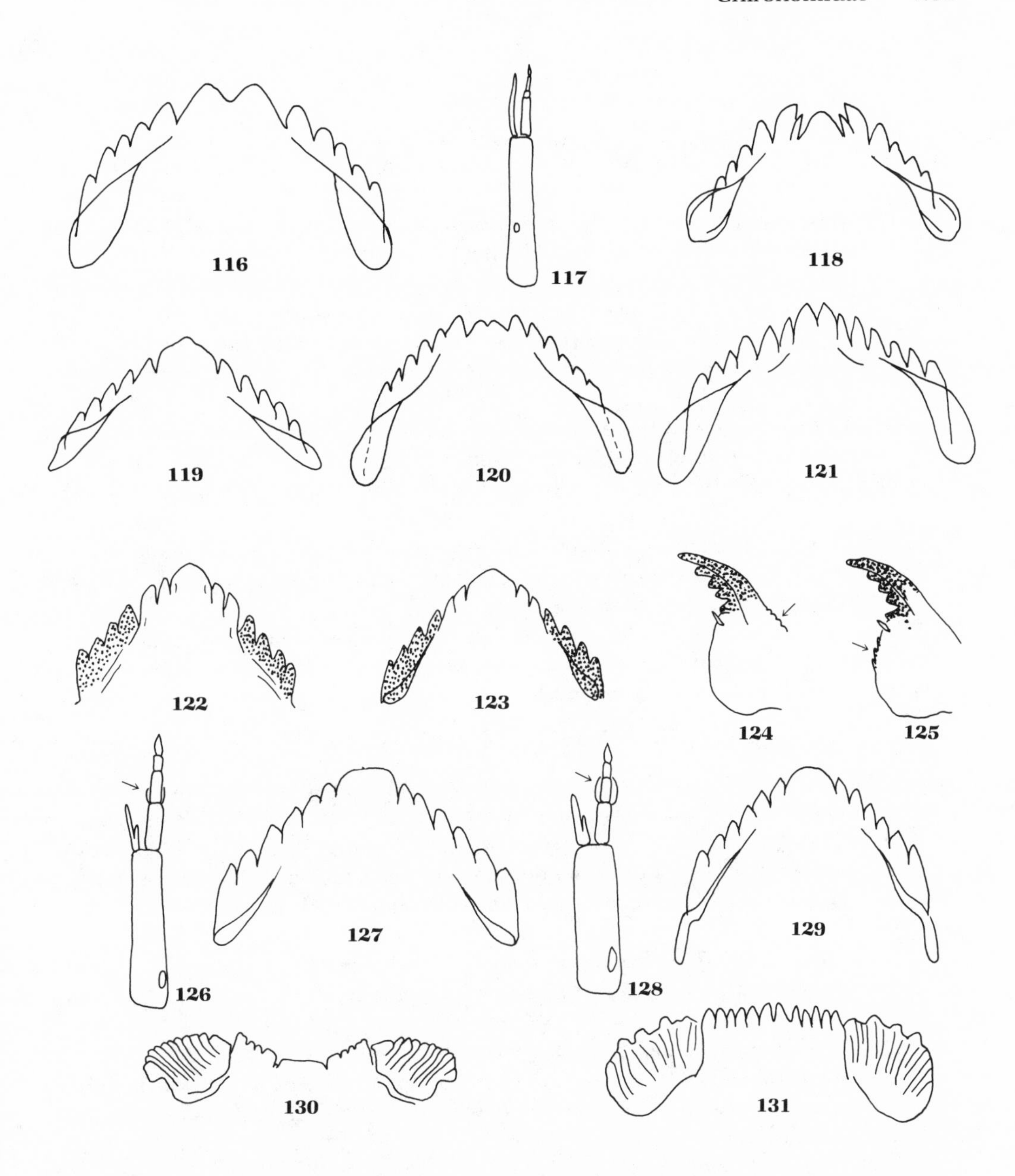

116. Mentum of *Paraphaenocladius*, dv. **117.** Antenna of *Unniella*. **118.** Mentum of *Unniella*, vv. **119.** Mentum of *Parakiefferiella*, vv. **120.** Mentum of *Trissocladius*, vv. **121.** Mentum of *Hydrobaenus*, vv. **122.** Mentum of *Orthocladius annectens*, vv. **123.** Mentum of *Cricotopus bicinctus*, vv. **124.** Mandible of *Orthocladius annectens*, vv. **125.** Mandible of *Cricotopus bicinctus*, vv. **126.** Antennae of *Cricotopus*. **127.** Mentum of *Cricotopus*, dv. **128.** Antenna of *Orthocladius*. **129.** Mentum of *Orthocladius*, dv. **130.** Mentum and ventromental plates of *Chernovskiia*, vv. **131.** Mentum and ventromental plates of *Robackia*, vv. dv, dorsal view; vv, ventral view. (120, 130 redrawn from Wiederholm 1983.)

111a (110b). Head capsule brown . . . . *Orthocladius* (in part)
111b. Head capsule yellow 112

112a (111b). Lauterborn organs usually weak and indistinct (Fig. 126); ventromental plate usually poorly developed (Fig. 127) . . . . *Cricotopus* (in part)
112b. Lauterborn organs usually distinct, as long as antennal segment 3 (Fig. 128); ventromental plate usually moderately well developed (Fig. 129) . . . . *Orthocladius* (in part)

There are no known characters for separating all larvae of *Cricotopus* from *Orthocladius*. Couplet 112 is designed to allow placement of most species to the correct genus.

113a (3a). **Chironominae:** Antenna arising from tubercle longer than wide, basal segment usually long and slightly curved; ventromental plates often close together; lauterborn organs large and conspicuous or placed on stalks . . . . Tribe Tanytarsini 157
113b. Antenna arising from tubercle not longer than wide, basal segment usually short and not curved; ventromental plates not usually close together; lauterborn organs small and sessile (Fig. 3) . . . . Tribes Chironomini and Pseudochironomini 114

114a (113b). Antenna with 6–8 segments 115
114b. Antenna with 5 segments 126

115a (114a). Antenna with 8 segments, elongate, almost as long as head, and weakly sclerotized; mentum pale and concave (Fig. 130); abdominal segments subdivided . . . . *Chernovskiia*
115b. Antenna with 6–7 segments, not as long as head, and normally sclerotized; mentum variable; abdominal segments not subdivided 116

116a (115b). Antenna with 7 segments; premandible with 4 teeth 117
116b. Antenna with 6 segments; premandible variable 119

117a (116a). Mentum with pale, dome-shaped median tooth and 7 pairs of darker lateral teeth (Fig. 132) . . . . *Demicryptochironomus*
117b. Mentum not with a pale dome-shaped median tooth and only 4–6 pairs of lateral teeth 118

118a (117b). Mentum with 4 pairs of lateral teeth; mandible with 2 inner teeth . . . . *Beckidia*
118b. Mentum with 5–6 pairs of lateral teeth (Fig. 131); mandible with 4 inner teeth . . . . *Robackia*

119a (116b). Mentum with single median tooth, often broad and dome-shaped and higher than lateral teeth (Figs. 133, 134) 120

119b. Mentum usually with paired median teeth (Figs. 135–138, 140), or if single, not higher than all lateral teeth (Fig. 141) 121

120a (119a). Ventromental plates about as wide as mentum and widely separated (Fig. 133); mandible with 2–3 flattened inner teeth; premandible with 3 teeth; SI simple .... *Saetheria*

120b. Ventromental plates wider than mentum and medially reaching median tooth of mentum (Fig. 134); mandible with 3 pointed inner teeth; premandible with 2 teeth; SI plumose .... *Paralauterborniella*

121a (119b). Ventromental plates nearly touching medially (Fig. 135); larva with movable case; mentum with 1st lateral tooth reduced (Fig. 135); mandible with 2 inner teeth 122

121b. Ventromental plates more widely separated (Figs. 136, 140); larva without movable case; mentum and mandible variable 123

122a (121a). Case with slitlike opening; submental setae plumose; mandibular subdental seta distally serrate .... *Lauterborniella*

122b. Case with circular opening; submental setae and subdental setae simple .... *Zavreliella*

123a (121b). Mentum with median 2 or 3 teeth larger than (Fig. 136) or subequal to (Fig. 137) adjacent pair of lateral teeth; central teeth paler than lateral teeth .. 124

123b. Mentum with median 2 teeth shorter and narrower than adjacent pair of lateral teeth; central teeth not paler than lateral teeth (Figs. 138, 140) 125

124a (123a). Mentum with 2 larger median teeth and often with a smaller or subequal tooth between them, central teeth paler than rest (Fig. 136); mandible with 3 inner teeth .... *Microtendipes*

124b. Mentum usually with 4 subequal central teeth, paler than rest (Fig. 137); mandible with 2 inner teeth .... *Paratendipes*

125a (123b). Mentum with 1st lateral tooth shorter than 2nd lateral tooth (Fig. 138); antenna longer than mandible .... *Omisus*

125b. Mentum with 1st lateral tooth longer than 2nd lateral tooth (Fig. 140); antenna subequal in length to mandible .... *Stictochironomus*

126a (114b). Antennal segment 2 with basal ⅔ unsclerotized (Fig. 139) ..... *Cyphomella*

126b. Antennal segment 2 fully sclerotized 127

127a (126b). Mentum with broad, dome-shaped median tooth, medially pale, and 5–7 pairs of darker oblique lateral teeth (Fig. 141) 128

127b. Not with above combination of characters 129

128a (127a). Pecten epipharyngis with 3 weak lobes; premandible with 6 narrow teeth

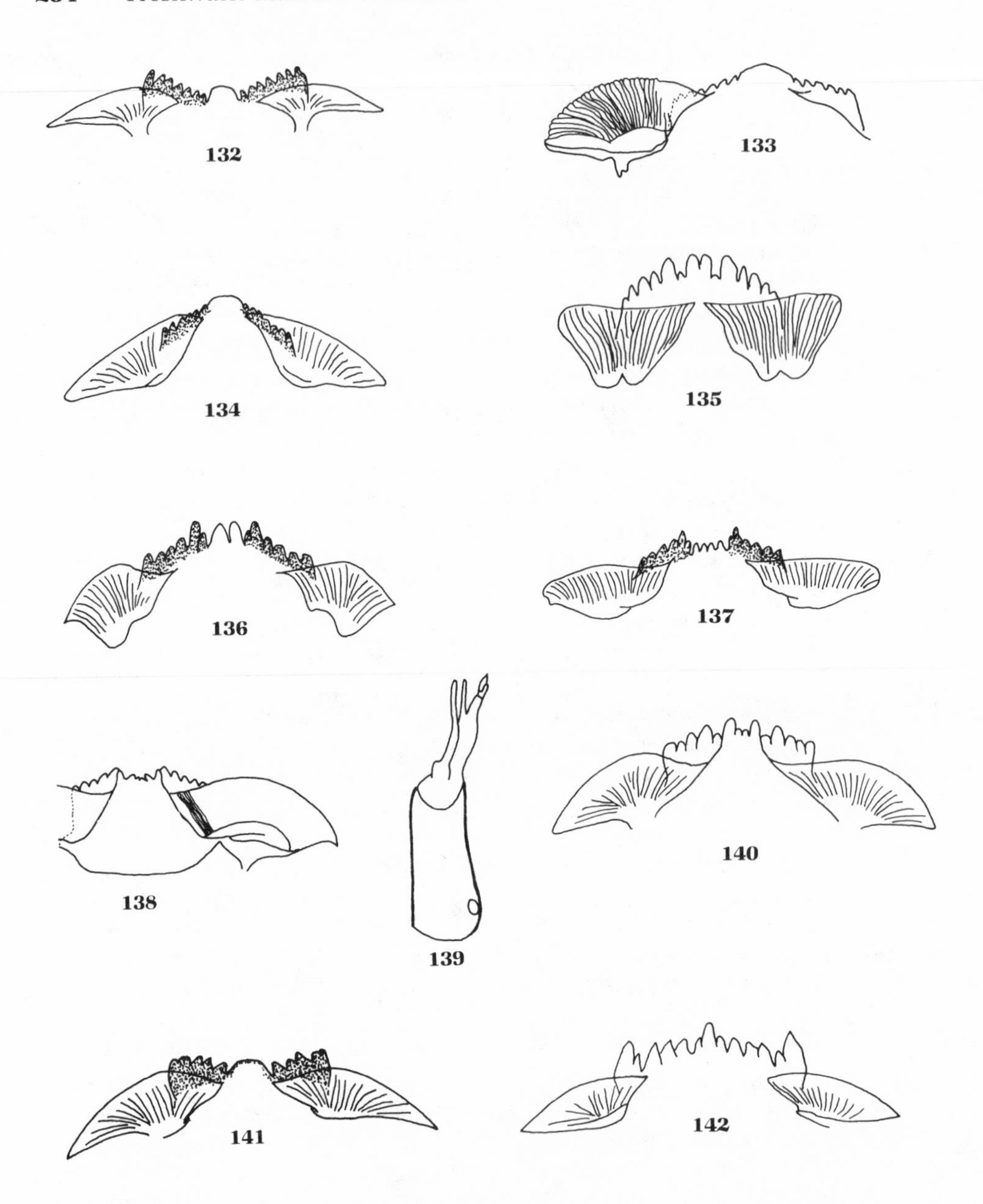

132. Mentum and ventromental plates of *Demicryptochironomus*, vv. **133.** Mentum and ventromental plates of *Saetheria*, vv. **134.** Mentum and ventromental plates of *Paralauterborniella*, vv. **135.** Mentum and ventromental plates of *Lauterborniella*, vv. **136.** Mentum and ventromental plates of *Microtendipes*, vv. **137.** Mentum and ventromental plates of *Paratendipes*, vv. **138.** Mentum and ventromental plates of *Omisus*, vv. **139.** Antenna of *Cyphomella*. **140.** Mentum and ventromental plates of *Stictochironomus*, vv. **141.** Mentum and ventromental plates of *Cryptochironomus*, vv. **142.** Mentum and ventromental plates of *Microchironomus*, vv. vv, ventral view. (133, 138, 139 redrawn from Wiederholm 1983.)

. . . . *Gillotia*

128b. Pecten epipharyngis triangular and serrate; premandible with 4–6 wide teeth . . . . *Cryptochironomus*

129a (127b). Mentum convex with outer pair of lateral teeth larger than preceding teeth (Figs. 142–144); premandible bifid 130

129b. Mentum usually not convex, outer lateral teeth not enlarged, or, if enlarged, premandible not bifid 132

130a (129a). Median tooth of mentum trifid and usually pointed (Fig. 142); antennal blade longer than flagellum . . . . *Microchironomus*

130b. Median tooth of mentum single or paired, mentum usually strongly convex; antennal blade shorter than flagellum 131

131a (130b). Median tooth of mentum broad and rounded, sometimes notched laterally (Fig. 143); antenna with basal segment subequal to flagellum . . . . *Cryptotendipes*

131b. Median teeth of mentum paired or, if a single tooth, notched medially (Fig. 144); antenna with basal segment distinctly longer than flagellum . . . . *Cladopelma*

132a (129b). Ventromental plates slender, nearly touching medially (Figs. 145, 146) 133

132b. Ventromental plates not slender, separated medially (Fig. 147) 135

133a (132a). Mandible with apical tooth pale; premandible bifid; mentum with 2nd pair of lateral teeth usually reduced (Fig. 147) . . . . *Pseudochironomus*

133b. Mandible with apical tooth darkened; premandible with at least 5 teeth; mentum not with 2nd pair of lateral teeth reduced 134

134a (133b). Mentum with median tooth trifid; mandibular teeth flattened; antennal pedestal without tubercle . . . . *Axarus*

134b. Mentum with 4 small median teeth; mandibular teeth not flattened; antennal pedestal with tubercle . . . . *Lipiniella*

135a (132b). Head capsule with prominent lamellar brush projecting anteriorly (Fig. 148); mentum with median tooth trifid and recessed below 1st lateral teeth (Fig. 147); mandible with 3 minute inner teeth . . . . *Xenochironomus*

135b. Head capsule without projecting lamellar brush; mentum and mandible not as above 136

136a (135b). Mandible with subdental seta serrate on inner margin (Fig. 149); mentum with median tooth trifid, some lateral teeth overlapping; 8th abdominal segment with 2 pairs of ventral tubules . . . . *Goeldichironomus*

136b. Mandible with subdental seta not serrate; mentum variable; 8th abdominal

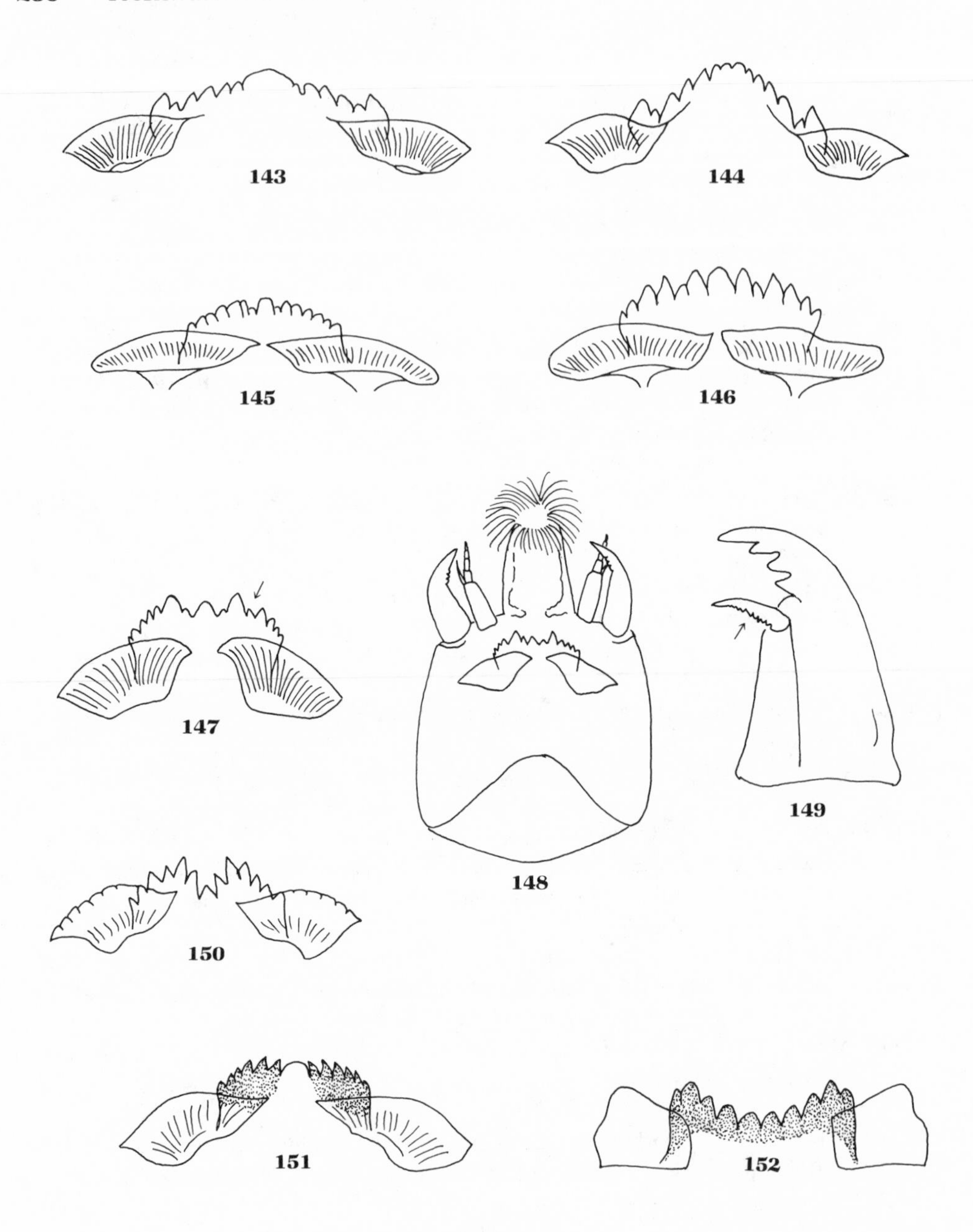

143. Mentum and ventromental plates of *Cryptotendipes*, vv. **144.** Mentum and ventromental plates of *Cladopelma*, vv. **145.** Mentum and ventromental plates of *Axarus*, vv. **146.** Mentum and ventromental plates of *Pseudochironomus*, vv. **147.** Mentum and ventromental plates of *Xenochironomus*, vv. **148.** Head capsule of *Xenochironomus*, vv. **149.** Mandible of *Goeldichironomus*, dv. **150.** Mentum and ventromental plates of *Stelechomyia*, vv. **151.** Mentum and ventromental plates of *Nilothauma*, vv. **152.** Mentum and ventromental plates of *Stenochironomus*, vv. dv, dorsal view; vv, ventral view.

segment usually not with 2 pairs of ventral tubules (except in *Chironomus*) 137

137a (136b). Midportion of mentum concave, with median tooth and 1st lateral teeth recessed (Fig. 150); mandible with 4 flattened inner teeth ...... *Stelechomyia*

137b. Mentum not as described above; mandibular teeth usually pointed 138

138a (137b). Mandible pale with 4 small dark inner teeth in a convex arrangement, apical tooth long and slender; mentum with convex or truncate median tooth and 6 pairs of lateral teeth (Fig. 151) .... *Nilothauma*

138b. Mandible not as described above; mentum variable 139

139a (138b). Mentum with single wide median tooth, usually pale, sometimes with notches, and at least 4 times as wide as lateral tooth; 6–7 pairs of lateral teeth present 140

139b. Mentum not with single wide median tooth 141

140a (139a). Antennal segment 2 subequal in length to segment 3; mandible with 1–2 flattened inner teeth; ventromental plate only weakly striate ..... *Harnischia*

140b. Antennal segment 2 longer than segment 3, segments 3–5 reduced; mandible with 2–3 pointed inner teeth; ventromental plate coarsely striate .... *Paracladopelma*

141a (139b). Mentum concave, with 10 large dark teeth (Fig. 152); mandible robust, triangular, distally dark; ventromental plates difficult to discern .... *Stenochironomus*

141b. Mentum and mandible not as described above; ventromental plates apparent 142

142a (141b). Mentum pale (no dark teeth), with paired median teeth and minute 1st lateral teeth (Fig. 153); premandible with 3 teeth, middle 1 broad and blunt and inner and outer both narrow and pointed; small, up to 4 mm long .... *Pagastiella*

142b. Not with above combination of characters 143

143a (142b). Median tooth of mentum appearing trifid (Fig. 154) 144

143b. Median tooth of mentum not appearing trifid although sometimes with shallow lateral notches 146

144a (143a). Premandible with 5–7 teeth; 8th abdominal segment with 1 pair of ventral tubules .... *Kiefferulus*

144b. Premandible bifid; 8th abdominal segment with 0–2 pairs of ventral tubules 145

145a (144b). Eighth abdominal segment usually with 2 pairs of ventral tubules, if ventral tubules absent, pecten epipharyngis with more than 7 teeth .... *Chironomus*

145b. Eighth abdominal segment with 0–1 pair of ventral tubules, if ventral tubules absent, pecten epipharyngis with less than 7 teeth . . . . *Einfeldia* (in part)

146a (143b). Mentum with single median tooth, rounded, pointed, or notched (Figs. 155, 156) 147
146b. Mentum with paired median teeth (Figs. 157–159) 151

147a (146a). Mentum with median tooth medially pointed or notched, with 7 pairs of narrow, pointed lateral teeth of decreasing size (Fig. 155); ventromental plate crenulate anteriorly; basal segment of antenna distinctly longer than flagellum . . . . *Parachironomus* (in part)
147b. Mentum with median tooth variable and 6 pairs of lateral teeth (Fig. 156); ventromental plate crenulate or smooth anteriorly; basal segment of antenna generally subequal to flagellum 148

148a (147b). Ventromental plate not as wide as mentum 149
148b. Ventromental plate wider than mentum (Fig. 156) 150

149a (148a). Mandible with 2 inner teeth; ventromental plate not crenulate anteriorly; 8th abdominal segment with 1 pair of ventral tubules . . . . *Demeijerea*
149b. Mandible with 3–4 inner teeth; ventromental plate crenulate anteriorly; 8th abdominal segment lacking ventral tubules . . . . *Dicrotendipes*

150a (148b). Ventromental plates usually crenulate anteriorly (Fig. 156); mandible usually with large, single pale dorsal tooth . . . . *Glyptotendipes*
150b. Ventromental plates usually smooth anteriorly; mandible usually with 2 small dorsal teeth . . . . *Einfeldia* (in part)

151a (146b). Mentum with median teeth partially fused, may appear as single tooth with medial notch (Fig. 160); 6–7 pairs of narrow, acutely pointed lateral teeth of decreasing size; mandible with 2 inner teeth . . . . *Parachironomus* (in part)
151b. Mentum not as described above, lateral teeth not so narrow and acutely pointed; mandibles usually with 3 inner teeth 152

152a (151b). Mentum with 1st lateral teeth shorter than 2nd lateral teeth (Fig. 157), or teeth of mentum gradually decreasing in size laterally 153
152b. Mentum with 1st lateral teeth not reduced; usually with 2nd lateral teeth reduced 154

153a (152a). Ventromental plates more than 3 times as wide as long (Fig. 161) . . *Asheum*
153b. Ventromental plates less than 3 times as wide as long (Fig. 157) . . . . *Polypedilum*

154a (152b). Median pair of teeth of mentum deeply recessed below 1st lateral teeth, 6 pairs of lateral teeth present (Fig. 158); basal inner tooth of mandible with 2–3 points . . . . *Hyporhygma*

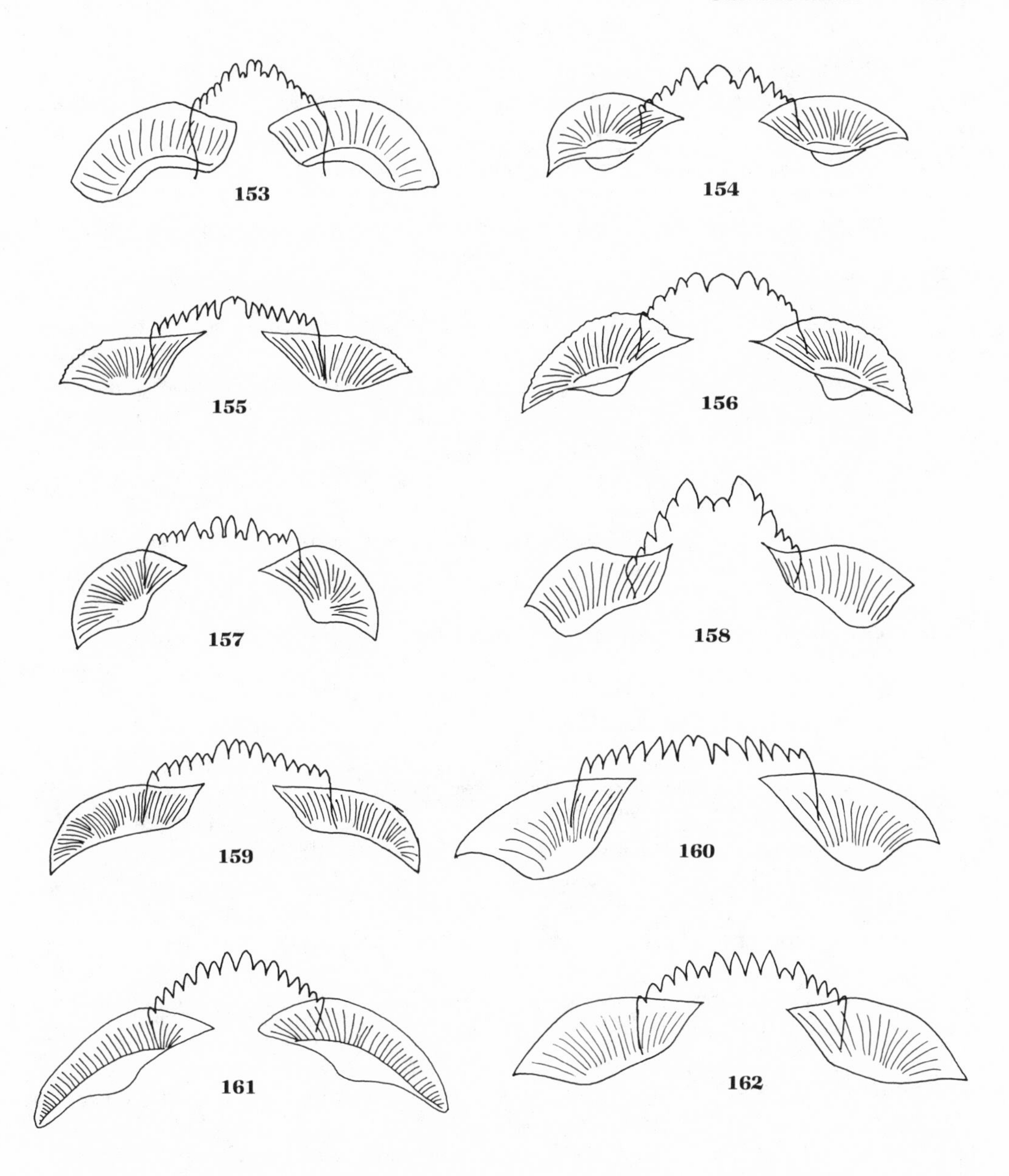

153. Mentum and ventromental plates of *Pagastiella*, vv. **154.** Mentum and ventromental plates of *Chironomus*, vv. **155.** Mentum and ventromental plates of *Parachironomus*, vv. **156.** Mentum and ventromental plates of *Glyptotendipes*, vv. **157.** Mentum and ventromental plates of *Polypedilum*, vv. **158.** Mentum and ventromental plates of *Hyporhygma*, vv. **159.** Mentum and ventromental plates of *Endochironomus*, vv. **160.** Mentum of *Parachironomus*, vv. **161.** Mentum of *Asheum*, dv. **162.** Mentum of *Phaenopsectra*, vv. dv, dorsal view; vv, ventral view. (158 redrawn from Wiederholm 1983.)

154b. Median teeth of mentum not so deeply recessed, 7 pairs of lateral teeth present (Fig. 159); basal inner tooth of mandible lacking additional points 155

155a (154b). Median teeth of mentum mostly fused; ventromental plates 3–4 times as wide as long (Fig. 159) .... *Endochironomus*

155b. Median teeth of mentum only partially fused (Fig. 162); ventromental plates less than 3 times as wide as long 156

156a (155b). Antennal blade distinctly longer than flagellum; premandible with 2 apical teeth and no basal tooth .... *Tribelos*

156b. Antennal blade subequal to flagellum; premandible with basal tooth in addition to 2 apical teeth .... *Phaenopsectra*

157a (113a). Ventromental plates nearly touching medially, inner margins usually truncate or rounded 158

157b. Ventromental plates widely separated, inner margins pointed; larva with portable case of sand or detritus 163

158a (157a). Lauterborn organs on stalks shorter than length of antennal segment 3 (Figs. 163, 164) 159

158b. Lauterborn organs on stalks longer than or equal to length of antennal segment 3 (Figs. 165, 166) 160

159a (158a). Antennal segment 2 shorter than segment 3 and distinctly widened distally (Fig. 163); lauterborn organs on short stalks .... *Cladotanytarsus*

159b. Antennal segment 2 longer than segment 3 and only slightly widened distally (Fig. 164); lauterborn organs sessile or on very short stalks .... *Paratanytarsus*

160a (158b). Lauterborn organs on stalks much longer than antennal segments 3–5 (Figs. 165, 166) 161

160b. Lauterborn organs on stalks no more than slightly longer than antennal segments 3–5 162

161a (160a). Premandible bifid (Fig. 167); posterior prolegs with more than 20 claws .... *Micropsectra*

161b. Premandible with 3–5 teeth (Fig. 168); posterior prolegs with less than 20 claws, usually only 10–12 .... *Tanytarsus*

162a (160b). Mentum strongly arched, with 3 central teeth subequal (Fig. 169); antennal segment 2 distinctly widened distally; lauterborn organs extending slightly beyond antennal apex; head capsule brown .... *Sublettea*

162b. Mentum not so strongly arched, median tooth often notched or trifid (Fig. 170); antennal segment 2 only slightly widened distally; lauterborn organs not extending beyond antennal apex; head capsule usually yellow .... *Rheotanytarsus*

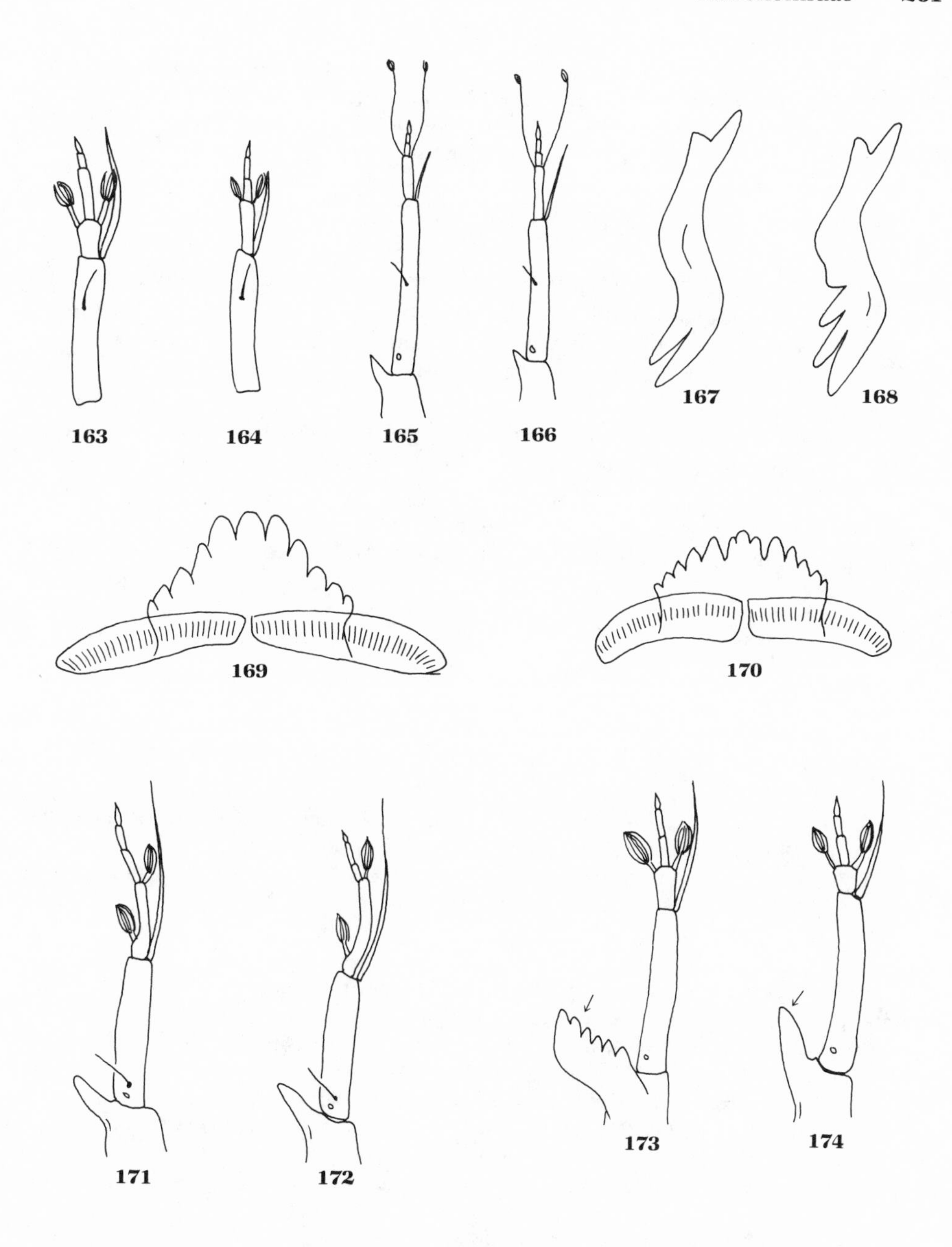

163. Antenna of *Cladotanytarsus*. **164.** Antenna of *Paratanytarsus*. **165.** Antenna of *Micropsectra*. **166.** Antenna of *Tanytarsus*. **167.** Premandible of *Micropsectra*, dv. **168.** Premandible of *Tanytarsus*, dv. **169.** Mentum and ventromental plates of *Sublettea*, vv. **170.** Mentum and ventromental plates of *Rheotanytarsus*, vv. **171.** Antenna of *Zavrelia*. **172.** Antenna of *Stempellinella*. **173.** Antenna of *Stempellina*. **174.** Antenna of *Constempellina*. dv, dorsal view; vv, ventral view.

163a (157b). Lauterborn organs originating at different heights on antennal segment 2, 1 proximally and 1 distally (Figs. 171, 172); antennal segment 2 subequal to or longer than combined lengths of segments 3–5 (Figs. 171, 172) 164

163b. Lauterborn organs both originating from distal end of antennal segment 2 (Figs. 173, 174); antennal segment 2 shorter than combined lengths of segments 3–5 (Figs. 173, 174) 165

164a (163a). Antennal segment 2 subequal in length to combined lengths of segments 3–5 (Fig. 171); premandible with 4–5 teeth .... *Zavrelia*

164b. Antennal segment 2 distinctly longer than combined lengths of segments 3–5 (Fig. 172); premandible with 2–3 teeth .... *Stempellinella*

165a (163b). Antennal tubercle with palmate process (Fig. 173); head capsule often with dorsal humps or projections; procercus and anal setae often strongly modified .... *Stempellina*

165b. Antennal tubercle with simple spur (Fig. 174); head capsule sometimes dorsally granular, but lacking projections; procercus and anal setae mostly normal .... *Constempellina*

References on Chironomidae Systematics

(*Used in construction of key.)

Ashe, P. 1983. A catalogue of chironomid genera and subgenera of the world including synonyms (Diptera: Chironomidae). Entomol. Scand. Suppl. 17:1–68.

Beckett, D.C., and P.A. Lewis. 1982. An efficient procedure for the slide mounting of larval chironomids. Trans. Amer. Microsc. Soc. 101(1):96–99.

Berg, C.O. 1950. Biology of certain Chironomidae reared from *Potamogeton*. Ecol. Monogr. 20:83–101.

Bode, R.W. 1980. Larvae and pupae of *Eukiefferiella* (Diptera: Chironomidae) found in New York State. M.S. thesis. Cornell Univ., Ithaca, N.Y. 178 pp.

——. 1983. Larvae of North American *Eukiefferiella* and *Tvetenia* (Diptera: Chironomidae). Bull. N.Y. State Mus. 452:1–40.

Boesel, M.W. 1983. A review of the genus *Cricotopus* in Ohio with keys to adults of species of the northeastern United States (Diptera: Chironomidae). Ohio J. Sci. 83:74–90.

——. 1985. A brief review of the genus *Polypedilum* in Ohio, with keys to known stages of species occurring in northeastern United States (Diptera: Chironomidae). Ohio J. Sci. 85:245–262.

Boesel, M.W., and R.W. Winner. 1980. Corynoneurinae of northeastern United States with a key to adults and observations on their occurrence in Ohio (Diptera: Chironomidae). J. Kans. Entomol. Soc. 53:501–508.

Borkent, A. 1984. The systematics and phylogeny of the *Stenochironomus* complex (*Xestochironomus, Harrisus,* and *Stenochironomus*) (Diptera: Chironomidae). Mem. Entomol. Soc. Can. 128:1–269.

Bryant, R.M., Jr., and R.J. Smith. 1985. Use of sodium dodecyl sulfate (SDS) as a clearing agent for chironomid larvae prior to slide mounting. Freshwat. Invert. Biol. 4:48–52.

Coffman, W.P. 1973. Energy flow in a woodland stream ecosystem. II. The taxonomic composition and phenology of the Chironomidae as determined by the collection of pupal exuviae. Arch. Hydrobiol. 71:281–322.

*Coffman, W.P., and L.C. Ferrington, Jr. 1984. Chironomidae. *In* R.W. Merritt and K.W. Cummins (eds.). An introduction to the aquatic insects of North America, 2nd ed., pp. 551–652. Kendall/Hunt Publ. Co., Dubuque, Iowa.

Coffman, W.P., and S.S. Roback. 1984. *Lopescladius (Cordiella) hyporheicus,* a new subgenus and species (Diptera: Chironomidae; Orthocladiinae). Proc. Acad. Nat. Sci. Phila. 136:130–144.

Coffman, W.P., and W.H. Walker. 1977. A selected bibliography of the family Chironomidae (Diptera: Nematocera). Spec. Publ. Pymatuning Lab. Ecol. Univ. Pittsburgh, Linesville, Penn. 50 pp.

Cranston, P.S. 1982. The metamorphosis of *Symposiocladius lignicola* (Kieffer) n. gen., n. comb., a wood-mining Chironomidae (Diptera). Entomol. Scand. 13:419–429.

Cranston, P.S., and D.R. Oliver. 1988. Additions and corrections to the Nearctic Orthocladiinae (Diptera: Chironomidae). Can. Entomol. 120:425–462.

Curry, L.L. 1958. Larvae and pupae of the species of *Cryptochironomus* (Diptera) in Michigan. Limnol. Oceanogr. 3:427–442.

——. 1961. A key for the larval forms of aquatic midges (Tendepedidae: Diptera) found in Michigan. Atom. Energy Comm. Contract (11–1)–350. Rept. no. 1. 145 pp.

Doughman, J.S. 1983. A guide to the larvae of Nearctic Diamesinae (Diptera: Chironomidae). The genera *Boreoheptagyia, Protanypus, Diamesa,* and *Pseudokiefferiella.* USGS Water Res. Inv. Rpt. no. 83–4006. 58 pp.

——. 1985a. Annotated keys to the genera of the tribe Diamesini (Diptera: Chironomidae). Inst. Wat. Res. Univ. Alaska-Fairbanks. Rept. 1WR–107. 37 pp.

——. 1985b. *Sympotthastia* Pagast (Diptera: Chironomidae), an update based on larvae from North Carolina, *S. diastena* (Sublette) comb. n., and other Nearctic species. Brimleyana 11:39–53.

Epler, J.H. 1987. Revision of the Nearctic *Dicrotendipes* Kieffer, 1913 (Diptera: Chironomidae). Evol. Monogr. 9. 102 pp. + plates.

——. 1988. Biosystematics of the genus *Dicrotendipes* Kieffer, 1913 (Diptera: Chironomidae: Chironominae) of the world. Mem. Amer. Entomol. Soc. 36:1–214.

Ferrington, L.C., Jr. 1987. Collection and identification of surface-floating pupal exuviae of Chironomidae for use in studies of surface water quality. US EPA Region VII SOP no. FW130A. 38 pp.

*Fittkau, E.J., F. Reiss, and O. Hoffrichter. 1976. A bibliography of the Chironomidae. Gunneria 26:1–177.

Grodhaus, G. 1987. *Endochironomus* Kieffer, *Tribelos* Townes, *Synendotendipes,* n. gen., and *Endotribelos,* n. gen. (Diptera: Chironomidae) of the Nearctic region. J. Kans. Entomol. Soc. 60:167–247.

Hansen, D.C., and E.F. Cook. 1976. The systematics and morphology of the Nearctic species of *Diamesa* Meigen, 1835 (Diptera: Chironomidae). Mem. Amer. Entomol. Soc. 30:1–203.

Harp, G., and R.S. Campbell. 1967. The distribution of *Tendipes plumosus* (Linne) in mineral acid water. Limnol. Oceanogr. 12:260–263.

Jackson, G.A. 1977. Nearctic and Palearctic *Paracladopelma* Harnish and *Saetheria* n. gen. (Diptera: Chironomidae). J. Fish. Res. Bd. Can. 34:1321–1329.

*Johannsen, O.A. 1937a. Aquatic Diptera. Part III. Chironomidae: subfamilies Tanypodinae, Diamesinae, and Orthocladiinae. Mem. Cornell Univ. Agric. Exp. Stn. 205:1–84.

*——. 1937b. Aquatic Diptera. Part IV. Chironomidae: subfamily Chironominae. Mem. Cornell Univ. Agric. Exp. Stn. 210:1–56.

——. 1946. Revision of the North American species of the genus *Pentaneura* (Tendipedidae: Chironomidae, Diptera). J. N.Y. Entomol. Soc. 54:267–289.

*——. 1952. Guide to the insects of Connecticut. Part VI. The Diptera or true flies of Connecticut. Fasc. 5. Midges and gnats. Family Tendipedidae (= Chironomidae) except Tendipedini. Bull. Conn. State Geol. Nat. Hist. Surv. 80:3–26, 232–250.

Maschwitz, D.E. 1976. Revision of the Nearctic species of the subgenus *Polypedilum* (Chironomidae: Diptera). Ph.D. dissertation. Univ. Minnesota, Minneapolis-St. Paul. 325 pp.

*Mason, W.T., Jr. 1973. An introduction to the identification of chironomid larvae. Div. Pollut. Surv., Fed. Water Pollut. Control Admin. Cincinnati. 89 pp.

*Murray, D.A. 1980. Chironomidae. Ecology, systematics, cytology and physiology. Pergamon, New York. 354 pp.

Murray, D.A., and E.J. Fittkau. 1985. *Hayesomyia,* a new genus of Tanypodinae from the Holarctic (Diptera: Chironomidae). Spixiana Suppl. 11:195–207.

Neumann, D. 1976. Adaptations of chironomids to intertidal environments. Ann. Rev. Entomol. 21:387–414.

Oliver, D.R. 1971. Life history of the Chironomidae. Ann. Rev. Entomol. 16:211–230.

*——. 1981a. Chironomidae. *In* J.F. McAlpine, B.V. Peterson, G.E. Shewell, H.J. Teskey, J.R. Vockeroth, and D.M. Wood (coords.). Manual of the Nearctic Diptera, vol. 1, pp. 423–458. Res. Branch Agric. Can. Monogr. no. 27. Ottawa. 674 pp.

——. 1981b. Description of *Euryhapsis* new genus including three new species (Diptera: Chironomidae). Can. Entomol. 113:711–722.

——. 1982. *Xylotopus,* a new genus of Orthocladiinae (Diptera: Chironomidae). Can. Entomol. 114:167–168.

Oliver, D.R., and R.W. Bode. 1985. Description of the larva and pupa of *Cardiocladius albiplumus* Saether (Diptera: Chironomidae). Can. Entomol. 117:803–809.

Oliver, D.R., and M.E. Dillon. 1988. Review of *Cricotopus* (Diptera: Chironomidae) of the Nearctic zone with description of two new species. Can. Entomol. 120:463–496.

Oliver, D.R., D. McClymont, and M.E. Roussel. 1978. A key to some larvae of Chironomidae (Diptera) from the Mackenzie and Porcupine River watersheds. Can. Fish. Mar. Serv. Tech. Rep. 791:1–73.

Oliver, D.R., and M.E. Roussel. 1982. The larvae of *Pagastia* Oliver (Diptera: Chironomidae) with descriptions of three Nearctic species. Can. Entomol. 114:849–854.

——. 1983. Redescription of *Brillia* Kieffer (Diptera: Chironomidae) with descriptions of Nearctic species. Can. Entomol. 114:257–279.

Pinder, L.C.V. 1986. Biology of freshwater Chironomidae. Ann. Rev. Entomol. 31:1–24.

Roback, S.S. 1962. The genus *Xenochironomus* (Diptera: Tendipedidae) Kieffer, taxonomy and immature stages. Trans. Amer. Entomol. Soc. 88:235–245.

——. 1969. The immature stages of the genus *Tanypus* Meigen. Trans. Amer. Entomol. Soc. 94:407–428.

——. 1971. The adults of the subfamily Tanypodinae (= Pelopiinae) in North America (Diptera: Chironomidae). Monogr. Acad. Nat. Sci. Phila. 17:1–410.

——. 1974. The immature stages of the genus *Coelotanypus* (Chironomidae: Tanypodinae: Coelotanypodini) in North America. Proc. Acad. Nat. Sci. Phila. 126:9–19.

——. 1976. The immature chironomids of the eastern United States. I. Introduction and Tanypodinae-Coelotanypodini. Proc. Acad. Nat. Sci. Phila. 127:147–201.

——. 1977. The immature chironomids of the eastern United States. II. Tanypodinae-Tanypodini. Proc. Acad. Nat. Sci. Phila. 128:55–87.

——. 1978. The immature chironomids of the eastern United States. III. Tanypodinae-Anatopyniini, Macropelopiini and Natarsiini. Proc. Acad. Nat. Sci. Phila. 129:151–202.

——. 1979. *Hudsonimyia karelena,* a new genus and species of Tanypodinae, Pentaneurini. Proc. Acad. Nat. Sci. Phila. 131:1–8.

——. 1980. The immature chironomids of the eastern United States. IV. Tanypodinae-Procladiini. Proc. Acad. Nat. Sci. Phila. 132:1–62.

——. 1981. The immature chironomids of the eastern United States. V. Pentaneurini-*Thienemannimyia* group. Proc. Acad. Nat. Sci. Phila. 133:73–128.

—— (ed.). 1983. Proc. 8th Int. Symp. on Chironomidae, July 25–28, 1982, Jacksonville, Fla. Mem. Amer. Entomol. Soc. 34:1–385.

——. 1985. The immature chironomids of the eastern United States. VI. Pentaneurini-Genus *Ablabesmyia*. Proc. Acad. Nat. Sci. Phila. 137:153–212.

——. 1986a. The immature chironomids of the eastern U.S.A. VII. Pentaneurini genus *Monopelopia*. New record with redescription of the male adults and description of some Neotropical material. Proc. Acad. Nat. Sci. Phila. 138:350–365.

——. 1986b. The immature chironomids of the eastern U.S.A. VIII. Pentaneurini genus *Nilotanypus* with the description of a new species from Kansas. Proc. Acad. Nat. Sci. Phila. 138:443–465.

——. 1987. The immature chironomids of the eastern United States. IX. Pentaneurini—Genus *Labrundinia*, with the description of some Neotropical material. Proc. Acad. Nat. Sci. Phila. 139:211–221.

Roback, S.S., D.J. Bereza, and M.F. Vidrine. 1980. Description of an *Ablabesmyia* (Diptera: Chironomidae; Tanypodinae) symbiont of unionid freshwater mussels (Mollusca: Bivalvia; Unionacea), with notes on its biology and zoogeography. Trans. Amer. Entomol. Soc. 105: 577–619.

Roback, S.S., and K.J. Tennessen. 1978. The immature stages of *Djalmabatista pulcher* [= *Procladius (Calotanypus) pulcher* (Joh.)]. Proc. Acad. Nat. Sci. Phila. 129:11–20.

Rosenberg, D.M., and A.P. Wiens. 1976. Community and species responses of Chironomidae (Diptera) to contamination of fresh waters by crude oil and petroleum products, with special reference to the Trail River, Northwest Territories. J. Fish. Res. Board Can. 33:1955–1963.

Rosenberg, D.M., A.P. Wiens, and B. Bilyj. 1980. Sampling emerging Chironomidae (Diptera) with submerged funnel traps in a new northern Canadian reservoir, Southern Indiana Lake, Manitoba. Can. J. Fish. Aquat. Sci. 37:927–936.

Saether, O.A. 1975a. Nearctic and Palearctic *Heterotrissocladius* (Diptera: Chironomidae). Bull. Fish. Res. Bd. Can. 193:1–67.

——. 1975b. Twelve new species of *Limnophyes* Eaton, with keys to Nearctic males of the genus (Diptera: Chironomidae). Can. Entomol. 107:1029–1056.

——. 1976. Revision of *Hydrobaenus, Trissocladius, Zalutschia, Paratrissocladius,* and some related genera (Diptera: Chironomidae). Bull. Fish. Res. Bd. Can. 195:1–287.

——. 1977a. Taxonomic studies on Chironomidae: *Nanocladius, Pseudochironomus,* and the *Harnischia* complex. Bull. Fish. Res. Bd. Can. 196:1–143.

——. 1977b. Female genitalia in Chironomidae and other Nematocera: morphology, phylogenies, keys. Bull. Fish. Res. Bd. Can. 197:1–209.

——. 1979a. Workshop on Chironomidae: introduction. 26th Ann. Mtg. N. Amer. Benthol. Soc., May 10, 1978, Winnipeg, Canada. Entomol. Scand. Suppl. 10:15–16.

—— (ed.). 1979b. Recent developments in chironomid studies (Diptera: Chironomidae). Entomol. Scand. Suppl. no. 10. 150 pp.

——. 1979c. Chironomid communities as water quality indicators. Holarct. Ecol. 2:65–74.

*——. 1980. Glossary of chironomid morphology and terminology. Entomol. Scand. Suppl. 14:1–51.

——. 1981. *Doncricotopus bicaudatus* n. gen., n. sp. (Diptera: Chironomidae, Orthocladiinae) from the Northwest Territories, Canada. Entomol. Scand. 12:223–229.

——. 1982. Orthocladiinae (Diptera: Chironomidae) from SE U.S.A., with descriptions of *Plhudsonia, Unniella* and *Platysmittia* n. genera and *Atelopodella* n. subgen. Entomol. Scand. 13:465–510.

——. 1983a. Three new species of *Lopescladius* Oliveira, 1967 (syn. "*Cordites*" Brundin, 1966, n. syn) with a phylogeny of the *Parakiefferiella* group. Mem. Amer. Entomol. Soc. 34:279–298.

——. 1983b. *Oschia dorsenna* n. gen. n. sp. and *Saetheria hirta* n. sp., two new members of the *Harnischia* complex (Diptera: Chironomidae). Entomol. Scand. 14:395–404.

——. 1983c. A review of Holarctic *Gymnometriocnemus* Goetghebuer, 1932, with the description of *Raphidocladius* subgen. n. and *Sublettiella* gen. n. (Diptera: Chironomidae). Aquat. Ins. 5:209–226.

——. 1985a. A review of *Odontomesa* Pagast, 1947 (Diptera: Chironomidae, Prodiamesinae). Spixiana Suppl. 11:15–29.

——. 1985b. The imagines of *Mesosmittia* Brundin, 1956, with description of seven new species (Diptera: Chironomiadae). Spixiana Suppl. 11:37–54.

——. 1985c. A review of *Rheocricotopus* Thienemann & Harnisch, 1932, with the description of three new species (Diptera: Chironomidae). Spixiana Suppl. 11:59–108.

Saether, O.A., and G.A. Halvorsen. 1981. Diagnoses of *Tvetenia* Kieff. emend., *Dratnalia* n. gen. and *Eukiefferiella* Thien. emend., with a phylogeny of the *Cardiocladius* group (Diptera: Chironomidae). Entomol. Scand. Suppl. 15:269–285.

Saether, O.A., and J.E. Sublette. 1983. A review of the genera *Doithrix* n. gen., *Georthocaldius* Strenzke, *Parachaetocladius* Wülker, and *Pseudorthocladius* Goetghebuer (Diptera: Chironomidae, Orthocladiinae). Entomol. Scand. Suppl. 20:1–100.

Sawedal, L. 1981. Taxonomy, morphology, phylogenetic relationships and distribution of *Micropsectra* Kierrer, 1909 (Diptera: Chironomidae). Entomol. Scand. 13:371–400.

*Simpson, K.W. 1982. A guide to basic taxonomic literature for the genera of North American Chironomidae (Diptera)—adults, pupae, and larvae. Bull. N.Y. State Mus. 447:1–43.

*Simpson, K.W., and R.W. Bode. 1980. Common larvae of Chironomidae (Diptera) from New York State streams and rivers. Bull. N.Y. State Mus. 439:1–105.

Simpson, K.W., R.W. Bode, and P. Albu. 1983. Keys for the genus *Cricotopus* adapted from "Revision der Gattung *Cricotopus* Vander Wulp und ihrer Verwandten (Diptera, Chironomidae)" by M. Hirvenoja. Bull. N.Y. State Mus. 450:1–133.

Smith, I., and D.R. Oliver. 1976. The parasitic associations of larval water mites with aquatic insects, especially Chironomidae. Can. Entomol. 108:1427–1442.

Soponis, A.R. 1977. A revision of the Nearctic species of *Orthocladius (Orthocladius)* Van Der Wulp (Diptera: Chironomidae). Mem. Entomol. Soc. Can. 102:1–187.

Steffan, A.W. 1968. On the epizoic associations of Chironomidae (Diptera) and their phyletic relationships. Proc. 12th Int. Congr. Entomol., July 8–16, 1964, London. 1:77–78.

Steiner, J.W., J.S. Doughman, and C.R. Moore. 1982. A generic guide to the larvae of the Nearctic Tanytarsini. USGS Open-File Rpt. 82–768.

Storey, A.W., and L.C.V. Pinder. 1985. Mesh-size and efficiency of sampling of larval Chironomidae. Hydrobiologia 124:193–198.

Sublette, J.E. 1967. Type specimens of Chironomidae (Diptera) in the Cornell University collection. J. Kans. Entomol. Soc. 40:477–564.

——. 1979. Scanning electron microscopy as a tool in taxonomy and phylogeny of Chironomidae (Diptera). Entomol. Scand. Suppl. 10:47–65.

Surber, E.W. 1959. *Cricotopus bicinctus,* a midgefly resistant to electroplating wastes. Trans. Amer. Fish. Soc. 88:111–116.

Townes, H.K., Jr. 1945. The Nearctic species of Tendipedini. Amer. Midl. Nat. 34:1–206.

——. 1952. Guide to the insects of Connecticut. Part VI. The Diptera of true flies of Connecticut. Fasc. 5. Midges and gnats. Tribe Tendipedini (= Chironomini). Bull. Conn. State Geol. Nat. Hist. Surv. 80:27–147, 232–250.

Walker, I.R., and C.G. Paterson. 1985. Efficient separation of subfossil Chironomidae from lake sediments. Hydrobiologia 122:189–192.

Walshe, B.M. 1951. The feeding habits of certain chironomid larvae (subfamily Tendipedinae). Proc. Zool. Soc. Lond. 121:63–79.

Warwick, W.F. 1980. Chironomidae (Diptera) responses to 2800 years of cultural influence: a palaeolimnological study with special reference to sedimentation, eutrophication, and contamination processes. Can. Entomol. 112:1193–1238.

*Wiederholm, T. (ed.). 1983. Chironomidae of the Holarctic region. Keys and diagnoses. Part I. Larvae. Entomol. Scand. Suppl. 19:1–457.

—— (ed.). 1986. Chironomidae of the Holarctic region. Keys and diagnoses. Part II. Pupae. Entomol. Scand. Suppl. 28:1–482.

Williams, C.J. 1985. A comparison of net and pump sampling methods in the study of chironomid larval drift. Hydrobiologia 124:243–250.

Wilson, R.S., and P.L. Bright. 1973. The use of chironomid pupal exuviae for characterizing streams. Freshwat. Biol. 3:282–302.

15 Freshwater Crustacea

Classification

The phylum Arthropoda contains 80% of all known species of animals. Although Insecta is the dominant class of arthropods on land, the class Crustacea dominates the water. Several features distinguish crustaceans from insects. Many, if not most, of the body segments of crustaceans bear jointed appendages that are fundamentally *biramous* (have two axes). Crustaceans have two pairs of antennae, often bearing flagella. In the first antenna, the flagellum begins with the fourth segment, while in the second antenna it begins with the sixth segment. In many amphipods the flagellum of the second antenna may appear to begin with the fifth segment because the basal segment of the peduncle is very short and is fused to the head. Crustaceans also have more than three pairs of jointed thoracic appendages. The thorax of crustaceans primitively had eight segments, each with a pair of legs. Amphipods and isopods have each had the most anterior segment incorporated into the head, so the thorax has seven segments. In mysidaceans and decapods, the thorax has been covered by the carapace. In crayfish, the part of the cephalothorax posterior to the cervical groove may be called the thorax. Another difference between crustaceans and insects is that crustaceans continue to molt after they become adults. Most of the 35,000 known species of crustaceans are marine; of the 26 orders that occur in the United States 13 are represented in freshwater, but only 5 of those are found exclusively in freshwater (Pennak 1978). Here we follow the classification scheme of Dexter (1959).

Many of the freshwater crustaceans rarely attain sizes greater than 1 mm and therefore are not readily collected with techniques effective for sampling aquatic insects. These microcrustaceans are the Cladocera (water fleas, in the subclass Branchiopoda) and the Calanoida, Cyclopoida, and Harpacticoida (all in the subclass Copepoda); the Caligoida and Lernaeopodoida, two orders of copepods composed entirely of parasites on freshwater fish (Wilson 1944), have modified morphologies and are not considered in this chapter. Members of two other microcrustacean orders attain sizes somewhat larger than 1 mm and may be entrained in aquatic dip nets, but they can easily be overlooked when the sample of aquatic insects is hand sorted. These orders are the Conchostraca (clam shrimp, subclass Branchiopoda) and the Podocopa (seed shrimp, subclass Ostracoda). Because identification of genera of the microcrustacean orders requires techniques different from those used for the collection, preservation, and identification of aquatic insects, the free-living forms are keyed to order but not to genera in the following key.

Seven remaining orders—Anostraca, Notostraca, Arguloida, Mysidacea, Isopoda, Amphipoda, and Decapoda—are considered macrocrustaceans and are treated similarly to the orders of aquatic insects. Notostracans (tadpole shrimp) do not occur in northeastern North America. We discuss the natural history of each of the other six macrocrustacean orders separately because they are so diverse. One feature common to all crustaceans is that they respire either through gills or through the general body integument. For more information on the biology of freshwater crustaceans, see Edmondson 1959.

Anostraca

The anostracans (fairy shrimp, subclass Branchiopoda) are restricted either to freshwater, where they occur only in temporary pools, or to very saline habitats (e.g., brine shrimp: *Artemia salina*). Along with two other orders of Branchiopoda (Notostraca and Conchostraca), they are noted for having life cycles specialized for survival in temporary habitats. Females retain either sexual or parthenogenetic eggs externally for a few days, then drop them in the mud. Thin-walled "summer" eggs hatch almost immediately. But thick-shelled resting, or "winter," eggs can withstand heat, cold, and long periods of desiccation. Viable eggs have been kept in dried pond mud on laboratory shelves for as long as 15 years (Pennak 1978). Eggs hatch within 30 hours after reflooding, and nauplii go through from 12 to 16 instars, adding more segments and more appendages at each successive molt, until sexual maturity is reached. The number of instars varies with species, temperature, and food conditions. Fairy shrimp have been recorded to complete their entire life cycles (from egg laying to egg laying) in as few as 16 days. This enables them to reproduce in extremely ephemeral pools, or to oviposit before predaceous aquatic insects colonize the habitat or reach sufficient size to decimate their populations.

Fairy shrimp swim on their backs (venter up), beating their appendages. This behavior not only propels them through the water, but it also serves as a mechanism for obtaining food. These shrimp consume phytoplankton, bacteria, protozoans, rotifers, and suspended detrital material.

Anostracans can be collected with a coarse mesh dip net and preserved in 70% ethanol. They can be distinguished from members of other orders of Crustacea by their lack of a carapace. Genera that occur in northeastern North America are separated on the basis of the structure of the second antenna. Most species of fairy shrimp, and all species of tadpole shrimp, are found in the southwestern United States, where temporary rain pools are more common. For more information on the natural history of this and related groups, see Dexter 1953, 1959 and Pennak 1978.

Arguloida

The order Arguloida, subclass Branchiura, is represented in North America by one genus, *Argulus* (fish lice). Adult *Argulus* are parasitic on freshwater fish, attaching either to the gill chamber or to the surface of the body. Their mouthparts are greatly modified or reduced for their parasitic existence; the second maxillae have been transformed into suction cups for attachment to their hosts and a piercing organ for obtaining host blood (Pennak 1978). Adult *Argulus* have four pairs of swimming legs, which they use when they leave their hosts, especially during the breeding season. Females oviposit on stones or sticks of the substrate (Pennak 1978). For more information on the natural history of this group, see the reviews by Meehean (1940) and Wilson (1944).

Mysidacea

The mysids (opossum shrimp) belong to the subclass Malacostraca. They are almost exclusively marine except for a few species that are native to deep, cold oligotrophic lakes, such as the Great Lakes and the Finger Lakes of New York State (Pennak 1978). Mysids have been introduced into other large lakes, such as Lake Tahoe in California and Flathead Lake in Montana, as a potential source of forage food for game fish. However, this management scheme has backfired because the mysids are predaceous on the native zooplankton (cladocerans). Thus, high mysid populations depress natural populations of food for fish. In addition, they are themselves a less desirable food source to fish than the native zooplankton because their diel migration behavior from the epilimnion to the sediments during daylight makes them less available to fish that choose food by sight (Cooper and Goldman 1980).

Only one species of mysid exists in northeastern North America (*Mysis relicta*). Females carry developing eggs and young in a marsupium for one to three months. After reaching 3–4 mm in length, young mysids leave the mother and undergo direct development, with no naupliar stages. At the latitudes of the region covered by this key, life cycles of *M. relicta* are about two years (Pennak 1978).

Mysids are generally restricted to cooler water and spend the summer in the hypolimnion of large lakes. As mentioned above, they also migrate vertically, remaining within a meter of the bottom during the day and migrating to the epilimnion at night. These changes in distribution should be taken into account when collecting mysids. The most popular method of collection is to use plankton tows at night on big lakes. Mysids can be stored in 70% ethanol. They are easily distinguished from members of the other orders of Crustacea by the shape of their carapace. Further information on this group can be found in Chace et al. 1959, Pennak 1978, and Morgan 1982.

Isopoda

Isopods (pill bugs and sow bugs), also in the subclass Malacostraca, are mostly terrestrial and marine; only 5% of the known North American species occur in freshwater. The freshwater sow bugs are restricted to springs, spring brooks, streams, and subterranean waters. A few species may be found in ponds or shallow bays of lakes. All freshwater species are scavengers on dead animal and plant material. They tend to remain under stones or in detritus.

Breeding occurs throughout the year, with little seasonal periodicity. However, few gravid females have been observed in October and February (Pennak 1978). Males clasp females, holding them beneath their bodies. Clasping and copulation can last as long as 24 hours. After the individuals separate, females shed the anterior half of their preadult exoskeleton, at which time the oostegites are transformed from buds to a functional marsupium ready to accept developing eggs. A few days after copulation, the eggs pass through the oviduct and through the seminal receptacles, where they are fertilized, then out the genital pores and into the marsupium. Females brood between 20 and 250 eggs, retaining the newly hatched young for 20–30 days. Young isopods go through at least 15 instars and are considered adults after five to eight molts. They live for about one year (Pennak 1978).

Isopods are collected along with other benthic invertebrates, and can be preserved in 70% ethanol. For detailed identification, appendages may be dissected and slide-mounted. Isopods are distinguished from members of other orders of crustaceans by the number of thoracic appendages and the degree of dorsoventral body flattening. Two genera of the family Asellidae occur in northeastern North America. They are separated by features of the head and the third pleopod. For more information on the biology of aquatic isopods, see Williams 1972 and Pennak 1978.

Amphipoda

Amphipods (scuds or side swimmers) also belong to the subclass Malacostraca and are primarily marine. Only about 50–90 species exist in freshwaters of North America. They are common in unpolluted lakes, ponds, streams, springs, and subterranean waters. Like isopods, amphipods are restricted primarily to the littoral benthos and are general scavengers. One species (*Pontoporeia affinis*) occurs in the profundal benthos of Cayuga Lake, in central New York State. It is also the only freshwater amphipod found regularly in the limnetic zone, since it migrates vertically to the epilimnion at night. As with mysids, this behavior reduces the risk of detection of amphipods by visual fish predators. Pond and stream amphipod species are often restricted to habitats too small to support natural fish populations.

Male amphipods carry the females on their backs during mating. Pairs may feed and swim for up to one week until the female molts into the first adult instar. The pairs separate while the female molts, then pair again and copulate within 24 hours. With their pleopods, females sweep ejected sperm into the marsupium. Then the females release their eggs from the oviducts into the marsupium, where they are fertilized. Brood sizes range from 15 to 50 eggs. Developing eggs hatch in one to three weeks, after which the young are retained in the marsupium for about a week. At the time of the mother's first molt after copulation, the young are released from the marsupium. Young amphipods molt eight (females) or nine (males) times before they are sexually mature (Pennak 1978).

Amphipods are collected along with other benthic invertebrates and can be stored in 70% ethanol. They can easily be distinguished from members of other orders of Crustacea by the number of thoracic appendages and the degree of lateral body compression. Characteristics used to separate families and genera of amphipods are features of the first and second antennae and accessory antennal flagellae. For more-detailed information on amphipods, see Chace et al. 1959, Holsinger 1972, Bousfield 1973, and Pennak 1978.

Decapoda

The decapods (subclass Malacostraca) are the best known of the freshwater crustaceans. Like the other malacostracans, most decapods are marine. A few species of freshwater prawns and river shrimp (family Palaemonidae) are found in northeastern North America, but the crayfish (family Cambaridae) are the most diverse group of freshwater decapods and are the only family covered in the key. Crayfish occur in a wide variety of habitats, including streams, large rivers, ponds, lakes, swamps, subterranean waters, and wet meadows. They are generally omnivorous, feeding on aquatic vegetation (generally detritus) and on animals (snails and other invertebrates).

There are two distinct types of adult male cambarids. Form I adults are breeding males. Form II males, the instars just preceding breeding males, lack the sculpturing of the first pleopods that is typical of breeding males. Thus, form II males are difficult to distinguish from juveniles. Male crayfish display some courtship behavior to reduce the aggressive tendencies of females (Ameyaw-Akumfi 1981). Copulation lasts from a few minutes to 10 hours. Spermatophores are stored in the annulus ventralis of the female until she lays her eggs, several weeks to several months after copulation. Just before the eggs are extruded, the female secretes from ventral glands a sticky substance that covers the ventral surface of the abdomen, tail fan, and pleopods. The sperm are released into this sticky substance while the female releases her eggs from the genital pore. As the female moves her body, the eggs become dispersed throughout the sticky substance, fertilized, and attached to the pleopods. Females brood from 10 to 700 eggs for 2–20 weeks (depending on temperature). Newly hatched young remain attached to the pleopods for three instars, after which they release their hold. After 6–10 molts many females are sexually mature. Copulation can occur in the fall, but is more common in the spring. After the first mating season, crayfish usually molt two to four times before they die. The lifespan of both males and females is usually less than two years (Pennak 1978).

Crayfish differ in their propensity to burrow (Hobbs 1974b). *Primary burrowers* are restricted to burrows; *secondary burrowers* generally stay in their burrows but may travel to open water when the mouths of their burrows are flooded; and *tertiary burrowers* are commonly in open waters, burrowing only in times of drought, breeding seasons, or as a defense against predators. This information should be taken into account when collecting crayfish; it is included in the key as an aid to identification.

Crayfish taxonomy is based mostly on the structure of the breeding male's (form I) copulatory stylets (first pleopods). Juveniles and form II males are more difficult to identify. However, the key is written so that all forms of crayfish can be identified. Couplet phrases refer to both sexes unless specified. Much terminology unique to crayfish is included in the key. Terms are defined in the glossary. To ensure correct identification, check additional references for more-detailed descriptions of the species (Crocker 1957; Crocker and Barr 1968; Bell 1971; Hobbs 1972, 1974a, b). For more information on the natural history of crayfish, see Chace et al. 1959, Pennak 1978, and Hobbs 1981.

Checklist of Freshwater Crustacea

Subclass Branchiopoda
- Order Anostraca
  - Chirocephalidae
    - *Dexteria floridans*
    - *Eubranchipus*
  - Streptocephalidae
    - *Streptocephalus*
- Order Cladocera
- Order Conchostraca

Subclass Branchiura
- Order Arguloida
  - Argulidae
    - *Argulus*

Subclass Copepoda
- Order Calanoida
- Order Cyclopoida
- Order Harpacticoida

Subclass Malacostraca
- Order Amphipoda
  - Crangonyctidae
    - *Crangonyx*
    - *Stygonectes*
  - Gammaridae
    - *Gammarus*
  - Haustoriidae
    - *Pontoporeia affinis*
  - Talitridae
    - *Hyalella*
- Order Decapoda
  - Cambaridae
    - *Cambarus bartonii bartonii*
    - *C. diogenes*
    - *C. robustus*
    - *Fallicambarus fodiens*
    - *Orconectes immunis*
    - *O. limosus*
    - *O. obscurus*
    - *O. propinquus*
    - *O. rusticus*
    - *O. virilis*
    - *Procambarus acutus acutus*
- Order Isopoda
  - Asellidae
    - *Caecidotea*
    - *Lirceus*
- Order Mysidacea
  - Mysidae
    - *Mysis relicta*

Subclass Ostracoda
- Order Podocopa

Key to Freshwater Crustacea

1a. A pair of prominent suction cups on ventral side of body (Fig. 1); usually ectoparasitic on fish . . . . Order ARGULOIDA, **Argulidae**, *Argulus*

1b. Suction cups absent on ventral side of body 2

2a (1b). A subcylindrical or bivalvelike carapace present (Figs. 2, 3, 36), obscuring all thoracic segments in dorsal view 3

2b. Carapace absent (Fig. 5), or reduced (Fig. 6) so that at least 2 posterior thoracic segments are exposed in dorsal view 6

3a (2a). Carapace bivalvelike (Fig. 2) 4

3b. Carapace subcylindrical, covering both the dorsal and lateral surfaces of the cephalothorax (Fig. 36) . . . . Order DECAPODA, **Cambaridae** 20

4a (3a). Bivalvelike carapace enclosing the entire body (Figs. 2, 3) 5

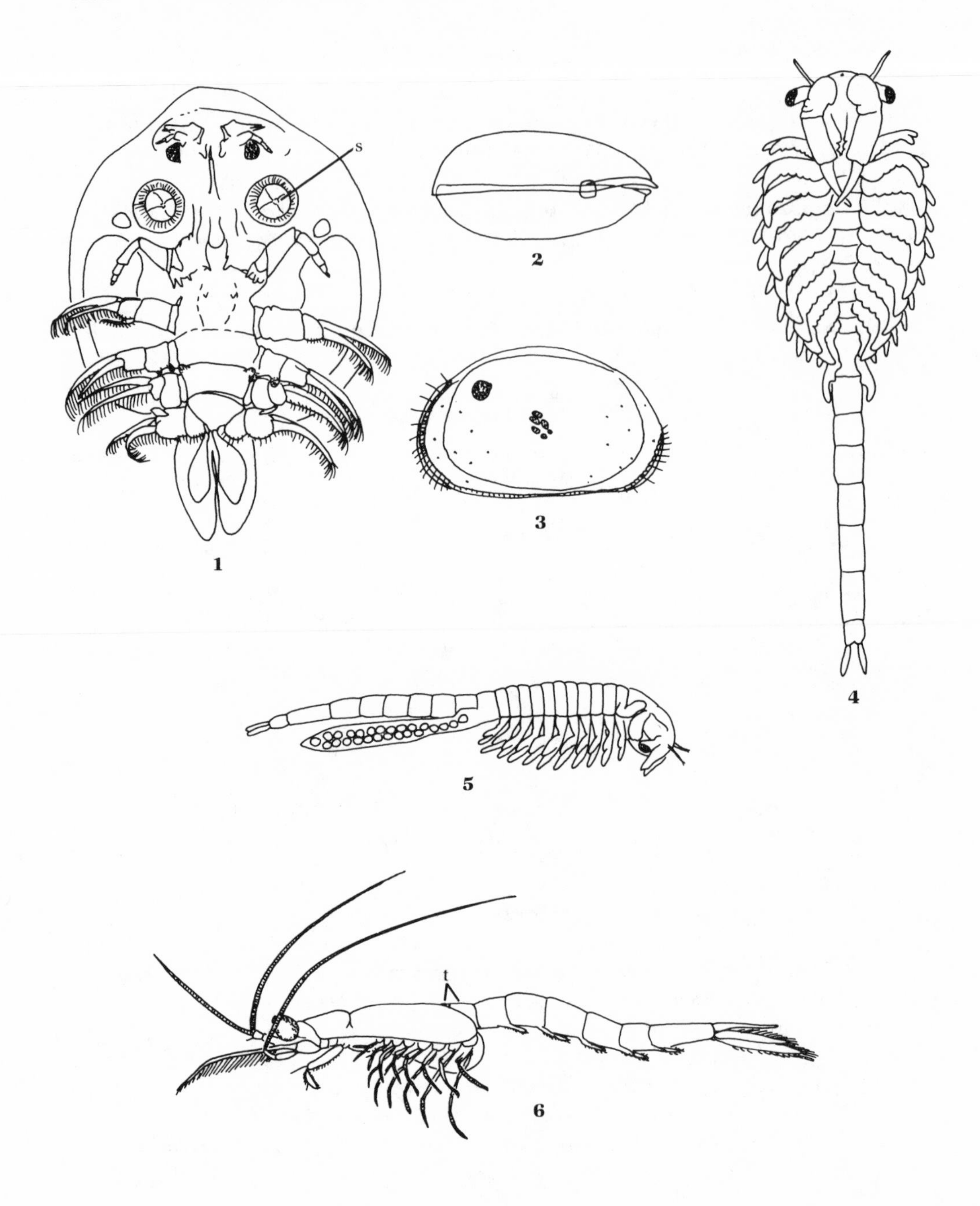

1. *Argulus* (Arguloida), vv: s, suction cup. **2.** *Physocypria* (Podocopa), dv. **3.** *Cypria* (Podocopa), lv. **4.** *Branchinecta* (Anostraca), vv. **5.** *Branchinecta* (Anostraca), lv. **6.** *Mysis* (Mysidacea), lv: t, thoracic segments. dv, dorsal view; lv, lateral view; vv, ventral view. (1–6 redrawn or modified from Pennak 1978.)

4b. Bivalvelike carapace not enclosing the head (Fig. 7) .... Order CLADOCERA (in part)

5a (4a). Thorax with 10–32 pairs of appendages (Fig. 8); carapace often with concentric rings .... Order CONCHOSTRACA

5b. Thorax with 2–3 pairs of appendages; carapace lacking concentric rings (Figs. 2, 3, 9) .... Order PODOCOPA

6a (2b). Eyes stalked (Fig. 4) 7

6b. Eyes sessile 8

7a (6a). Carapace absent; thoracic appendages flattened and leaflike (Figs. 4, 5) .... Order ANOSTRACA 13

7b. Carapace present, covering all but 2 posterior thoracic segments in dorsal view; abdominal appendages cylindrical and thin (Fig. 6) .... Order MYSIDACEA, **Mysidae,** *Mysis relicta*

8a (6b). Seven thoracic segments, each with paired appendages (Fig. 11) 9

8b. Four to six thoracic segments, each with paired appendages (Fig. 10) 10

9a (8a). Body dorsoventrally depressed (Figs. 11, 12); pleopods forming broad plates (Figs. 13–16); 6 or 7 pairs of legs, and with or without a pair of gnathopods (Fig. 11) .... Order ISOPODA, **Asellidae** 15

9b. Body laterally compressed (Fig. 17); pleopods slender; 5 pairs of legs, and 2 pairs of gnathopods (Fig. 17) .... Order AMPHIPODA 16

10a (8b). Second antennae biramous (Figs. 7, 10) .... Order CLADOCERA (in part)

10b. Second antennae uniramous 11

11a (10b). Anterior part of body subequal in width to the posterior part of the body (Fig. 18) .... Order HARPACTICOIDA

11b. Anterior part of body much wider than posterior part of body (Figs. 19, 20) 12

12a (11b). First antennae as long as entire body; 1st antennae of female with 23–25 segments; ovigerous female with 1 egg sac, carried medially (Fig. 19); only right 1st antenna of male may be geniculate (elbowed) (Fig. 21) .... Order CALANOIDA

12b. First antennae no longer than anterior portion of body; 1st antennae of female with 6–17 segments; ovigerous female with 2 egg sacs, carried laterally (Fig. 20); both right and left 1st antennae of male geniculate .... Order CYCLOPOIDA

13a (7a). ANOSTRACA: Second antenna of male with a scissorlike terminal segment (Fig. 22) .... **Streptocephalidae,** *Streptocephalus*

13b. Second antenna of male without a scissorlike terminal segment (Fig. 23) . . 14

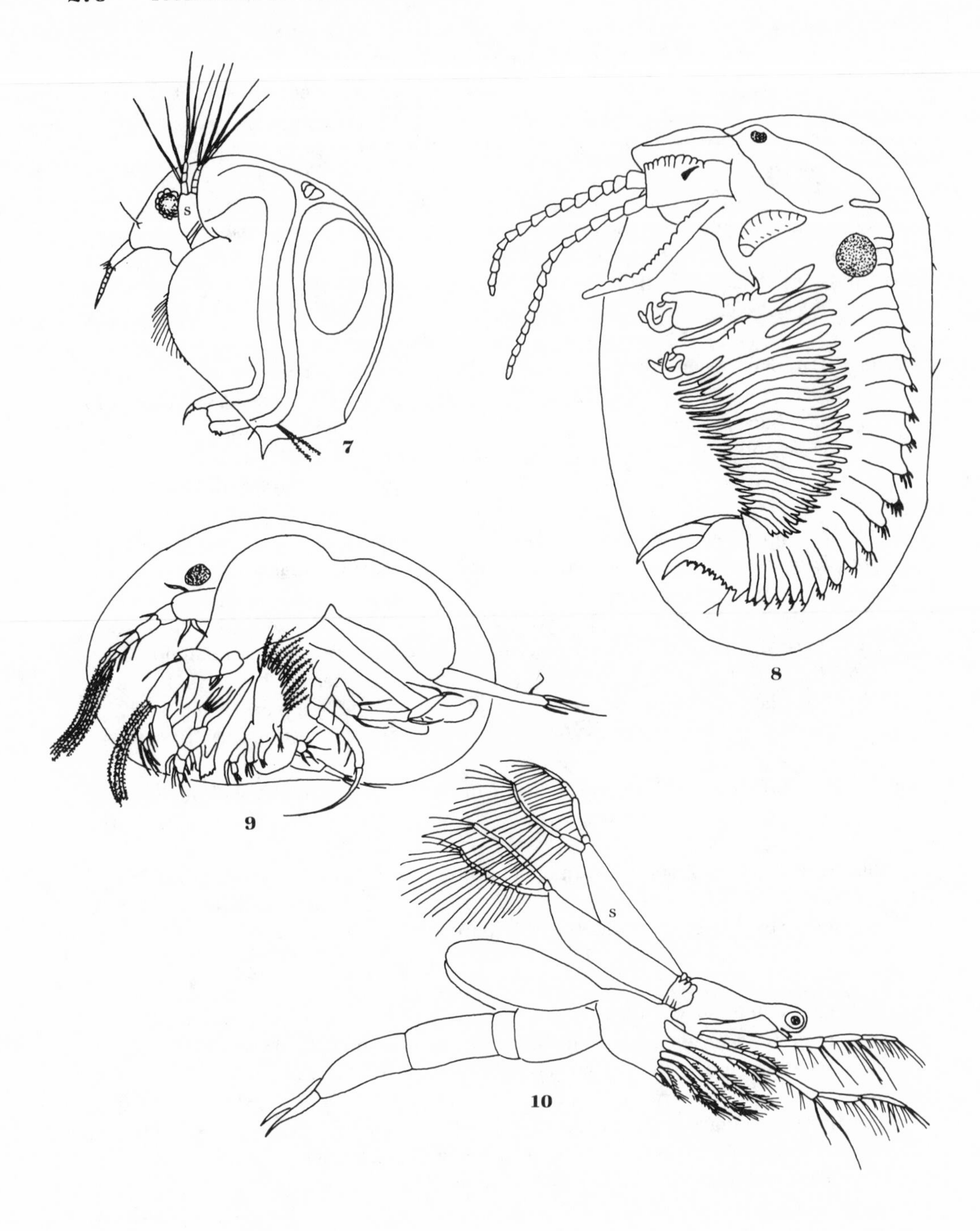

7. *Bosmina* (Cladocera), lv: s, second antenna. **8.** Male *Cyzicus* (Conchostraca), cutaway, lv. **9.** Female ostracod (Podocopa), cutaway, lv. **10.** *Leptodora* (Cladocera), lv: s, second antenna. lv, lateral view. (7, 9, 10 modified from Pennak 1978; 8 modified from Mattox 1959.)

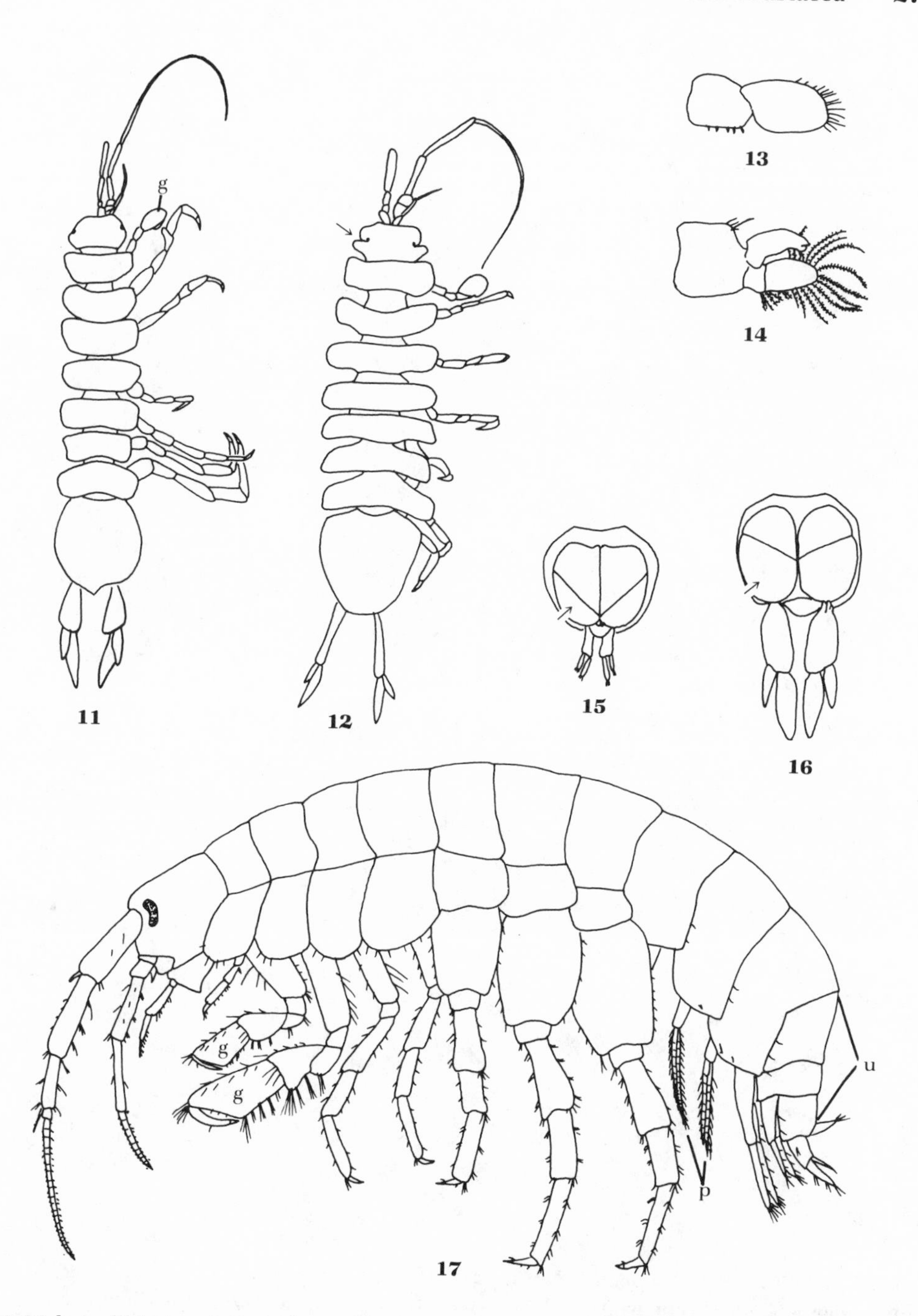

11. Male asellid isopod, *Caecidotea*, dv: g, gnathopod. **12.** Male asellid isopod, *Lirceus*, dv. **13.** First pleopod of male isopod, vv. **14.** Second pleopod of male isopod, vv. **15.** Third pleopods and uropods of *Lirceus* (Asellidae), vv. **16.** Third pleopods and uropods of *Caecidotea* (Asellidae), vv. **17.** Generalized gammarid amphipod, lv: g, gnathopods; p, pleopods; u, uronites. dv, dorsal view; lv, lateral view; vv, ventral view. (11, 12 redrawn from Williams 1972; 13–16 redrawn or modified from Cole 1957; 17 modified from Holsinger 1972.)

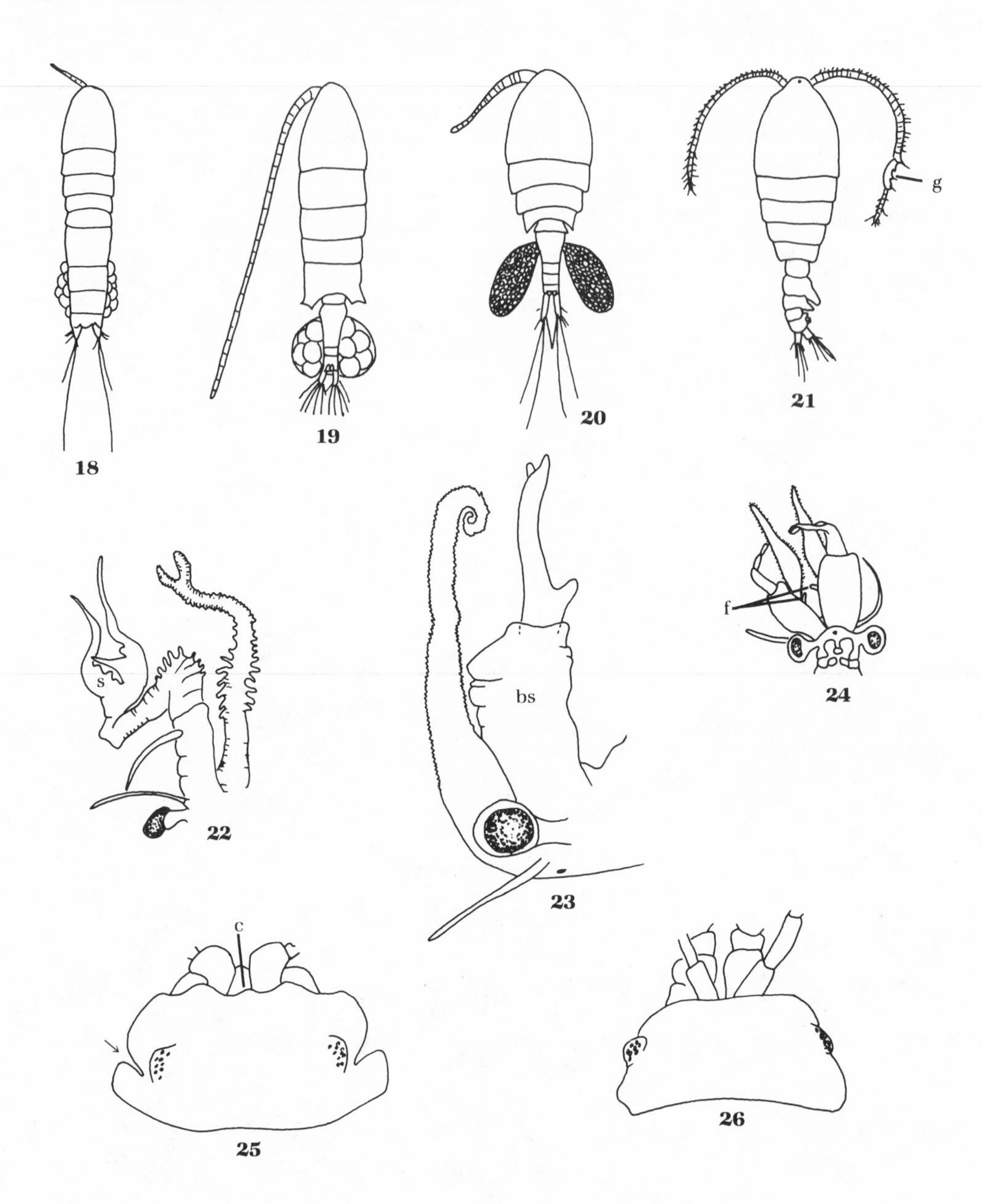

18. Female harpacticoid copepod, dv. **19.** Female calanoid copepod with 1 ventral egg sac, dv. **20.** Female cyclopoid copepod with 2 lateral egg sacs, dv. **21.** Male *Epischura* (Copepoda), dv: g, geniculate antenna. **22.** Left part of head of male *Streptocephalus* (Streptocephalidae), dv: s, scissorlike segment. **23.** Head of *Eubranchipus bundyi* (Chirocephalidae) male, lv: bs, basal segment of second antenna. **24.** Head of male *Dexteria* (Chirocephalidae), dv: f, fingerlike process. **25.** Head of *Lirceus* (Asellidae), dv: c, carina. **26.** Head of *Caecitodea* (Asellidae), dv. dv, dorsal view; lv, lateral view. (18–20 redrawn from Wilson and Yeatman 1959; 21 redrawn from Cole 1957; 22 modified from Pennak 1978; 23 redrawn from Linder 1941; 24 redrawn from Dexter 1953; 25, 26 redrawn from Williams 1972.)

14a (13b). Basal segment of 2nd antenna (male) with a fingerlike process on the median surface near midlength (Fig. 24) .... **Chirocephalidae,** *Dexteria floridans*

14b. Basal segment of 2nd antenna (male) without a fingerlike process (Fig. 23) ... . *Eubranchipus*

15a (9a). ISOPODA, **Asellidae:** Lateral margin of head produced to form a thin plate that covers or overhangs the base of the mandible, plate sometimes incised (Figs. 12, 25); anterior margin of head with a pointed median protuberance (carina) between bases of 1st pair of antennae (Fig. 25); terminal segment of exopod of 3rd pleopod triangular or half-moon shaped (Fig. 15) (suture between exposed segments of 3rd pleopod running from the median posterior angle very obliquely toward the lateral margin [Fig. 15]); up to 25 mm long; eyes present .... *Lirceus*

15b. Lateral margin of head not produced to cover or overhang base of mandible (Figs. 11, 26); anterior margin of head without a carina (Fig. 26); terminal segment of 3rd pleopod quadrangular (Fig. 16) (suture between exopod segments of 3rd pleopod running from the median margin less obliquely toward the lateral margin [Fig. 16]); up to 20 mm long; eyes present or absent .... *Caecidotea*

16a (9b). AMPHIPODA: Antenna 1 shorter than antenna 2 (Fig. 27) 17

16b. Antenna 1 longer than antenna 2 (Fig. 17) 18

17a (16a). Accessory flagellum present on antenna 1 (Figs. 28, 29); pereiopod 5 (last pair) much shorter than 4 (Fig. 27); telson deeply cleft (as in Figs. 30, 31) .... **Haustoriidae,** *Pontoporeia affinis*

17b. Accessory flagellum absent; pereiopod 5 longer than or subequal to 4; telson entire .... **Talitridae,** *Hyalella*

18a (16b). Accessory flagellum of antenna 1 with 2–7 segments (but usually with 3 or more) (Fig. 29); uronites (Fig. 17) with prominent dorsal spines; telson cleft nearly to base (Fig. 30) .... **Gammaridae,** *Gammarus*

18b. Accessory flagellum of antenna 1 never with more than 2 segments (Fig. 28); uronites without prominent dorsal spines; telson cleft or not (Fig. 32), if cleft then no more than ¾ way to base (Fig. 31) .... **Crangonyctidae** 19

19a (18b). **Crangonyctidae:** Antenna 2 of mature male with paddleshaped calceoli (Fig. 33); eyes usually present and pigmented; pereiopod 6 longer than 7; apical margin of telson distinctly cleft (Fig. 31) .... *Crangonyx*

19b. Antenna 2 of mature male without paddle-shaped calceoli (as in Fig. 28); eyes never present (if subterranean); pereiopod 7 sometimes longer than 6; apical margin of telson entire (Fig. 32) or with a shallow cleft .... *Stygonectes*

20a (3b). DECAPODA, **Cambaridae:** First pleopod of male terminating in more than 2 elements (Figs. 34, 35); carapace, posterior to cervical groove, covered with tubercles of such height that surface feels granular; tertiary burrower .... *Procambarus acutus acutus*

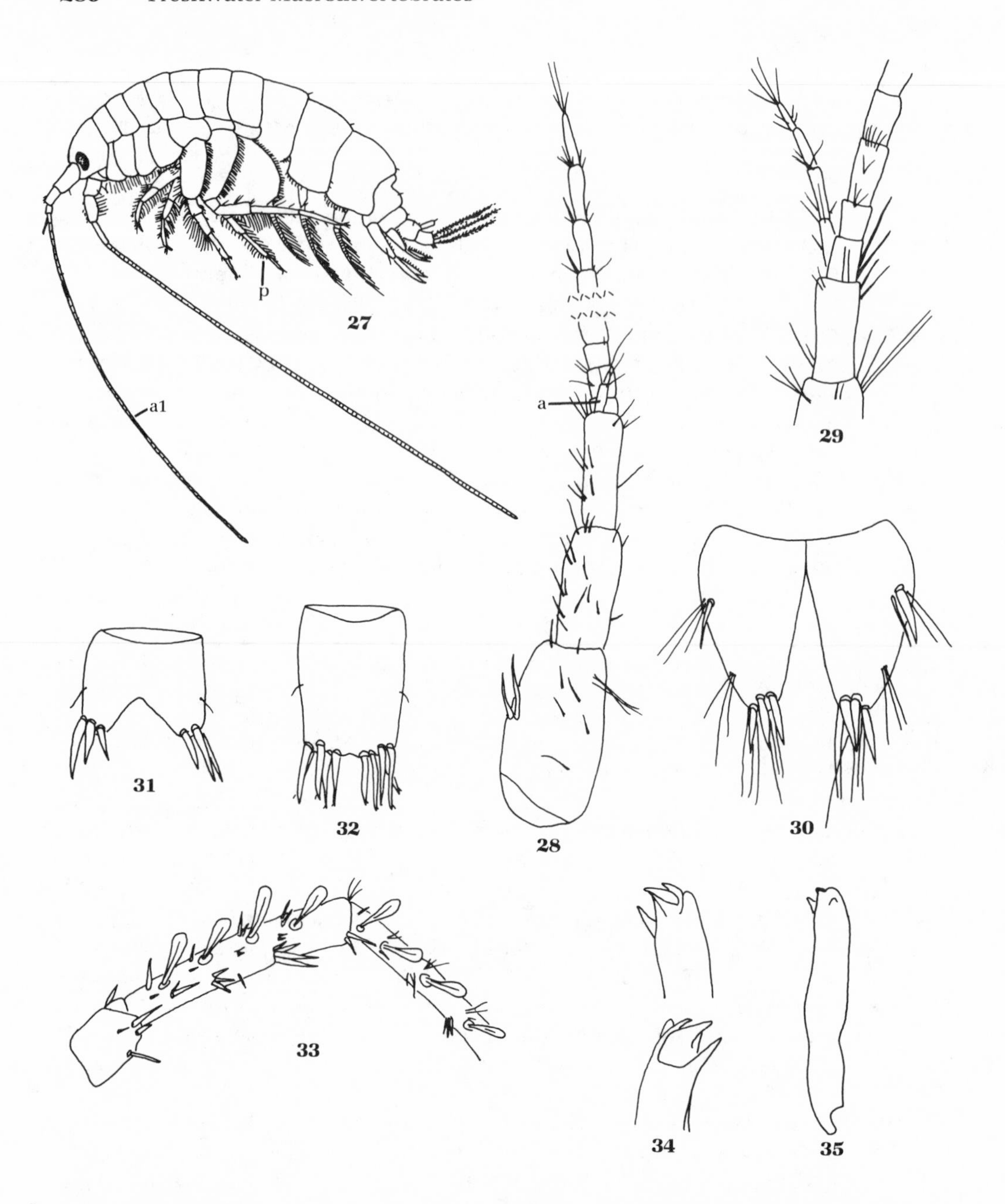

27. Male *Pontoporeia* (Haustoriidae), lv: a1, antenna 1; p, 5th pereiopod. **28.** Antenna 1 of crangonyctid: a, accessory flagellum. **29.** Accessory flagellum of *Gammarus* (Gammaridae). **30.** Telson of *Gammarus* (Gammaridae), vv. **31.** Telson of *Crangonyx* (Crangonyctidae), vv. **32.** Telson of crangonyctid, vv. **33.** Antenna 2 of *Crangonyx* (Crangonyctidae). **34.** First pleopod of form I male *Procambarus acutus acutus* (Cambaridae), dv (upper), mv (lower). **35.** First pleopod of form II male *Procambarus acutus acutus* (Cambaridae), lv. dv, dorsal view; lv, lateral view; mv, mesal view; vv, ventral view. (27 redrawn from Pennak 1978; 28, 32 redrawn from Holsinger 1978; 29–31, 33 redrawn from Holsinger 1972; 34, 35 redrawn from Hobbs 1974a.)

20b. First pleopod of male terminating in 2 elements (Fig. 37); carapace, posterior to cervical groove, not covered with tubercles 21

21a (20b). The ratio (length of areola)/(width of areola) > 9.6 (e.g., Figs. 38–41) (see Fig. 36 for length and width); a relatively narrow areola 22

21b. The ratio (length of areola)/(width of areola) < 9.6 (e.g., Figs. 51–56) (see Fig. 36 for length and width); a relatively wide areola 25

22a (21a). Areola linear or obliterated at its narrowest portion (Figs. 38, 40) 23

22b. Areola not linear or obliterated at its narrowest portion (Figs. 39, 41) 24

23a (22a). Carapace with triangular suborbital projection (Fig. 36); cervical groove continuous laterally (Fig. 42); inner margin of dactyl of chela without a notch at its base (Fig. 43); primary burrower .... *Cambarus diogenes*

23b. Carapace without suborbital projection; cervical groove interrupted laterally; inner margin of dactyl of chela with a notch at its base (Fig. 44); primary or secondary burrower .... *Fallicambarus fodiens*

24a (22b). Dactyl of large chela with a notch at its base on the inner margin (Fig. 45); ventral margin of joint between dactyl and propodus with only 1 tubercle; terminal processes of 1st pleopod in males strongly curved caudally and subequal in length (Fig. 46); fossa of female seminal receptacle off to one side (Fig. 47); tertiary burrower .... *Orconectes immunis*

24b. Dactyl of large chela with inner margin nearly straight (Fig. 48); ventral margin of joint between dactyl and propodus often with 2 small tubercles; terminal processes of 1st pleopod in males weakly curved caudally, the central longer than the mesial (Fig. 49); fossa of female seminal receptacle large and central (Fig. 50); tertiary burrower .... *O. virilis*

25a (21b). Marginal spines or tubercles on rostrum present (Figs. 36, 51–54) 26

25b. Marginal spines or tubercles on rostrum absent (Figs. 55, 56) 29

26a (25a). Lateral surface of carapace anterior to cervical groove with 2 or more hepatic spines (Figs. 36, 51); in males, terminal processes of 1st pleopod distinctly divergent (Fig. 57); tertiary burrower .... *Orconectes limosus*

26b. Lateral surface of carapace anterior to cervical groove with no hepatic spines; terminal processes of 1st pleopod in males not distinctly divergent 27

27a (26b). Sides of rostrum slightly concave (Fig. 52); dactyls of large chelae with distinct subterminal black bands (Fig. 58); central process of 1st pleopod in males longer than mesial process (Fig. 59); seminal receptacle oval (Fig. 60); tertiary burrower .... *O. rusticus*

27b. Sides of rostrum straight (Figs. 53, 54); dactyls of large chelae without distinct black bands; terminal processes of 1st pleopod in males subequal in length (Figs. 61, 62) or, if unequal, central process may or may not be longer than mesial; seminal receptacle diamond-shaped (Figs. 63, 64) 28

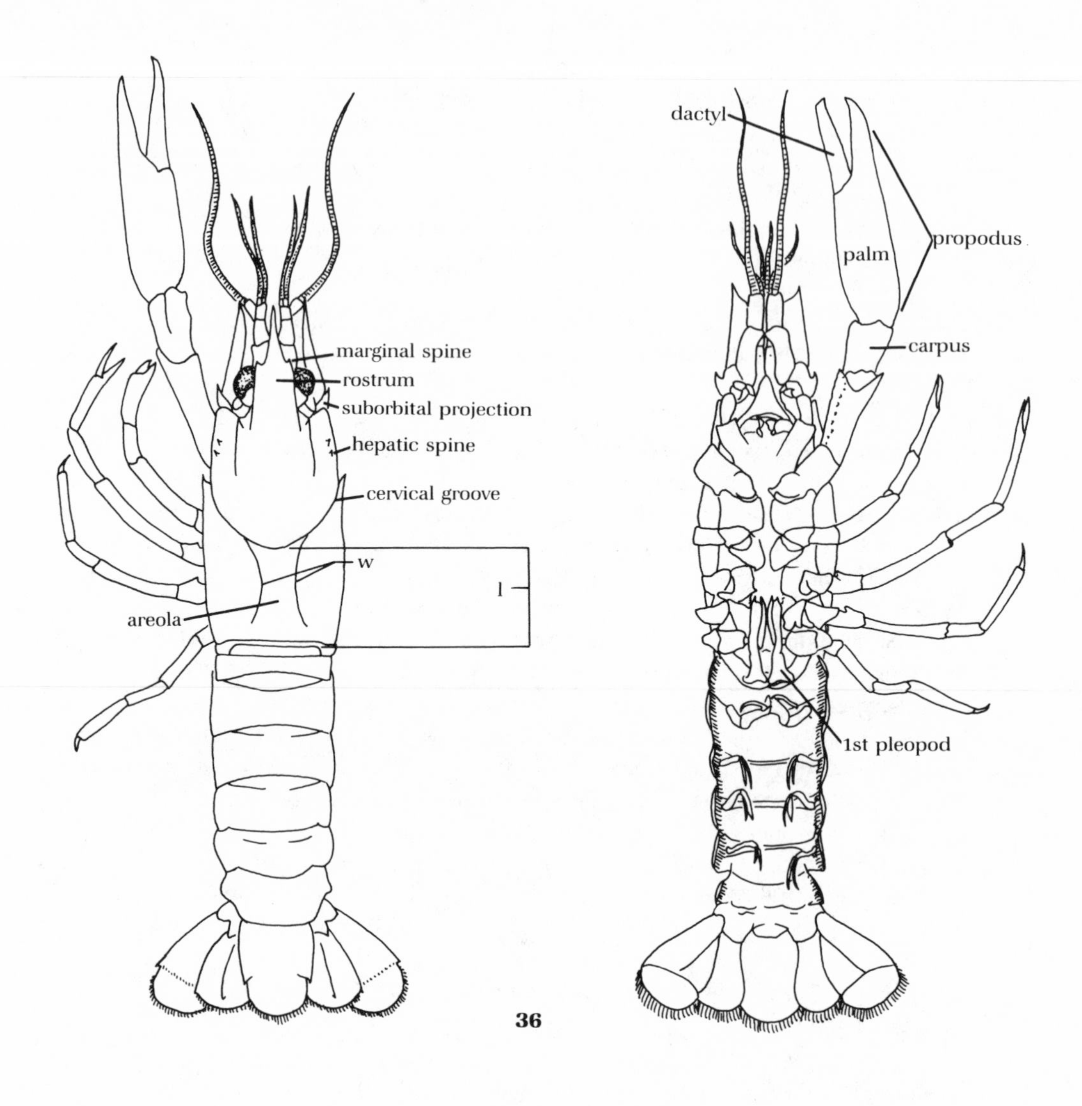

36. Generalized male crayfish, dv (left), vv (right): l, length of areola; w, width of areola. dv, dorsal view; vv, ventral view. (36 redrawn from Hobbs 1972.)

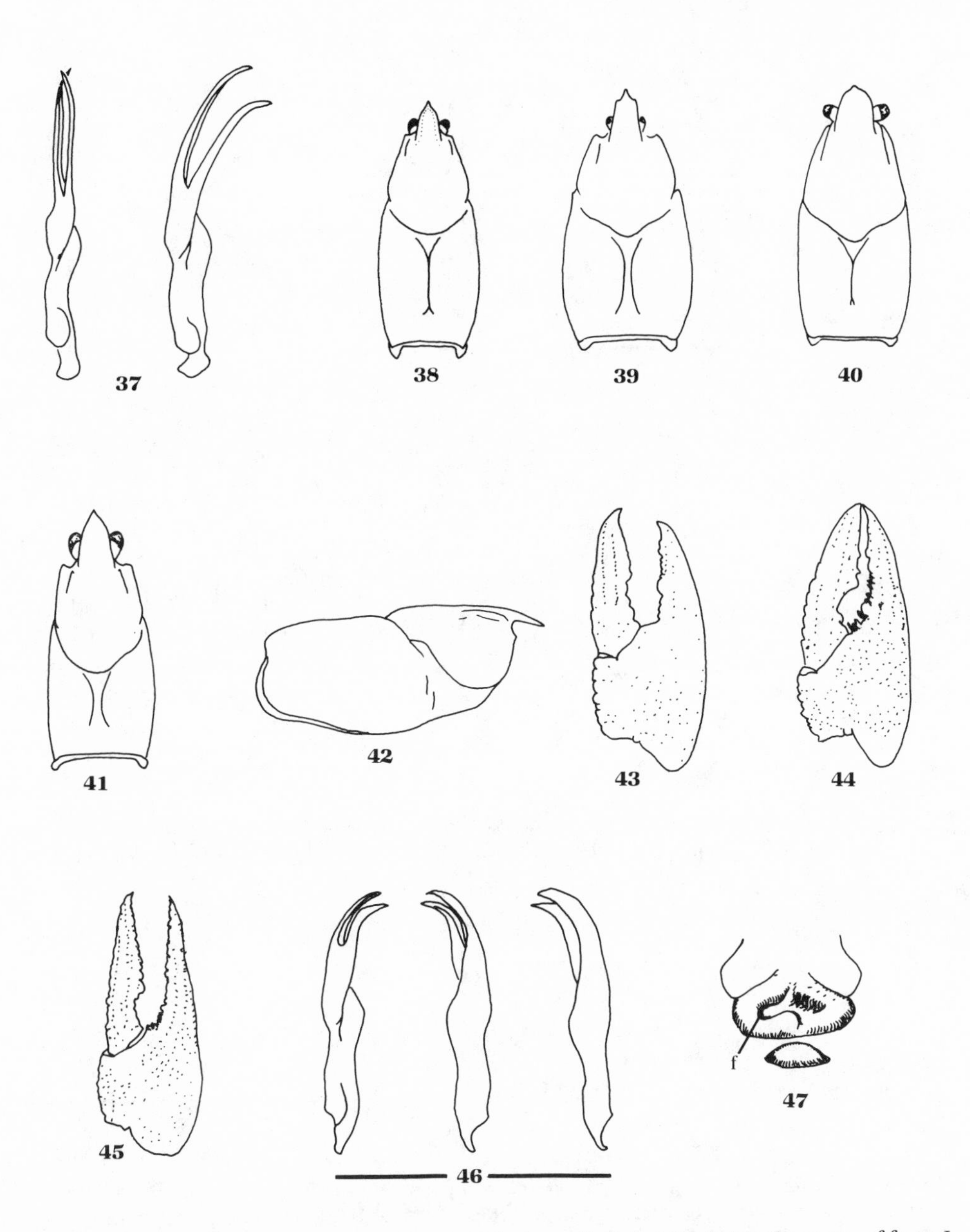

37. First pleopod of male *Orconectes* (Cambaridae), mv (left), lv (right). **38.** Carapace of form I male *Cambarus diogenes* (Cambaridae), dv. **39.** Carapace of form I male *Orconectes virilis* (Cambaridae), dv. **40.** Carapace of form I male *Fallicambarus fodiens* (Cambaridae), dv. **41.** Carapace of form I male *Orconectes immunis* (Cambaridae), dv. **42.** Carapace of form I male *Cambarus diogenes* (Cambaridae), lv. **43.** Chela of form I male *Cambarus diogenes* (Cambaridae), dv. **44.** Chela of form I male *Orconectes rusticus* (Cambaridae), dv. **45.** Chela of form I male *Orconectes immunis* (Cambaridae), dv. **46.** First pleopod of form I (mv, left; lv, middle) and form II (lv, right) male *Orconectes immunis* (Cambaridae). **47.** Seminal receptacle of *Orconectes immunis* (Cambaridae), dv: f, fossa. dv, dorsal view; lv, lateral view; mv, mesal view. (37 redrawn from Hobbs 1972; 38–47 redrawn from Hobbs 1974a.)

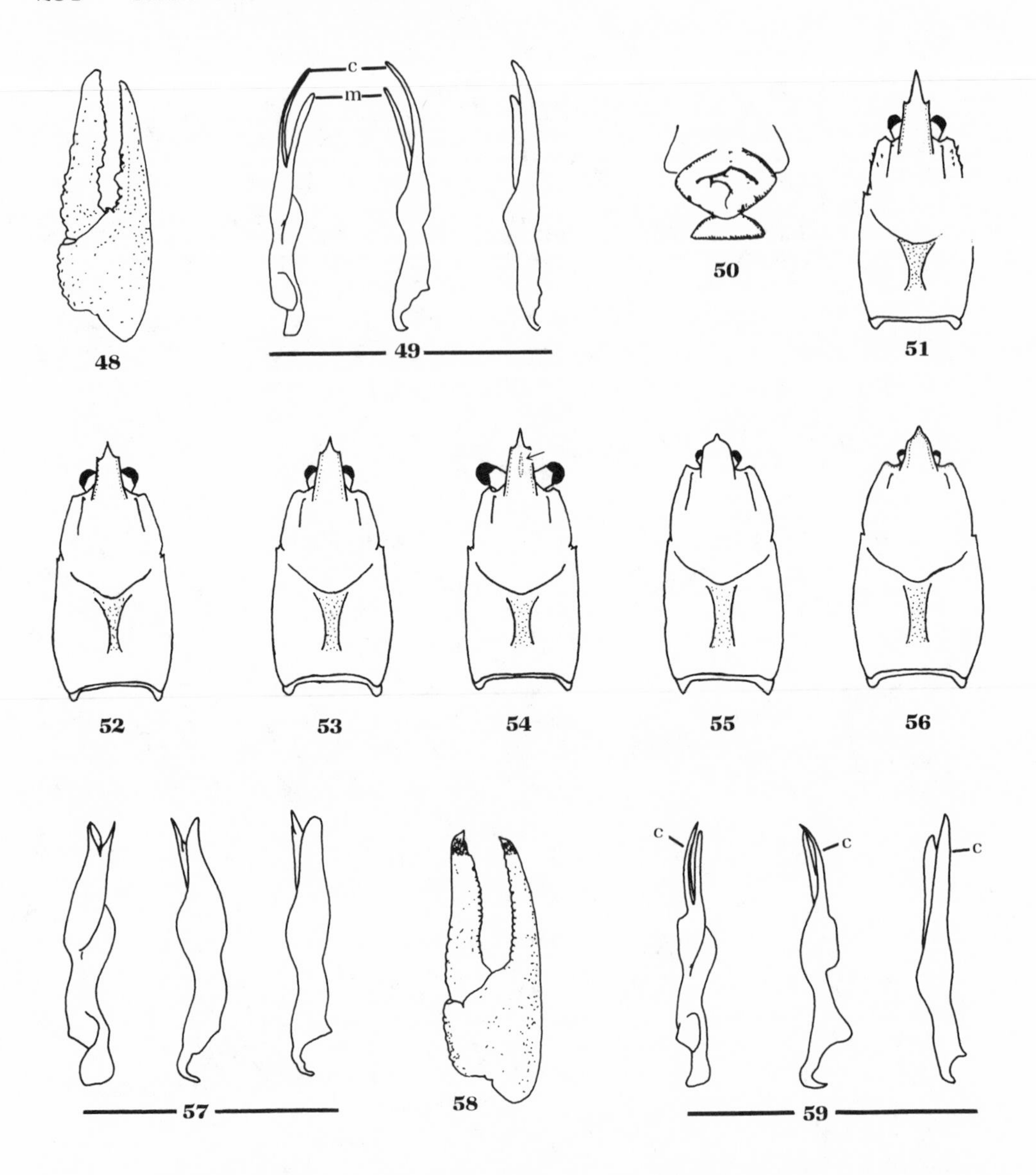

48. Chela of form I male *Orconectes virilis* (Cambaridae), dv. **49.** First pleopod of form I (mv, left; dv, middle) and form II (lv, right) male *Orconectes virilis* (Cambaridae); c, central process; m, mesal process. **50.** Seminal receptacle of *Orconectes virilis* (Cambaridae), vv. **51.** Carapace of form I male *Orconectes limosus* (Cambaridae), dv. **52.** Carapace of form I male *Orconectes rusticus* (Cambaridae), dv. **53.** Carapace of form I male *Orconectes obscurus* (Cambaridae), dv. **54.** Carapace of form I male *Orconectes propinquus* (Cambaridae), dv. **55.** Carapace of form I male *Cambarus bartonii bartonii* (Cambaridae), dv. **56.** Carapace of form I male *Cambarus robustus* (Cambaridae), dv. **57.** First pleopod of form I (mv, left; lv, middle) and form II (lv, right) male *Orconectes limosus* (Cambaridae). **58.** Chela of form I male *Orconectes rusticus* (Cambaridae), dv. **59.** First pleopod of form I (mv, left; lv, middle) and form II (lv, right) male *Orconectes rusticus* (Cambaridae): c, central process. dv, dorsal view; lv, lateral view; mv, mesal view; vv, ventral view. (48–59 redrawn from Hobbs 1974a.)

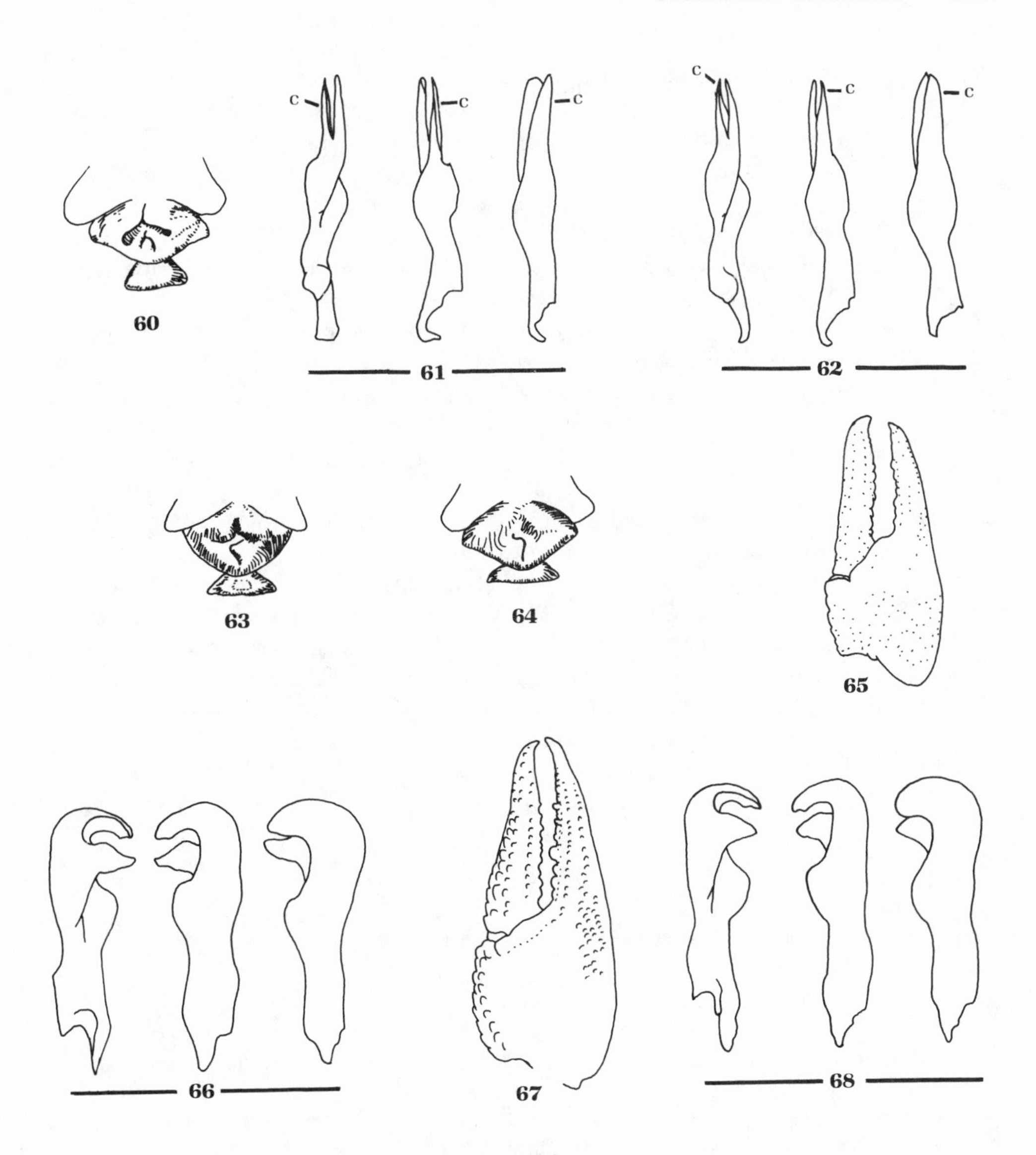

60. Seminal receptacle of *Orconectes rusticus* (Cambaridae), vv. **61.** First pleopod of form I (mv, left; lv, middle) and form II (lv, right) male *Orconectes propinquus* (Cambaridae): c, central process. **62.** First pleopod of form I (mv, left; lv, middle) and form II (lv, right) male *Orconectes propinquus* (Cambaridae): c, central process. **63.** Seminal receptacle of *Orconectes obscurus* (Cambaridae), vv. **64.** Seminal receptacle of *Orconectes propinquus* (Cambaridae), vv. **65.** Chela of form I male *Cambarus bartonii bartonii* (Cambaridae), dv. **66.** First pleopod of form I (mv, left; lv, middle) and form II (lv, right) male *Cambarus bartonii bartonii* (Cambaridae). **67.** Chela of *Cambarus robustus* (Cambaridae), dv. **68.** First pleopod of form I (mv, left; lv, middle) and form II (lv, right) male *Cambarus robustus* (Cambaridae). lv, lateral view; mv, mesal view; vv, ventral view. (60–66, 68 redrawn from Hobbs 1974a; 67 modified from Crocker and Barr 1968.)

28a (27b). Usually without rostral carina (Fig. 53); ventral anterior margin of carpus with a median tubercle or low spine; 1st pleopod on form I males with a distinct, right-angled shoulder on anterior margin in lateral view (Fig. 61); seminal receptacle with 2 high tubercles, which are fused along the midline, on anterior margin (Fig. 63); tertiary burrower .... *O. obscurus*

28b. Usually with rostral carina (Fig. 54) (Bell [1971] suggests that the rostrum can be completely dry before determining the presence of a rostral carina. It is easily overlooked when the rostrum is wet.); ventral anterior margin of carpus (Fig. 36) usually without median tubercle or spine; 1st pleopod on form I males without a distinct, right-angled shoulder on anterior margin in lateral view, although there may be a shallow indentation there (Fig. 62); seminal receptacle with 2 low tubercles, which are never fused along the midline, on anterior margin (Fig. 64); tertiary burrower .... *O. propinquus*

29a (25b). Inner margin of palm with a single row of low tubercles (Fig. 65); palm inflated, without conspicuous depression near outer margin; rostrum often square, tapering acutely to its tip (Fig. 55); tip of mesial process of form I male copulatory stylets (1st pleopods) generally pointing ventrally (Fig. 66); tertiary burrower .... *Cambarus bartonii bartonii*

29b. Inner margin of palm with 2 rows of low tubercles (Fig. 67); palm with depression that is visible from both dorsal and ventral sides near its outer margin; rostrum rectangular, not tapering acutely to its tip (Fig. 56); tip of mesial process of form I male copulatory stylets (1st pleopods) generally pointing dorsally (Fig. 68); tertiary burrower .... *C. robustus*

References on Freshwater Crustacea Systematics

(*Used in construction of key.)

Ameyaw-Akumfi, C. 1981. Courtship in the crayfish *Procambarus clarkii* (Girard) (Decapoda, Astacidae). Crustaceana 40:57–64.

Belk, G.D. 1974. Zoogeography of the Arizona Anostraca with a key to the North American species. Ph.D. dissertation. Arizona State Univ., Tempe. 100 pp.

Belk, [G.]D. 1982. Branchiopoda. *In* S.P. Parker (ed.). Synopsis and classification of living organisms, vol. 2, pp. 174–180. McGraw Hill, New York.

*Bell, R.T. 1971. Handbook of the Malacostraca of Vermont and neighboring regions (crayfish, sowbugs and their relatives). R.T. Bell, Burlington, Vt. 65 pp.

Bouchard, R.W. 1978. Taxonomy, distribution, and general ecology of the genera of North American crayfishes. Fisheries 3:11–19.

Bousfield, E.L. 1973. Shallow-water gammaridean Amphipoda of New England. Cornell Univ. Press, Ithaca, N.Y. 312 pp.

Bowman, T.E. 1975. Three new troglobitic asselids from North America (Crustacea: Isopoda: Asellidae). Int. J. Speleol. 7:339–356.

Carter, J.C.H., M.J. Dadswell, J.C. Roff, and W.G. Sprules. 1980. Distribution and zoogeography of planktonic crustaceans and dipterans in glaciated eastern North America. Can. J. Zool. 58:1355–1387.

*Chace, F.A., J.G. Mackin, L. Hubricht, A.H. Banner, and H.H. Hobbs, Jr. 1959. Malacostraca. *In* W.T. Edmondson (ed.). Fresh-water biology, 2nd ed., pp. 869–901. John Wiley and Sons, New York.

*Cole, G.A. 1957. Some epigean isopods and amphipods from Kentucky. Trans. Ky. Acad. Sci. 18:29–39.

——. 1959. A summary of our knowledge of Kentucky crustaceans. Trans. Ky. Acad. Sci. 20:60–81.

*——. 1983. Textbook of limnology, 3rd ed. C.V. Mosby Co., St. Louis. 426 pp.

Cooper, S.D., and C.R. Goldman. 1980. Opossum shrimp (*Mysis relicta*) predation on zooplankton. Can. J. Fish. Aquat. Sci. 37:909–919.

Creaser, E.P. 1929. The Phyllopoda of Michigan. Pap. Mich. Acad. Sci. Arts Lett. 11:381–388.

——. 1930. The Michigan decapod crustaceans. Pap. Mich. Acad. Sci. Arts Lett. 13:257–276.

*Crocker, D.W. 1957. The crayfishes of New York State (Decapoda, Astacidae). Bull. N.Y. State Mus. 355:1–97.

*——. 1979. The crayfishes of New England. Proc. Biol. Soc. Wash. 92:225–252.

*Crocker, D.W., and D.W. Barr. 1968. Handbook of the crayfishes of Ontario. R. Ont. Mus. Life Sci. Misc. Publ. Univ. of Toronto Press, Toronto. 158 pp.

*Dexter, R.W. 1953. Studies on North American fairy shrimps with the description of two new species. Amer. Midl. Nat. 49:751–771.

——. 1956. A new fairy shrimp from the western United States, with notes on other North American species. J. Wash. Acad. Sci. 46:159–165.

*——. 1959. Anostraca. *In* W.T. Edmondson (ed.). Fresh-water biology, 2nd ed., pp. 558–571. John Wiley and Sons, New York.

Dodds, G.S. 1923. A new species of phyllopod. Occas. Pap. Univ. Mich. Mus. Zool. Mich. 141:1–3.

Edmondson, W.T. (ed.). 1959. Fresh-water biology, 2nd ed. John Wiley and Sons, New York.

Fincham, A.A. 1980. Eyes and classification of malacostracan crustaceans. Nature 287:729–731.

Forest, J., I. Gordon, L.B. Holthuis, A.G. Humes, H.K. Schminke, J.H. Stock, and J.C. Von Vaupel-Klein (eds.). 1979. Studies on Decapoda (biology, ecology, morphology, and systematics). Crustaceana, suppl. 5. E.J. Brill, Leiden, Netherlands. 242 pp.

Ginsburger-Vogel, T. 1985. Sex determination in Amphipoda. Bull. Soc. Zool. Fr. 110:49–62.

Goldman, C.R. (ed.). 1983. Freshwater crayfish. Pap. 5th Int. Symp. Freshwat. Crayfish, Davis, Calif., 1981. Avi Publ. Co., Westport, Conn. 569 pp.

Hartland-Rowe, R. 1965. The Anostraca and Notostraca of Canada with some new distribution records. Can. Field Nat. 79:185–189.

Hobbs, H.H., Jr. 1942. A generic revision of the crayfishes of the subfamily Cambarinae (Decapoda, Astacidae) with the description of a new genus and species. Amer. Midl. Nat. 28:334–357.

*——. 1972. Crayfishes (Astacidae) of North and Middle America. Biota Freshwat. Ecosyst. Ident. Man. 9:1–173.

*——. 1974a. A checklist of the North and Middle American crayfishes (Decapoda: Astacidae and Cambaridae). Smithson. Contrib. Zool. 166:1–161.

——. 1974b. Synopsis of the families and genera of crayfishes (Crustacea: Decapoda). Smithson. Contrib. Zool. 164.

——. 1981. The crayfishes of Georgia. Smithson. Contrib. Zool. 318:1–549.

Hobbs, H.H., Jr., H.H. Hobbs III, and M.A. Daniel. 1977. A review of the troglobitic decapod crustaceans of the Americas. Smithson. Contrib. Zool. 244:1–183.

*Holsinger, J.R. 1972. The freshwater amphipod crustaceans (Gammaridae) of North America. Biota Freshwat. Ecosyst. Ident. Man. 5:1–89.

——. 1977. A review of the systematics of the holarctic amphipod-family Crangonyctidae. Proc. 3rd Int. Colloq. on *Gammarus* and *Niphargus*. Schlitz, Germany, 1975. Crustaceana (suppl.) 4:244–281.

*——. 1978. Systematics of the subterranean amphipod genus *Stygobromus* (Crangonyctidae). Part II. Species of the eastern United States. Smithson. Contrib. Zool. 266:1–144.

Holthius, L.B. 1952. The subfamily Palaemoninae. A general revision of the Palaemonidae (Crustacea: Decapoda: Natantia) of the Americas. Allan Hancock Found. Publ. Occas. Pap. 12:1–396.

Jezerinac, R.F., and R.F. Thomas. 1984. An illustrated key to the Ohio *Cambarus* and *Fallicambarus* (Decapoda, Cambaridae) with comments and a new subspecies record. Ohio J. Sci. 84:120–124.

Kaestner, A. 1980. Invertebrate zoology. Vol. 3. Crustacea. Robert E. Krieger Publ. Co., Huntington, N.Y. 523 pp.

Larsen, A.A. 1959. A study of the fresh-water Crustacea (exclusive of the Copepoda) of the Rochester area. Proc. Rochester Acad. Sci. 10:183–240.

Lewis, J.J. 1980. A comparison of *Pseudobaicalasellus* and *Caecidotea*, with a description of *Caecidotea bowmani*, n. sp. (Crustacea: Isopoda: Asellidae). Proc. Biol. Soc. Wash. 93:314–326.

*Linder, F. 1941. Contributions to the morphology and the taxonomy of the Branchiopoda Anostraca. Zool. Bidr. Uppsala 20:101–302.

——. 1952. Contributions to the morphology and taxonomy of the Branchipoda Notostraca, with special reference to the North American species. Proc. U.S. Natl. Mus. 102:1–69.

*Mattox, N.T. 1959. Conchostraca. *In* W.T. Edmondson (ed.). Fresh-water biology, 2nd ed., pp. 577–586. John Wiley and Sons, New York.

Meehean, O.L. 1940. A review of the parasitic crustacea of the genus *Argulus* in the collections of the United States National Museum. Proc. U.S. Natl. Mus. 88:459–522.

Morgan, M.D. (ed.). 1982. Ecology of Mysidacea. Hydrobiologia. 93:1–222.

*Needham, J.G., and P.R. Needham. 1962. A guide to the study of freshwater biology, 5th ed. Holden-Day, Inc., San Francisco. 107 pp.

Norrocky, M.J. 1984. Burrowing crayfish trap. Ohio J. Sci. 84:65–66.

Page, L.M. 1985. The crayfishes and shrimps (Decapoda) of Illinois. Ill. Nat. Hist. Surv. Bull. 33:335–448.

Payne, J.F. 1978. Aspects of the life histories of selected species of North American crayfishes. Fisheries 3:5–7.

*Pennak, R.W. 1978. Freshwater invertebrates of the United States, 2nd ed. John Wiley and Sons, New York. 803 pp.

*Pickett, J.F., J. Nasca, and W. Gall. 1982. A new subspecies of crayfish in New York State. The Conservationist 36:48.

Richardson, H. 1904. Contributions to the natural history of the Isopoda. Proc. U.S. Natl. Mus. 27:1–89.

Sanders, H.L. 1963. The Cephalocarida: functional morphology, larval development, comparative external anatomy. Mem. Conn. Acad. Arts Sci. 15:1–80.

Smith, D.S. 1979. Variability of crayfish of the *virilis* section (Cambaridae: *Orconectes*) introduced into New England and eastern New York. Amer. Midl. Nat. 102:388–391.

Somers, K.M., and D.P.M. Stechey. 1986. Variable trappability of crayfish associated with bait type, water temperature and lunar phase. Amer. Midl. Nat. 116:36–44.

Strenth, N.W. 1976. A review of the systematics and zoogeography of the freshwater species of *Palaemonetes* Heller of North America (Crustacea: Decapoda). Smithson. Contrib. Zool. 228:1–27.

*Tressler, W.L. 1959. Ostracoda. *In* W.T. Edmondson (ed.). Fresh-water biology, 2nd ed., pp. 657–734. John Wiley and Sons, New York.

Wildish, D.J. 1982. Talitroidea (Crustacea, Amphipoda) and the driftwood ecological niche. Can. J. Zool. 60:3071–3074.

——. 1988. Ecology and natural history of aquatic Talitroidea. Can. J. Zool. 66:2340–2359.

Williams, W.D. 1970. A revision of North American epigean species of *Asellus* (Crustacea: Isopoda). Smithson. Contrib. Zool. 49:1–80.

*——. 1972. Freshwater isopods (Asellidae) of North America. Biota Freshwat. Ecosyst. Ident. Man. 7:1–45.

Wilson, C.B. 1944. Parasitic copepods in the United States National Museum. Proc. U.S. Natl. Mus. 94:529–582.

*Wilson, M.S., and H.C. Yeatman. 1959. Free-living Copepoda. *In* W.T. Edmondson (ed.). Freshwater biology, 2nd ed., pp. 735–861. John Wiley and Sons, New York.

16 Hydrachnidia

Bruce P. Smith

The Hydrachnidia (true water mites) are not the only aquatic mites, but they are the most successful group of mites found in freshwater. They inhabit almost every aquatic habitat, and densities frequently exceed 200 mites/m^2 (see, e.g., Efford 1966). In northeastern North America, probably more than 95% of all species and well over 95% of all individuals of freshwater aquatic mites encountered belong to the subcohort Hydrachnidia. Water mites are frequently brightly colored and sometimes relatively large (over 2 mm), making them more conspicuous than other aquatic mites, and thus exaggerating their already significant predominance.

Life History

The basic life cycle of a water mite is characteristic of the cohort Parasitengona, to which the subcohort Hydrachnidia belongs, and is an interesting parallel to the complete metamorphosis of holometabolous insects. There is an egg, prelarva, six-legged larva, nymphochrysalis, eight-legged deutonymph, teleiochrysalis (= imagochrysalis), and an eight-legged adult (Prasad and Cook 1972, Krantz 1978). The larva is morphologically very different from the deutonymph and the adult, whereas the deutonymph is smaller than, but often quite similar to, the adult. The nymphochrysalis and teleiochrysalis are pupalike resting stages that occur within the cuticle of the previous stage. The prelarva is also a resting stage in almost all species of mites and is frequently overlooked because it occurs within the egg. The ancestral acariform mite life cycle consists of an egg, six-legged prelarva, six-legged larva, eight-legged protonymph, eight-legged deutonymph, eight-legged tritonymph, and eight-legged adult. All but the egg are active stages and are similar in appearance (Krantz 1978).

Most species of water mites in the more "primitive" families (within the Eylaoidea, Hydrovolzioidea, and Hydryphantoidea) have aerial larvae; they are weakly sclerotized and have long, six-segmented legs and long setae (Figs. A, B; Mitchell 1957d, Prasad and Cook 1972). This form of larva breaks through the surface and encounters a potential host out of water. Larvae of the "advanced" families (within the superfamilies Arrenuroidea, Hygrobatoidea, and Lebertioidea) are heavily sclerotized, have reduced body setation and stout, five-segmented legs, (Figs. C, D) and swim or crawl in the water in search of hosts (Mitchell 1957d, Prasad and Cook 1972). Aquatic larvae have also evolved independently in the family Hydrachnidae (Mitchell 1957d), the genus *Rhyncholimnochares* (family Limnocharidae; B. P. Smith 1988), and in the subfamily Wandesiinae (family Hydryphantidae) (Prasad and Cook 1972).

The larva is typically a parasite of aquatic insects. The majority of species are parasites on adult chironomid midges (Diptera; I. M. Smith and Oliver 1976, 1986); larvae of most remaining species parasitize most other aquatic insects, including nymphs and adults of aquatic and semiaquatic hemipterans, adult aquatic beetles, adult odonates, and both adult and immature trichopterans and plecopterans (I. M. Smith and Oliver 1976, 1986). Most water mite larvae will parasitize any species of from one to several families of insects (see, e.g., B. P. Smith and McIver 1984a, b; B. P. Smith 1988). Despite this wide spectrum of acceptable hosts, one species of host will usually bear the majority of one species of mite within a given habitat.

Certain water mites, for example, members of the Eylaidae and Hydrachnidae families, modify their life cycle by extending the association with the host through the nymphochrysalis stage (Mitchell 1957d). A few other species do not feed as larvae, such as some *Piona* spp. (Prasad and Cook 1972, I. M. Smith 1976a), or they bypass an active larval stage altogether and emerge from the egg as a deutonymph, as do some species of the genus *Thyas* and of the family Pionidae (Prasad and Cook 1972, I. M. Smith 1976a). Usually the parasitic larvae are permanently affixed to their host for the duration of the engorgement period, although some parasites of juvenile hemipterans, for example *Limnochares aquatica* (Böttger 1972a), will detach and transfer to the teneral insect when the host molts.

Some, if not all, larval water mites secrete a feeding tube called the stylostome into the host's tissues (Davids 1973; Åbro 1979, 1982, 1984; Redmond and Hochberg 1981; Redmond and Lanciani 1982). The stylostome remains in the host after the mite has departed, leaving a permanent record of parasitism (Lanciani 1979b). The nutrient drain of parasitic larval mites can reduce the fecundity and survivorship of adult hosts (see, e.g., Lanciani and Boyt 1977; Lanciani 1982, 1983; B. P. Smith and McIver 1984c), or it can increase the duration of instars and decrease survivorship of parasitized immature insects (Lanciani 1982, Lanciani and May 1982).

The deutonymph of the "primitive" water mites is a smaller version of the adult. In the more "advanced" superfamilies (e.g., *Arrenurus* spp.; Arrenuroidea), the ventral coxal plates may be similar to those of the adult even though the deutonymph lacks most of the sclerotization associated with the adult. For some species of more "advanced" water mites (e.g., *Aturus* spp.; Hygrobatoidea), even the coxal plates of the deutonymph are quite unlike those of the adult.

Whether or not the deutonymph and the adult are morphologically similar, their life styles are the same. Both typically are free-living predators, frequently feeding on chironomid larvae or on small aquatic crustaceans (Böttger 1970, Davids et al. 1981, Reissen 1982a). In the genus *Hydrachna*, however, adults and deutonymphs special-

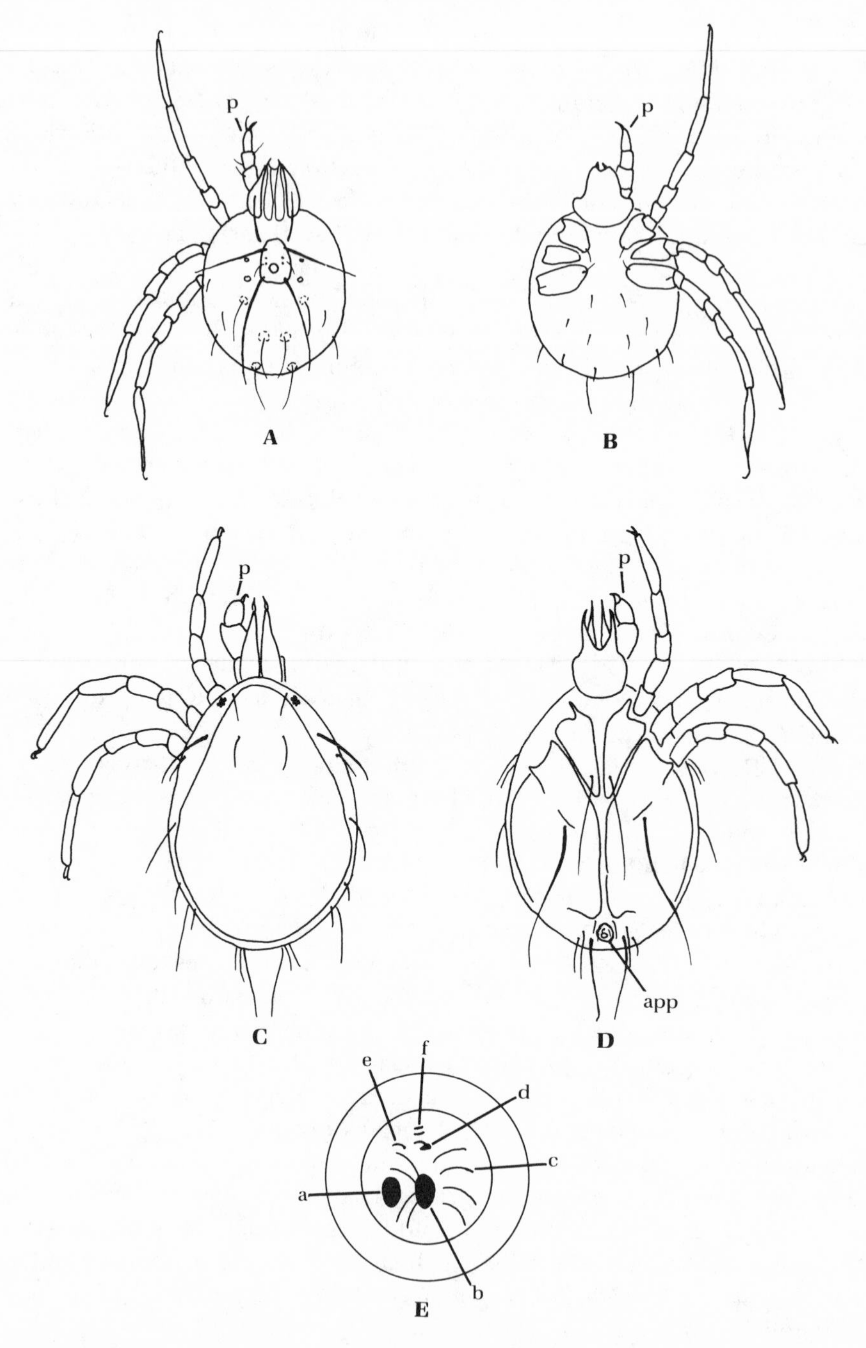

A. Hydryphantoid larva, dv: p, palp. **B.** Hydryphantoid larva, vv: p, palp. **C.** Hygrobatoid larva, dv: p, palp. **D.** Hygrobatoid larva, vv: app, anal pore plate; p, palp. **E.** Slide-mounted adult mite (museum mount): a, dorsal surface; b, ventral surface and legs (anterior face up); c, legs (posterior face up); d, capitulum with one palp showing lateral surface; e, one palp showing medial surface; f, chelicerae. dv, dorsal view; vv, ventral view. (A–D modified from Mitchell 1957d; E modified from Cook 1974a.)

ize on eggs, often feeding on the eggs of the same species that served as host for the larval stage (Davids 1973). Many species of *Unionicola* have parasitic deutonymphs and adults, which feed on freshwater mussels, snails, or sponges. Most larval *Unionicola* are parasites of adult chironomids. So species may have one parasitic relationship as larvae and a different one as deutonymphs and adults. *Najadicola ingens* (Pionidae) is also a parasite of freshwater mussels in its adult and deutonymphal stages, and apparently evolved this association independently (Simmons and I. M. Smith 1984). A few unusual species of water mite have been reported that have multiple deutonymph and adult stages, for example *Limnochares aquatica* (Böttger 1972a).

Characteristic of Chelicerata, most water mites do not use direct internal sperm transfer. Some, such as *Hydrachna* spp., will randomly deposit spermatophores in the absence of the female (Davids and Belier 1979). Species in other genera may indirectly transfer sperm using modified legs or other body parts, a process often associated with elaborate behaviors and profound sexual dimorphism (Mitchell 1957a, Cassagne-Méjean 1966). Among the species of *Eylais* there is a complete range, from species that randomly deposit spermatophores to those that use direct internal transfer (Lanciani 1972). Male water mites have a characteristic internal cuticular structure, the ejaculatory complex (Barr 1972), which for some species functions as a penis for copulation (Cook 1974a). Sexual selection appears to be intense for at least some species of *Unionicola;* the males are strongly territorial and defend harems of up to 50 or more females (Dimock 1983, 1985).

Many water mites exhibit bright color patterns, containing greens, blues, oranges, or reds. Red predominates among species in the "primitive" Hydrovolzioidea, Hydrachnoidea, Eylaoidea, and Hydryphantoidea, and often the term *the red water mites* is used to denote this ancestral stem (Prasad and Cook 1972). Fish and invertebrates will eat water mites, but the brightly colored species are apparently distasteful and predators learn to reject them (Kerfoot 1982).

Systematics

The higher categories of the subclass Acari have changed dramatically since 1970, and hence a confusing array of names is applied to the water mites. To summarize, the water mites are *either* called the Hydrachnellae and represent a sister group to the Parasitengona (used commonly in European literature), *or* they are called the Hydrachnidia (= Hydrachnida, = Hydracarina) and are grouped within the Parasitengona (= Parasitengonae) as a sister group to the terrestrial mites of the Trombidia (= Trombida, consisting of the Calyptostomatoidea, Erythreoidea, and Trombidioidea). The water mites and their terrestrial relatives belong to the order Acariformes, suborder Prostigmata (= suborder Trombidiformes, = suborder Actinedida). While it is common practice to cite orders and families in literature pertaining to insects, the suborder, supercohort, cohort, and subcohort can be more useful categories for references to mites.

The species of the subcohort Hydrachnidia are organized into 7 superfamilies and some 47 families of mites worldwide. More than 300 genera are currently recognized

(K. O. Viets 1987). One or more new genera are added each year. It is expected that new genera will regularly be discovered in North America, as well as species within genera not previously known from this continent. An estimated 425 species have been found in Canada, with an estimated 533 yet to be discovered (I. M. Smith and Lindquist 1979), a fair indication of our knowledge of the North American fauna.

Comparisons of the morphological and species diversity of the Hydrachnidia with their closest relatives, the Trombidia, indicate that the Hydrachnidia underwent rapid speciation and diversification when they invaded the aquatic environment. Although the species of water mites can usually be grouped into distinctive and discrete genera, they are often so highly modified as adults that it is difficult to trace relationships among genera or to associate the water mites with their terrestrial relatives. Convergence of body form for certain life styles necessitates complex descriptions to outline some family and generic groupings. Larval morphology is more conservative and is quite useful for clarifying evolutionary relationships. Unfortunately, the larval stage is still unknown for approximately 75% of the genera and many of the species (based on data in I. M. Smith and Oliver 1986).

The Hydrachnidia are among the most numerically abundant and taxonomically diverse of the various freshwater aquatic mites. Several factors have probably contributed to their success (I. M. Smith and Oliver 1986).

1. Their parasitic association with insects greatly increases their dispersal.
2. Presumably, when the Chironomidae underwent rapid species radiation, so did the species of mites that parasitized them.
3. Mite larva feed on a different food source (either a different species or a different life history stage) from that of the deutonymph and adult. This trait adds energy from a different resource and minimizes competition between instars. (An analogous argument has been advanced as one of the factors contributing to the success of holometabolous insects.)
4. The resting stages are ideally suited for diapause, and several groups (e.g., *Arrenurus planus;* Wiggins et al. 1980) use these stages to avoid unfavorable conditions. (Again, an analogous argument has been applied to holometabolous insects.)
5. The ability to swim, restricted among mites to species of Hydrachnidia, greatly contributes to their dispersal and allows for a broader diversity of life styles.

Habitats and Collection

Species of Hydrachnidia are common in such lentic waters as swamps, marshes, ponds, and the littoral and profundal zones of lakes (Barr 1973). They are often associated with vegetation or with the top few millimeters of substrate, but they can also lead a planktonic existence. Water mites are common, too, in the erosional and depositional zones of rivers, and the air-water interface at the margins of various water bodies harbors a variety of these mites. Some species are adapted to live in such extreme environments as thermal springs, glacial meltwater rivers, temporary pools, waterfalls, and in groundwater buried within gravel banks of streams (interstitial habitats). A few species can inhabit oceans and inland saline waters, although most are limited to freshwater.

Water mites in lentic waters are often free-swimming and conspicuously colored, but many interesting species are cryptic, clinging to vegetation or buried in the substrate. The best method for collecing mites from lakes, ponds, and slow (depositional) rivers is to vigorously sweep the vegetation and skim the surface of the substrate with an aquatic net (Cook and Mitchell 1952, Barr 1973). The contents of the net can then be sorted in a shallow white tray filled with water, and the mites can be removed with an eyedropper. Many species in lentic and slow lotic habitats are photopositive and can be collected with aquatic light traps (Barr 1979), although this method is highly biased for certain taxa. In streams and erosional zones of rivers, water mites are found crawling on rocks or buried in the gravel. Such habitats can be sampled by kicking up the gravel upstream of a net. Care should be taken to disturb the gravel as deeply as possible, since there is a distinctive fauna deep within gravel beds. Wet moss and other vegetation in bogs, springs, seepage areas, and at the margins of various water bodies harbor a characteristic mite fauna. One effective method for extracting this fauna is to place the vegetation in water with a temperature gradient (i.e., 35–10°C) and wait for the mites to move to cooler water (Fairchild et al. 1987).

Preservation and Preparation

Water mites are best preserved in Koenike's solution (also called GAW; 5 parts Glycerine:1 part *A*cetic acid:4 parts *W*ater). Preservation in alcohol is effective, but it makes it more difficult to clear specimens for slide-mounting. Water mites should never be preserved in solutions that contain formaldehyde. Andre's solution (acetic corrosive; 50 g chloral hydrate:50 ml acetic acid:50 ml water), lactic acid, or mild KOH solution is used for clearing specimens before slide-mounting. Hoyer's solution is the standard mounting medium for routine slide preparation, although glycerine jelly is preferred for permanent museum specimens. For accurate identification, specimens should be dissected before mounting on slides and should be positioned to clearly show all body parts (Fig. E). Methods for preservation and preparation of specimens are discussed in greater depth by Mitchell and Cook (1952), Barr (1973), and Cook (1974a).

Keys for water mites typically rely heavily on characters of the mouthparts and placement of setae and glands to distinguish major groups. Use of those keys requires that specimens be cleared and mounted on slides for examination with a compound microscope. Yet, with a little experience, one can recognize many specimens to genus using characters that require minimal magnification. I have used such characters wherever possible in the following key.

Novices should be able to identify larger live and preserved whole specimens with only a good quality dissecting microscope capable of 100× magnification (requiring 20× eyepieces). While most specimens should be identifiable without being cleared and slide-mounted, some material may require slide-mounting if diagnostic characters are not visible. As students gain experience, they should need to slide-mount only a few specimens within a sample.

Several techniques are handy for examining live material. Specimens can be restrained by trapping them in a thin film of water on a concavity slide. Mites can also be slowed by cooling them in a refrigerator or anesthetized by placing them in an aqueous Alka Seltzer solution (one tablet in a glass of water) for approximately one minute.

Morphology

What is known of the internal morphology of Hydrachnidia can be found in articles such as those written by Wesenberg-Lund (1939), Mitchell (1954b, 1957b, c), Cassagne-Méjean (1966), Barr (1972), and Wiles (1984). External morphology has been outlined by a number of authors (e.g., Cook 1974a, Krantz 1978, I. M. Smith 1976a) and is briefly covered below.

Idiosoma. The term *idiosoma* refers to the unsegmented body of the mite. To the uninitiated some Hydrachnidia appear to be nothing more than a head with legs. Body shape is commonly globular, although it can be a flattened teardrop, circular disk, concave saucer-shaped disk, elongate disk, American football, laterally compressed American football, or without definite form, much like a beanbag. The body can be entirely membranous (see Fig. 9), or it can bear a dorsal plate (Fig. 21), a dorsal plate and marginal platelets (Fig. 107), or a series of small plates (Fig. 28; plates may be knoblike). The entire body can also be encased in a hardened integument, with only a thin membranous line around a single dorsal plate (Fig. 134). In some species this dorsal plate is joined to the body at the anterior or posterior margin, so that the membranous line forms an arching dorsal furrow (see Fig. 135). Deviations from a spherical membranous body are presumably adaptations for more efficient locomotion. Dorsal plates allow for stronger muscle attachments and are commonly seen on crawling species found in marginal waters or in fast-flowing water (Mitchell 1957c). Disk-, saucer-, and football-shaped mites are common in dense vegetation, a distribution that suggests that these body forms are adaptations for living in densely overgrown habitats. Many stream-dwelling species are flattened, presumably an adaptation to minimize drag. The beanbag form of limnocharids appears to be an adaptation for squeezing into crevices and between objects (Mitchell 1964).

The coxa is flattened into a plate and fused with the body (coxal plate) in all species of Hydrachnidia. There can be varying degrees of fusion among these plates. In the most ancestral state there is fusion between coxal plates I and II and coxal plates III and IV, no fusion between plates from opposite sides (Fig. 6), no enlargement of the plates, and distinct sutures. The other extreme is represented by *Frontipoda,* in which all coxal plates are enlarged and fused into a shell that almost encloses the entire body, and sutures are indistinct (Figs. 1–3).

Almost all species bear clusters of genital acetabula (also called papillae) around the gonopore. These acetabula appear to be osmoregulatory organs and are papilla-like or cup-shaped depressions. It is common for species to have three pairs of acetabula, although some have more than 50 pairs. In most species these structures are obvious (as in Fig. 52), but in some they are very small and cannot be seen under a dissecting microscope (Fig. 10). Genital acetabula may be borne on genital plates (as in Fig. 18), covered by a pair of doorlike genital flaps (Fig. 17), or loosely scattered over the body wall (Fig. 10). The gonopore, genital acetabula, and associated flaps or plates are frequently referred to collectively as the genital field.

Water mites typically have two pairs of laterally borne simple eyes, either located individually on the body wall (see Fig. 37) or borne in pairs on plates (Fig. 84). Rarely, both pairs of lateral eyes are borne together on a central eyebridge (Figs. 7, 11). A primitive characteristic is the presence of a fifth medianly placed simple eye, often borne on a central plate (Fig. 21).

Gnathosoma. The gnathosoma (mouthparts) of Hydrachnidia consist of a capsule (the capitulum), a pair of chelicerae enclosed by the capitulum, and a pair of leglike structures (palps) that are attached to the sides of the capitulum (Figs. 31, 81). In many terrestrial mites, the chelicerae are chelate (the subterminal segment extends beyond the terminal segment forming an apposable digit, like a lobster claw), but this extension has been lost in the water mites. Palps are typically five-segmented, and in the preferred method of identification the segments are called the trochanter, femur, genu, tibia, and tarsus (Fig. 81). The palp can be simple (Fig. 31), chelate (Fig. 81), or uncate (apex of the tibia broadened so that the tarsus can fold against it; Fig. 139).

Legs. Adult and deutonymph Hydrachnidia have four pairs of legs (Fig. 1), with six segments on each leg (Fig. 42). The preferable method for distinguishing legs and segments is to use roman numerals to refer to specific legs, and to use the terms *trochanter, basifemur, telofemur, genu, tibia,* and *tarsus* to refer to the segments. (For example, IIITa denotes the tarsus of leg III).

Males of species in certain families have highly modified leg segments (Figs. 65, 69, 72) or specialized setae (Figs. 66, 123). These adaptations are associated with mating. Mites that swim typically bear rows of long, thin setae on their legs (swimming setae; Figs. 1, 69). Most species of water mites have stout whorls of setae around the base of leg segments. The whorls aid in crawling and in burrowing through vegetation and organic debris. There are normally two retractable claws on the tip of each leg, but some groups have no claws on leg IV (Fig. 20). The lack of claws is used in the following key to separate some groups, but it may be difficult to see claws that are retracted. Typically a leg with claws is blunt at the tip, often with the tip as wide or wider than the rest of the leg segment (Figs. 42, 123), while legs without claws usually taper to a point (Fig. 20).

Identification: Use of the Key

Most Hydrachnidia are distinctively different from other mites. They are highly modified for an aquatic existence, whereas most other aquatic and semiaquatic mites are typically quite similar to their terrestrial relatives. Only the Hydrachnidia include species that swim. Species of Halacaridae are largely marine, but a number of freshwater species occur; these usually have three claws on each leg, while Hydrachnidia never have more than two. Some species of Halacaridae have fewer than three claws on each leg and can be confused with *Hydrovolzia* spp. (Hydrachnidia), but the two can usually be distinguished because halacarids are typically much smaller (< 500 μm long) than *Hydrovolzia* spp., lack genital flaps, and in life are usually transparent or straw-colored, whereas *Hydrovolzia* spp. are bright red. Stygothrombiid mites are also occasionally encountered in interstitial water and can be confused with *Wandesia* spp. (Hydrachnidia). But stygothrombiids have three claws on each leg, and the tarsus of the first leg is swollen as in terrestrial trombidiids. A few species of Gamasida are semiaquatic, such as some members of the Ascidae. But these mites have seven leg segments, unlike Hydrachnidia species, which have six. Aquatic or semiaquatic species of Gamasida, Oribatida, and Acaridida are very similar to their terrestrial relatives—when in doubt, refer to Krantz 1978.

This key was designed for use with adult water mites and should not be used for juvenile stages. The larval stage is obvious, since it has only six legs. Keys to genera for larval Hydrachnidia are presented by Prasad and Cook (1972) and Wainstein (1980). Although the deutonymphs are often very similar to adults, they lack a genital opening (*gonopore*) and have fewer genital acetabula than adults. In many species, the deutonymphs have only two pairs of acetabula, while adults will always have either more than two pairs or, rarely, none at all. K. Viets (1936) published diagnoses for deutonymphs of some genera, but there is no comprehensive key currently available for deutonymphs of Hydrachnidia.

I have attempted to put into words many of the conspicuous characters an expert would use to recognize adult members of distinctive genera of Hydrachnidia. I sometimes use long and qualified couplets, but I believe it is necessary to retain the reliability of the keys devised by such authors as Cook (1974a) and Newell (1959). Because I have had to generalize on such characters as body shape, a few of the less common species may not be identifiable. Nevertheless, I estimate that 95% of the species and 99% of the individual specimens collected in northeastern North America could be accurately identified using this key. In this key, northeastern North America is defined as Connecticut, Maine, Massachusetts, New Hampshire, New York, Rhode Island, and Vermont, and New Brunswick, Nova Scotia, southern and central Ontario, Prince Edward Island, and southern Quebec. In the couplets, I frequently list several criteria in order of importance. Supplemental characteristics are therefore available in case of uncertainty. Additional information that can aid identification, such as habitat, behavior, likelihood of encounter, and color of living specimens, is sometimes also included in square brackets.

There are many examples of parallel acquisition of morphological characters among taxa (e.g., multiplication of genital acetabula, development of dorsal plates), as well as strong sexual dimorphism in some species. The key is thus artificial, and some genera occur several times. In the few cases where specimens cannot be identified to genus unless slide-mounted, the genera involved are listed under the couplet. If in doubt, refer to Cook's book (1974a), which has the most comprehensive and reliable keys.

Checklist of Hydrachnidia

Subclass Acari
Order Acariformes
Suborder Prostigmata
Cohort Parasitengona
Subcohort Hydrachnidia
Superfamily Arrenuroidea
Acalyptonotidae
Paenecalyptonotus
Arrenuridae
Arrenurus
Athienemanniidae
Chelomideopsis[a,b]
Chappuisididae
Chappuisides
Krendowskiidae
Geayia
Krendowskia
Laversiidae
Laversia
Mideidae
Midea
Mideopsidae
Mideopsis
Nudomideopsis
Momoniidae
Momonia
Stygomomonia
Neoacaridae
Neoacarus
Volsellacarus[c]
Uchidastygacaridae
Uchidastygacarus
Superfamily Eylaoidea
Eylaidae
Eylais
Limnocharidae
Limnochares
Neolimnochares
Rhyncholimnochares
Piersigiidae
Piersigia
Superfamily Hydrachnoidea
Hydrachnidae
Hydrachna
Superfamily Hydrovolzioidea
Hydrovolziidae
Hydrovolzia
Superfamily Hydryphantoidea
Hydrodromidae
Hydrodroma
Hydryphantidae
Euthyas
Hydryphantes
Panisopsis
Panisus
Protzia
Pseudohydryphantes
Tartarothyas[a]
Thyas
Thyasides
Thyopsella
Thyopsis
Trichothyas[a]
Wandesia
Zschokkea
Rhynchohydracaridae
Clathrosperchon
Superfamily Hygrobatoidea
Aturidae
Albia
Aturus
Axonopsis
Brachypoda
Estellacarus
Kongsbergia
Ljania
Woolastookia
Feltriidae
Feltria
Hygrobatidae
Atractides
Hygrobates
Limnesiidae
Limnesia
Meramecia
Tyrrellia
Pionidae
Forelia
Huitfeldtia
Hydrochoreutes
Najadicola
Nautarachna
Neotiphys
Piona
Pionacercus
Pionopsis
Pseudofeltria
Tiphys
Wettina
Unionicolidae
Koenikea
Neumania
Unionicola
Superfamily Lebertioidea
Anisitsiellidae
Bandakia
Lebertiidae
Lebertia

Oxidae
Frontipoda
Oxus
Sperchonidae
Sperchon
Sperchonopsis
Teutoniidae
Teutonia
Torrenticolidae
Testudacarus
Torrenticola

Source: Derived and updated from Cook 1974a.
[a]Possibly occurs in, but has not yet been recorded from, northeastern North America.
[b]Not included in the key.
[c]Present, but no published records (I. M. Smith, pers. comm., 1987).

Key to Genera of Hydrachnidia Adults

1a. All legs inserted far forward on the body in a whorl around the anterior end, octopus-like, with leg IV inserted dorsolaterally to leg I (Fig. 1); leg IV tapers to a clawless point (Fig. 20); extreme fusion and expansion of the coxal plates; sutures between adjacent plates are reduced and do not reach the genital field (Figs. 3, 5) [elongate-oval, shaped like an American football; sometimes laterally compressed] 2

1b. Legs inserted along the longitudinal axis of the body, with insertion of leg IV clearly posterior to insertions of other legs (Figs. 14, 18, 66, 123); tip of leg IV generally blunt and with claws (Figs. 42, 53, 123; claws may be difficult to see if retracted) or, if clawless, then there is a complete division between coxal plates II and III (Figs. 17, 18); coxal plates may be small and well defined (Figs. 6, 27) or may show various stages of fusion and expansion (Figs. 25, 123, 141) 3

2a (1a). Body laterally compressed; coxal plates greatly expanded, extending behind and enclosing genital field, also extending up the sides of the body onto the dorsum, leaving only a narrow median membranous strip (Figs. 1–3); narrow dorsomedian membranous strip, usually containing 1 large or several smaller plates (Fig. 2), but in some species these plates are absent [dark olive-green or orange swimmers; common in permanent standing waters and slow streams] . . . . *Frontipoda*

2b. Body usually not compressed; coxal plates usually not greatly expanded, never enclosing genital field (posterior margin is open) (Fig. 5), usually not extending up onto the dorsal side of the body; dorsum membranous and without plates; if body is laterally compressed and coxal plates do extend up onto the dorsum then there is a moderately wide dorsal membranous strip (Fig. 4) [typically reddish brown; strong swimmers; fairly common in permanent standing water, occasionally found in slow streams] . . . . *Oxus*

3a (1b). Dorsum of body mostly membranous with no more than ¼ of its surface covered with a plate or plates (Figs. 13, 37, 84) 4

3b. Dorsum of body with more than ¼ of its surface covered with a plate or plates (Figs. 59, 87, 89, 113) often to the extent that the body is encased in a rigid cuticle shell with, at most, a thin membranous line around a single dorsal plate (Fig. 134) 41

4a (3a). Eyes attached to an eyebridge (Figs. 7, 11); mouth opening appearing to be circular (Figs. 8, 12); numerous very small genital acetabula (not visible under a dissecting microscope) scattered over the body wall and not clustered in groups (Fig. 10); no genital plates or flaps [large (> 3 mm long) red mites] . . 5

4b. Eyebridge absent (eyes rarely may also be absent); if anterodorsal plate is present then eyes are not attached to it (Figs. 21, 33); mouth opening not circular; 3 to numerous pairs of genital acetabula, usually conspicuous, and always grouped (Figs. 15, 38, 43); usually with either genital plates (Figs. 43, 49, 83) or genital flaps (Figs. 25, 80) 7

5a (4a). Eyebridge wider than long, spectacle-like (Fig. 7); coxal plates III and IV fused only at median tip (Fig. 6) [in life the mite is red, and a firm disk shape; swims dragging its 4th pairs of legs; very common, found in temporary and permanent lentic waters] . . . . *Eylais*

5b. Eyebridge much longer than wide (Fig. 11); coxal plates III and IV fused together along most of their length into a strap-shaped plate (Fig. 10) [in life these mites are pillow-shaped] 6

6a (5b). Gnathosoma attached to a long protrusible tube of soft integument; gnathosoma is well removed from coxal plate I [with gnathosoma extended, mite appears elephant-like, while retracted, integument tube folds like a turtleneck; fairly common, found in streams and never swims; dorsum often with a few straplike or circular dorsal plates, although often not visible on live specimens] . . . . *Rhyncholimnochares*

6b. Gnathosoma not attached to a long protrusible tube of soft integument; gnathosoma always located near coxal plate I [common, found in lakes, streams, marshes and bogs; 1 species of *Limnochares* swims using only legs III and IV, while other species do not swim; no North American species of *Neolimnochares* can swim; most *Neolimnochares* species have a few straplike or circular plates on the dorsum (Fig. 13), although often not visible on live specimens] . . . . *Limnochares, Neolimnochares*

Palp is 5-segmented in *Limnochares,* 4-segmented in *Neolimnochares.* Specimens must be slide-mounted for this distinction to be seen.

7a (4b). Genital plates and flaps absent or minute and not visible without a compound microscope; genital acetabula present but may be inconspicuous (Figs. 14–16) [crawlers; found in lotic waters] 8

7b. Genital plates (Figs. 18, 19, 40, 41, 85) or genital flaps (Figs. 17, 30) present and conspicuous, or the genital acetabula are grouped on the body wall (Fig. 76); genital acetabula present and typically conspicuous 9

8a (7a). Lateral eyes either absent or lying below the integument, without a lens or

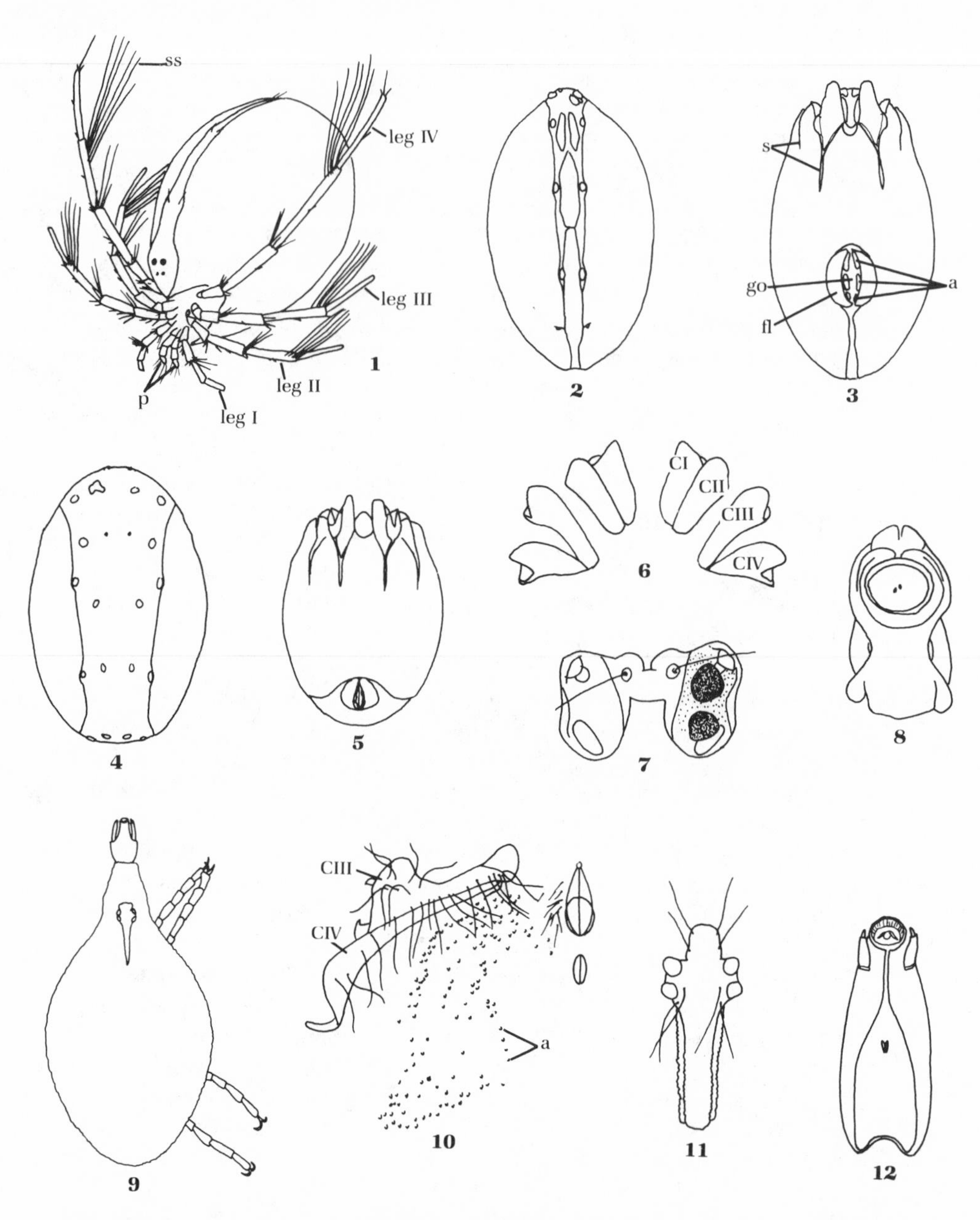

1. *Frontipoda* (Oxidae), dav: p, palps; ss, swimming setae. **2.** *Frontipoda* (Oxidae), dv. **3.** *Frontipoda* (Oxidae), vv: a, genital acetabula; fl, genital flap; go, gonopore; s, sutures. **4.** *Oxus* (Oxidae), dv. **5.** *Oxus* (Oxidae), vv. **6.** Venter of *Eylais* (Eylaidae): CI–CIV, coxal plates I–IV. **7.** Eyebridge of *Eylais* (Eylaidae), dv. **8.** Gnathosoma of *Eylais* (Eylaidae). **9.** *Rhyncholimnochares* (Limnocharidae), dv. **10.** *Limnochares* (Limnocharidae) genital field and coxal plates (C)III and IV (one side only), vv: a, acetabula. **11.** Eyebridge of *Limnochares* (Limnocharidae), dv. **12.** Gnathosoma of *Limnochares* (Limnocharidae). dav, diagonal anterolateral view; dv, dorsal view; vv, ventral view. (1 modified from Barr 1973; 2–11 modified from Cook 1974a; 12 modified from Lundblad 1941.)

capsule (as in Fig. 26); body very soft and usually greatly elongated (Fig. 14); 3–4 pairs of genital acetabula (Figs. 14, 15) [reduced body pigment, partially transparent; fairly uncommon] . . . . *Wandesia*

8b. Lateral eyes present, borne in capsules with lenses (e.g., Figs. 21, 84); body not greatly elongated (Fig. 16); with numerous pairs of genital acetabula [firm-bodied, red; uncommon] . . . . *Protzia*

9a (7b). Leg IV without claws and tapering to a point (Fig. 20); leg segments unmodified, cylindrical [found in permanent lentic waters, springs, and slow streams] 10

9b. Leg IV with claws, the tip of the leg is blunt and not tapering to a point (Figs. 42, 53, 123); if leg segments modified and not cylindrical then claws may be reduced and not conspicuous (Figs. 58, 69) 11

10a (9a). Coxal plate IV square or blocklike; possessing well-developed genital flaps covering 3 pairs of acetabula; no fusion between coxal plates I (Fig. 17) [uncommon; found in streams, springs, or in far northern ponds] . . . . . *Teutonia*

10b. Coxal plate IV triangular in shape (Figs. 18, 19); 3 (less commonly 4 to many) pairs of acetabula borne on a pair of genital plates (genital plates are fused medially on males); coxal plates I often fused at median edge [very common, frequently in permanent ponds and lakes, but can be found in most types of freshwater; typically not red; strong swimmers] . . . . *Limnesia* (in part)

11a (9b). Dorsum with a single anteromedian plate with posterolateral projections present (Figs. 21–23); variable genital field (Fig. 24), the 3 paired genital acetabula configuration is most commonly encountered; like a red flattened pillow in life; swimming mites, with long swimming setae on legs III and IV (e.g., Fig. 69) [very common in temporary and permanent ponds, can be found in lakes and sluggish streams] . . . . *Hydryphantes*

11b. Dorsum without dorsal plates, with several plates, or if a single dorsal plate is present, it lacks posterolateral projections 12

12a (11b). Genital plates or flaps present and genital acetabula not borne on them (acetabula either in a median row between genital flaps or plates (Figs. 25, 27) or distributed around genital plates in at least 2 groups per plate (Fig. 85); 3 pairs of genital acetabula 13

12b. Either genital plates present and acetabula borne on plates (Figs. 38, 39, 46–48) or genital plates and flaps absent and acetabula are grouped on the body wall (Fig. 76); 3 to numerous pairs of genital acetabula 17

13a (12a). Without dorsal plates (Fig. 26); genital acetabula forming a median row between genital flaps or plates (Figs. 25, 27) 14

13b. A series of dorsal plates present (Figs. 84, 86), which may be raised and knoblike (Fig. 28); genital acetabula may form a median row between genital flaps or plates (Fig. 80), or be borne in at least 2 groups around each genital plate (Figs. 83, 85) [crawlers; almost always orange-red] 38

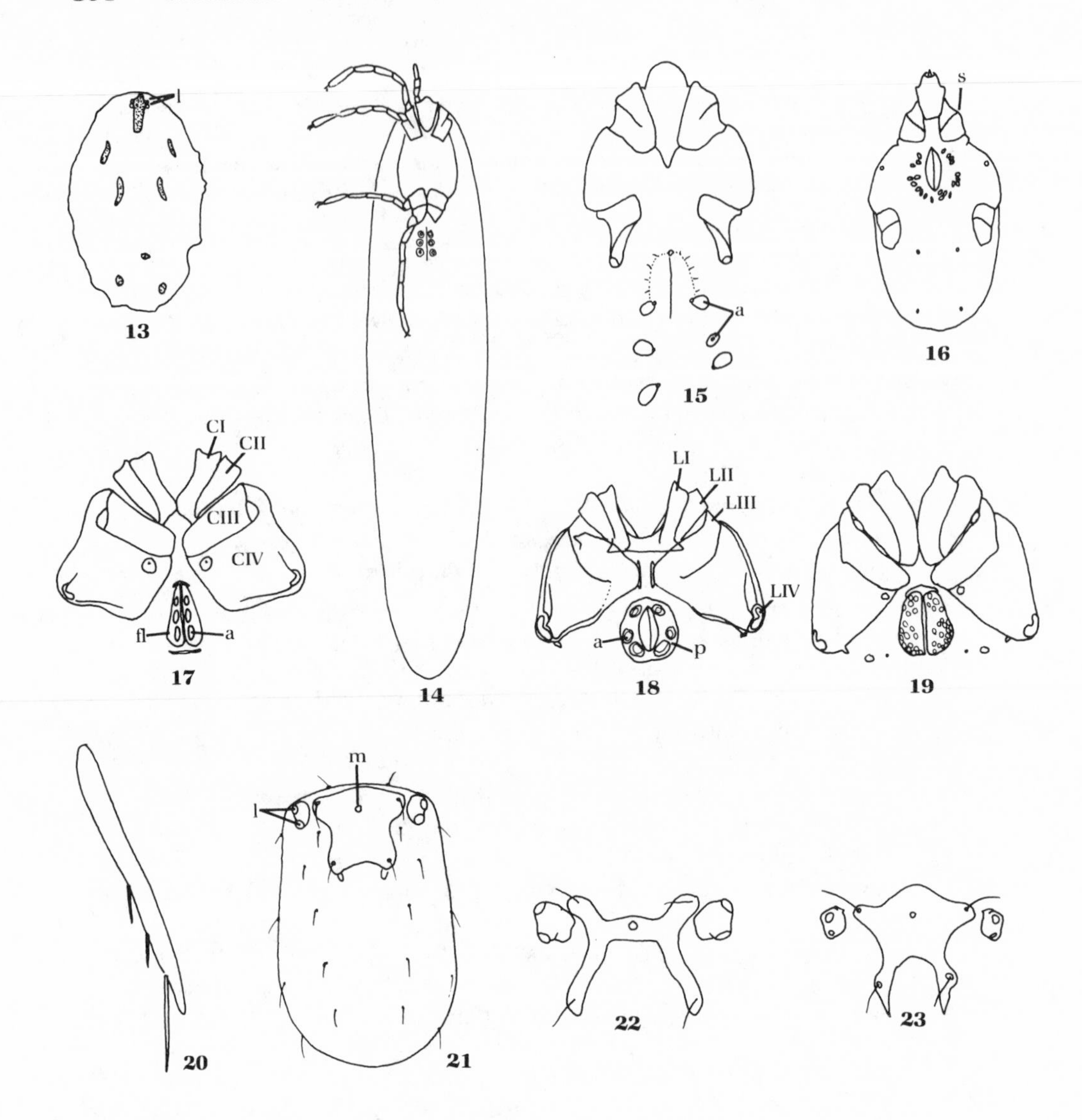

13. *Neolimnochares* (Limnocharidae), dv: l, lateral eyes. **14.** *Wandesia* (Hydryphantidae), vv. **15.** *Wandesia* (Hydryphantidae), vv: a, acetabula. **16.** *Protzia* (Hydryphantidae), vv: s, suture. **17.** *Teutonia* (Teutoniidae), vv: a, acetabulum; CI–IV, coxal plates I–IV; fl, genital flap. **18.** Male *Limnesia* (Limnesiidae), vv: a, acetabulum; LI–IV, insertion sites for legs I–IV; p, genital plate. **19.** Female *Limnesia* (Limnesiidae), vv. **20.** Leg IV tarsus of *Limnesia* (Limnesiidae). **21.** *Hydryphantes* (Hydryphantidae), dv: l, lateral eyes; m, median eye. **22.** Dorsal plate of *Hydryphantes* (Hydryphantidae), dv. **23.** Dorsal plate of *Hydryphantes* (Hydryphantidae), dv. dv, dorsal view; vv, ventral view. (13–23 modified from Cook 1974a.)

14a (13a). Coxal plates I fused at median edge, produced anteriorly on either side of the gnathosoma (Fig. 25); sutures separating coxal plates I and II coming together and joining the median suture between coxal plates II to form a Y configuration; partial fusion between coxal plates II and III; coxal plates IV closely flanking genital field, usually to the posterior limit [rarely entirely red; typically globular body shape; very common in lotic waters, but infrequently found in lakes; typically crawlers, but species in sluggish streams and in lakes are weak swimmers .... *Lebertia*

14b. Coxal plates I not fused medially and not produced anteriorly (Fig. 27); coxal sutures not forming a Y configuration; no fusion between coxal plates II and III; coxal plates IV never closely flanking genital field [red mites; body a flattened pillow shape] 15

15a (14b). Lateral eyes either absent or lying below the integument, without a lens or capsule (Fig. 26) [crawlers; very uncommon; found in springs and within gravel beds of streams] .... *Tartarothyas*

15b. Lateral eyes present and in capsules with lenses (e.g., Fig. 21) 16

16a (15b). Numerous long setae (swimming setae) inserted in longitudinal series on tibia and tarsus of legs III and IV; palps chelate, not prominent (e.g., Fig. 81) [swimmers; found in permanent and temporary ponds, sluggish streams] .... *Pseudohydryphantes*

16b. Without long setae in longitudinal series on tibia and tarsus of legs III and IV; palps simple, prominent (Fig. 31) [crawlers; leg setae arranged predominantly in whorls; found in permanent ponds, springs, seeps, and streams] .... *Sperchon*

17a (12b). Genital field begins adjacent to coxal plate III, with the posterior margin of the genital field typically anterior or adjacent to the posterior limit of coxal plate IV (Figs. 36, 38) 18

17b. Genital field begins no farther forward than the posteromedial corner of coxal plate III, and extends well beyond the posterior limit of coxal plate IV (Figs. 39, 40, 52) 19

18a (17a). Numerous genital acetabula borne on a single heart-shaped plate (Fig. 36); coxal plates I not fused medially; dorsal plates may be absent, there may be 1 to several pairs of small anterodorsal plates (Figs. 34, 35) or a single anterodorsal plate (Fig. 33), but never with a posterodorsal plate [large, spherical mites, usually red but sometimes blue or green; strong swimmers; very common, found in temporary and permanent ponds] .... *Hydrachna* (in part)

18b. Three pairs of genital acetabula borne on a genital plate (not heart-shaped) (Fig. 38); coxal plates I fused medially; 1 small posterior and either 1 or 2 small anterior dorsal plates (Fig. 37) [small, reddish mites; crawlers, and the only water mite capable of walking on dry land; fairly common at margins of streams, springs, lakes, and ponds] .... *Tyrrellia*

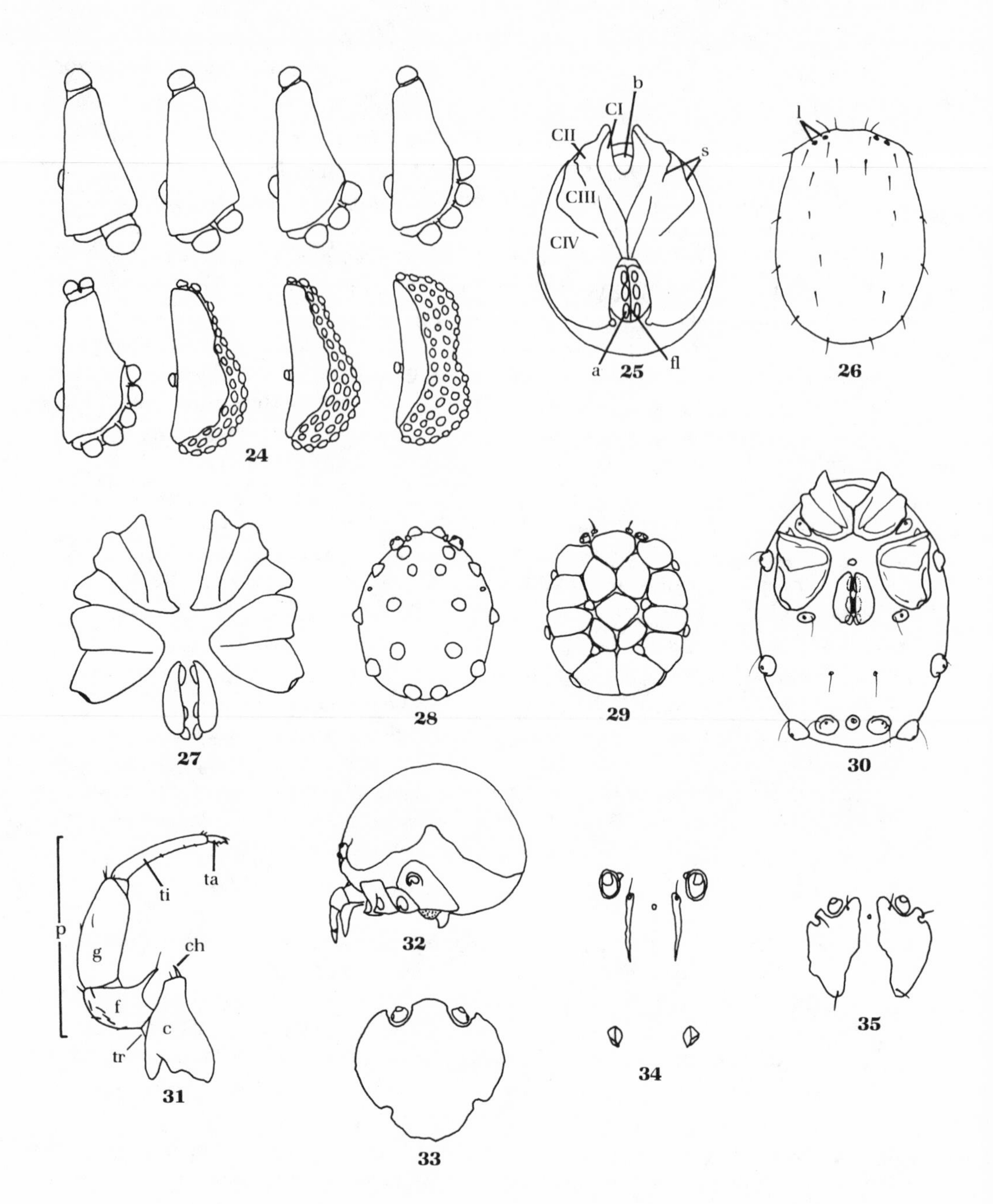

24. Various genital fields (one side only) of *Hydryphantes* (Hydryphantidae), vv. **25.** *Lebertia* (Lebertiidae), vv: a, acetabulum; b, gnathosomal bay; CI–IV, coxal plates I–IV; fl, genital flap; s, sutures. **26.** *Tartarothyas* (Hydryphantidae), dv: l, lateral eyes. **27.** *Pseudohydryphantes* (Hydryphantidae), vv. **28.** *Sperchon* (Sperchonidae), dv. **29.** *Sperchon* (Sperchonidae), dv. **30.** *Sperchon* (Sperchonidae), vv. **31.** Gnathosoma of *Sperchon* (Sperchonidae), lv: c, capitulum; ch, chelicerae; f, femur; g, genu; p, palp; ta, tarsus; ti, tibia; tr, trochanter. **32.** *Hydrachna* (Hydrachnidae), lv. **33.** Dorsal plate of *Hydrachna* (Hydrachnidae), dv. **34.** Dorsal plates of *Hydrachna* (Hydrachnidae), dv. **35.** Dorsal plates of *Hydrachna* (Hydrachnidae), dv. dv, dorsal view; vv, ventral view. (24–35 modified from Cook 1974a.)

19a (17b). Numerous (more than 40) pairs of genital acetabula located on a pair of raised, bean-shaped genital plates (Fig. 39); straplike coxal plates, with coxal plate IV with the width at least twice the length; no dorsal plate or plates [an orange-red, flattened pillow-shaped, swimming mite; very common, found in permanent standing waters, can be found in slow streams] .... *Hydrodroma*

19b. From 3 to numerous pairs of genital acetabula (Figs. 40, 49, 76), but if more than 40 pairs then not on a pair of raised genital plates, and plates often not bean-shaped; at least coxal plate IV is not strap-shaped, and its width is not twice its length; dorsal plate or plates present or absent 20

20a (19b). Not found within freshwater mollusks 21

20b. Found within freshwater mollusks (parasite) 23

21a (20a). Coxal plates III and IV blocklike, with anterior margin of coxal plates III and posterior margins of coxal plates IV approximately straight and colinear with margins from opposite side (Figs. 40, 45); median margins of coxal plates III and IV adjacent with their opposite pair [typically not red, with long legs and long, rigid, swordlike leg setae that are inserted into a moveable socket (Fig. 42)] 22

21b. Coxal plates III and IV not blocklike, with at least either the anterior margins of coxal plates III or the posterior margins of coxal plates IV not straight and colinear (Figs. 49, 55, 63, 64, 70, 75; in *Hydrochoreutes* [Figs. 63, 64] and some *Tiphys* [Fig. 70] coxal plates III and IV similar to above, but posterior margins of coxal plates IV not straight); median margins of coxal plates III and IV may be widely separated from their opposite pair (Fig. 49) 24

22a (21a). Sutures between coxal plates III and IV are incomplete (Fig. 45); usually with 5 or 6 pairs of acetabula (Figs. 46, 47), less commonly numerous pairs (Fig. 48); acetabula borne on 1 or 2 pairs of plates, if 2 pairs of plates then lower plates are never wider than long (Fig. 47) 23

22b. Sutures between coxal plates III and IV complete (Fig. 40); typically 10 or more pairs of genital acetabula (rarely fewer); acetabula typically borne on 1 pair of plates, if 2 pairs of plates then lower plates are approximately 2 times as wide as long and with numerous acetabula (Fig. 41; only 1 such species in North America) [very common in permanent lentic water, uncommon in streams; strong swimmers, typically with stout, dark blue-green or violet legs] .... *Neumania* (in part)

23a (20b, 22a). With numerous pairs of genital acetabula (Figs. 43, 44); genital acetabula located on 1 pair of winglike plates that are clearly wider than long; only found within freshwater mollusks [large] .... *Najadicola*

23b. Usually 5 or 6 pairs of genital acetabula (Figs. 46, 47), occasionally numerous (Fig. 48); genital acetabula borne on 1 or 2 pairs of plates that are never winglike and never wider than long; free-living or found within freshwater mollusks [many free-living species have very long legs and are strong swimmers; very common, in permanent standing water and slow streams] .. *Unionicola*

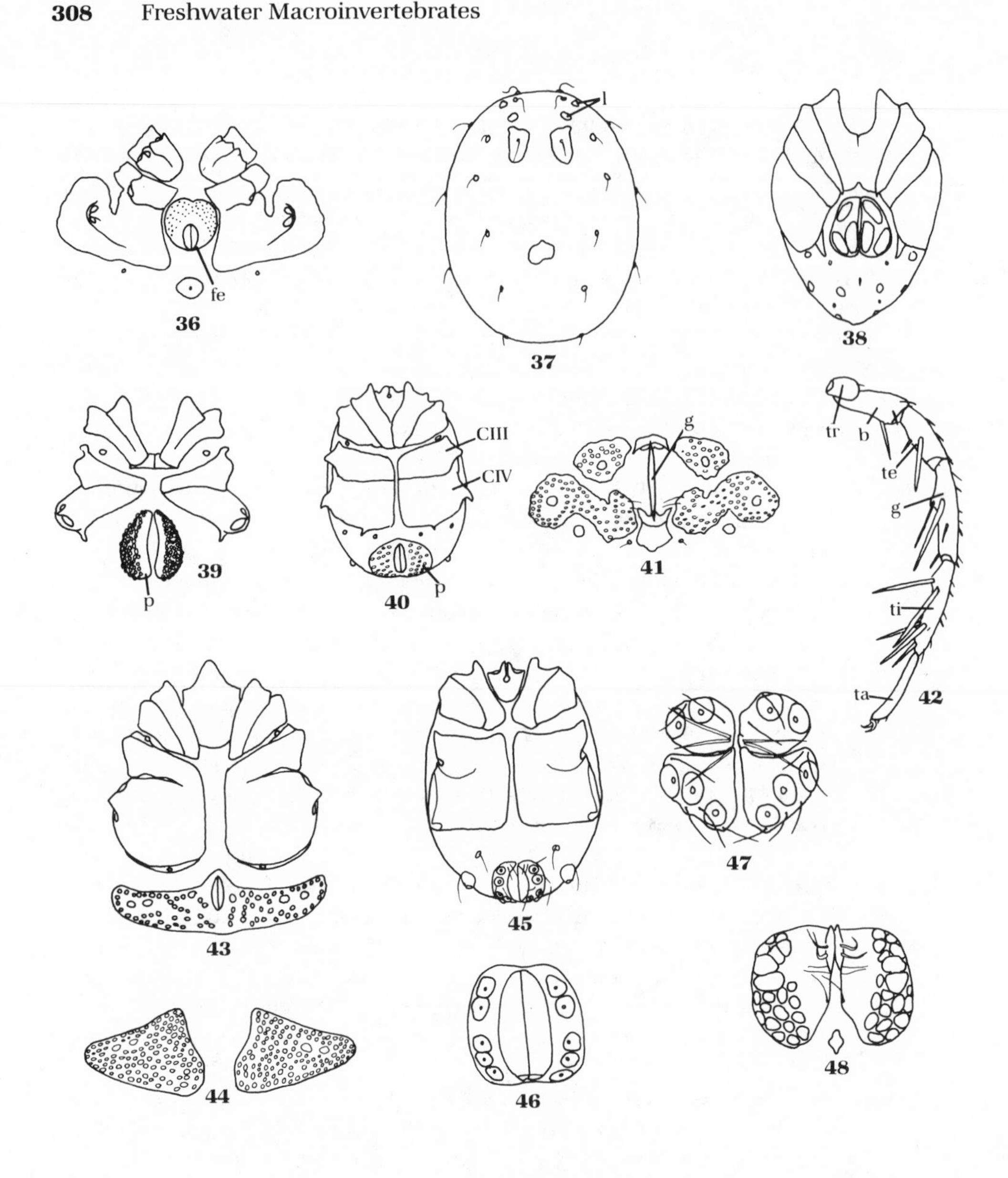

36. *Hydrachna* (Hydrachnidae), vv: fe, genital field. **37.** *Tyrrellia* (Limnesiidae), dv: l, lateral eyes. **38.** *Tyrrellia* (Limnesiidae), vv. **39.** *Hydrodroma* (Hydrodromidae), vv: p, genital plate. **40.** *Neumania* (Unionicolidae), vv: CIII–IV, coxal plates III–IV; p, genital plate. **41.** Genital field of *Neumania* (Unionicolidae), vv: g, gonopore. **42.** Leg I of *Neumania* (Unionicolidae), lv: b, basifemur; g, genu; ta, tarsus; te, telofemur; ti, tibia; tr, trochanter. **43.** Male of *Najadicola* (Pionidae), vv. **44.** Genital field of female *Najadicola* (Pionidae), vv. **45.** *Unionicola* (Unionicolidae), vv. **46.** Genital field of *Unionicola* (Unionicolidae), vv. **47.** Genital field of *Unionicola* (Unionicolidae), vv. **48.** Genital field of *Unionicola* (Unionicolidae), vv. dv, dorsal view; vv, ventral view. (36–48 modified from Cook 1974a.)

24a (21b). Leg IV is inserted at the posterolateral angle of coxal plate IV, close to the posterior limit of the plate (Figs. 49, 52); coxal plates I fused together; posterior margin of coxal plate IV straight or gently convex, the coxal plate appearing triangular [body globular, seldom if ever red; commonly with 3 pairs of acetabula (can be numerous); very common, found typically in fast-flowing water but can be found in lakes; typically crawlers, but species within lakes can swim] 25

24b. Leg IV is inserted near the midpoint of the lateral margin of coxal plate IV, often with ½ or more of the plate posterior to the insertion (Fig. 66); coxal plates I usually not fused together (Figs. 63, 71); posterior margin of coxal plate IV angled, the posteromedial portion often being concave, typically giving the plate a 4- or 5-sided appearance (Figs. 60, 63, 71) (*Wettina* spp. [Fig. 55] are exceptions, check other characters) 26

25a (24a). Gnathosoma fused to coxal plate I (Fig. 49); ITa (tarsus of leg I) straight (Fig. 50); palpal tibia with few (< 10) setae dorsally (visible only on slide-mounted specimens) (Fig. 51) [very common] . . . . *Hygrobates*

25b. Gnathosoma not fused to coxal plate I (Fig. 52); ITa (tarsus of leg I) usually curved (Fig. 53); palpal tibia with many (≥ 10) setae dorsally (visible only on slide-mounted specimens) (Fig. 54) [common] . . . . *Atractides*

26a (24b). Suture between coxal plates III and IV obliquely posteromedially directed such that coxal plate IV is triangular and its median margin is bounded by coxal plate III, so that coxal plates IV meet, at most, at a medial point (Figs. 55–57) 27

26b. Suture between coxal plates III and IV nearly transverse such that coxal plate IV is nearly quadrangular (often has a point or spur on its posterior margin so the plate appears 5-sided), and coxal plates IV face each other medially for at least part of their length (Figs. 63, 73) 29

27a (26a). Three of 4 pairs of genital acetabula (Fig. 55) . . . . *Wettina* (in part)

27b. Five to numerous pairs of genital acetabula (Figs. 56, 57, 60) 28

28a (27b). Integument is smooth, thin, and membranous; posteromedial angle of coxal plate IV is concave for females (Fig. 57), mildly concave for males (Fig. 56); IVTa of male with a dorsal concavity bearing numerous peglike setae (Fig. 58), and IVTa of female is cylindrical, unmodified [from various water bodies, but not cold springs and seepage areas] . . . . *Forelia* (in part)

28b. Integument consists of small papillae (Fig. 59, enlargement), is thickened and tough; posteromedial angle of coxal plate IV is widely rounded (Fig. 60); IVTa is cylindrical, unmodified [rare; restricted to cool slow streams and cold rheocene springs, but has been reported from the littoral zone of the ocean in Europe] . . . . *Nautarachna* (in part; some females)

29a (26b). Typically 3 pairs of genital acetabula (rarely more; *Tiphys* 3–6) 30

29b. More than 6 pairs of genital acetabula 37

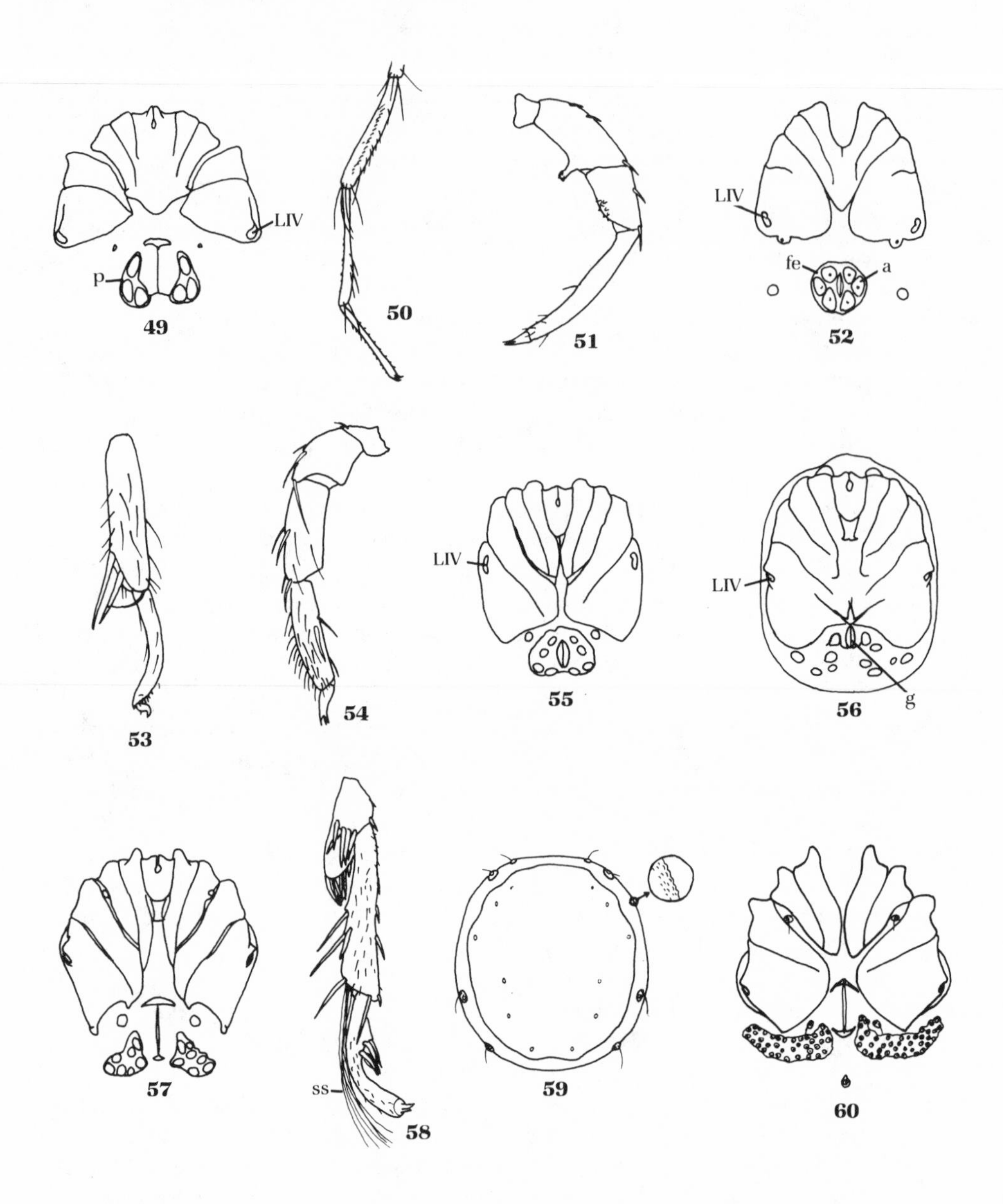

49. *Hygrobates* (Hygrobatidae), vv: LIV, insertion site for leg IV; p, genital plate. **50.** Leg I of *Hygrobates* (Hygrobatidae), lv. **51.** Palp of male *Hygrobates* (Hygrobatidae), lv. **52.** *Atractides* (Hygrobatidae), vv: a, acetabulum; fe, genital field; LIV, insertion site for leg IV. **53.** Leg I of *Atractides* (Hygrobatidae), lv. **54.** Palp of female *Atractides* (Hygrobatidae), lv. **55.** *Wettina* (Pionidae), vv: LIV, insertion site for leg IV. **56.** Male *Forelia* (Pionidae), vv: g, gonopore; LIV, insertion site for leg IV. **57.** Female *Forelia* (Pionidae), vv. **58.** Leg IV of male *Forelia* (Pionidae), lv: ss, swimming setae. **59.** Male *Nautarachna* (Pionidae), dv. **60.** Female *Nautarachna* (Pionidae), vv. dv, dorsal view; lv, lateral view; vv, ventral view. (49–60 modified from Cook 1974a.)

30a (29a). Genital plates widely separated medially, at least anteriorly (Figs. 67, 71, 73); all leg segments nearly cylindrical and body never with a petiole .... Females 31

30b. Genital plates are fused medially with one another, surrounding the gonopore (Figs. 64, 66); typically with either some leg segments modified (Figs. 66, 72, 74) or with a petiole at the posterior end of the body (Fig. 64) .... Males 33

31a (30a). Suture between coxal plates III and IV nearly transverse (Fig. 63) .... *Hydrochoreutes* (in part)

31b. Suture between coxal plates III and IV obliquely posteromedially directed (Figs. 67, 71) 32

32a (31b). Genital plates as wide or wider than long (Fig. 67); concave posterior edge of coxal plate IV nearly transverse .... *Neotiphys* (in part)

32b. Genital plates longer than wide (Figs. 71, 73); concave posterior edge of coxal plate IV obliquely posterolaterally directed [typically red and ovoid; strong swimmers] .... *Tiphys* (in part), *Pionacercus* (in part), *Pionopsis* (in part)

Tiphys and *Pionopsis* are very common, found in permanent and temporary standing water, or in slow, cool, flowing water. To separate further, slide-mount specimens and identify using the key in I. M. Smith 1976a.

33a (30b). With a petiole at the posterior edge of the body (Fig. 64); IIIGe modified, with a distoventral concavity bearing a long, curved, heavy seta at each of the proximal and distal extremities (Fig. 65); IVGe cylindrical, unmodified .... *Hydrochoreutes* (in part)

33b. Without a petiole; IIIGe cylindrical, unmodified; IVGe usually with some modification, e.g., flattened and expanded (Fig. 72) or bearing obvious specialized setae (Fig. 66) 34

34a (33b). IVTa with a dorsal concavity bearing numerous peglike setae (Fig. 69); IVGe cylindrical, may be produced distally, not bearing specialized paddle-like setae .... *Pionacercus* (in part)

34b. IVTa cylindrical, unmodified (Figs. 66, 72, 74); IVGe often modified, e.g., flattened and expanded (Fig. 72), or bearing paddle-like setae (Fig. 66) 35

35a (34b). IVGe dorsoventrally expanded and flattened laterally, 1½ to 3 times as great in height as IVTFe (Fig. 72) (see bracketed notes in 32b and note following 32b) .... *Tiphys* (in part)

35b. IVGe not or only slightly dorsoventrally expanded (Figs. 66, 74), not flattened laterally, less than 1½ times as great in height as IVTFe 36

36a (35b). IVGe with 2 conspicuous paddle-like setae; IVTi not concave dorsally (Fig. 66) .... *Neotiphys* (in part)

36b. IVGe without paddle-like setae; IVTi broadly concave dorsally (Fig. 74) (see bracketed notes in 32b and note following 32b) .... *Pionopsis* (in part)

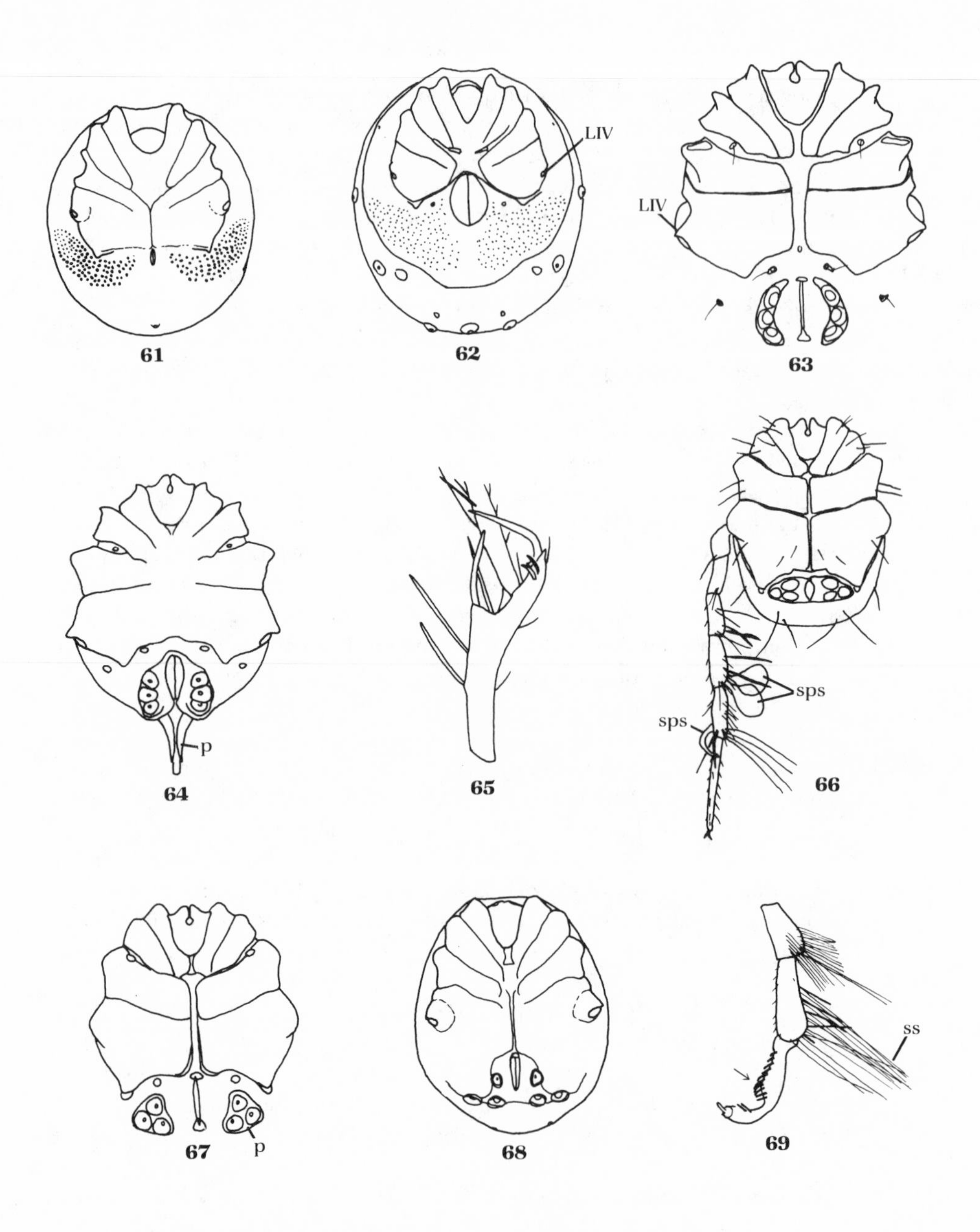

61. Male *Nautarachna* (Pionidae), vv. **62.** Female *Nautarachna* (Pionidae), vv: LIV, insertion site for leg IV. **63.** Female *Hydrochoreutes* (Pionidae), vv: LIV, insertion site for leg IV. **64.** Male *Hydrochoreutes* (Pionidae), vv: p, petiole. **65.** Leg III genu of male *Hydrochoreutes* (Pionidae), lv. **66.** Male *Neotiphys* (Pionidae), vv: sps, specialized setae. **67.** Female *Neotiphys* (Pionidae), vv: p, genital plate. **68.** Male *Pionacercus* (Pionidae), vv. **69.** Leg IV of male *Pionacercus* (Pionidae), lv: ss, swimming setae. lv, lateral view; vv, ventral view. (61–69 modified from Cook 1974a.)

37a (29b). Eight to ten genital acetabula (Fig. 75); IVGe of males never with a dorsal concavity [restricted to cold waters of profundal depths of oligotrophic lakes] . . . . *Huitfeldtia*

Huitfeldtia females are difficult to separate from females of some species of *Piona*. Specimens should be slide-mounted and identified using keys in I. M. Smith 1976a and Cook 1974a.

37b. Seven to numerous genital acetabula (Figs. 76, 77); IVGe of males with a deep dorsal concavity bearing numerous peglike setae (Fig. 78) [very common; usually in lentic water (including temporary pools) and rarely in slow lotic waters; typically red or purple and ovoid; strong swimmers] . . . . *Piona*

38a (13b). Second acetabulum midway along median length of genital plate, midway between 1st and 3rd acetabula (Figs. 30, 80); all acetabula lying in a paired row between genital flaps 39

38b. Second acetabulum ⅔ or more posterior along median length of genital plate, much closer to 3rd acetabulum than to 1st (Figs. 83, 85); 3rd pair of acetabula typically lying posterior to the genital plate (Fig. 83) or laterally displaced relative to the 2nd pair of acetabula (Fig. 85) 40

39a (38a). With a characteristic spindle-shaped median plate between pairs of lateral eyes (Fig. 79); dorsal plates flat, never raised and knoblike [orange-red, with the coxal and dorsal plates similar color to the rest of the body; flattened pillow-shape; palps small, not prominent (Fig. 81); common in temporary ponds, also known from streams] . . . . *Euthyas*

39b. Without a spindle-shaped median plate between pairs of lateral eyes; dorsal plates flat or may be raised and knoblike (Fig. 28) [typically red or salmon-colored, typically with the coxal and dorsal plates darker than the rest of the body; flattened pillow shape; typically with large, prominent palps (Fig. 31); very common in streams, springs, seepage areas, but never in temporary ponds] . . . . *Sperchon* (in part), *Sperchonopsis* (in part)

To separate, slide-mount specimens and refer to Cook 1974a. *Sperchonopsis* spp. have raised knoblike dorsal plates with numerous papillae, while *Sperchon* spp. have flattened dorsal plates or raised, knoblike dorsal plates without papillae (Fig. 28).

40a (38b). Fifth plates in the dorsomedial series of paired plates (termed dorsocentralia 5) fused medially into a single plate (Fig. 82) . . . . *Panisus* (in part)

40b. Fifth plates in the dorsomedial paired rows of dorsal plates (termed dorsocentralia 5) not fused medially (Figs. 84, 86) [flattened pillow-shaped; typically orange-red; very common, in temporary ponds (*Thyas* and *Thyasides*), springs, seepage areas, and cold streams] . . . . *Panisopsis* (in part), *Thyas, Thyasides, Zschokkea*

To separate genera, slide-mount specimens and follow key in Cook 1959 or Cook 1974a.

41a (3b). Genital flaps present, small (Fig. 88); genital field located far forward, terminating before the anterior limit of coxal plate IV; genital acetabula absent; su-

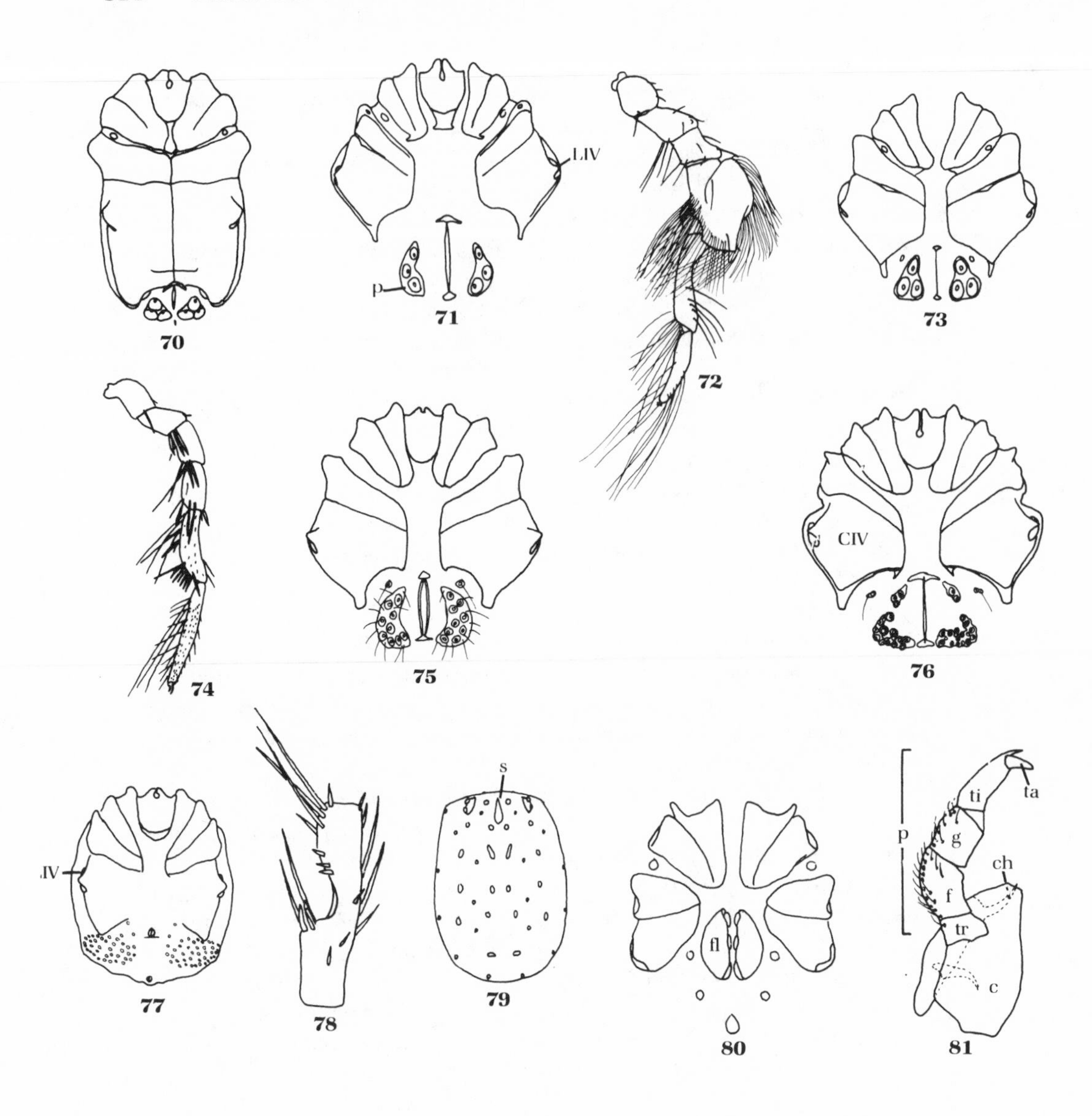

70. Male *Tiphys* (Pionidae), vv. **71.** Female *Tiphys* (Pionidae), vv: LIV, insertion site for leg IV; p, genital plate. **72.** Leg IV of male *Tiphys* (Pionidae), lv. **73.** Female *Pionopsis* (Pionidae), vv. **74.** Leg IV of male *Pionopsis* (Pionidae), lv. **75.** *Huitfeldtia* (Pionidae), vv. **76.** *Piona* (Pionidae), vv: CIV, coxal plate IV. **77.** Male *Piona* (Pionidae), vv: LIV, insertion site for leg IV. **78.** Leg IV genu of *Piona* (Pionidae), lv. **79.** *Euthyas* (Hydryphantidae), dv: s, spindle-shaped plate. **80.** *Euthyas* (Hydryphantidae), vv: fl, genital flap. **81.** Gnathosoma of *Euthyas* (Hydryphantidae), lv: c, capitulum; ch, chelicerae; f, femur; g, genu; p, palp; ta, tarsus; ti, tibia; tr, trochanter. dv, dorsal view; lv, lateral view; vv, ventral view. (70–76, 78–81 modified from Cook 1974a; 77 modified from Cook 1966.)

ture between coxal plates III and IV extending posterolaterally or more or less laterally [legs are held in a characteristic posture (Fig. 87); uncommon; red; found crawling in moss of cold streams and springs] ...... *Hydrovolzia*

41b. Genital flaps present or absent; genital field either extending beyond or originating posterior to anterior limit of coxal plate IV (Figs. 30, 36, 105, 138); genital acetabula present and typically conspicuous; suture between coxal plates III and IV typically anterolateral or lateral 42

42a (41b). Dorsum of mite bearing more than 1 dorsal plate, such that the largest plate is clearly smaller in area than the sum of the remaining dorsal and marginal plates, and the plates are separated by membranous integument (Figs. 29, 89) 43

42b. Dorsum of mite bearing a single dorsal plate and without marginal plates (Fig. 59), or a dominant dorsal plate of equal or greater area than the sum of the remaining dorsal and marginal plates (Figs. 104, 107, 113), or the body is entirely encased in a rigid cuticle shell with, at most, a thin membranous line around a single dorsal plate (Figs. 134, 140) 54

43a (42a). Eyes attached to an eyebridge (Fig. 90); mouth opening appearing to be circular; numerous acetabula borne in 4 groups (Fig. 91) [an uncommon red mite, found crawling at the margins of marshes, swamps, temporary ponds, sphagnum seeps] .... *Piersigia*

43b. Eyebridge absent, if anterodorsal plate is present then eyes are not attached to it (Figs. 29, 82, 92); mouth opening not circular; 3 to numerous pairs of acetabula, typically not borne in 4 groups (Figs. 30, 93, 96) 44

44a (43b). Dorsal and ventral plates with radiating (wheel-like) reticulations (Figs. 92, 93); reticulate patterning on the coxal plates (Fig. 93); capitulum attached to a long protrusible tube of soft integument (may only be apparent on dead specimens, unmistakable if extended, but inconclusive if not) [red crawlers, uncommon, found in streams] .... *Clathrosperchon*

44b. Dorsal and ventral plates typically not reticulate, if reticulate then not in a radiating pattern; coxal plates not reticulate; capitulum not attached to a long protrusible tube of soft integument 45

45a (44b). Three or 4 pairs of genital acetabula 46

45b. Five to numerous pairs of genital acetabula 50

46a (45a). Suture between coxal plates III and IV obliquely posteromedially directed such that the median margin of coxal plate IV is bounded by coxal plate III (Fig. 55); 3 or 4 pairs of acetabula on genital plates [found in springs, cold streams, and occasionally in cold, oligotrophic lakes; with swimming setae on some leg segments (e.g., Fig. 69)] .... *Wettina* (in part)

46b. Suture between coxal plates III and IV approximately transverse, such that coxal plate III does not bound the median margin of coxal plate IV (Figs. 30, 80); 3 pairs of acetabula situated between (Fig. 80) or around (Figs. 83, 95) genital flaps or plates [without swimming setae] 47

47a (46b). All 3 pairs of acetabula lie between the genital flaps, with the 2nd pair approximately midway between the 1st and 3rd pairs (Fig. 30); dorsal sclerites flat and platelike (Fig. 29) or raised and knoblike (Fig. 28) (see bracketed notes in 39b and note following 39b) . . . *Sperchon* (in part), *Sperchonopsis* (in part)

47b. The 3rd pair of genital acetabula lie posterior to the genital plates, often the 2nd pair of acetabula is closer to the 3rd pair than to the 1st pair (Figs. 83, 95); dorsal sclerites platelike (Figs. 82, 86, 94), never raised and knoblike [orange-red; flattened pillow shape; legs, coxal and dorsal plates are similar color to the body; palps are not prominent (Fig. 81); found in seepage areas, springs, cold streams] 48

48a (47b). Series of large dorsal plates forming 3 longitudinal rows (Fig. 94); the 2nd pair of acetabula is midway between the 1st and 3rd pair (Fig. 95) . . . . *Trichothyas*

48b. Series of large or small dorsal plates forming 4 longitudinal rows (Figs. 82, 86); the 2nd pair of acetabula is closer to the 3rd pair than to the 1st pair (Figs. 83, 85) 49

49a (48b). Fifth plates in the dorsomedial series of paired plates (termed dorsocentralia 5) and fused medially into a single plate (Fig. 82) . . . . *Panisus* (in part)

49b. Fifth plates in the dorsomedial series of paired plates not fused medially (Fig. 86) (see bracketed notes in 40b and note following 40b) . . *Panisopsis* (in part)

50a (45b). With 2 dorsal plates laterally paired lying behind the lateral eyes (Fig. 35); genital field located between coxal plates III and IV, anterior to the posterior margin of coxal plate IV (Fig. 36); numerous (more than 40) pairs of genital acetabula located on a single heart-shaped plate, with most of the acetabula anterior to the gonopore (see bracketed notes in 18a) . . . . *Hydrachna* (in part)

50b. Dorsal plates not laterally paired, or more than 1 pair present; genital field not located between coxal plates III and IV, extending beyond posterior limit of coxal plates IV (Fig. 40, 96, 98); variable number of genital acetabula, not borne on a single heart-shaped plate, most of acetabula not anterior to gonopore 51

51a (50b). Coxal plates III and IV blocklike (Fig. 40), with anterior margin of coxal plates III and posterior margin of coxal plates IV approximately straight and colinear with margins from opposite side (see bracketed notes in 22b) . . . . *Neumania* (in part)

51b. Coxal plates III and IV not blocklike, at least anterior margin of coxal plates III (Figs. 96, 97) and often the posterior margin of coxal plate IV (Figs. 56, 57, 98) not colinear with margin from opposite side 52

52a (51b). Posterior margin of coxal plates IV approximately colinear, forming a relatively transverse (Fig. 96), weakly convex, or weakly concave margin (Fig. 97); insertion of leg IV usually near the posterolateral limit of coxal plate IV; medial edge of coxal plates IV typically adjacent for at least a short longitudinal section (sometimes the plates are fused along this length); IVTa of males cy-

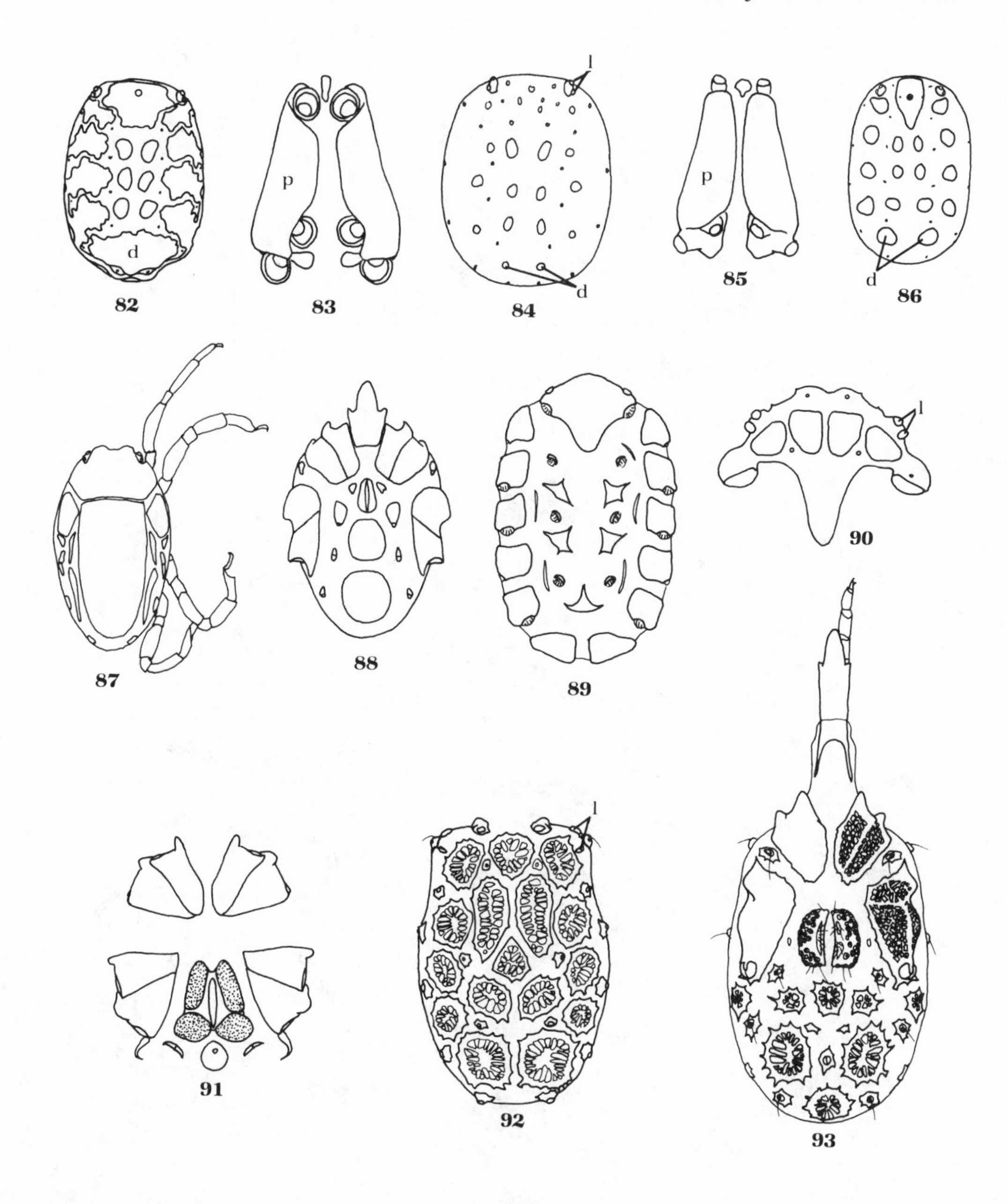

82. *Panisus* (Hydryphantidae), dv: d, dorsocentralia 5. **83.** Genital field of *Panisus* (Hydryphantidae), vv: p, genital plate. **84.** *Thyas* (Hydryphantidae), dv: d, dorsocentralia 5; l, lateral eyes. **85.** Genital field of *Thyas* (Hydryphantidae), vv: p, genital plate. **86.** *Panisopsis* (Hydryphantidae), dv: d, dorsocentralia 5. **87.** *Hydrovolzia* (Hydrovolziidae), dv. **88.** *Hydrovolzia* (Hydrovolziidae), vv. **89.** *Piersigia* (Piersigiidae), dv. **90.** Eyebridge of *Piersigia* (Piersigiidae), dv: l, lateral eyes. **91.** *Piersigia* (Piersigiidae), vv. **92.** *Clathrosperchon* (Rhynchohydracaridae), dv: l, lateral eyes. **93.** *Clathrosperchon* (Rhynchohydracaridae), vv. dv, dorsal view; vv, ventral view. (82–93 modified from Cook 1974a.)

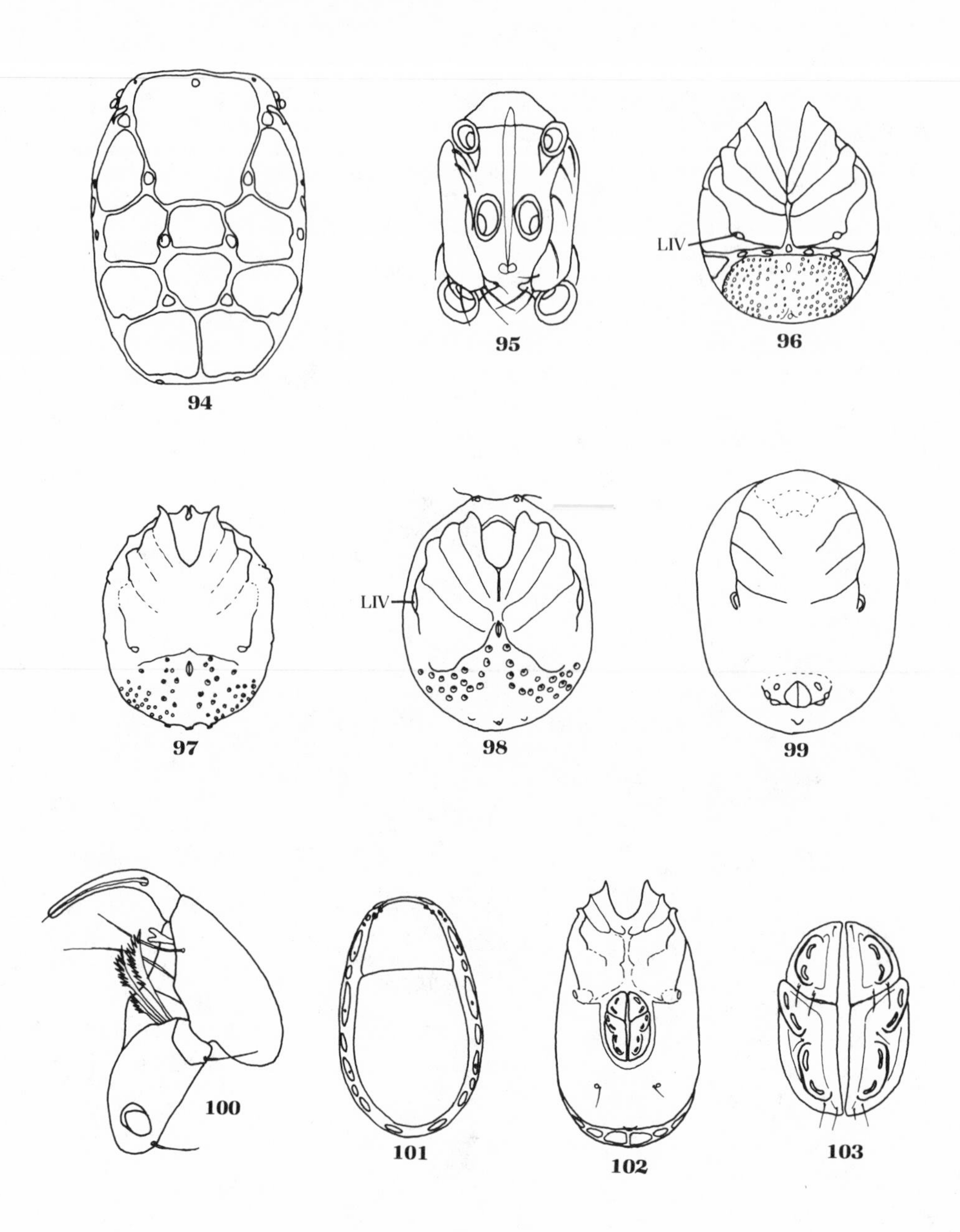

94. *Trichothyas* (Hydryphantidae), dv. **95.** Genital field of *Trichothyas* (Hydryphantidae), vv. **96.** *Feltria* (Feltriidae), vv: LIV, insertion site for leg IV. **97.** *Feltria* (Feltriidae), vv. **98.** *Pseudofeltria* (Pionidae), vv: LIV, insertion site of leg IV. **99.** *Uchidastygacarus* (Uchidastygacaridae), vv. **100.** Palp of *Uchidastygacarus* (Uchidastygacaridae), vv. **101.** *Meramecia* (Limnesiidae), dv. **102.** *Meramecia* (Limnesiidae), vv. **103.** Genital field of *Meramecia* (Limnesiidae), vv. dv, dorsal view; vv, ventral view. (94 modified from Mitchell 1953; 95–98, 101–103 modified from Cook 1974a; 99, 100 modified from I.M. Smith 1983.)

lindrical, unmodified [typically found in mosses and aquatic vegetation in swift cold streams, cataracts, and waterfalls] .... *Feltria* (in part)

52b. Posterior margin of coxal plates IV clearly not colinear, forming a strongly concave (Fig. 57) or angular (Fig. 98) margin; insertion of leg IV anterior to midpoint of lateral margin of coxal plate IV; coxal plates IV triangular, adjacent at most only at a median point, the diagonal anteromedian margin bounded by coxal plates III; IVTa of males with a dorsal concavity bearing numerous peglike setae (Fig. 58) 53

53a (52b). Swimming setae present (Fig. 58); dorsoventral expansion of leg segments, if present, confined to legs I and II; claws on legs III and IV of male relatively small and reduced (see bracketed notes in 28a) .... *Forelia* (in part)

53b. Swimming setae absent, all setae short and relatively heavy; all leg segments short and dorsoventrally expanded; claws on legs III and IV large and unmodified [typically found in cold springs and seepage areas] .... *Pseudofeltria* (in part)

54a (42b). Gnathosomal bay absent (Fig. 99); palp rotated so that tarsi fold medially toward each other on a horizontal plane; palp tibia greatly expanded basally (but not apically) relative to palp tarsus (Fig. 100) [flattened body; eyes reduced or absent; tip of leg IV blunt and without claws; 3 pairs of genital acetabula; found in gravel beds of streams in New Hampshire, Maine, New Brunswick, and Nova Scotia] .... *Uchidastygacarus*

54b. Gnathosomal bay present (Figs. 25, 120, 130); palp typically not rotated, the tarsi fold ventrally on a vertical plane (Fig. 81), except *Chappuisides*, which has a rotated uncate palp; palp tibia may be apically expanded or expanded along its entire length relative to palp tarsus (uncate, Figs. 132, 139), or not expanded (Figs. 31, 81, 128) 55

55a (54b). IVTa clawless and tapering to a point (Fig. 20) 56

55b. IVTa with claws and not tapering to a point (Figs. 123, 133) (One claw may be reduced in Axonopsinae or claws may be reduced but present in *Forelia* [Fig. 58], but leg never tapers to a point.) 57

56a (55a). Dorsal plate split into anterior and posterior plates (Fig. 101); with 6 pairs of crescent-shaped acetabula (Figs. 102, 103); with the genital plates split between the 2nd and 3rd acetabula .... *Meramecia*

56b. Dorsal plate entire; from 3 to numerous pairs of acetabula, never crescent-shaped (Figs. 18, 19); with the genital plates entire (see bracketed notes in 10b) .... *Limnesia* (in part)

57a (55b). Genital acetabula lying in a median row beneath a pair of genital flaps (Figs. 30, 105, 108); 3–6 pairs of genital acetabula 58

57b. Genital acetabula never covered by a pair of genital flaps, either lying on genital plates (Fig. 18), distributed around genital plates (Figs. 110, 112), apparently free on the body wall (Fig. 68), or lying within the gonopore (Fig. 117); 3 to numerous pairs of genital acetabula 60

58a (57a). Coxal plates I fused at median edge, strongly produced anteriorly on either side of the gnathosoma (Figs. 105, 106); sutures separating coxal plates I and II either coming together and joining the median suture between coxal plates II to form a Y configuration (Fig. 105), or sutures between coxal plates I and II extending to the genital field (Fig. 106); partial or total obliteration of suture lines between coxal plates II and III; body shape flattened tear-drop [very common in gravel bottoms of rivers and streams, occasionally present in lakes; commonly with a black, white, and sometimes red pattern in life; typically crawlers, but species in lakes are weak swimmers] .... *Testudacarus, Torrenticola*

In most cases, the specimens will need to be cleared in order to separate these genera. *Torrenticola* has only 1 or 2 pairs of anterior marginal plates (Fig. 104) and has 6 pairs of genital acetabula, while *Testudacarus* has numerous marginal plates surrounding the dorsal plate (Fig. 107) and has 3 pairs of genital acetabula.

58b. Coxal plates I may be fused on the medial edge, but not strongly produced anteriorly (Figs. 30, 108); no median suture between coxal plates II, no Y configuration, and sutures between coxal plates I and II not extending to the genital field; coxal plates II and III either well separated or with well-defined sutures; body shape a flattened ovoid 59

59a (58b). Coxal plates forming 4 distinct groups (grouped as plates I and II, III and IV, separate on each side), typically with a wide separation between coxal plates II and III (Fig. 30); coxal plate III not subdivided by a longitudinal suture (see bracketed notes in 39b and note following 39b) .... *Sperchon* (in part)

59b. Coxal plates fused into a single ventral plate, with coxal plates II and III separated by only a suture (Fig. 108); coxal plate III subdivided by a longitudinal suture (Fig. 108) [uncommon; reddish brown; found in springs, streams, and mossy seeps] .... *Bandakia*

60a (57b). Three to 6 pairs of genital acetabula, located on (Fig. 18) or around (Figs. 110, 112) a genital plate, or lying within the gonopore (Fig. 117) (*Neoacarus* females typically have 6–7 pairs of acetabula, but range from 4 to 9, and will key out from either choice of this couplet.) 61

60b. Seven to numerous pairs of genital acetabula, either lying clustered on genital plates (Figs. 57, 96) or lying free on the body wall (Figs. 61, 77), or sometimes lying inconspicuously along the posterior margin of the body (Fig. 123), rarely lying within the gonopore 73

61a (60a). Three pairs of genital acetabula (Figs. 110, 112); the 1st and 2nd pair are located between a pair of genital plates, with the 3rd pair located posterior to the plates, spaced more widely apart than the other pairs [orange-red crawlers; leathery, not encased in a rigid cuticle] 62

61b. Three to 6 pairs of genital acetabula; if more than 3 pairs then with various configurations, if only 3 pairs then they are located in 2 rows with similar spacing between partners, either placed on a pair of genital plates (Fig. 18),

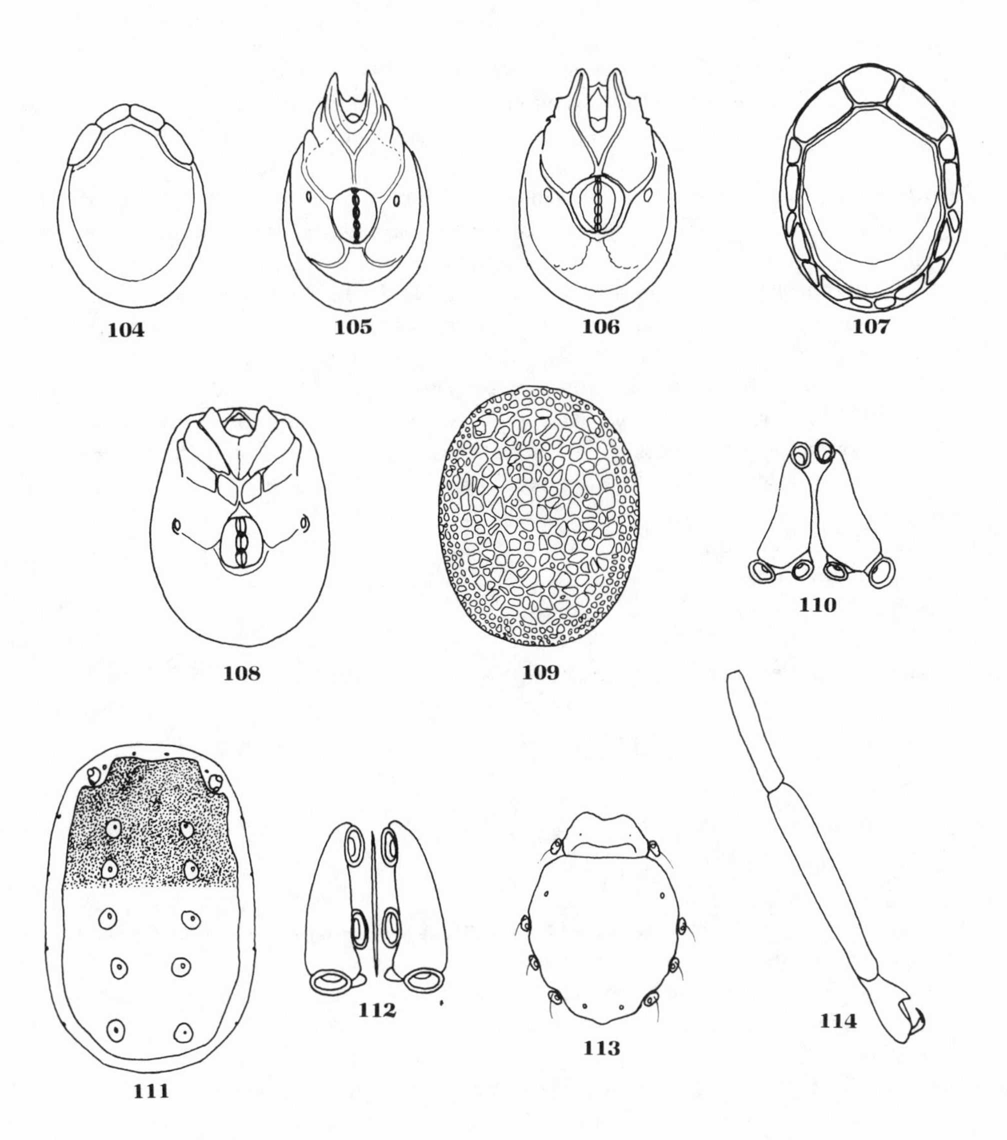

104. *Torrenticola* (Torrenticolidae), dv. **105.** *Torrenticola* (Torrenticolidae), vv. **106.** *Torrenticola* (Torrenticolidae), vv. **107.** *Testudacarus* (Torrenticolidae), dv. **108.** *Bandakia* (Anisitsiellidae), vv. **109.** *Thyopsis* (Hydryphantidae), dv. **110.** Genital field of *Thyopsis* (Hydryphantidae), vv. **111.** *Thyopsella* (Hydryphantidae), dv. **112.** Genital field of *Thyopsella* (Hydryphantidae), vv. **113.** Dorsal shield of *Momonia* (Momoniidae), dv. **114.** Leg I of *Momonia* (Momoniidae), lv. dv, dorsal view; lv, lateral view; vv, ventral view. (104, 105, 107–114 modified from Cook 1974a; 106 modified from Cook 1966.)

or as a median row in the gonopore (Fig. 117), or if posterior pair is spaced more widely apart than the other pairs then genital plates are absent (Fig. 68) 63

62a (61a). Second acetabulum at the posteromedial corner of the genital plate, 3rd acetabulum lateral to it (Fig. 110); dorsal plate with a conspicuous raised reticulate pattern (Fig. 109); dorsal plate covering almost the entire area of the dorsum; eye capsules always surrounded by dorsal plate [found occasionally, around temporary ponds, seepage areas, and springs] . . . . *Thyopsis*

62b. Second acetabulum located approximately midway along medial edge of genital plate, 3rd acetabulum posterior to it (Fig. 112); dorsal plate either reticulate or punctate (Fig. 111); dorsal plate not covering the entire area of the dorsum; eye capsules sometimes surrounded by dorsal plate [found around springs, seepage areas, and waterfalls] . . . . *Thyopsella*

63a (61b). ITa highly modified for grasping in both sexes (Fig. 114) [uncommon, found in streams] 64

63b. ITa not modified 65

64a (63a). Two dorsal plates, approximately ¼ (anterior) and ¾ (posterior) of the dorsum in area (Fig. 113) [darkly pigmented] . . . . *Momonia*

64b. Single dorsal plate [lightly pigmented; found in gravel beds of streams] . . . . *Stygomomonia*

65a (63b). IIITi is broadly expanded (Fig. 133) 82

65b. IIITi cylindrical, not expanded 66

66a (65b). Body not entirely encased in rigid cuticle, with a wide membranous band separating the dorsal plate from the rest of the armor (Fig. 115) 67

66b. Body entirely encased in rigid cuticle, with at most a thin line of membrane separating the dorsal plate from the rest of the armor (Fig. 134) 68

67a (66a). IVTa with a dorsal concavity bearing peglike setae (Fig. 69); 3rd genital acetabulum lateral to 2nd acetabulum (Fig. 68) . . *Pionacercus* (males, in part)

67b. IVTa cylindrical, unmodified; 3rd genital acetabulum posterior to 2nd acetabulum (Fig. 116) [collected from algae and mosses in cold springs and seepage areas; quite uncommon] . . . . *Paenecalyptonotus*

68a (66b). Body is strongly dorsoventrally compressed, body depth less than ½ width 69

68b. Body is not dorsoventrally compressed (but may be flat on the dorsal surface), body depth more than ½ width 70

69a (68a). Three to 5 genital acetabula, lying in the gonopore arranged in 2 parallel rows (Fig. 117); body has a saucerlike shape (dorsum is flat or slightly concave), and is approximately circular in dorsal view; gonopore is located subterminally on the ventral plate . . . . *Mideopsis, Nudomideopsis*

Mideopsis is fairly common in lakes and slow streams, and is often blue in life. *Nudomideopsis* is quite rare and is found in springs, seepage areas, and interstitial waters.

69b. Three or 4 genital acetabula, lying on genital plates typically arranged in 2 triangular or rectangular groups (sutures delimiting plates are often indistinct; Fig. 119); body is typically a dorsoventrally flattened ellipsoid; gonopore is usually located terminally, on the posterior edge of the ventral plate [fairly uncommon, found in streams and lakes; often yellowish with black markings at either end (Fig. 118); small, ca. 1 mm long or less] . . . . Axonopsinae (in part, including *Axonopsis, Brachypoda, Estellacarus, Ljania, Woolastookia*)

Specimens must be cleared and slide-mounted for further identification. See key in Cook 1974a.

70a (68b). Sutures between coxal plates III and IV angled acutely posteromedially so that if extended, they would meet within the gonopore (Fig. 130) 82

70b. Sutures between coxal plates III and IV angled slightly posteromedially so that if extended, they would meet before or at the anterior edge of the gonopore (Figs. 120–122) 71

71a (70b). Body 1½ times or more as long as wide (Fig. 120); palp rotated, palpal tarsi fold medially on a horizontal plane [found in streams, in interstitial gravel; quite uncommon] . . . . *Chappuisides*

71b. Body length approximately equal to width, length clearly less than 1½ times width (Figs. 121, 122); palp not rotated, palpal tarsi fold ventrally on a vertical plane [the margin of the dorsal plate is recessed into the marginal groove, not smoothly contoured into the shape of the body; uncommon] 72

72a (71b). Four or 5 pairs of acetabula present (Fig. 121); gnathosoma on a long eversible tube (Fig. 121), often everted on preserved or recently killed specimens (Eversible tube is unmistakable when extended, but cannot be reliably identified when the tube is not extended.) . . . . *Geayia*

72b. Three pairs of acetabula present (Fig. 122); gnathosoma not borne on a long eversible tube . . . . *Krendowskia*

73a (60b). Body dorsoventrally compressed, depth less than ½ width 74

73b. Body not dorsoventrally compressed (although the dorsal plate may be flattened), depth greater than ½ width 78

74a (73a). Gonopore is terminal (may not be visible, or only visible as a notch) (Fig. 123); genital acetabula are distributed as a thin band around the posterior margin of the body [males have heavy bladelike setae on IVTi or IVTa; small, often reddish; very common in lotic habitats] . . . . *Aturus, Kongsbergia*

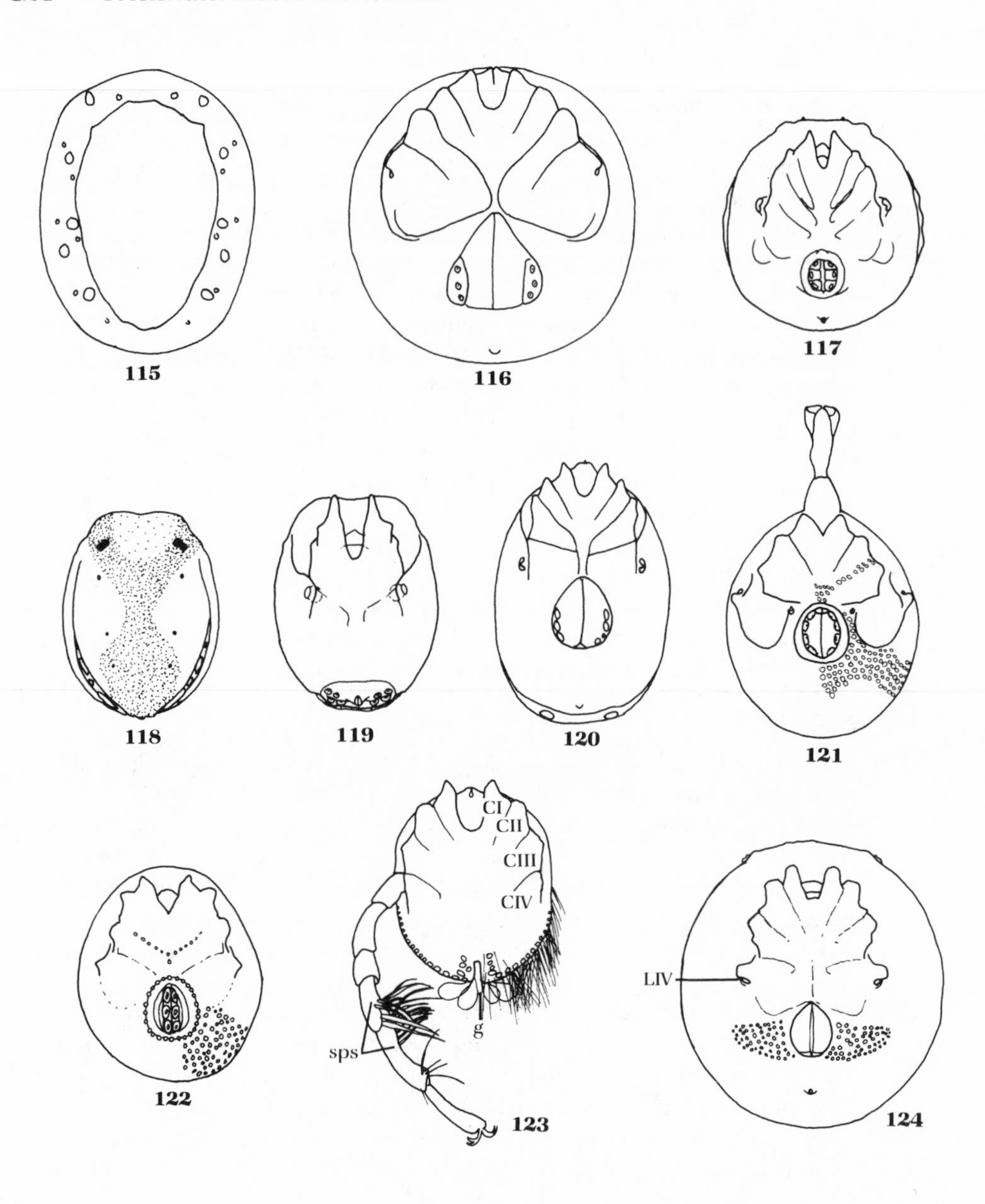

115. *Paenecalyptonotus* (Acalyptonotidae), dv. **116.** *Paenecalyptonotus* (Acalyptonotidae), vv. **117.** *Mideopsis* (Mideopsidae), vv. **118.** *Axonopsis* (Axonopsinae), dv. **119.** *Axonopsis* (Axonopsinae), vv. **120.** *Chappuisides* (Chappuisididae), vv. **121.** *Geayia* (Krendowskiidae), vv. **122.** *Krendowskia* (Krendowskiidae), vv. **123.** *Aturus* (Aturidae), vv: CI–IV, coxal plates I–IV; g, gonopore; sps, specialized setae. **124.** *Koenikea* (Unionicolidae), vv: LIV, insertion site for leg IV. dv, dorsal view; vv, ventral view. (115 modified from I.M. Smith 1976c; 116–124 modified from Cook 1974a.)

Kongsbergia spp. have a ventral projection associated with the opening for the 4th leg, while *Aturus* spp. do not. Specimens must be viewed with a compound microscope to see this character.

74b. Gonopore is subterminal (Figs. 56, 62, 77, 98); genital acetabula are not distributed as a thin band around the posterior margin of the body 75

75a (74b). Body entirely encased in rigid cuticle, never more than a peripheral line of membrane around the dorsal plate (Fig. 140) 76

75b. Body is not entirely encased in rigid cuticle, at least a band of peripheral membrane around dorsal plate (Fig. 59) (dorsal and ventral plates may be joined at their anterior edge) 84

76a (75a). Dorsal plate is very flat, even slightly concave, the body appearing saucer-shaped [in life, these mites tend to be yellow with black, white, and red marks on the dorsum; strong swimmers, common in lakes, and can be found in slow streams] . . . . *Koenikea*

76b. Dorsal plate at least convex at margins, the body not appearing saucer-shaped 77

77a (76b). Medial edge of coxal plates IV adjacent for most of its length (Fig. 125); leg IV appearing to emerge from within coxal plate IV [elongate-oval, swimming mites; typically bluish green in life without magnification; fairly common, found in permanent standing water and slow-flowing water] . . . . *Albia*

77b. Medial edge of coxal plates IV meet only at a medial point (Fig. 141); leg IV clearly emerges at the lateral margin of coxal plate IV [found in cold springs] . . . . *Laversia* (in part)

78a (73b). Genital field located between coxal plates III and IV, anterior to the posterior margin of coxal plates IV, with medial margins of coxal plates IV either parallel or converging posterior to the genital field (Fig. 36); genital acetabula borne on a single heart-shaped plate, with most of the acetabula anterior to the gonopore; dorsal plate ranging from an anterior plate on an otherwise membranous dorsum, to extending down the sides of the body almost to the coxal plates, encasing most of the body (Fig. 32) (See bracketed notes in 18a) . . . . *Hydrachna* (in part)

78b. Genital field not located between coxal plates III, and if located between coxal plates IV then these plates diverge posteriorly and the genital field extends back to at least the posterior limit of coxal plates IV (Figs. 126, 138, 141); genital acetabula not borne on a single heart-shaped plate, most of the genital acetabula are not anterior to the gonopore 79

79a (78b). Body is entirely encased in rigid cuticle, never more than a peripheral line of membrane around the dorsal plate (Figs. 134, 135) 80

79b. Body is not entirely encased in rigid cuticle, at least a band of peripheral membrane around the dorsal plate (Fig. 59) (Dorsal and ventral plates may be joined at their anterior edge.) 84

80a (79a). Six to 16 pairs of genital acetabula, flanking the gonopore in 1 or 2 rows on each of 2 crescent-shaped plates (female, Fig. 126), or lying in 2 rows within the gonopore, or genital field with posterolaterally extending winglike sclerites (male, Fig. 127) 81

80b. Numerous (more than 25) pairs of genital acetabula clustered in 2 wing-shaped or band-shaped fields (Figs. 138, 141) 83

81a (80a). Palp simple (Fig. 128); 9–16 pairs of genital acetabula, flanking the gonopore either in 1 or 2 rows on each of two crescent-shaped plates (female, Fig. 126), or 1–2 rows within the gonopore, sometimes on a genital field with posterolaterally extending winglike sclerites (male, Fig. 127) [uncommon; dorsal plate almost covering the entire dorsal surface of the body; found in standing and slow-flowing water] . . . . *Midea*

81b. Palp uncate (Fig. 132) or chelate (Fig. 129); 9 or fewer pairs of genital acetabula (range, 4–9) flanking the gonopore in single row on each of 2 crescent-shaped plates (Fig. 130) or lying within the gonopore in 2 rows (Fig. 131) 82

82a (65a, 70a, 81b). Palp chelate, like a crab claw, with tibiae extending far beyond insertion of the tarsi (Fig. 129) [quite uncommon; found in gravel beds of streams] . . . . *Volsellacarus*

82b. Palp uncate, not clawlike, with tibiae not extending beyond insertion of the tarsi (Fig. 132) [quite uncommon; typically found within gravel beds of streams, but 1 unusual species found in mesotrophic lakes, in association with aquatic vascular plants] . . . . *Neoacarus*

83a (80b). Genital acetabula located behind the posterior margin of coxal plates IV, never between these plates (Fig. 138); dorsal plate clearly does not cover the entire dorsal surface, ⅞ or less the length of the body in dorsal view (Figs. 134, 136); female is round or ovoid, male can be any of an assortment of unusual or bizarre shapes (Figs. 135–137) and may have the dorsal plate partially fused to the ventral plate [very common; frequently greenish, but may be red, brown, dark purple, or blue; strong swimmers, typically in temporary or permanent lentic or slow lotic waters] . . . . *Arrenurus*

83b. Genital acetabula located, in part, between coxal plates IV (Fig. 141); the dorsal plate almost covers the entire dorsal surface and is more than ⅞ the length of the body in dorsal view (Fig. 140); both sexes are ovoid, never unusual or bizarre shapes, and the dorsal plate is never fused to the ventral plate [found in cold springs] . . . . *Laversia* (in part)

84a (75b, 79b). IVGe with a deep, dorsal concavity bearing numerous, peglike setae (Fig. 78); IVTa cylindrical, unmodified 85

84b. IVGe without a deep, dorsal concavity; IVTa either cylindrical, unmodified, or modified, with a dorsal concavity bearing peglike setae (Fig. 58) 86

85a (84a). Posterior margin of coxal plates IV clearly not colinear, forming a strongly angular margin (Fig. 77) (see bracketed notes in 37b) . . . . . *Piona* (males, in part)

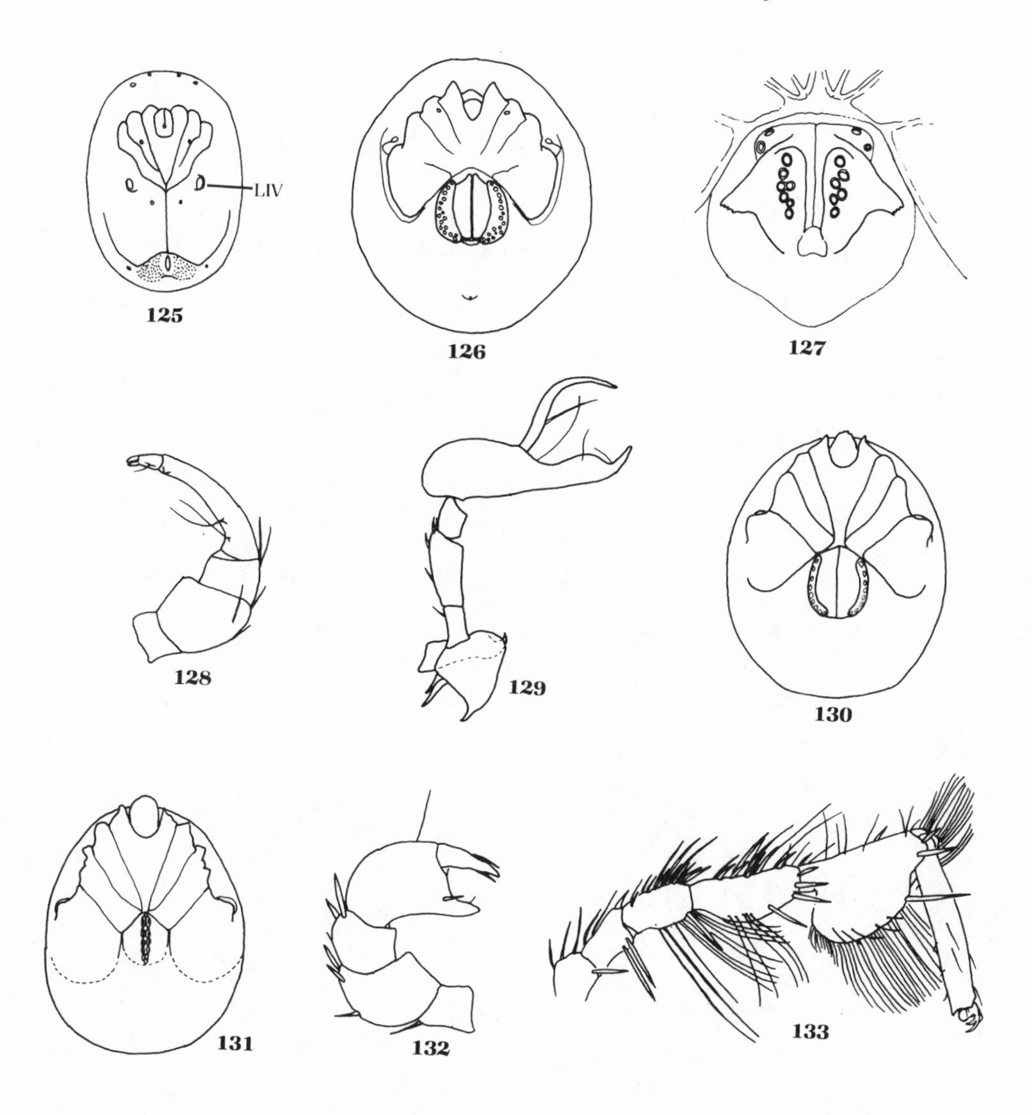

125. *Albia* (Aturidae), vv: LIV, insertion site for leg IV. **126.** Female *Midea* (Mideidae), vv. **127.** Genital field of male *Midea* (Mideidae), vv. **128.** Palp of *Midea* (Mideidae), vv. **129.** Gnathosoma of *Volsellacarus* (Neoacaridae), lv. **130.** Female *Neoacarus* (Neoacaridae), vv. **131.** Male *Neoacarus* (Neoacaridae), vv. **132.** Palp of *Neoacarus* (Neoacaridae). **133.** Leg III of *Neoacarus* (Neoacaridae), lv. lv, lateral view; vv, ventral view. (125–129 modified from Cook 1974a; 130–133 modified from I.M. Smith 1976b.)

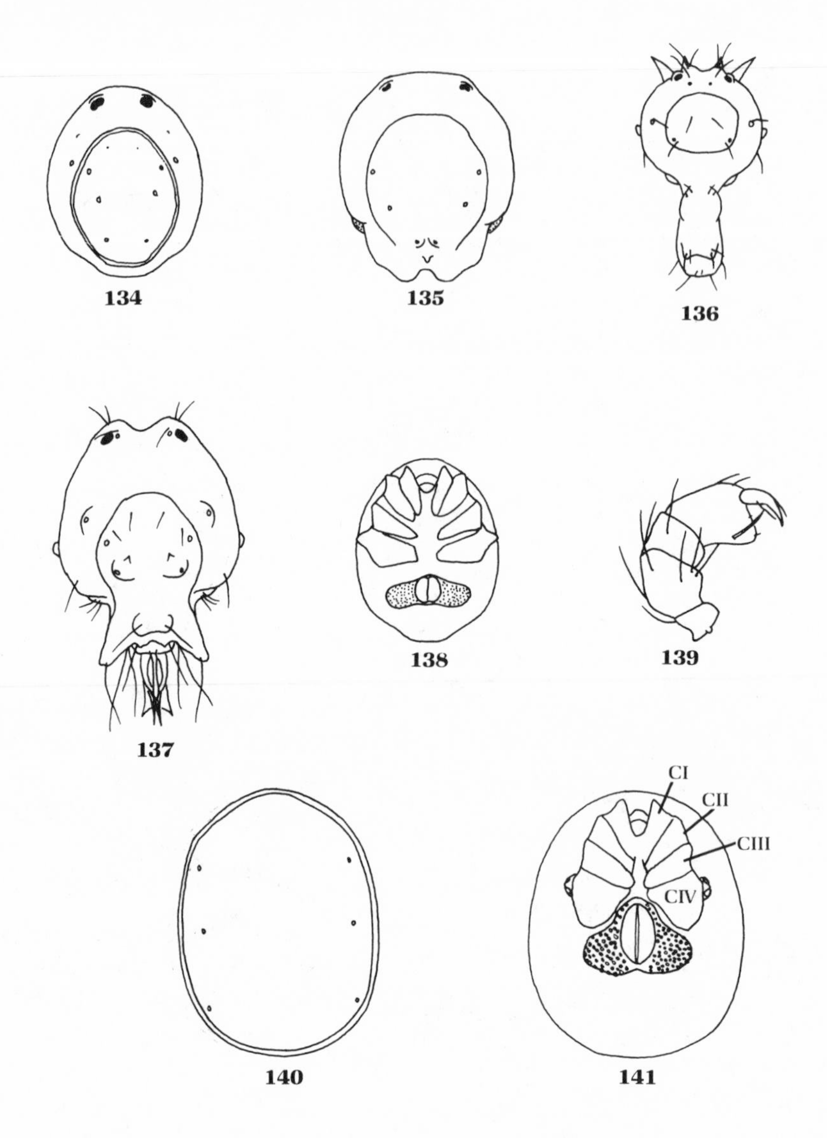

134. Female *Arrenurus* (Arrenuridae), dv. **135.** Male *Arrenurus* (Arrenuridae), dv. **136.** Male *Arrenurus* (Arrenuridae), dv. **137.** Male *Arrenurus* (Arrenuridae), dv. **138.** Female *Arrenurus* (Arrenuridae), vv. **139.** Palp of *Arrenurus* (Arrenuridae), dv. **140.** Dorsal plate of *Laversia* (Laversiidae), dv. **141.** *Laversia* (Laversiidae), vv: CI–IV, coxal plates I–IV. dv, dorsal view; vv, ventral view. (134–141 modified from Cook 1974a.)

85b. Posterior margin of coxal plates IV approximately colinear, forming a relatively transverse margin (Fig. 61) (see bracketed notes in 28b) .... *Nautarachna* (males)

86a (84b). Posterior margin of coxal plates IV approximately colinear, forming a relatively transverse (Fig. 96), mildly convex or mildly concave margin (Fig. 97); insertion of leg IV within the posterior half of coxal plate IV; medial edge of coxal plates IV meeting along at least a short longitudinal section (sometimes the plates are fused along this length); IVTa cylindrical, unmodified (see bracketed notes in 52a) .... *Feltria* (in part)

86b. Posterior margin of coxal plates IV clearly not colinear, forming a strongly concave (Fig. 56) or angular margin (Fig. 62); insertion of leg IV at or anterior to the midpoint of coxal plate IV; coxal plates IV approximately triangular, if meeting medially then only as a median point, the diagonal anteromedian margin bounded by coxal plates III; IVTa of females unmodified, of males with a dorsal concavity bearing numerous peglike setae (Fig. 58) 87

87a (86b). IVTa cylindrical, unmodified 88

87b. IVTa with a dorsal concavity bearing numerous peglike setae (Fig. 58) 89

88a (87a). Seven to 30 pairs of genital acetabula; 1–4 dorsal plates (see bracketed notes in 53b) .... *Pseudofeltria* (females, in part)

88b. Eighty-five to 110 pairs of genital acetabula (Fig. 62); 1 dorsal plate (see bracketed notes in 28b) .... *Nautarachna* (females, in part)

89a (87b). Swimming setae present (Fig. 69); dorsoventral expansion of leg segments, if present, confined to legs I and II; claws on legs III and IV relatively small and reduced (see bracketed notes in 28a) .... *Forelia* (males, in part)

89b. Swimming setae absent, all setae short and relatively heavy; all leg segments short and dorsoventrally expanded; claws on legs III and IV large and unmodified (see bracketed notes in 53b) .... *Pseudofeltria* (males, in part)

References on Hydrachnidia

(*Used in construction of key.)

Åbro, A. 1979. Attachment and feeding devices of water-mite larvae (*Arrenurus* spp.) parasitic on damselflies (Odonata, Zygoptera). Zool. Scr. 8:221–234.

——. 1982. The effects of parasitic water mite larvae (*Arrenurus* spp.) on zygopteran imagoes (Odonata). J. Invert. Pathol. 39:373–381.

——. 1984. The initial stylostome formation by parasitic larvae of the water-mite genus *Arrenurus* on zygopteran imagines. Acarologia 25:33–46.

Angelier, E., M.-L. Angelier, and J. Lauga. 1985. Recherches sur l'écologie des hydracriens (Hydrachnellae, Acari) dans les eaux courantes. Ann. Limnol. 21:25–64.

Baker, R.A. 1982. Unionicolid mites from central New York. J. N.Y. Entomol. Soc. 90:176–180.

Balogh, J. 1972. The oribatid genera of the world. Akademiai Kiado, Budapest. 188 pp.

Barr, D.W. 1970. Tiny wolves of the water. Nat. Hist. 79:40–45.

——. 1972. The ejaculatory complex in water mites (Acari: Parasitengona): morphology and potential value for systematics. R. Ont. Mus. Life Sci. Contrib. 81:1–87.

*——. 1973. Methods for the collection, preservation and study of water mites. R. Ont. Mus. Life Sci. Misc. Publ. 28 pp.

——. 1979. Water mites (Acari, Parasitengona) sampled with chemoluminescent bait in underwater traps. Int. J. Acarol. 5:187–194.

——. 1982. Comparative morphology of genital acetabula of aquatic mites (Acari, Prostigmata): Hydrachnoidea, Eylaoidea, Hydryphantoidea and Lebertioidea. J. Nat. Hist. 16:147–160.

Barr, D.W., and B.P. Smith. 1979. The contribution of setal blades to effective swimming in the aquatic mite *Limnochares americana* (Acari: Prostigmata: Limnocharidae). Zool. J. Linn. Soc. 65:55–69.

——. 1980. Stable swimming by diagonal phase synchrony in arthropods. Can. J. Zool. 58:782–795.

Böttger, K. 1970. Die Ernährungsweise der Wassermilben (Hydrachnellae, Acari). [Feeding in water mites.] Int. Rev. ges. Hydrobiol. 55:895–912.

——. 1972a. Vergleichend biologisch-ökologische Studien zum Entwicklungszyklus der Süsswassermilben (Hydrachnellae, Acari). I. Der Entwicklungszyklus von *Hydrachna globosa* und *Limnochares aquatica.* Int. Rev. ges. Hydrobiol. 57:109–152.

——. 1972b. Vergleichend biologisch-ökologische Studien zum Entwicklungszyklus der Süsswassermilben (Hydrachnellae, Acari). II. Der Entwicklungszyklus von *Limnesia maculata* und *Unionicola crassipes.* Int. Rev. ges. Hydrobiol. 57:263–319.

——. 1976. Types of parasitism by larvae of water mites (Acari: Hydrachnellae). Freshwat. Biol. 6:497–500.

——. 1977. The general life cycle of fresh water mites (Hydrachnellae, Acari). Acarologia 18:496–502.

Cassagne-Méjean, F. 1966. Contribution à l'étude des Arrenuridae (Acari, Hydrachnellae) de France. Acarologia 8: (fasc. suppl.): 1–186.

Conroy, J.C. 1968. The water-mites of western Canada. Natl. Mus. Can. Bull. 223:23–42.

*Cook, D.R. 1955. Studies on the Thyasinae of North America (Acarina: Hydryphantidae). Amer. Midl. Nat. 62:402–428.

——. 1960. Water mites of the genus *Piona* in the United States (Acarina: Pionidae). Ann. Entomol. Soc. Amer. 53:35–60.

——. 1963a. Studies on the phreaticolous water mites of North America: new or unreported genera of Axonopsidae. Amer. Midl. Nat. 70:110–125.

——. 1963b. Studies on the phreaticolous water mites of North America: the genus *Feltria* (Acarina: Feltriidae). Ann. Entomol. Soc. Amer. 56:488–500.

*——. 1966. The water mites of Liberia. Mem. Amer. Entomol. Inst. 6:1–418.

——. 1968. Water mites of the genus *Stygomomonia* in North America (Acarina, Momoniidae). Proc. Entomol. Soc. Wash. 70:210–224.

——. 1969. The zoogeography of interstitial water mites. *In* G.O. Evans (ed.). Proc. 2nd Int. Congr. Acarol., July 19–24, 1967, Sutton Bonington, England, pp. 81–87. Akademiai Kiado, Budapest.

——. 1970. North American species of the genus *Hydrochoreutes* (Acarina: Pionidae). Mich. Entomol. 3:108–117.

*——. 1974a. Water mite genera and subgenera. Mem. Amer. Entomol. Inst. 21:1–860.

——. 1974b. North American species of the genus *Axonopsis* (Acarina: Aturidae: Axonopsinae). Great Lakes Entomol. 7:55–80.

——. 1975. North American species of the genus *Brachypoda* (Acarina: Aturidae: Axonopsinae). Proc. Entomol. Soc. Wash. 77:278–289.

——. 1976b. Contributions to the water mite fauna of North America. North American species of the genus *Mideopsis* (Acarina: Mideopsidae). Contrib. Amer. Entomol. Inst. 11:101–148.

Cook, D.R., and R.D. Mitchell. 1952. Notes on collecting water mites. Turtox News 30:122–125.

Corbet, P.S. 1963. Reliability of parasitic water mites (Hydracarina) as indicators of physiological age in mosquitoes (Diptera: Culicidae). Entomol. Exp. Appl. 6:215–233.

Crowell, R.M. 1960. The taxonomy, distribution and developmental stages of Ohio water mites. Bull. Ohio Biol. Surv. 1(2):1–77.

——. 1961. Catalogue of the distribution and ecological relationships of North American Hydracarina. Can. Entomol. 93:321–359.

Davids, C. 1973. The water mite *Hydrachna conjecta* Koenike, 1895 (Acari: Hydrachnellae), bionomics and relation to species of Corixidae (Hemiptera). Neth. J. Zool. 23:363–429.

Davids, C., and R. Belier. 1979. Spermatophores and sperm transfer in the water mite *Hydrachna conjecta* Koen. Reflections of the descent of water mites from terrestrial forms. Acarologia 21:84–90.

Davids, C., C.F. Heijnis, and J.E. Weekenstroo. 1981. Habitat differentiation and feeding strategies in water mites in Lake Maarsseveen. 1. Hydrobiol. Bull. 15:87–91.

Dewez, A., and G. Wauthy. 1984. Description and efficiency of a catching system for water mites in lotic environments. Arch. Hydrobiol. 101:461–467.

Dimock, R.V., Jr. 1983. In defense of the harem: intraspecific aggression by male water mites (Acari: Unionicolidae). Ann. Entomol. Soc. Amer. 76:463–465.

——. 1985. Population dynamics of *Unionicola formosa* (Acari: Unionicolidae), a water mite with a harem. Amer. Midl. Nat. 114:168–179.

Efford, I. 1963. The parasitic ecology of some watermites. J. Anim. Ecol. 32:141–156.

——. 1966. Observations on the life-history of three stream-dwelling water mites. Acarologia 8:86–93.

Fairchild, W.L., M.C.A. O'Neill, and D.M. Rosenberg. 1987. Quantitative evaluation of the behavioral extraction of aquatic invertebrates from samples of *Sphagnum* mosses. J. N. Amer. Benthol. Soc. 6:281–287.

Gledhill, T., J. Cowley, and R. Gunn. 1982. Some aspects of the host:parasite relationships between adult blackflies (Diptera: Simuliidae) and larvae of the water mite, *Sperchon setiger* (Acari; Hydrachnellae) in a small chalk stream in southern England. Freshwat. Biol. 12:345–357.

Gordon, M.J., B.K. Swan, and C.G. Paterson. 1979. The biology of *Unionicola formosa* (Dana and Whelpley): a water mite parasitic in the unionid bivalve, *Anodonta cataracta* (Say) in a New Brunswick lake. Can. J. Zool. 57:1748–1756.

Habeeb, H. 1967. A checklist of North American water-mites. Leafl. Acad. Biol. 43:1–8. [Almost every issue of this publication is devoted to water mites.]

Harris, S.C. 1981. An illustrated key to the species of *Tyrrellia* (Prostigmata: Limnesiidae). Proc. Entomol. Soc. Wash. 83:524–531.

Kerfoot, W.C. 1982. A question of taste: crypsis and warning coloration in freshwater zooplankton communities. Ecology 63:538–554.

*Krantz, G. 1971. A manual of acarology. Oreg. State Univ. Bookstores, Inc., Corvallis. 335 pp.

——. 1978. A manual of acarology, 2nd ed. Oreg. State Univ. Bookstores, Inc., Corvallis. 509 pp.

Lanciani, C. 1969. Three species of *Eylais* (Acari: Eylaidae) parasitic on aquatic Hemiptera. Trans. Amer. Microscop. Soc. 88:356–365.

——. 1970. New species of *Eylais* (Acari: Eylaidae) parasitic on aquatic Coleoptera. Trans. Amer. Microscop. Soc. 89:169–188.

——. 1971. Host exploitation and synchronous development in a water mite parasite of the marsh treader *Hydrometra myrae* (Hemiptera: Hydrometridae). Ann. Entomol. Soc. Amer. 64:1254–1259.

——. 1972. Mating behaviour of water mites of the genus *Eylais*. Acarologia 14:631–637.

——. 1979a. Water mite-induced mortality in a natural population of the mosquito *Anopheles crucians* (Diptera: Culicidae). J. Med. Entomol. 15:529–532.

——. 1979b. Detachment of parasitic water mites from the mosquito *Anopheles crucians* (Diptera: Culicidae). J. Med. Entomol. 15:99–102.
——. 1982. Parasite-mediated reductions in survival and reproduction of the backswimmer *Buenoa scimitra* (Hemiptera: Notonectidae). Parasitology 85:593–603.
——. 1983. Overview of the effects of water mite parasitism on aquatic insects. *In* M.A. Hoy, G.L. Cunningham, and L. Knutson (eds.). Biological control of pests by mites, pp. 86–90. Univ. Calif. Div. Agric. Nat. Resour. Spec. Publ. no. 3304.
——. 1984. Crowding in the parasitic stage of the water mite *Hydrachna virella* (Acari, Hydrachnidae). J. Parasitol. 70:270–272.
Lanciani, C., and A.D. Boyt. 1977. The effect of a parasitic water mite, *Arrenurus pseudotenuicollis* on the survival and reproduction of the mosquito *Anopheles crucians* (Diptera: Culicidae). J. Med. Entomol. 14:10–15.
Lanciani, C., and P.G. May. 1982. Parasite-mediated reductions in growth of nymphal backswimmers. Parasitology 85:1–7.
*Lundblad, O. 1941. Die hydracarinenfauna Südbrasiliens und Paraguays. Kungl. Svenska Vetenskapsakademiens Handlingar. Tredje Serien. Band 19. no. 7.
Marshall, R. 1929. Canadian Hydracarina. Univ. Toronto Stud. Biol. Ser. 33:57–93.
McDaniel, B. 1979. How to know the mites and ticks. W.C. Brown Publ. Co., Dubuque, Iowa. 355 pp.
*Mitchell, R.D. 1953. A new species of *Lundbladia* and remarks on the family Hydryphantidae (water mites). Amer. Midl. Nat. 49:159–170.
——. 1954a. Check list of North American water-mites. Fieldiana: Zool. 35:29–70.
——. 1954b. The structure and evolution of water mite mouthparts. J. Morphol. 110:41–59.
——. 1954c. Water-mites of the genus *Aturus* (Family Axonopsidae). Trans. Amer. Microscop. Soc. 73:350–367.
——. 1957a. The mating behaviour of pionid water-mites. Amer. Midl. Nat. 58:360–366.
——. 1957b. Anatomy, life history, and evolution of the mites parasitizing fresh-water mussels. Misc. Publ. Univ. Mich. Mus. Zool. 89:1–28.
——. 1957c. Locomotor adaptations of the family Hydryphantidae (Hydrachnellae: Acari). Abh. Naturw. Ver. Bremen 35:75–100.
——. 1957d. Major evolutionary lines in water mites. Syst. Zool. 6:137–148.
——. 1957e. On the mites parasitizing *Anodonta* (Unionidae; Mollusca). J. Parasitol. 43:101–104.
——. 1960. The evolution of thermophilus water mites. Evolution 14:361–377.
——. 1962. The structure and evolution of water mite mouthparts. J. Morphol. 110:41–59.
——. 1964. An approach to the classification of water mites. Acarologia (fasc. hors sér): 75–79. C.R. 1er Cong. Int. Acarologie, 1963.
Mitchell, R.D., and D.R. Cook. 1952. The preservation and mounting of water-mites. Turtox News 30:169–172.
Mullen, G.R. 1974. Acarine parasites of mosquitoes. II. Illustrated larval key to the families and genera of mites reportedly parasitic on mosquitoes. Mosq. News 34:183–195.
——. 1976. Water mites of the subgenus *Truncaturus* (Arrenuridae: *Arrenurus*) in North America. Search: Agric. 6:1–35.
——. 1977. Acarine parasites of mosquitoes. IV. Taxonomy, life history and behavior of *Thyas barbigera* and *Thyasides sphagnorum* (Hydrachnellae: Thyasidae). J. Med. Entomol. 13:475–485.
*Newell, I.M. 1959. Acari. *In* W.T. Edmondson (ed.). Ward and Whipple, Fresh-water biology, pp. 1080–1116. John Wiley and Sons, New York.
OConnor, B.M. 1981. A systematic revision of the family-group taxa in the non-psoroptidid Astigmata (Acari: Acariformes). Ph.D. dissertation. Cornell Univ., Ithaca, N.Y. 594 pp.
Oliver, D.R., and I.M. Smith. 1980. Host associations of some pionid water mite larvae (Acari:

Prostigmata: Pionidae) parasitic on chironomid imagos (Diptera: Chironomidae). Acta Univ. Carol. Biol. 1978: 157–162.

Paterson, C.G., and R.K. MacLeod. 1979. Observations on the life history of the water mite, *Unionicola formosa* (Acari: Hydrachnellae). Can. J. Zool. 57:2047–2049.

Prasad, V., and D.R. Cook. 1972. The taxonomy of water mite larvae. Mem. Amer. Entomol. Inst. 18:1–326.

Redmond, B.L., and J. Hochberg. 1981. The stylostome of *Arrenurus* spp. (Acari: Parasitengona) studied with the scanning electron microscope. J. Parasitol. 67:308–313.

Redmond, B.L., and C.A. Lanciani. 1982. Attachment and engorgement of a water mite *Hydrachna virella* (Acari: Parasitengona), parasitic on *Buenoa scimitra* (Hemiptera: Notonectidae). Trans. Amer. Microscop. Soc. 101:388–394.

Riessen, H.P. 1980. Diel vertical migration of pelagic water mites. *In* W.C. Kerfoot (ed.). Evolution and ecology of zooplankton communities, pp. 129–137. Univ. Press of New England, Hanover, N.H.

——. 1982a. Predatory behavior and prey selectivity of the pelagic mite *Piona constricta.* Can. J. Fish. Aquat. Sci. 39:1569–1579.

——. 1982b. Pelagic water mites: their life history and seasonal distribution in the zooplankton community of a Canadian lake. Arch. Hydrobiol. Suppl. 62:410–439.

Robinson, J.V. 1983. Effects of water mite parasitism on the demographics of an adult population of *Ischnura posita* (Hagen) (Odonata: Coenagrionidae). Amer. Midl. Nat. 109:169–174.

Simmons, T.W., and I.M. Smith. 1984. Morphology of larvae, deutonymphs, and adults of the water mite *Najadicola ingens* (Prostigmata: Parasitengona: Hygrobatoidea) with remarks on phylogenetic relationships and revision of taxonomic placement of Najadicolinae. Can. Entomol. 116:691–701.

Smith, B.P. 1983. The potential of mites as biological control agents of mosquitoes. *In* M.A. Hoy, G.L. Cunningham, and L. Knutson (eds.). Biological control of pests by mites, pp. 79–85. Univ. Calif. Div. Agric. Nat. Resour. Spec. Publ. no. 3304.

Smith, B.P. 1988. Host-parasite interaction and impact of larval water mites on insects. Ann. Rev. Entomol. 33:487–507.

Smith, B.P., and D.W. Barr. 1977. Swimming by the water mite, *Limnochares americana* Lundblad (Acari, Parasitengona, Limnocharidae). Can. J. Zool. 55:2050–2059.

Smith, B.P., and S. McIver. 1984a. The patterns of mosquito emergence (Diptera: Culicidae; *Aedes* spp.): their influence on host selection by parasitic mites (Acari: Arrenuridae; *Arrenurus* spp.). Can. J. Zool. 62:1106–1113.

——. 1984b. Factors influencing host selection and successful parasitism of *Aedes* spp. mosquitoes by *Arrenurus* spp. mites. Can. J. Zool. 62:1114–1120.

——. 1984c. The impact of *Arrenurus danbyensis* Mullen (Acari: Prostigmata; Arrenuridae) on a population of *Coquillettidia perturbans* (Walker) (Diptera:Culicidae). Can. J. Zool. 62:1121–1134.

*Smith, I.M. 1972. A review of the water mite genus *Nautarachna* (Acari: Parasitengona: Pionidae). Life Sci. Cont. R. Ont. Mus. 86:1–17.

*——. 1976a. A study of the systematics of the water mite family Pionidae (Prostigmata: Parasitengona). Mem. Entomol. Soc. Can. 98:1–249.

*——. 1976b. An unusual new species of *Neoacarus* (Acari: Parasitengona: Neoacaridae) from a lake in Ontario. Can. Entomol. 108:993–995.

*——. 1976c. *Paenecalyptonotus fontinalis* n. gen., n. sp., with remarks on the family Acalyptonotidae (Acari: Parasitengona: Arrenuroidea). Can. Entomol. 108:997–1000.

*——. 1977. A new species of *Mideopsis* (*Nudomideopsis*) (Acari: Parasitengona: Mideopsidae) from North America. Can. Entomol. 109:533–535.

——. 1978. Descriptions and observations on host associations of some larval Arrenuroidea

(Prostigmata: Parasitengona), with comments on phylogeny in the superfamily. Can. Entomol. 110:957–1001.

——. 1979. A review of water mites of the family Anisitsiellidae (Prostigmata: Lebertioidea) from North America. Can. Entomol. 111:529–550.

——. 1982. Larvae of water mites of the genera of the superfamily Lebertioidea (Prostigmata: Parasitengona) in North America with comments on phylogeny and higher classification of the superfamily. Can. Entomol. 114:901–990.

——. 1983. Description of adults of a new species of *Uchidastygacarus* (s.s.) (Acari: Parasitengona: Arrenuroidea) from eastern North America, with comments on distribution of mites of the genus. Can. Entomol. 115:1177–1179.

——. 1984. Larvae of water mites of some genera of Aturidae (Prostigmata: Hygrobatoidea) in North America with comments on phylogeny and classification of the family. Can. Entomol. 116:307–374.

Smith, I.M., and E. Lindquist. 1979. Prostigmata. *In* H. Danks (ed.). Canada and its insect fauna, pp. 267–277. Mem. Entomol. Soc. Can. no. 108.

Smith, I.M., and D.R. Oliver. 1976. The parasitic associations of larval water mites with imaginal aquatic insects, especially Chironomidae. Can. Entomol. 108:1427–1442.

——. 1986. Review of parasitic associations of larval water mites (Acari: Parasitengona; Hydrachnida) with insect hosts. Can. Entomol. 118:407–472.

Stechmann, D.-H. 1977. Zur Phänologie und zum Wirtsspektrum einiger an Zygopteren (Odonata) und Nematoceran (Diptera) ektoparasitisch auftretenden *Arrenurus*-Arten (Hydrachnellae, Acari). Z. Ang. Entomol. 82:349–355.

——. 1978. Eiablage, Parasitismus und postparasitische Entwicklung von *Arrenurus*-Arten (Hydrachnellae, Acari). Z. Parasitkde. 57:169–188.

——. 1979. Zum Wirtskreis synoptischer *Arrenurus*-Arten (Hydrachnellae, Acari) mit parasitischer Entwicklung an Nematocera (Diptera). Z. Parasitkde. 62:267–283.

Viets, K. 1936. Wassermilben oder Hydracarina (Hydrachnellae und Halacaridae). *In* Die Tierwelt Deutschlands, vols. 31–32. F. Dahl, Jena. 574 pp.

——. 1956. Die Milben des Süsswassers und des Meeres. Hydrachnellae et Halacaridae (Acari). Zweiter und dritter Teil: Katalog und Nomenclator. Gustav Fischer Verlag, Jena. 870 pp.

Viets, K.O. 1982. Die Milben des Süsswassers (Hydrachnellae und Halacaridae [part.], Acari). 1: Bibliographie. Sonderbände des Naturwissenschaftlichen Vereins in Hamburg. Verlag Paul Parey, Hamburg and Berlin. 116 pp.

——. 1987. Die Milben des Süsswassers (Hydrachnellae und Halacaridae [part.], Acari). 2: Katalog. Sonderbände des Naturwissenschaftlichen Vereins in Hamburg. Verlag Paul Parey, Hamburg and Berlin. 1,012 pp. [Includes update of bibliography to 1987.]

Wainstein, B.A. 1980. Opredelitel lichinok vodjanych kleshchei. [The water mite larvae.] Inst. Biol. Vnutrenn. Vod., Nauka. 238 pp.

Wesenberg-Lund, C.J. 1939. Biologie der Süsswassertiere. Wirbellose Tiere. J. Springer, Vienna. 817 pp.

Wiggins, G.B., R.J. Mackay, and I.M. Smith. 1980. Evolutionary and ecological strategies of animals in annual temporary pools. Arch. Hydrobiol. Suppl. 58:97–206.

Wiles, P.R. 1984. Watermite respiratory systems. Acarologia 25:27–32.

Wilson, J. 1981. Two new species of water mites (Acarina: Arrenuridae, genus *Arrenurus*, subgenus *Micruracarus*) with redescriptions of two closely related species. J. Tenn. Acad. Sci. 56:78–83.

17 Freshwater Mollusca

David Strayer

Mollusks, snails (class Gastropoda) and clams (class Bivalvia), are well represented in freshwater habitats throughout the northeastern United States. The molluscan fauna of this region contains about 135 species distributed among 13 families. The fauna can be divided into five groups: the prosobranch and pulmonate snails and the corbiculid, sphaeriid, and unionacean clams.

Gastropoda

The prosobranchs are gill-breathing snails derived from marine ancestors. Because prosobranchs depend on oxygen dissolved in the water for respiration, they are intolerant of sites where dissolved oxygen is scarce, such as sites of organic pollution. They also are absent from temporary waters. Some prosobranchs are long-lived, with life spans of three to five years. Although most prosobranch species have separate sexes and lay eggs, there are some conspicuous exceptions to this pattern. Members of the family Valvatidae are hermaphrodites, and members of the family Viviparidae (the "mystery snails") are ovoviviparous, including one common species, *Campeloma decisum*, that is parthenogenetic.

Pulmonate snails are descended from terrestrial snails, so they have lungs and breathe air. Although this feature frees them from a dependence on oxygen dissolved in the water (some pulmonates inhabit grossly polluted sites), most pulmonates must come to the water's surface to breathe. A few pulmonates have developed some kind of secondary "gills," which enable them to remain submerged indefinitely. The limpets (family Ancylidae) have an external *pseudobranch* that functions as a gill, and in some of the Lymnaeidae, the highly vascularized mantle cavity can be filled with water and used as a gill. Many pulmonates are short-lived and are able to complete their life cycles in a year or less. All of the freshwater pulmonates in the northeastern

I am grateful to D. G. Smith, E. H. Jokinen, A. H. Clarke, W. N. Harman, and J. Barnes for their helpful comments on earlier versions of this chapter. Kurt Jirka was especially helpful with final revisions. This is a contribution to the program of the Institute of Ecosystem Studies of The New York Botanical Garden.

United States are egg-laying hermaphrodites. Pulmonates are common in all kinds of freshwater habitats.

Most of the freshwater snails in the northeastern United States, prosobranch and pulmonate, feed by scraping algae and organic debris from stones, leaves, and other substrates. Major predators of snails include fish, waterfowl, crayfish, leeches, and sciomyzid flies.

Bivalvia

The corbiculid clams (superfamily Sphaeracea, family Corbiculidae) are represented in North America by a single species, *Corbicula fluminea*. It was introduced from Asia into the Pacific Northwest in the early 1900s. A hermaphrodite, *C. fluminea* lives two to three years. The young are released from adults as tiny free-living benthic larvae. Since its initial introduction, *C. fluminea* has spread throughout the entire continental United States except for the northernmost tier of states. The species is said to be intolerant of low winter temperatures, so it is not clear how widely it will be able to establish itself in the northeastern United States. It has only recently been reported from Long Island (Foehrenbach and Raihle 1984). *C. fluminea* often occurs at spectacularly high density (more than 10,000 individuals/m^2). It has been accused of competitively displacing native bivalves and can cause serious economic problems by attaching to the inside of and clogging pipes that carry cooling water to power plants.

The sphaeriid clams (superfamily Sphaeracea, family Sphaeriidae) are tiny (3–20 mm) bivalves known as fingernail clams or pea clams. Sphaeriids are hermaphrodites also, but differ from corbiculid clams in that young sphaeriids are brooded by the parent and are not released into the environment until they are relatively large (1/20–1/4 the size of the parent), well-developed juveniles. Sphaeriids live for a year or two. These clams are especially abundant in standing waters, both permanent and temporary; a few species are common in running waters.

The two families of clams of the superfamily Unionacea that occur in the northeastern United States, the Unionidae and Margaritiferidae, have similar biological characteristics. The unionaceans are the large (3–20 cm) pearly mussels of lakes, streams, and rivers. The life cycle of the unionaceans is bizarre, with ecological and evolutionary implications that are not fully understood. Typically, sexes are separate, although a few hermaphroditic species are known. The developing young are held within their mother's gills for 1–10 months, and when they are released as tiny (0.2 mm), specialized larvae called *glochidia*, they are obligate parasites of fish (or, in one case, of the amphibian *Necturus*). The glochidia are nonmotile, and to survive they must encounter and attach to an appropriate host fish (not all fish species are suitable hosts for each clam species) within a few days of their release from their mother. A glochidium that beats the long odds and attaches to a correct host remains attached to the fish for 1–10 weeks, during which time it undergoes a metamorphosis into a juvenile mussel. Most glochidia apparently derive little nutrition from their hosts; the chief purpose of the parasitic period appears to be dispersal. After metamorphosing, the juvenile clam breaks free of the fish, drops to the bottom, and begins its long development (one to eight years) into an adult clam. Some unionaceans are extraordinarily long-lived. Many species live for 20 years or more, and one species, *Margaritifera margaritifera*, has been reported to live for more than a century (Hendelberg 1960).

All freshwater clams are filter feeders, subsisting on phytoplankton, zooplankton, detritus, and bacteria. Some species of fish consume clams regularly, and several species of mammals (most notably muskrats and raccoons) prey heavily on unionaceans. In the rivers of the midwestern United States, an active commercial fishery for unionaceans still exists; the shells are used in the Japanese cultured-pearl industry.

Collection and Preservation

In general, no special techniques are needed to collect freshwater mollusks; Brinkhurst (1974) and Merritt and Cummins (1984) review apparatuses and techniques suitable for the collection of benthic animals, including mollusks. For many purposes, it is sufficient to keep only the empty shell of the mollusk and to discard the soft parts. Although most of the molluscan species that occur in the northeastern United States can be identified by the shell alone, it is preferable to preserve both the shell and the soft parts. Entire animals can be killed and preserved in 70% ethanol (do not use formaldehyde, as it hardens the soft parts).

The best specimens are obtained if the animals are relaxed before they are killed. A variety of narcotics have been used successfully with mollusks, including sodium nembutal (van der Schalie 1953), menthol crystals (Smith 1982), and benzocaine. Menthol crystals can be sprinkled on the surface of the water in the containers holding the snails. The animals should be killed only after they do not react to gentle prodding. The amount of narcotic needed and the length of time needed for narcotization depend on many factors (e.g., water temperature, species of mollusk) and are learned by trial and error.

As an alternative to narcotization, Clarke (1982) recommends putting mollusks, uncrowded, into a plastic bag partly full of water and then freezing them. The mollusks die in a relaxed position. They should then be thawed, fixed in 10% formalin, and transferred to 70% ethanol for storage. I have had only limited success with this method, but it is worth trying because of its simplicity.

The fine structure of the gastropod radula is also of taxonomic importance. To prepare a radula for examination, remove the buccal mass (Fig. 1) and place it in a few drops of household bleach. After the tissues have been softened for a few minutes, remove the ribbonlike radula, wash it several times in clear water, and mount it on a microscope slide, either in water (temporary mount) or in CMC or a similar medium (permanent mount). See Jokinen 1983 for additional information.

Further details on preparation of mollusk specimens are also available in papers by van der Schalie (1953), Jacobson (1974), LaMarche et al. (1982), and Smith (1982).

Notes on Use of the Keys

These keys include all of the species of mollusks known to occur in the freshwaters of New York and New England. Collections made in brackish waters, or in freshwaters south or west of New York State, will probably contain species that are not in the keys.

The keys proceed to species except for the family Sphaeriidae, for which only generic identifications are made. The sphaeriids (especially the species of *Pisidium*) are difficult for beginners to identify. In addition, the northeastern United States contains nearly all of the North American sphaeriids, so a regional key would be only a little simpler than existing keys to the entire North American sphaeriid fauna. Readers who wish to identify sphaeriids to species should consult Herrington 1962, Burch 1975a, Mackie et al. 1980, and Clarke 1981a.

Several problems are responsible for many of the misidentifications of mollusks. First, it is unfortunately inevitable that some key couplets are based on subtle, subjectively determined characters of shell shape. When you are confronted with such a couplet, read it carefully, look at the illustrations, and, if you have access to a good series of correctly identified reference shells, compare your specimen with those shells. Keep in mind that it is far better to report a correct generic identification than an incorrect specific identification. Second, the keys for the identification of snails work only for adult specimens. If you have an immature snail, you may misidentify it. To determine whether a specimen is an adult, count the number of whorls (complete turns of the spiral). An adult shell usually has at least four whorls. Third, snails of the subclass Prosobranchia (families Bithyniidae, Hydrobiidae, Pleuroceridae, Valvatidae, Viviparidae) have a distinguishing feature—the operculum—that may be missing. The operculum is a plate that covers the aperture when the snail has withdrawn into the shell; it is obvious in a living snail or a specimen with soft parts. If the snail shell is empty, you have no way of knowing whether it was operculate. Finally, the snail key includes only truly aquatic species; it does not include species of land snails, which may be common along the margins of lakes and streams. A live specimen, or one in which the soft parts have been preserved, can easily be identified as either a land or an aquatic snail. A land snail has two pairs of tentacles in its head, while an aquatic snail has but one pair (Figs. E, F) ("two if by land and one if by sea"—Paul Revere was wrong!). There are no easy ways of telling land snails from aquatic snails if you have only an empty shell. Do not try to force a species name onto a shell that will not key out; it may be a land snail.

The taxonomy and identification of snails of the family Physidae present special problems. This confusing family has been revised recently by George Te (1978), who has, to date, published only small parts of his revision (Te 1975, 1980). The species of the subfamily Physinae (the former genus *Physa*) are difficult to distinguish using shell characters, and records published before 1978 must be viewed with caution. The Physidae section of the following key, which follows Te's system and uses features of the male reproductive system to distinguish species, is much more reliable than keys based on shell characters (available in Harman and Berg 1971 and

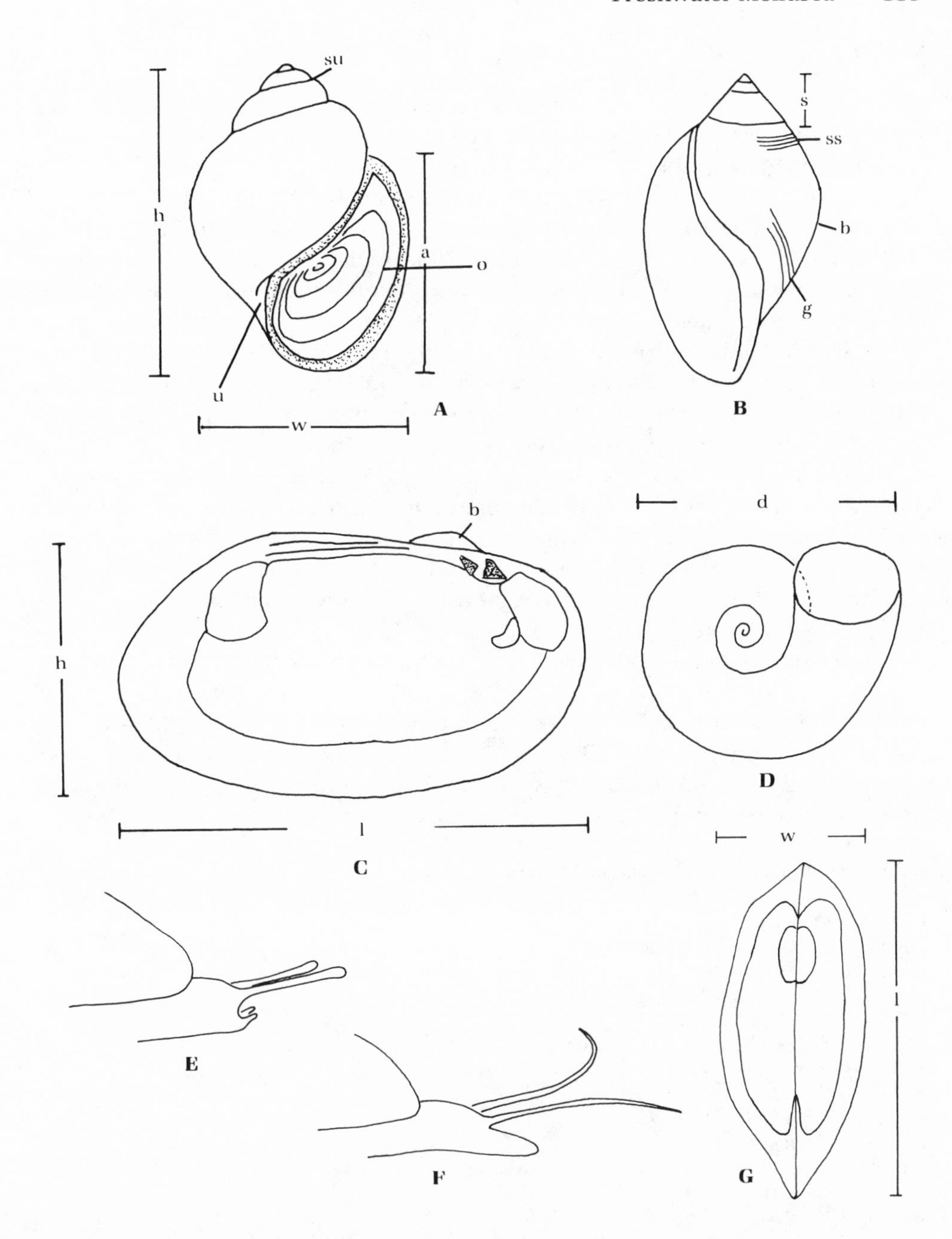

A. Snail shell with dextral orientation, vv: a, aperture; h, height; o, operculum; su, suture; u, umbilicus; w, width. **B.** Snail shell with sinistral orientation, vv: b, body whorl; g, growth lines; s, spire; ss, spiral striae. **C.** Interior of a unionacean valve, lv: b, beak; h, height; l, length. **D.** Schematic illustrating diameter dimension of a snail shell, lv: d, diameter. **E.** Schematic of the head of a land snail, lv. **F.** Schematic of the head of an aquatic snail, lv. **G.** Bivalve shell, dv: l, length; w, width. dv, dorsal view; lv, lateral view; vv, ventral view. (A modified from Jokinen 1983.)

Te 1975). The male reproductive system is on the left side of the animal, just behind the tentacles (Fig. 1). Animals used for dissection should be killed in an extended condition, either by drowning them in a closed vial full of water or by narcotizing them overnight by sprinkling crystals of menthol, benzocaine, or chloretone on the surface of the water in a shallow dish. Animals should then be stored in 70% ethanol. The male system is exposed by cutting or tearing open the top of the head along its midline and teasing out with a needle the twisted penial complex (which will appear as a thick white thread). The dissection and identification are not difficult. See Jokinen 1983 for further details.

These keys contain several morphological terms with which the user may be unfamiliar. The majority of these terms are defined in the glossary, and many of them are illustrated in Figures A–G. Furthermore, to identify bivalves, you will need to distinguish the anterior and posterior ends of shells. The anterior end of the shell is where the foot protrudes. On a unionid, the anterior end is the shorter end of the shell as measured from the beaks. To determine which end of an empty sphaeriid shell is anterior, hold the shell with the beaks up and with the left valve (with two cardinal and two lateral teeth) in your left hand and the right valve (with one cardinal and two pairs of lateral teeth) in your right hand. The anterior end of the shell is now facing away from you.

The classification and nomenclature of the North American freshwater mollusks have historically been very unstable. As a result, it is common (and frustrating) to find a species referred to by several different names in the literature. Burch (1975a, b, 1982) offers some fine synonymies, which are helpful in untangling nomenclatural problems. As an aid to users of this book, I have compiled a brief summary that shows the most common names used over the past 30 or so years.

binneyana (*Amnicola*) = *Probythinella lacustris*
Bulimus = *Bithynia*
calceolus (*Alasmidonta*) = *viridis*
coccineum (*Pleurobema*) = *cordatum*
costata (*Amblema*) = *plicata*
Elimia = *Goniobasis*
elliptica (*Physella*) = *gyrina*
emarginata (*Stagnicola*) = *catascopium*
Epioblasma = *Dysnomia*
Helisoma (in part) = *Planorbella*
hirsutus (*Gyraulus*) = *deflectus*
hudsonicus (*Promenetus*) = *exacuous*
humilis (*Lymnaea*) = various species of *Fossaria*
Hydrobia (in part) = *Fontigens*
hypnorum (*Aplexa*) = *elongata*
integra (*Amnicola*) = *Cincinnatia cincinnatiensis*
jenksii (*Planorbella*) = *armigera*
ligamentina (*Actinonaias*) = *carinata*
lustrica (*Marstonia, Amnicola*) = *decepta*
luteola (*Lampsilis*) = *siliquoidea*
Lyogyrus = *Amnicola* (in part)
Margaritana = *Margaritifera*
Micromya = *Villosa*
Nitocris = *Leptoxis*

palustris (*Stagnicola*) = *elodes*
Plagiola = *Dysnomia* or (in older literature) *Truncilla*
Potamilus = *Proptera*
rugosus (*Strophitus*) = *undulatus*
sayii (*Physella*) = *gyrina*
Simpsonaias = *Simpsoniconcha*
skinneri (*Physa*) (in part) = *vernalis*
Somatogyrus (in part) = *Birgella*
Spirodon = *Leptoxis*
tarda (*Ferrissia*) = *rivularis*
Toxolasma = *Carunculina*
ventricosa (*Lampsilis*) = *ovata*
viridus (*Lasmigona*) = *compressa*

Checklist and Distributional Information of Mollusca

Species	Drainage[a]					
	Erie-Niagara/ Allegheny	Lake Ontario	St. Lawrence/ Lake Champlain	Delaware/ Susquehanna	Hudson	New England (excl. Lake Champlain)
Class Gastropoda						
Ancylidae						
Ferrissia fragilis	?	+	?	?	+	+
F. parallelus	+	+	?	+	+	+
F. rivularis	+	+	+	+	+	+
F. walkeri						+
Laevapex fuscus	+	+	?	+	+	+
Bithyniidae						
Bithynia tentaculata	+	+	+		+	
Hydrobiidae						
Amnicola grana					+	+
A. limosa	+	+	+	+	+	+
A. pupoidea					+	+
A. walkeri	+	+				
Birgella subglobosa	?	?	+		+	
Cincinnatia cincinnatiensis	+	+	?		+	
Fontigens nickliniana	+					
Gillia altilis		+	+		+	
Marstonia decepta	+	+	?	?	+	+
Probythinella lacustris	+	+	?		+	
Pyrgulopsis letsoni	+					
Lymnaeidae						
Acella haldemani	?	+	+	+		
Bulimnea megasoma			+		?	+
Fossaria[a]	+	+	+	+	+	+
F. cyclostoma						
F. dalli						
F. galbana						
F. obrussa group						
F. parva						
Lymnaea stagnalis	+	+	+	+	+	+
Pseudosuccinea columella	+	+	+	+	+	+
Radix auricularia		+	?		+	?
Stagnicola caperata	+					

	Drainage[a]					
Species	Erie-Niagara/ Allegheny	Lake Ontario	St. Lawrence/ Lake Champlain	Delaware/ Susquehanna	Hudson	New England (excl. Lake Champlain)
S. catascopium	+	+	+	+	+	+
S. elodes	+	+	+	+	+	+
Physidae						
Aplexa elongata	+	+	+	+	+	+
Physa vernalis	?	?	?	?	+	+
Physella ancillaria		?	?	?	+	+
P. gyrina	+	+	+	+	+	+
P. heterostropha	+	+	+	+	+	+
P. integra	+	+	+	?	+	?
P. magnalacustris	?	?	?			
P. vinosa	?	?	?		+	
Planorbidae						
Armiger crista		+				+
Gyraulus circumstriatus	?	?	?	?	+	+
G. deflectus	+	+	+	+	+	+
G. parvus	+	+	+	+	+	+
Helisoma anceps	+	+	+	+	+	+
Menetus dilatatus	+	+	?	?	+	+
Planorbella campanulata	+	+	+	+	+	+
P. trivolvis	+	+	+	+	+	+
Planorbula armigera	+	+	+	+	+	+
Promenetus exacuous	+	+	+	+	+	+
Pleuroceridae						
Goniobasis livescens	+	+	+		+	
G. virginica		+		+	+	+
Leptoxis carinata				+		
Pleurocera acuta	+	+			+	
Valvatidae						
Valvata bicarinata	?					
V. lewisi	?	+	?		+	?
V. perdepressa	+	+				
V. piscinalis	+	+	+		+	
V. sincera	?	+	+		+	+
V. tricarinata	+	+	+	+	+	+
Viviparidae						
Campeloma decisum	+	+	+	+	+	+
Cipangopaludina chinensis	+	+	+		+	+
Lioplax subcarinata				?	+	
Viviparus georgianus	+	+	+	+	+	+
Class Bivalvia						
Superfamily Sphaeracea						
Corbiculidae						
Corbicula fluminea				(Long Island)		
Sphaeriidae						
Musculium	+	+	+	+	+	+
Pisidium	+	+	+	+	+	+
Sphaerium	+	+	+	+	+	+
Superfamily Unionacea						
Margaritiferidae						
Margaritifera margaritifera		+	+			+
Unionidae						
Actinonaias carinata	+	?				

Species	Drainage[a]					
	Erie-Niagara/ Allegheny	Lake Ontario	St. Lawrence/ Lake Champlain	Delaware/ Susquehanna	Hudson	New England (excl. Lake Champlain)
Alasmidonta heterodon				?		+
A. marginata	+	+	+	+	+	
A. undulata		+	+	+	+	+
A. varicosa				+	+	+
A. viridis	+	+			?	
Amblema plicata	+	+				
Anodonta cataracta		+	+	+	+	+
A. grandis	+	+	+		+	
A. imbecilis	+	+			+	
A. implicata				+	+	+
Anodontoides ferussacianus	+	+	+	+	+	
Carunculina parva	+	+				
Cyclonaias tuberculata	?					
Dysnomia triquetra	+					
Elliptio complanata		+	+	+	+	+
E. dilatata	+	+	+			
Fusconaia flava	+	+			+	
Lampsilis cariosa			+	+	+	+
L. fasciola	+					
L. ochracea				?	+	+
L. ovata	+	+	+		+	
L. radiata		+	+	+	+	+
L. siliquoidea	+	+	+		+	
Lasmigona compressa	+	+	+		+	
L. costata	+	+	+		+	
L. subviridus		+		+	+	
Leptodea fragilis	+	+	+		+	
L. laevissima	+	+				
Ligumia nasuta		+	+	+	+	+
L. recta	+	+	+			
Obovaria olivaria	+		+			
Pleurobema cordatum	+					
Proptera alata	+	+	+		+	
Ptychobranchus fasciolare	+					
Quadrula pustulosa	+					
Q. quadrula	+					
Simpsoniconcha ambigua	+					
Strophitus undulatus	+	+	+	+	+	+
Truncilla donaciformis	?					
T. truncata	+					
Villosa fabalis	+					
V. iris	+	+			?	

[a]Distributions of species of *Fossaria* identified in the key are unclear, hence distributional information is provided on the generic level only.

Sources of distributional information: Johnson 1915, Ortmann 1919, Robertson and Blakeslee 1948, Clench and Turner 1955, Clarke and Berg 1959, Jacobson and Emerson 1961, Clench 1962a, Harman 1970b, Harman and Berg 1971, Buckley 1977, Te 1978, Clarke 1981a, Burch 1982, Jokinen 1982, 1983, Taylor and Jokinen 1984, Thompson 1984, Smith 1985, and Strayer 1987.

Note: Occurrence of a species in a drainage is indicated by a +. A ? indicates that there is no record of the species in the drainage but it possibly occurs there. A blank indicates that there is no record of the species and it is unlikely that it occurs in the drainage.

Key to Species of Freshwater Gastropoda

1a. Shell not coiled, but shaped like a short, wide cone (Figs. 2–6) . . **Ancylidae** 2
1b. Shell coiled into a spiral (Figs. 7–65) 5

2a (1a). **Ancylidae**: Tentacles with a black core; apex of shell unsculptured (difficult to see) (Fig. 2) . . . . *Laevapex fuscus*
2b. Tentacles with a white core (so they are white all the way through); apex of shell finely sculptured (difficult to see) (Fig. 5) . . . . *Ferrissia* 3

3a (2b). Apex of shell very strongly off center, reaching almost to the edge of the shell (Fig. 3); this rare species, limited to coastal ponds, may be an ecophenotype of *F. fragilis* . . . . *F. walkeri*
3b. Apex not strongly off center; widely distributed and common 4

4a (3b). (Triplet) Shell elongate, L/W ≥ 1.8, with almost parallel sides (Fig. 4); shell length to 9 mm . . . . *F. parallelus*
4b. Shell sturdy, elliptical (Fig. 5), L/W = 1.3–1.7; up to 7 mm long; found in streams . . . . *F. rivularis*
4c. Shell small (≤ 4 mm long) and fragile, elliptical; L/W variable (Fig. 6); usually in quiet waters . . . . *F. fragilis*

5a (1b). Shell shaped like a flat disk (Figs. 56–65); without a raised spire or operculum . . . . **Planorbidae** 52
5b. Shell with a raised spire (Figs. 7–55) (or, in a few rare cases, without a raised spire but with an operculum) 6

6a (5b). Shell sinistral (Figs. 49–55) . . . . **Physidae** 45
6b. Shell dextral (Figs. 7–48) (A snail shell held with the aperture facing you and the spire pointing away from you is dextral if the aperture is on your right.) 7

7a (6b). Shell not operculate . . . . **Lymnaeidae** 33
7b. Shell operculate 8

8a (7b). Shell as wide as or wider than high, and small (≤ 5 mm high) (Figs. 8–14) . . . . **Valvatidae**, *Valvata* 12
8b. Shell higher than wide 9

9a (8b). Operculum concentric (Figs. 7, 66), whorls inflated; adult shell fairly large (> 8 mm high) 10
9b. Operculum spiral (Figs. 26, 27, 32), whorls flattened or inflated; adult shell small to large 11

10a (9a). Adult shell small (8–15 mm high) . . . . **Bithyniidae**, *Bithynia tentaculata*
10b. Adult shell large (> 15 mm high) . . . . **Viviparidae** 17

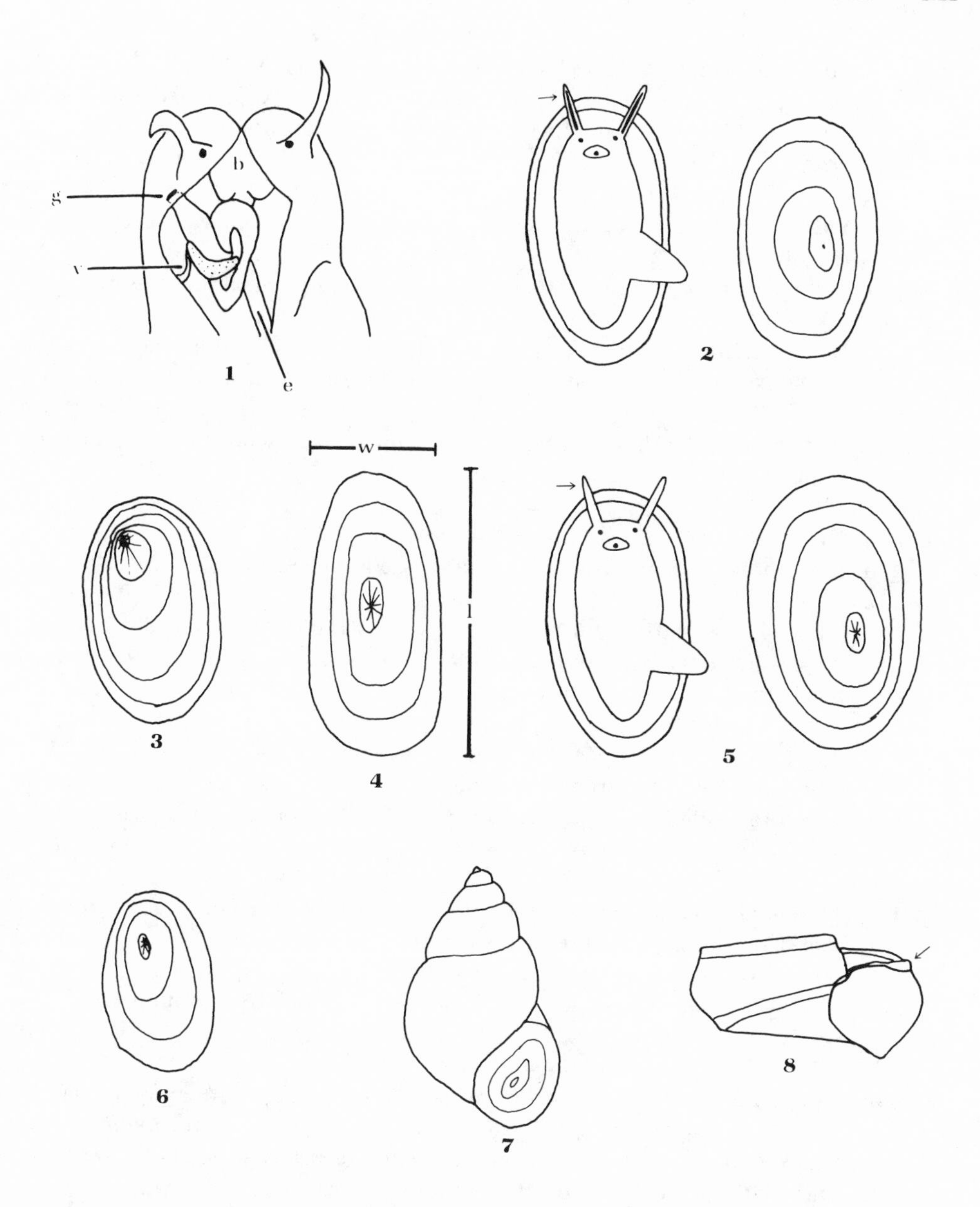

1. Head of dissected *Physa* s.l. (Physidae) showing male reproductive tract, dv: b, buccal mass; e, esophagus; g, gonopore; v, vas deferens. **2.** *Laevapex fuscus* (Ancylidae), vv (left), dv. **3.** *Ferrissia walkeri* (Ancylidae), dv. **4.** *Ferrissia parallelus* (Ancylidae), dv: l, length; w, width. **5.** *Ferrissia rivularis* (Ancylidae), vv (left), dv. **6.** *Ferrissia fragilis* (Ancylidae), dv. **7.** *Bithynia tentaculata* (Bithyniidae), vv. **8.** *Valvata bicarinata* (Valvatidae), vv. dv, dorsal view; vv, ventral view. (1, 4, 6 modified from Jokinen 1983; 2 [right], 3, 5 [right], 7, 8 redrawn or modified from Burch 1982.)

11a (9b). Adult shell small ($\leq$ 9 mm high); whorls inflated (Figs. 19–28, 32) .... **Hydrobiidae** 20

11b. Adult shell large (> 9 mm high); whorls flattened (Figs. 33–36) .... **Pleuroceridae** 30

12a (8a). **Valvatidae**: Whorls angular in cross section (Figs. 8–10); marginal carinae usually present 13

12b. Whorls rounded in cross section (Figs. 11–14); marginal carinae absent ... 14

13a (12a). Dorsal carina extending beyond the body whorl dorsally; spire very low, usually sunken below the body whorl; with 2–3 carinae (Fig. 8); rare .... *V. bicarinata*

13b. Body whorl extending dorsally above the dorsal carina (if present); spire elevated; 0–3 carinae (Figs. 9, 10); common .... *V. tricarinata*

14a (12b). Shell with coarse growth lines; depressed (H/W < 0.75) (Fig. 11) ..... *V. lewisi*

14b. Shell without coarse growth lines; depressed or elevated 15

15a (14b). Shell depressed (H/W < 0.70) (Fig. 12); apical whorls of shell usually dull purple or pink; a rare species of the Great Lakes .... *V. perdepressa*

15b. Shell elevated (Figs. 13, 14) 16

16a (15b). Aperture very large (Ap/W > 0.50; Ap = maximum diameter of aperture); whorls inflated and rapidly expanding (Fig. 13) .... *V. piscinalis*

16b. Aperture moderately large (Ap/W = 0.3–0.5); whorls not especially inflated or rapidly expanding (Fig. 14) .... *V. sincera*

17a (10b). **Viviparidae**: Shell greenish; relatively slender (H/W $\geq$ 1.35); whorls shouldered; umbilicus absent or nearly closed 18

17b. Shell dark olive-green or brownish, often with spiral stripes (Fig. 16); H/W usually $\leq$ 1.35; whorls flattened or rounded; umbilicus present (may be partly closed over) 19

18a (17a). Operculum entirely concentric (Fig. 66); shell large (up to 40 mm high) and solid, without spiral ridges (Fig. 17); common .... *Campeloma decisum*

18b. Operculum concentric with a spiral center (Fig. 67); shell small (H < 25 mm) and relatively fragile, often with low spiral ridges (Fig. 18); a rare species found in the northeastern United States only in the tidal Hudson River .... *Lioplax subcarinata*

19a (17b). Adult shell large (H > 35 mm) but thin, dark olive-green (Fig. 15) .... *Cipangopaludina chinensis*

19b. Adult shell usually < 35 mm high, solid, and usually with prominent brown spiral stripes (Fig. 16) .... *Viviparus georgianus*

20a (11a). **Hydrobiidae**: Shell relatively large (to 9 mm high), subglobose (H/W = 1.08–1.25), with a very large aperture (Ap/H > 0.5) (Figs. 19, 20) 21

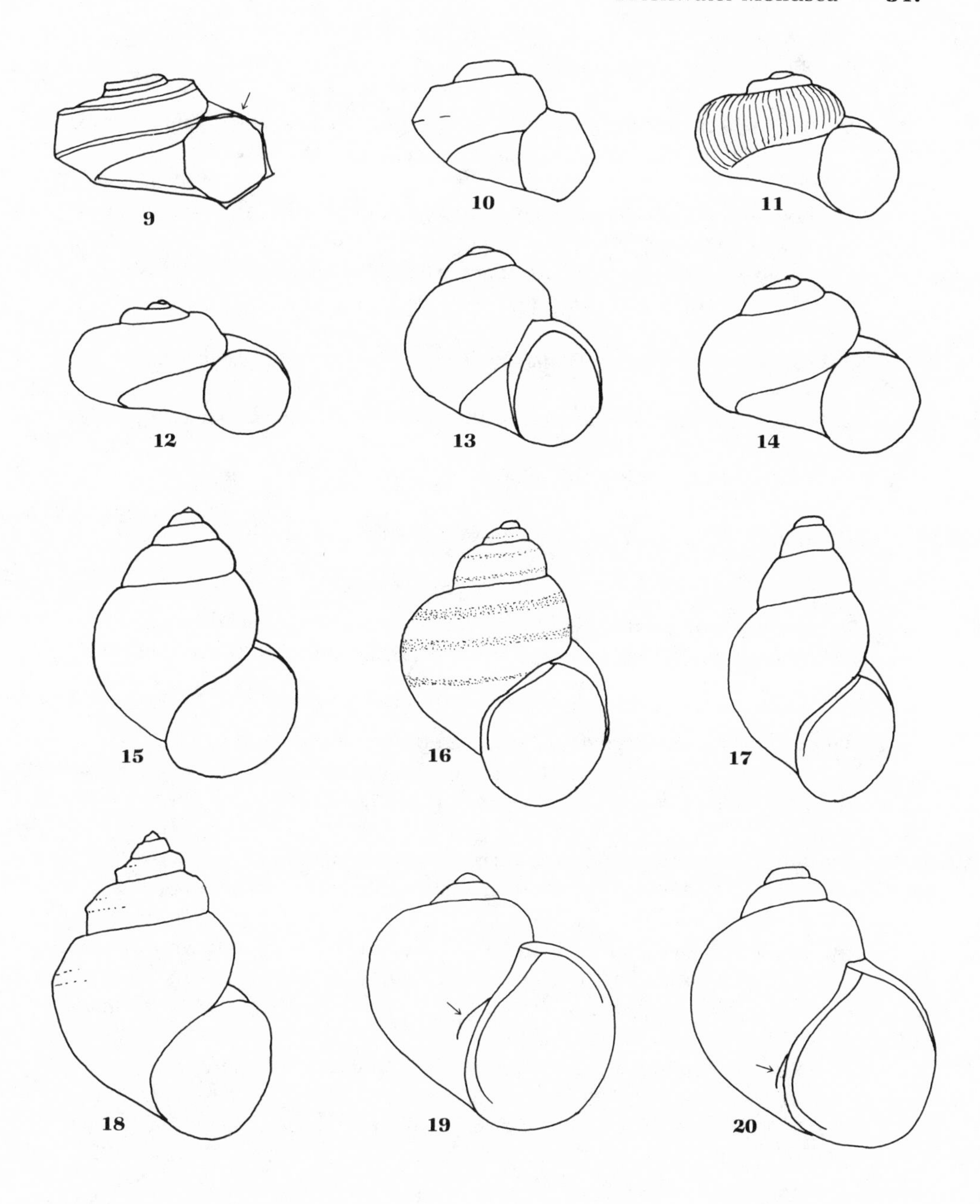

9. *Valvata tricarinata* (morph *tricarinata*) (Valvatidae), vv. **10.** *Valvata tricarinata* (morph *simplex*) (Valvatidae), vv. **11.** *Valvata lewisi* (Valvatidae), vv. **12.** *Valvata perdepressa* (Valvatidae), vv. **13.** *Valvata piscinalis* (Valvatidae), vv. **14.** *Valvata sincera* (Valvatidae), vv. **15.** *Cipangopaludina chinensis* (Viviparidae), vv. **16.** *Viviparus georgianus* (Viviparidae), vv. **17.** *Campeloma decisum* (Viviparidae), vv. **18.** *Lioplax subcarinata* (Viviparidae), vv. **19.** *Birgella subglobosa* (Hydrobiidae), vv. **20.** *Gillia altilis* (Hydrobiidae), vv. vv, ventral view. (9, 11–18 redrawn from Burch 1982; 10 redrawn from Harman and Berg 1971; 19, 20 modified from Thompson 1984.)

20b. Not as above; < 6 mm high 22

21a (20a). Body whorl typically somewhat shouldered; shell usually distinctly umbilicate (Fig. 19); male organ bearing an accessory lobe (Fig. 68) . . . . *Birgella subglobosa*

21b. Body whorl typically evenly rounded; shell usually without an open umbilicus (Fig. 20); male organ without accessory lobe (Fig. 69) . . . . . . *Gillia altilis*

B. subglobosa and *G. altilis* are extremely difficult to distinguish by shell characteristics.

22a (20b). Umbilicus absent or very narrow (Figs. 21, 22); rare species of western New York 23

22b. Umbilicus obvious (Figs. 23–28, 32); common, widespread species 24

23a (22a). Whorls inflated (Fig. 21); shell of moderate size (to 6 mm high) . . . . *Fontigens nickliniana*

23b. Whorls flattened (Fig. 22); shell very small (≤ 3 mm high) . . . . *Pyrgulopsis letsoni*

24a (22b). Nuclear whorl raised above the following whorls (Fig. 29) 25

24b. Nuclear whorl level with (Fig. 30) or sunken below (Fig. 31) the following whorls 27

25a (24a). Adult shell very small (≤ 2.5 mm high); aperture round (Fig. 23) . . . . *Amnicola walkeri*

25b. Adult shell 4–6 mm high; aperture ovate (Figs. 24, 25) 26

26a (25b). Adult shell small (< 4.5 mm high), usually relatively narrow (W/H ≤0.68); apex of shell blunt (Fig. 24), with a narrow, partially closed umbilicus . . . . *Marstonia decepta*

26b. Adult shell moderately small (up to 6 mm high) usually relatively wide (W/H ≥ 0.70); apex of shell pointed, with a broad and deep umbilicus (Fig. 25) . . . . . *Cincinnatia cincinnatiensis*

27a (24b). Adult shell minute (≤ 2.5 mm high); aperture almost round; operculum multispiral (Figs. 26, 27) . . . . *Amnicola* (in part) 28

27b. Adult shell small (to 5 mm high); aperture ovate; operculum paucispiral (Fig. 32) 29

28a (27a). Aperture margin continuous with body whorl for a very short distance (Fig. 27) . . . . *A. pupoidea*

28b. Aperture margin continuous with body whorl for a long distance (Fig. 26) . . . *A. grana*

29a (27b). Nuclear whorl level with following whorls (Fig. 30) *Amnicola limosa* (Fig. 32)

29b. Nuclear whorl sunken below following whorls (Fig. 31) . . . . *Probythinella lacustris* (Fig. 28)

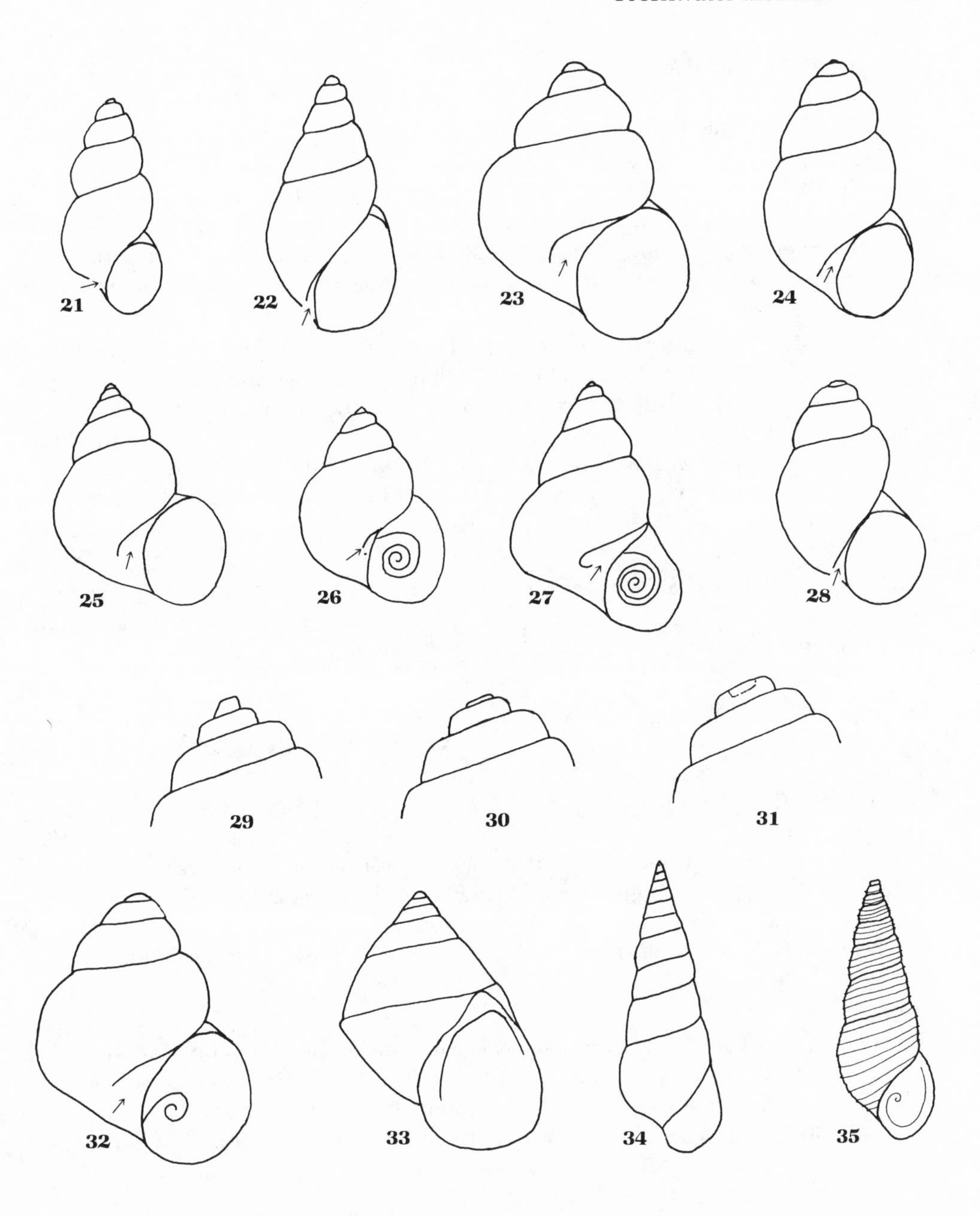

21. *Fontigens nickliniana* (Hydrobiidae), vv. **22.** *Pyrgulopsis letsoni* (Hydrobiidae), vv. **23.** *Amnicola walkeri* (Hydrobiidae), vv. **24.** *Marstonia decepta* (Hydrobiidae), vv. **25.** *Cincinnatia cincinnatiensis* (Hydrobiidae), vv. **26.** *Amnicola grana* (Hydrobiidae), vv. **27.** *Amnicola pupoidea* (Hydrobiidae), vv. **28.** *Probythinella lacustris* (Hydrobiidae), vv. **29.** Shell apex with a raised nuclear whorl. **30.** Shell apex with a nearly level nuclear whorl. **31.** Shell apex with a sunken nuclear whorl. **32.** *Amnicola limosa* (Hydrobiidae), vv. **33.** *Leptoxis carinata* (Pleuroceridae), vv. **34.** *Pleurocera acuta* (Pleuroceridae), vv. **35.** *Goniobasis virginica* (Pleuroceridae), vv. vv, ventral view. (21, 23–25, 28, 32–35 redrawn from Burch 1982; 26, 27 redrawn from Jokinen 1983.)

30a (11b). **Pleuroceridae**: Shell globose ($H/W \leq 2.5$) (Fig. 33); Susquehanna basin .... *Leptoxis carinata*

30b. Shell elevated ($H/W \geq 2.5$) 31

31a (30b). Shell extremely elevated (H/Ap usually ≥ 3.33); whorls very flat-sided; sutures shallow (Fig. 34); rare except in western New York .... *Pleurocera acuta*

31b. Shell elevated (H/Ap usually ≤ 3.33); whorls moderately flat-sided; sutures fairly deep; widespread .... *Goniobasis* 32

32a (31b). Adult shells large (ca. 30 mm high), often with coarse spiral striae on body whorl (Fig. 35); yellow to chestnut, often with 2 brown spiral stripes; in large streams of the Atlantic drainage northeast to the Connecticut River .... *G. virginica*

32b. Adult shells smaller, rarely exceeding 20 mm, never striated and rarely striped (Fig. 36); western New York, Lake Champlain, and scattered in the Oswego River and Hudson River basins .... *G. livescens*

33a (7a). **Lymnaeidae**: Shell extremely elevated ($H/W \geq 5$); rare (Fig. 37) .... *Acella haldemani*

33b. Shell not extremely elevated ($H/W < 5$); common 34

34a (33b). $Ap/H > 0.6$ 35

34b. $Ap/H < 0.6$ 36

35a (34a). $Ap/H < 0.75$; adult $H < 15$ mm (Fig. 38); shell amber and very fragile; very common .... *Pseudosuccinea columella*

The semiaquatic land snail *Succinea* will key out here. It can be distinguished from *Pseudosuccinea* by its lack of spiral striae, which are well developed in *Pseudosuccinea*.

35b. $Ap/H > 0.75$; adult $H > 15$ mm (Fig. 39); shell whitish; sporadic and uncommon .... *Radix auricularia*

36a (34b). Shell small ($H < 15$ mm), usually lacking spiral striae; inner lip of aperture flat .... *Fossaria* 37

The nominal species of *Fossaria* are difficult to distinguish.

36b. Shell often > 15 mm high, with obvious spiral striae; inner lip of aperture twisted 41

37a (36a). Adult shell (with ca. 5 whorls) < 7 mm high; whorls rounded to slightly shouldered; aperture ovate 38

37b. Adult shell (with ca. 5 whorls) > 5 mm high; whorls rounded to strongly shouldered; aperture round, ovate, or squarish 39

38a (37a). Lateral teeth of radula bicuspid (Fig. 70) .... *F. dalli*

38b. Lateral teeth of radula tricuspid (Fig. 40) .... *F. parva*

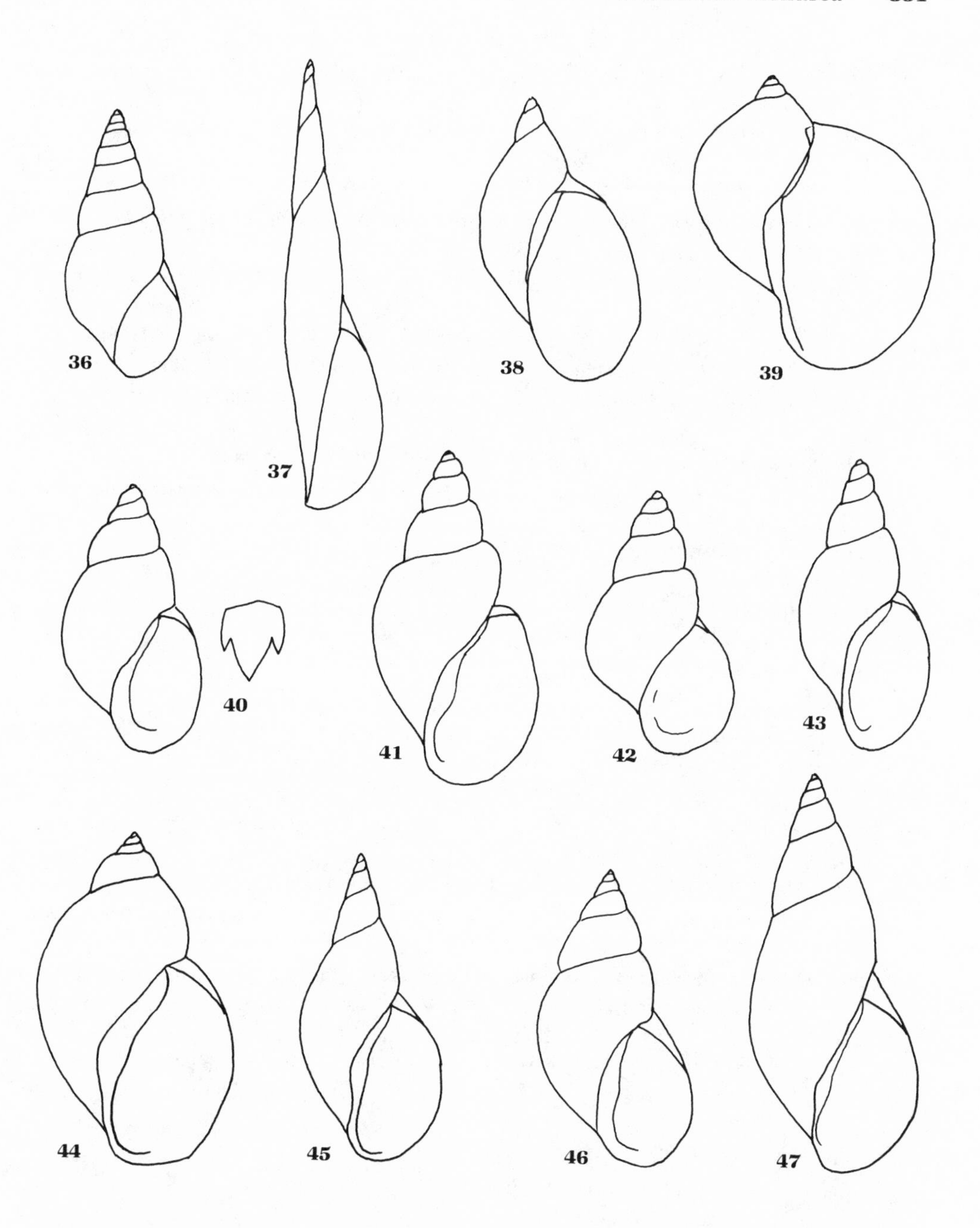

36. *Goniobasis livescens* (Pleuroceridae), vv. **37.** *Acella haldemani* (Lymnaeidae), vv. **38.** *Pseudosuccinea columella* (Lymnaeidae), vv. **39.** *Radix auricularia* (Lymnaeidae), vv. **40.** Shell and lateral tooth of *Fossaria parva* (Lymnaeidae), vv. **41.** *Fossaria galbana* (Lymnaeidae), vv. **42.** *Fossaria cyclostoma* (Lymnaeidae), vv. **43.** *Fossaria obrussa* (Lymnaeidae), vv. **44.** *Bulimnea megasoma* (Lymnaeidae), vv. **45.** *Lymnaea stagnalis* (Lymnaeidae), vv. **46.** *Stagnicola caperata* (Lymnaeidae), vv. **47.** *Stagnicola elodes* (Lymnaeidae), vv. vv, ventral view. (36–47 redrawn from Burch 1982.)

39a (37b). Shell thick and often whitish; whorls strongly shouldered; aperture squarish (Fig. 41) . . . . *F. galbana*
39b. Shell relatively thin, usually brown; whorls rounded or slightly shouldered; aperture rounded to ovate 40

40a (39b). Spire noticeably longer than aperture; aperture almost round (Fig. 42) . . . . *F. cyclostoma*
40b. Spire not much longer than aperture; aperture ovate (Fig. 43) . . . . *F. obrussa* group

Species included in *F. obrussa* group are *F. exigua*, *F. modicella*, *F. obrussa*, and *F. rustica*. These may not all be "good" species and are difficult to distinguish. See Burch 1982 for keys and illustrations.

41a (36b). Shell large (often > 35 mm high), inflated (W/H > 0.6; Ap/H > 0.5), brown with streaks of green, orange, or purple; interior of shell copper-colored; rare (Fig. 44) . . . . *Bulimnea megasoma*
41b. Not as above; common 42

42a (41b). Adult shell large (H usually > 35 mm), slender (Ap/H often < 0.55), and usually whitish (Fig. 45) . . . . *Lymnaea stagnalis*
42b. Adult shell < 35 mm high, often brown . . . . *Stagnicola* 43

43a (42b). Shell sculptured with fine raised spiral lines; rare . . . . *S. caperata* (Fig. 46)
43b. Shell with malleations or spiral striae, but not raised spiral lines 44

44a (43b). W/H $\leq$ 0.5; Ap/H usually $\leq$ 0.5 (Fig. 47); whorls increasing in size gradually; usually in ponds and marshes . . . . *S. elodes*
44b. W/H $\geq$ 0.5; Ap/H $\geq$ 0.5 (Fig. 48); whorls increasing in size rapidly; usually in large lakes or rivers . . . . *S. catascopium*

45a (6a). **Physidae**: Shell elongate and highly polished (Fig. 49); mantle edge (of body) without fingerlike projections . . . . *Aplexa elongata*
45b. Shell more or less broad, glossy, but not highly polished; mantle edge (of body) with fingerlike projections on one or both sides (Fig. 50) . . . . **Physinae** (= *Physa* s.l.) 46

46a (45b). Penial sheath having 2 parts (Figs. 71, 72) . . . . *Physella* (*Physella*) 47
46b. Penial sheath having 1 part (Figs. 73, 74) 50

47a (46a). Terminal nonglandular section of the penial sheath not transparent, longer than glandular section; glandular section not swollen distally (Fig. 71); shell W/H usually < 0.62 . . . . *P.* (*P.*) *gyrina*
47b. Terminal nonglandular section of the penial sheath transparent, shorter and narrower than glandular section (Fig. 72); glandular section swollen distally; shell W/H usually > 0.62 48

48a (47b). Shell with very small spire and strongly shouldered body whorl (Fig. 51) .... *P.* (*P.*) *ancillaria*
48b. Shell with moderately small spire, body whorl not shouldered (Figs. 52, 53) 49

49a (48b). Shell whitish and very thick; whorls shouldered; spire very short and broad (Fig. 52) .... *P.* (*P.*) *magnalacustris*
49b. Shell green or brown and not especially thick; whorls rounded; spire relatively higher and narrower than above (Fig. 53) .... *P.* (*P.*) *vinosa*

50a (46b). Penial sheath glandular and very long (about 3 times as long as the preputium) (Fig. 73); fingerlike projections on one side of the mantle (Fig. 50) ... . *Physa vernalis*
50b. Penial sheath nonglandular, about 1–1½ times as long as the preputium (Fig. 74); fingerlike projections present on both sides of the mantle .... *Physella* (*Costatella*) 51

51a (50b). Shell spire wide and short (0.18–0.33 of the shell height; see Clarke 1981a); shell fairly thin .... *P.* (*C.*) *heterostropha* (Fig. 54)
51b. Shell spire narrow and long (0.27–0.43 of the shell height; see Clarke 1981a); shell thick .... *P.* (*C.*) *integra* (Fig. 55)

52a (5a). **Planorbidae**: Shell large (> 7 mm diameter), or, if smaller (immatures), then quite high (H/diam. ≥ 0.6) 53
52b. Shell small (< 7 mm diameter) and usually flat (H/diam. ≤ 0.6) 55

53a (52a). Body whorl usually flared (in mature specimens) into a bell-like expansion (Fig. 56); sutures deeply impressed on both sides of the shell (Fig. 56) .... *Planorbella campanulata*
53b. Body whorl without bell-like expansion; sutures deeply impressed on one or neither side of the shell (Figs. 57, 58) 54

54a (53b). Sutures shallow on both sides of the shell, forming smooth-sided depressions (Fig. 57) .... *Helisoma anceps*
54b. Sutures shallow on one side of the shell and deep on the other, forming 1 smooth-sided and 1 rough-sided depression (Fig. 58) .... *Planorbella trivolvis* group

There are many named "species," "subspecies," and "morphs" that belong to the *P. trivolvis* group. The true taxonomic relationships among them are not clear, and there is not even agreement on how to separate the named taxa. Descriptions, discussions, and keys to the taxa are given in Clarke 1973, 1981a and Burch 1982; they should be used cautiously.

55a (52b). Shell very small (< 3 mm diam.), with large raised ribs externally (Fig. 59); rare .... *Armiger crista*

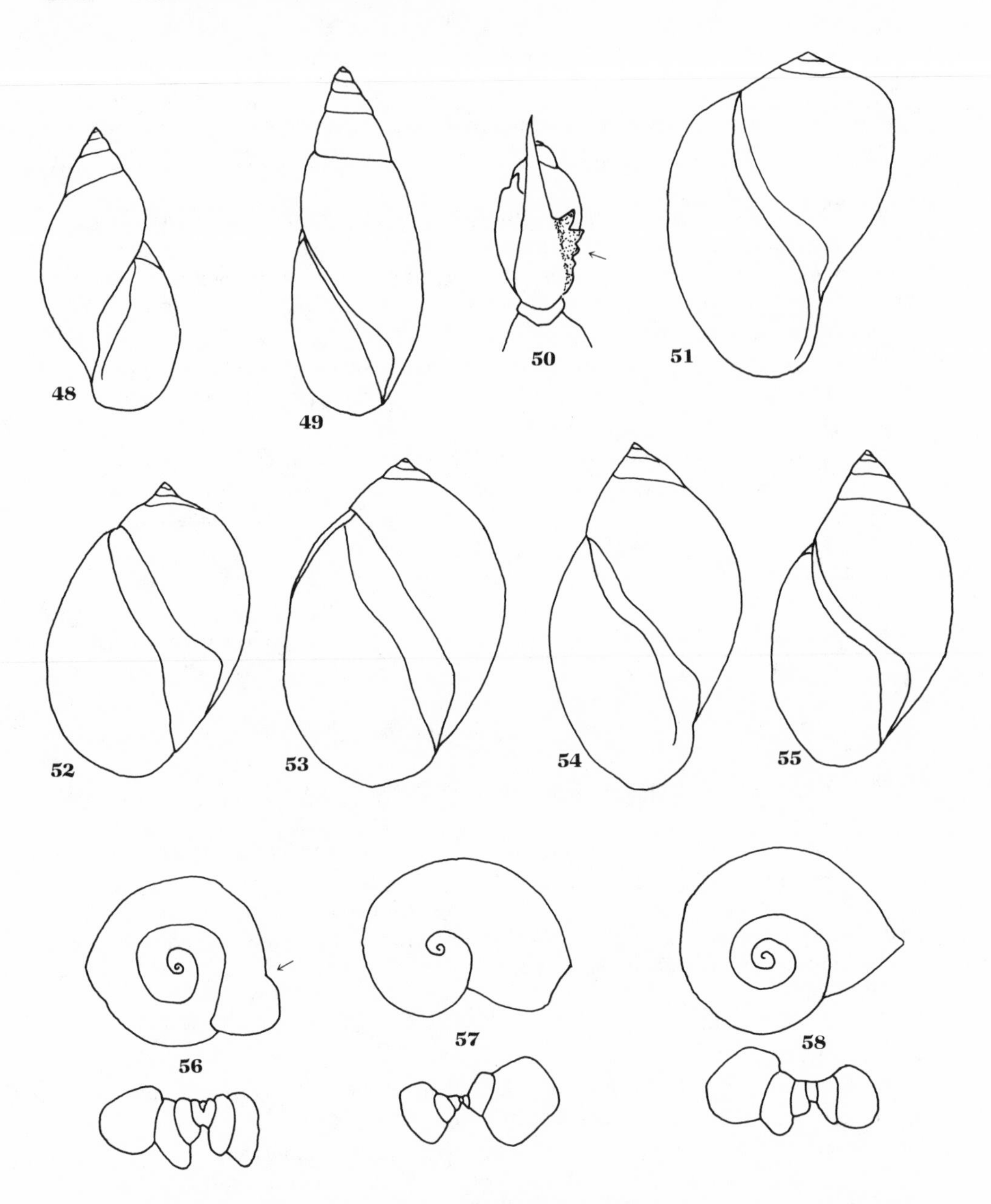

48. *Stagnicola catascopium* (Lymnaeidae), vv. **49.** *Aplexa elongata* (Physidae), vv. **50.** Underside of living animal of *Physa* (Physidae), showing fingerlike mantle projections. **51.** *Physella ancillaria* (Physidae), vv. **52.** *Physella magnalacustris* (Physidae), vv. **53.** *Physella vinosa* (Physidae), vv. **54.** *Physella heterostropha* (Physidae), vv. **55.** *Physella integra* (Physidae), vv. **56.** *Planorbella campanulata* (Planorbidae) in umbilical view (upper) and cross section. **57.** *Helisoma anceps* (Planorbidae) in umbilical view (upper) and cross section. **58.** *Planorbella trivolvis* (Planorbidae), in umbilical view (upper) and cross section. vv, ventral view. (48, 49, 52–55, 58 [upper] redrawn from Burch 1982; 50, 56–58 [lower] modified or redrawn from Harman and Berg 1971; 51, 56, 57 [upper] redrawn from Jokinen 1983.)

55b. Shell small to large, but without ribs; common and widespread 56

56a (55b). Adult shell moderately large (≥ 4.5 mm diam.), with internal lamellae ("teeth") projecting from the inner walls of the body whorl (look *deep* into the aperture!) (Fig. 60); typically in temporary pools and marshes .... *Planorbula armigera*

56b. Adult shell small to large, but without lamellae 57

57a (56b). Shell very flat (H/diam. ≤ 0.33), with carina on shell circumference; sutures shallow (Fig. 61) .... *Promenetus exacuous*

57b. Shell not with this combination of characters 58

58a (57b). Adult shell very small (usually ≤ 3 mm diameter), relatively high (H/diam. usually ≥ 0.4), flattened on top and rounded on bottom (Fig. 62) .... *Menetus dilatatus*

58b. Adult shell ≥ 3 mm in diameter (usually ca. 5 mm), flat (H/diam. usually ≤ 0.4), and slightly rounded on top and bottom (Figs. 64, 65) ...... *Gyraulus* 59

59a (58b). Shell relatively large (4–7 mm diam. as adults), often with spiral lines of periostracal hairs (most apparent on dry shells), malleations, or a marginal carina (Fig. 63) .... *G. deflectus*

59b. Shell relatively small (usually ≤ 5 mm diam.), without spiral lines of periostracal hairs, malleations, or carina 60

60a (59b). Whorls increasing in size slowly (Fig. 64); shell usually white or yellow and semitransparent; apical and umbilical sides of the shell nearly identical; usually in temporary waters .... *G. circumstriatus*

60b. Whorls increasing in size rapidly (Fig. 65); shell usually brown and translucent but not transparent; apical side of the shell raised and obviously different from the umbilical side; in permanent and sometimes in temporary waters .... *G. parvus*

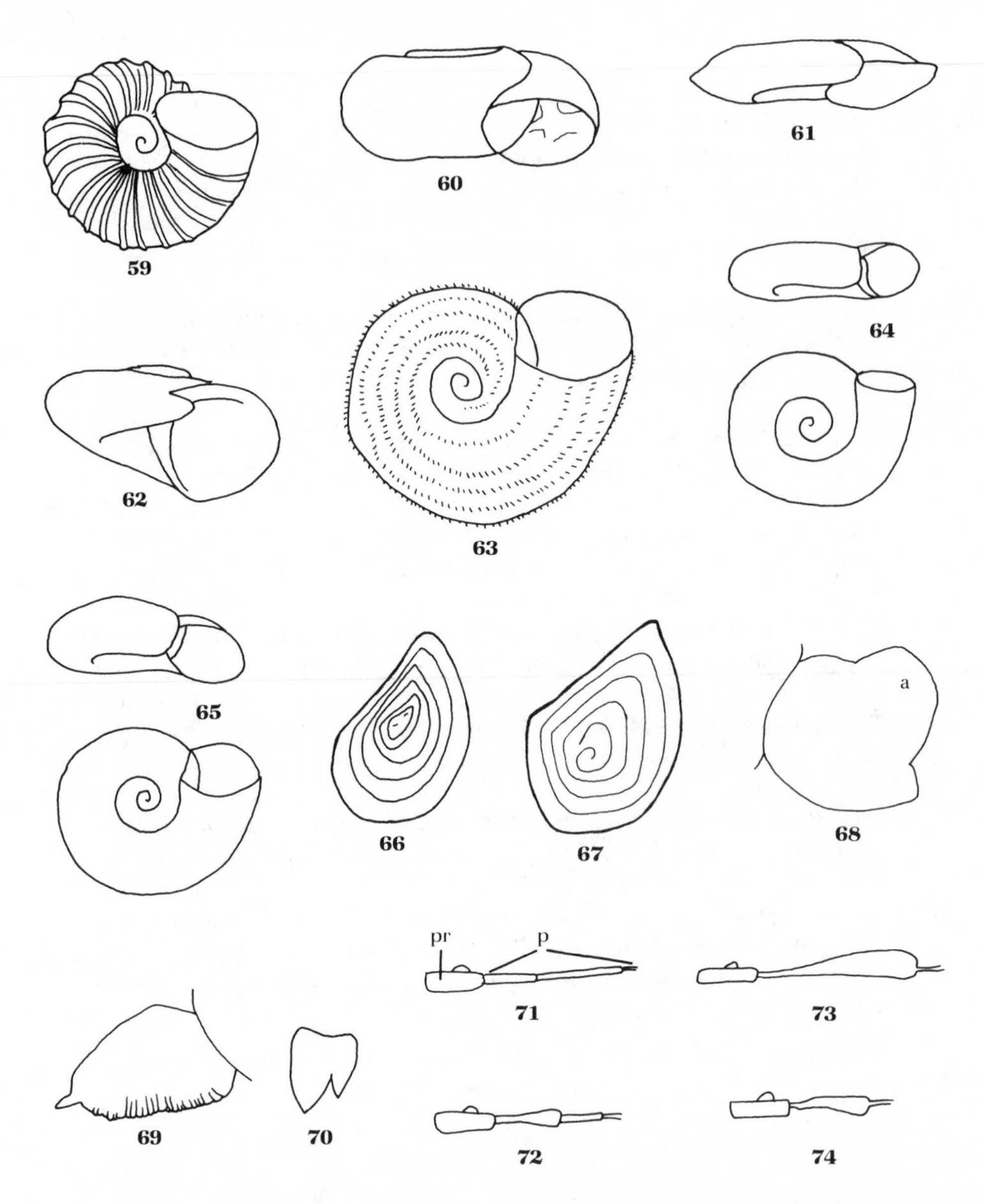

59. *Armiger crista* (Planorbidae), uv. **60.** *Planorbula armigera* (Planorbidae), vv. **61.** *Promenetus exacuous* (Planorbidae), vv. **62.** *Menetus dilatatus* (Planorbidae), vv. **63.** *Gyraulus deflectus* (Planorbidae), uv. **64.** *Gyraulus circumstriatus* (Planorbidae), vv (upper), uv. **65.** *Gyraulus parvus* (Planorbidae), uv (lower), vv. **66.** Operculum of *Campeloma decisum* (Viviparidae), vv. **67.** Operculum of *Lioplax subcarinata* (Viviparidae), vv. **68.** Male organ of *Birgella subglobosa* (Hydrobiidae), lv: a, accessory lobe. **69.** Male organ of *Gillia altilis* (Hydrobiidae), lv. **70.** Lateral tooth of *Fossaria dalli* (Lymnaeidae). **71.** Penial complex of *Physella gyrina* (Physidae), lv: p, penial sheath; pr, preputium. **72.** Penial complex of *Physella vinosa* (Physidae), lv. **73.** Penial complex of *Physa vernalis* (Physidae), lv. **74.** Penial complex of *Physella integra* (Physidae), lv. lv, lateral view; uv, umbilical view; vv, ventral view. (59–62, 70 redrawn from Burch 1982; 63–65 redrawn from Jokinen 1983; 68, 69 modified from Thompson 1984; 71–74 redrawn from Te 1975.)

Key to Species of Freshwater Bivalvia

1a. Shell always < 50 mm long and almost always < 20 mm long; hinge teeth consist of a set of cardinal teeth near the beaks, flanked on *both* sides by lateral teeth (Fig. 75); interior of shell usually dull bluish grey . . . . Superfamily SPHAERACEA 2

1b. Shell often > 50 mm long; hinge teeth consist of a set of pseudocardinal teeth near the beaks, flanked on *one* side by lateral teeth (Fig. 76), either or both of these sets of teeth may be vestigial or lacking; interior of shell usually a lustrous white, purple, orange, or pink . . . . Superfamily UNIONACEA 5

2a (1a). SPHAERACEA: Lateral teeth serrated (Fig. 77); exterior of shell with coarse growth lines (Fig. 78); shell often > 20 mm long; an introduced species that has just recently reached the northeastern United States . . . . **Corbiculidae**, *Corbicula fluminea*

2b. Lateral teeth smooth; growth lines usually not especially coarse; shell never > 20 mm long; very common and widespread . . . . **Sphaeriidae** 3

3a (2b). **Sphaeriidae**: Beaks posterior to center of shell, usually obviously so (Fig. 79); shell small (< 12 mm long) . . . . *Pisidium*

3b. Beaks near center of shell or slightly anterior; shell up to 20 mm long 4

4a (3b). Beaks "capped" (separated from the remainder of the shell by a distinct suture) (Fig. 80); shell often yellowish and translucent . . . . *Musculium*

4b. Beaks not capped (Fig. 81); shell usually brown or gray . . . . *Sphaerium*

5a (1b). UNIONACEA: Shell with obvious pustules (Figs. 83, 84) or ridges (Fig. 82); species of central and western New York . . . . **Unionidae** (in part) 6

5b. Shell smooth or a little rough, but without distinct pustules or ridges; widespread 9

6a (5a). **Unionidae** (in part): Nacre purple . . . . *Cyclonaias tuberculata*

6b. Nacre white 7

7a (6b). Shell with ridges (Fig. 82) . . . . *Amblema plicata*

7b. Shell with pustules (Figs. 83, 84) . . . . *Quadrula* 8

8a (7b). Shell with prominent median sulcus, giving the shell a squarish shape (Fig. 83) . . . . *Q. quadrula*

8b. Shell evenly rounded, usually with a broad green color ray on beaks (Fig. 84) . . . . *Q. pustulosa*

9a (5b). Obvious hinge teeth absent . . . . **Unionidae** (in part) 10

9b. Obvious hinge teeth present 15

10a (9a). **Unionidae** (in part): Beaks not projecting above the hinge line (Fig. 85); a species of western and central New York . . . . *Anodonta imbecilis*

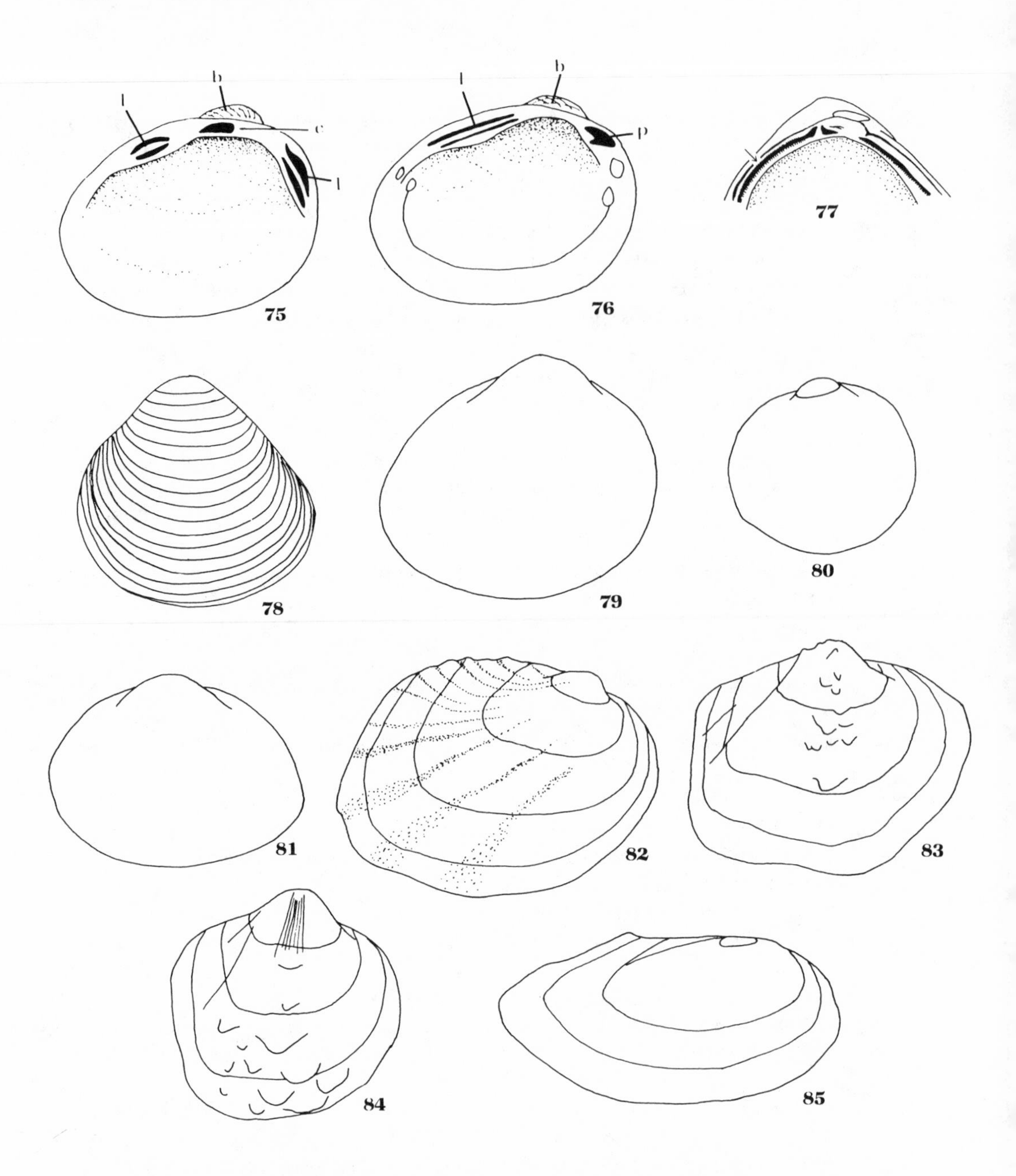

75. Interior of a sphaeriid shell: b, beak; c, cardinal teeth; l, lateral teeth. **76.** Interior of a unionid shell: b, beak; l, lateral teeth; p, pseudocardinal teeth. **77.** *Corbicula fluminea* (Corbiculidae), with detail of hinge teeth, lv. **78.** *Corbicula fluminea* (Corbiculidae), lv. **79.** *Pisidium* (Sphaeriidae), lv. **80.** *Musculium* (Sphaeriidae), lv. **81.** *Sphaerium* (Sphaeriidae), lv. **82.** *Amblema plicata* (Unionidae), lv. **83.** *Quadrula quadrula* (Unionidae), lv. **84.** *Quadrula pustulosa* (Unionidae), lv. **85.** *Anodonta imbecilis* (Unionidae), lv. lv, lateral view. (77–81 modified from Burch 1975a; 82–85 modified from Burch 1975b.)

10b. Beaks projecting above hinge line (Fig. 86); widespread 11

11a (10b). Beak sculpture double-looped (seagull-shaped) (Figs. 86, 87) . . . . *Anodonta* (in part) 12

11b. Beak sculpture concentric (Figs. 88, 89) 14

12a (11a). Nacre salmon or copper-colored; shell thick; in or near tidewaters . . . . *A. implicata*

12b. Nacre bluish or white; widespread 13

13a (12b). Beak sculpture nodulous (Fig. 86); periostracum usually brown or yellowish; common in the Allegheny, Lake Erie, Lake Champlain, Lake Ontario, and parts of the Hudson basins, rare or absent elsewhere in the northeastern United States . . . . *A. grandis*

13b. Beak sculpture not nodulous (Fig. 87); periostracum usually greenish; absent from the Allegheny and Lake Erie basins, but common and widespread elsewhere in the northeastern United States . . . . *A. cataracta*

A. grandis and *A. cataracta* are difficult to distinguish and have been shown by Kat (1983a) to hybridize in our area.

14a (11b). Beak sculpture coarse (Fig. 88); nacre usually orange in the beak cavity; pseudocardinal teeth usually represented by a faint thickening of the nacre near the beaks . . . . *Strophitus undulatus*

14b. Beak sculpture fine (Fig. 89); nacre usually bluish white; pseudocardinal teeth absent . . . . *Anodontoides ferussacianus*

15a (9b). Lateral teeth well developed, functional and interlocking . . . . **Unionidae** (in part) 22

15b. Lateral teeth absent or reduced, not functional and not interlocking 16

16a (15b). Shell elongate (H/L ca. 0.50) and often arched (Fig. 90); periostracum without color rays; lateral teeth entirely or almost entirely absent 17

16b. Shell shape variable (H/L = 0.48–0.70); periostracum usually with numerous fine color rays; vestigial lateral teeth present . . . . **Unionidae** (in part) 18

17a (16a). Shell large (adults $>$ 75 mm long); periostracum blackish; scattered in cool, softwater streams from Oneida Lake northeast (Fig. 90) . . . . **Margaritiferidae**, *Margaritifera margaritifera*

17b. Shell small (adults $<$ 60 mm long); periostracum grey to brown; a very rare species formerly found in the northeastern United States near Buffalo, N.Y. . . . . **Unionidae** (in part), *Simpsoniconcha ambigua*

18a (16b). **Unionidae** (in part): Transverse ridges on the posterior slope prominent (Fig. 91); shell compressed (W/H $<$ 0.6) . . . . *Lasmigona costata*

18b. Transverse ridges on the posterior slope absent or fine; shell inflated (W/H usually $>$ 0.6) . . . . *Alasmidonta* (in part) 19

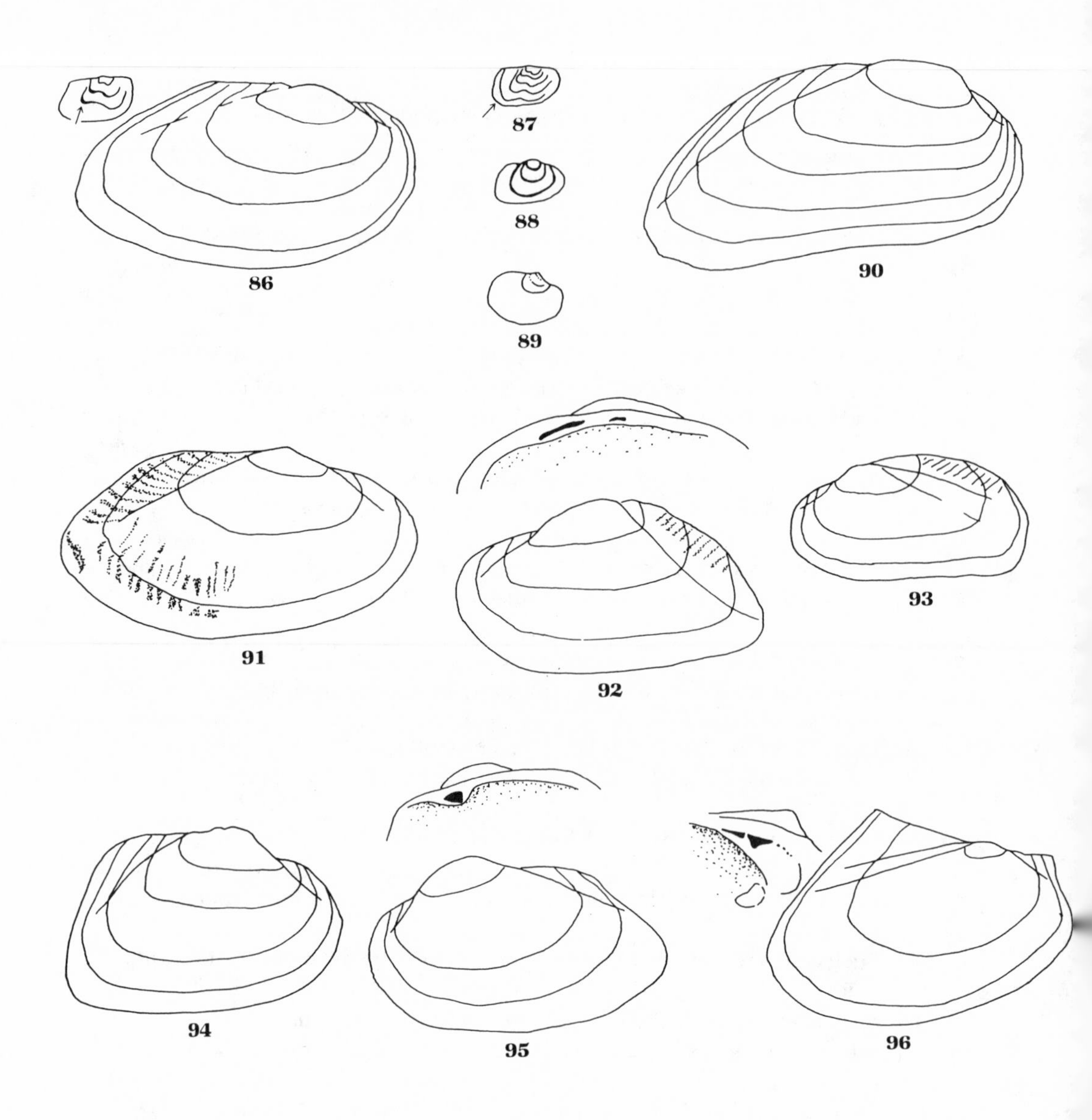

86. *Anodonta grandis* (Unionidae) and beak sculpture, lv. **87.** Beak sculpture of *Anodonta cataracta* (Unionidae), lv. **88.** Beak sculpture of *Strophitus undulatus* (Unionidae), lv. **89.** Beak sculpture of *Anodontoides ferussacianus* (Unionidae), lv. **90.** *Margaritifera margaritifera* (Margaritiferidae), lv. **91.** *Lasmigona costata* (Unionidae), lv. **92.** *Alasmidonta marginata* (Unionidae), and hinge, lv. **93.** *Alasmidonta varicosa* (Unionidae), lv. **94.** *Alasmidonta viridis* (Unionidae), lv. **95.** *Alasmidonta undulata* (Unionidae) and hinge, lv. **96.** *Leptodea fragilis* (Unionidae) and hinge, lv. lv, lateral view. (86 [lower], 90, 91, 94, 96 modified from Burch 1975b; 86 [upper], 87–89 modified from Clarke and Berg 1959; 92 [lower], 93 modified from Ortmann 1919; 92 [upper], 95 modified from Clarke 1981b.)

19a (18b). Fine transverse ridges present on the posterior slope (Figs. 92, 93); pseudocardinal teeth reduced and elongate, with smooth surfaces (Fig. 92) 20

19b. Transverse ridges on posterior slope absent (Figs. 94, 95); pseudocardinal teeth strong and triangular, with rough surfaces (Fig. 95) 21

20a (19a). Posterior ridge angular and prominent; shell truncate and small to large (Fig. 92); found in scattered sites east to Albany, N.Y. . . . . *A. marginata*

20b. Posterior ridge rounded; shell rounded and usually < 70 mm long (Fig. 93); found in streams of the Atlantic drainage . . . . *A. varicosa*

21a (19b). Shell < 50 mm long; subrhomboid (Fig. 94); uncommon and found in the northeastern United States only in western and central New York . . *A. viridis*

21b. Shell small to large; triangular to ovate (Fig. 95); widespread and common in the northeastern United States except in western New York . . . . . *A. undulata*

22a (15a). **Unionidae** (in part): Right valve with two lateral teeth; rare . . . . *Alasmidonta heterodon*

22b. Right valve with one lateral tooth; common 23

23a (22b). Shell compressed, with prominent dorsal wing (Fig. 96) ("heel-splitters"); species of larger lakes and rivers, northeast to Lake Champlain 24

23b. Shell without dorsal wing 25

24a (23a). Pseudocardinal teeth reduced, thin, and bladelike (Fig. 96); nacre white to pink or purple . . . . *Leptodea* 52

24b. Pseudocardinal teeth strong and triangular (Fig. 97); nacre purple or pink . . . . *Proptera alata*

25a (23b). H/L ≤ 0.48 . . . . *Ligumia* 26

25b. H/L > 0.48 27

26a (25a). Posterior ridge prominent; posterior end of shell subangular; pseudocardinal teeth elongate (Fig. 98); shell usually < 110 mm long . . . . *L. nasuta*

26b. Posterior ridge not prominent; posterior end of shell rounded; pseudocardinal teeth triangular (Fig. 99); shell often > 110 mm long . . . . *L. recta*

27a (25b). Nacre purple . . . . *Elliptio* (in part) 28

27b. Nacre white or colored, but not purple 29

28a (27a, 39b). Shell subrhomboid with well-defined posterior ridge and slope (Fig. 100); very common and widespread in the northeastern United States, but absent from western New York . . . . *E. complanata*

28b. Shell subelliptical with posterior ridge rounded and following the hinge line so that the posterior slope is small and poorly defined (Fig. 101); common in western New York, but rare elsewhere in the northeastern United States . . . . *E. dilatata*

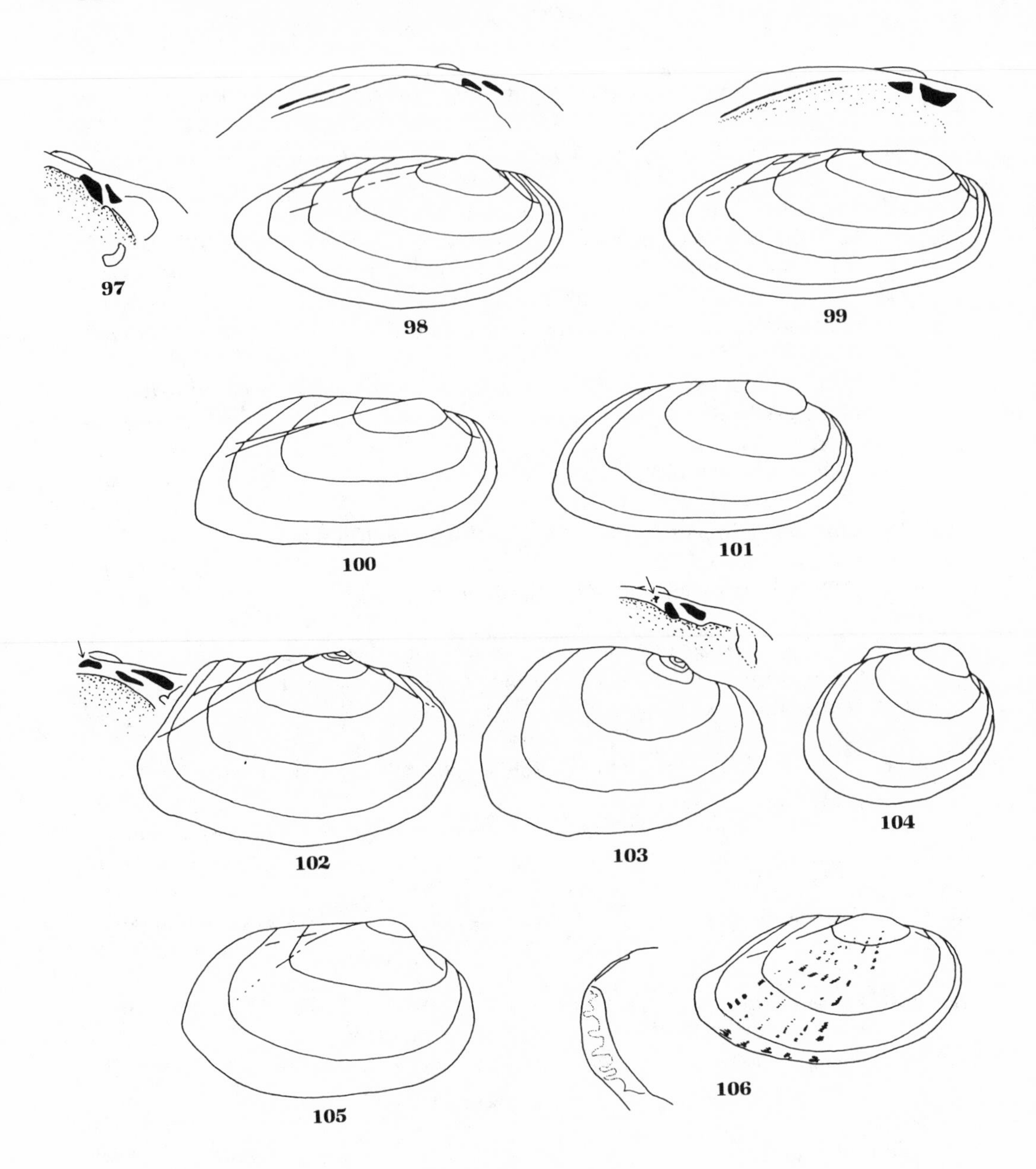

97. Hinge of *Proptera alata* (Unionidae), lv. **98.** *Ligumia nasuta* (Unionidae) and hinge, lv. **99.** *Ligumia recta* (Unionidae) and hinge, lv. **100.** *Elliptio complanata* (Unionidae), lv. **101.** *Elliptio dilatata* (Unionidae), lv. **102.** *Lasmigona compressa* (Unionidae) and hinge, lv. **103.** *Lasmigona subviridus* (Unionidae) and hinge, lv. **104.** *Obovaria olivaria* (Unionidae), lv. **105.** *Carunculina parva* (Unionidae), lv. **106.** *Villosa iris* (Unionidae) and mantle edge, lv. lv, lateral view. (97–106 modified from Burch 1975b.)

29a (27b). With interdental tooth, so that the left valve appears to have 3 pseudocardinal teeth (Figs. 102, 103); shell more or less compressed and subrhomboid; periostracum dark green with numerous color rays, or brown .... *Lasmigona* (in part) 30

29b. No interdental tooth, so that the left valve appears to have 2 pseudocardinal teeth 31

30a (29a). Adult shell usually > 70 mm long; interdental tooth prominent (Fig. 102); widespread east to Lake Champlain and the Hudson River basin .... *L. compressa*

30b. Adult shell < 65 mm long; interdental tooth small (Fig. 103); from sites in the Atlantic drainage and along the Erie Canal west to Syracuse, N.Y. .... *L. subviridus*

31a (29b). Shell ovate to elliptical with beaks very near the anterior end (distance from beaks to anterior end less than ⅕ of the shell length) (Fig. 104); shell with color rays; only parts of outer demibranchs swollen in gravid females (see Fig. 125); hinge teeth massive; a rare species of the St. Lawrence and Niagara rivers .... *Obovaria olivaria*

31b. Not with the above combination of characters; common and widespread 32

32a (31b). Adult shell small (< 35 mm long); periostracum green to black without color rays (Fig. 105); a species of western New York .... *Carunculina parva*

32b. Not with the above combination of characters; widespread 33

33a (32b). Collected from western New York (Allegheny River, Lake Erie, and Niagara River basins) 40

33b. Collected from other drainage systems 34

34a (33b). Outer 2 demibranchs swollen in gravid females (see Fig. 125); periostracum usually smooth and glossy, yellow or green, often with color rays 35

34b. All 4 demibranchs swollen in gravid females (see Fig. 125); periostracum not smooth and glossy, usually brown, color rays usually absent, but if present they are fine 39

35a (34a). Adult shell small (< 75 mm long) and delicate; shell subelliptical; yellow-green with broad color rays; nacre white, often with an iridescent sheen; posterior mantle margins with fingerlike projections (Fig. 106); St. Lawrence basin .... *Villosa iris*

35b. Adult shell often > 75 mm long and not delicate; shell subovate (Figs. 107–114); nacre white, bluish, or pink; posterior mantle margins without fingerlike projections; widespread .... *Lampsilis* (in part) 36

36a (35b). Shell small (usually < 80 mm long) and thin, hardly thicker anteriorly than posteriorly; periostracum dull yellow without rays or with fine rays all over the shell (Figs. 107, 108); in or near tidewaters .... *L. ochracea*

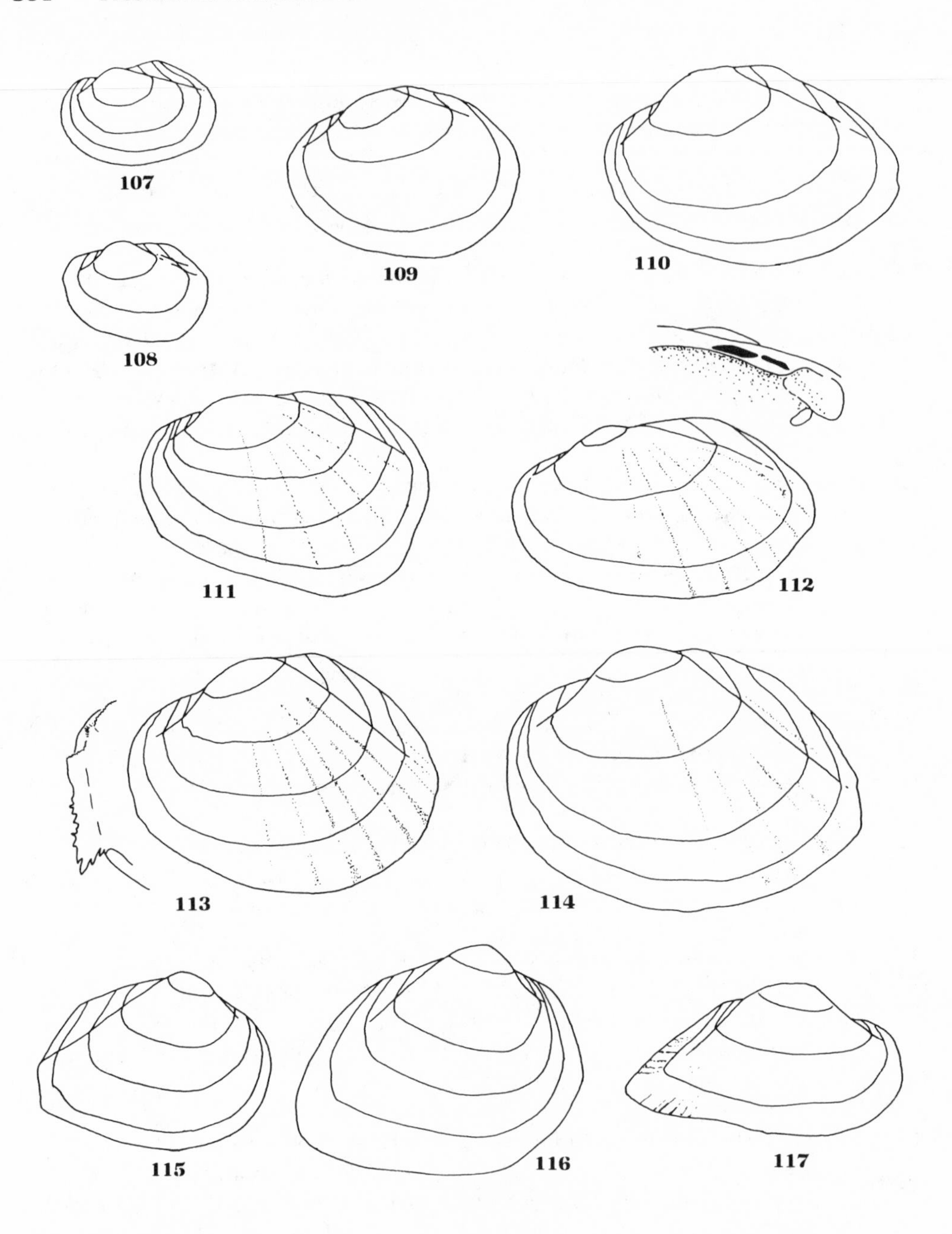

107. *Lampsilis ochracea* (male) (Unionidae), lv. **108.** *Lampsilis ochracea* (female) (Unionidae), lv. **109.** *Lampsilis cariosa* (female) (Unionidae), lv. **110.** *Lampsilis cariosa* (male) (Unionidae), lv. **111.** *Lampsilis radiata* (female) (Unionidae), lv. **112.** *Lampsilis radiata* (male) (Unionidae) and hinge, lv. **113.** *Lampsilis ovata* (female) (Unionidae) and mantle, lv. **114.** *Lampsilis ovata* (male) (Unionidae), lv. **115.** *Fusconaia flava* (Unionidae), lv. **116.** *Pleurobema cordatum* (Unionidae), lv. **117.** *Dysnomia triquetra* (female) (Unionidae), lv. lv, lateral view. (107–111, 112 [lower], 113 [right], 114 modified from Ortmann 1919; 112 [upper], 113 [left], 115–117 modified from Burch 1975b.)

36b. Shell often > 80 mm long, much thicker anteriorly than posteriorly; may have obvious broad color rays; widely distributed 37

37a (36b). Shell without color rays (Figs. 109, 110), or with color rays only on the posterior slope .... *L. cariosa*

37b. Shell with color rays all over the shell (although they may be obscure in old specimens) 38

38a (37b). H/L < 0.60 in males and in most females; posterior ridge low and rounded; beaks not prominent (Figs. 111, 112); common and widespread .... *L. radiata* group

L. siliquoidea is a species or subspecies closely allied to and difficult to distinguish from *L. radiata*. See Clarke and Berg 1959 and Kat 1986 for notes on the distribution, identification, and taxonomic status of these two taxa.

38b. H/L > 0.60 in both sexes; posterior ridge usually well developed; beaks often prominent (Figs. 113, 114); western New York, the Lake Champlain and St. Lawrence River basins, and rarely in the Lake Ontario and Hudson River basins .... *L. ovata*

39a (34b). Shell subtriangular (H/L > 0.70) (Fig. 115) and thick, with massive hinge teeth; western New York .... *Fusconaia flava*

39b. H/L < 0.70, shell subrhomboid to subelliptical (Figs. 100, 101) and not especially thick; widespread and common .... *Elliptio* (in part) 28

40a (33a). Periostracum usually dull, brown, and without color rays (rarely, dark green with many fine color rays); entire outer demibranchs swollen in gravid females (see Fig. 125); inner demibranchs may be swollen as well 41

40b. Periostracum often shiny, usually yellow or light green, usually with obvious green color rays; only parts of outer demibranchs swollen in gravid females (see Fig. 125) 43

41a (40a). Shell elongate (H/L < 0.60) .... *Elliptio dilatata*

41b. Shell short and high (H/L > 0.60) 42

42a (41b). All 4 demibranchs swollen in gravid females (see Fig. 125); shell subtriangular to subquadrate, usually flattened or with a slight median sulcus and posterior slope often sharply set off; beaks relatively central (distance from beaks to anterior margin of shell usually ⅕–⅖ of shell length) (Fig. 115) .... *Fusconaia flava*

42b. Outer 2 demibranchs swollen in gravid females (see Fig. 125); shell subtriangular to subovate, usually rounded and posterior slope not sharply set off; beaks relatively anterior (distance from beaks to anterior margin of shell usually 1/20–¼ of shell length) (Fig. 116) .... *Pleurobema cordatum*

It is difficult for a novice to distinguish the shell of *F. flava* from that of *P. cordatum*; gravid females should be examined for a positive identification.

43a (40b). Shell small (usually < 60 mm long); usually subtriangular with a prominent and sometimes angular posterior ridge (Figs. 117–120); rare species of Lake Erie and large streams 44

43b. Shell small to large; subovate, subelliptical, or subrhomboid (Figs. 111–114); posterior ridge rarely prominent and never angular; widespread 46

44a (43a). Posterior slope finely corrugated (Figs. 117, 118); posterior ridge of female ending in raised radiating lines (Fig. 117) . . . . . *Dysnomia triquetra*

44b. Posterior slope smooth, not corrugated . . . . . *Truncilla* 45

45a (44b). Shell subtriangular, sharp, straight, posterior ridge extends to the shell margin; posterior slope small and steep (Fig. 120) . . . . . *T. truncata*

45b. Shell subtriangular to subovate; posterior ridge fades out before it reaches the margin of the shell and often is bowed downward; posterior slope well developed (Fig. 119) . . . . . *T. donaciformis*

46a (43b). Adult shell small (< 40 mm long) and thick; lateral teeth short and stout; periostracum usually green or brown with numerous fine color rays (Fig. 121); posterior mantle margins with long fingerlike projections (as in Fig. 106); rare . . . . . *Villosa fabalis*

46b. Not with the above combination of characters 47

47a (46b). Shell subrhomboid to subelliptical, compressed (W/L = 0.18–0.33); lateral teeth short, stout, and swollen posteriorly (Fig. 122); entire ventral part of demibranch swollen in gravid females . . . . . *Ptychobranchus fasciolare*

47b. Shell subovate to subelliptical (Figs. 111–114), compressed to inflated (W/L = 0.20–0.50); lateral teeth not as above; posterior part of outer demibranch swollen in gravid females 48

48a (47b). Shell small (< 75 mm), compressed (W/L < 0.35); subelliptical and relatively elongate (H/L < 0.60) and thin; hinge teeth fine and delicate; posterior mantle margins with fingerlike projections (Fig. 106) . . . . . *Villosa iris*

48b. Shell small to large, compressed to inflated (W/L = 0.25–0.50); subovate (H/L = 0.52–0.80) and thin to thick; hinge teeth fine to stout; posterior mantle margins without fingerlike projections 49

49a (48b). Pseudocardinal teeth large and triangular, with ragged surfaces; shell thick; posterior mantle margins without ribbonlike projections; shell relatively compressed (W/L = 0.25–0.40) and elongate (H/L = 0.54–0.70) (Fig. 123); in rivers and large creeks . . . . . *Actinonaias carinata*

49b. Pseudocardinal teeth not triangular, but flexed upward and with relatively smooth surfaces (Fig. 112); posterior mantle margins with ribbonlike flaps that may resemble a small fish (Fig. 113); shell various shapes (W/L = 0.25–0.50; H/L = 0.52–0.80), thin to thick; common and widespread in streams and lakes . . . . . *Lampsilis* (in part) 50

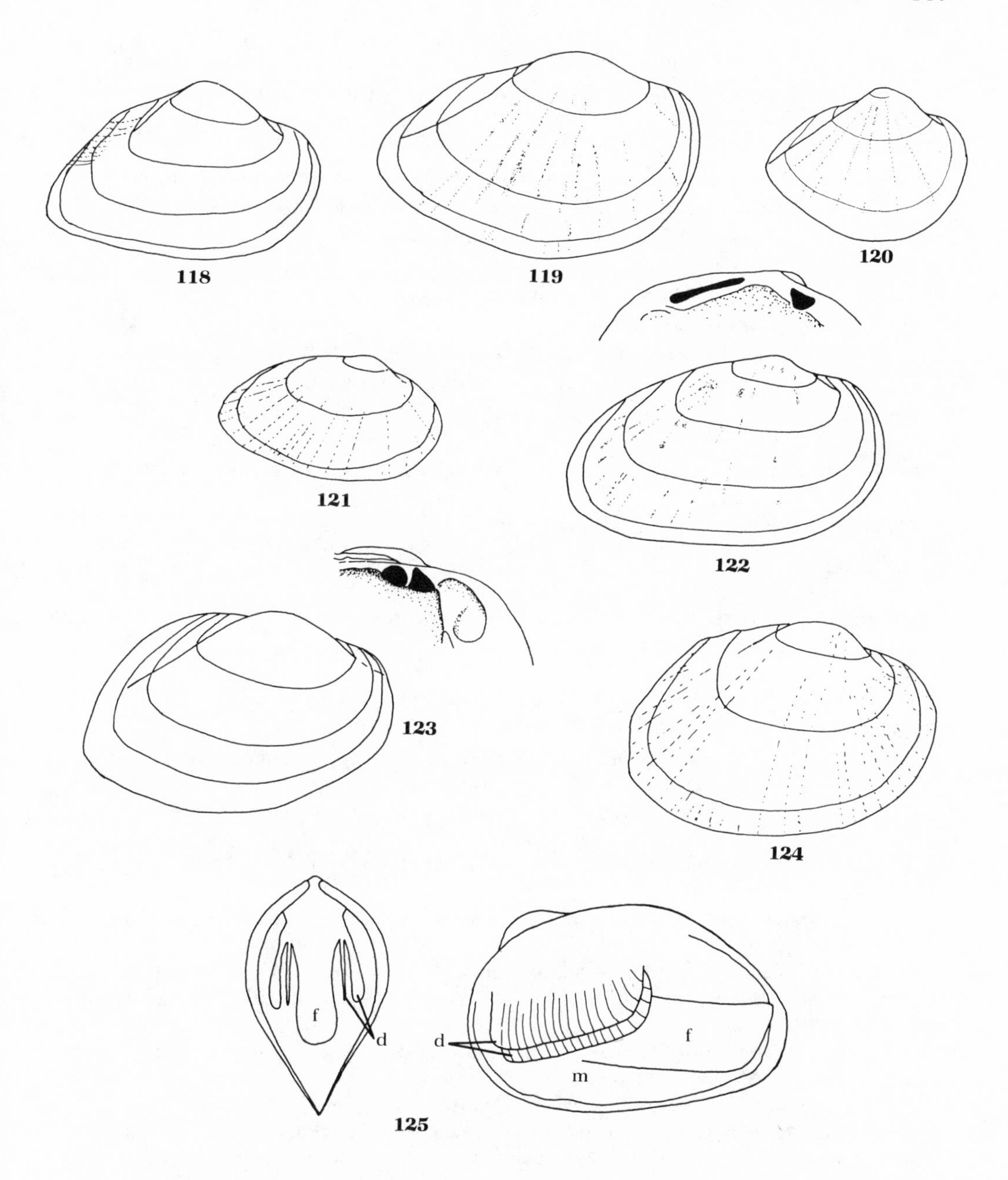

118. *Dysnomia triquetra* (male) (Unionidae), lv. **119.** *Truncilla donaciformis* (Unionidae), lv. **120.** *Truncilla truncata* (Unionidae), lv. **121.** *Villosa fabalis* (Unionidae), lv. **122.** *Ptychobranchus fasciolare* (Unionidae) and hinge, lv. **123.** *Actinonaias carinata* (Unionidae) and hinge, lv. **124.** *Lampsilis fasciola* (Unionidae), lv. **125.** Diagrammatic sketch of the internal anatomy of unionid in cross section (left) and lateral view (right). The outer demibranchs on the left are swollen and filled with developing glochidia: d, demibranchs; f, foot; m, mantle. lv, lateral view. (118, 119, 121–124 modified from Burch 1975b.)

50a (49b). Shell elongate and relatively compressed (H/L $\leq$ 0.625 in almost all males and some females; W/L = 0.25–0.40); posterior ridge low and rounded; beaks not prominent (Figs. 111, 112) . . . . *L. radiata*
See comments under 38a.

50b. Shell relatively short and inflated (H/L $\geq$ 0.625 in all males and females; W/L = 0.30–0.50); posterior ridge may be well developed; beaks often prominent 51

51a (50b). Shell small (< 80 mm), with numerous fine green color rays; beaks not especially prominent (Fig. 124) . . . . *L. fasciola*

51b. Shell small to large, with color rays usually present, but not especially numerous or fine; beaks and posterior ridge often prominent (Figs. 113, 114) . . . . *L. ovata*

52a (24a). Nacre white (sometimes pink in the beak cavity); periostracum smooth but dull . . . . *L. fragilis*

52b. Nacre pink or purple; periostracum smooth and glossy . . . . *L. laevissima*

References on Freshwater Mollusca Systematics

(*Used in construction of keys.)

Athearn, H.D., and A.H. Clarke. 1962. The freshwater mussels of Nova Scotia. Natl. Mus. Can. Bull. 183:11–41.

Baker, F.C. 1911. The Lymnaeidae of North and Middle America. Spec. Publ. Chicago Acad. Sci. 3:1–539.

——. 1916. The relation of mollusks to fish in Oneida Lake. N.Y. State Coll. For. Tech. Publ. 4:1–366.

——. 1928a. The freshwater Mollusca of Wisconsin. Part I. Gastropoda. Bull. Wisc. Acad. Sci. Arts Lett. 70(1):1–507.

——. 1928b. The freshwater Mollusca of Wisconsin. Part II. Pelecypoda. Bull. Wisc. Acad. Sci. Arts. Lett. 70(2):1–495.

——. 1945. The molluscan family Planorbidae. Univ. Ill. Press, Urbana. 530 pp.

*Basch, P.F. 1963. A review of the recent limpet snails of North America. Harv. Univ. Mus. Comp. Zool. Bull. 129:401–461.

*Berry, E.G. 1943. The Amnicolidae of Michigan: distribution, ecology, and taxonomy. Misc. Publ. Univ. Mich. Mus. Zool. 57:1–68.

Brinkhurst, R.O. 1974. The benthos of lakes. St. Martin's Press, New York. 190 pp.

Buckley, D.E. 1977. The distribution and ecology of the aquatic molluscan fauna of the Black River drainage basin in northern New York. Occas. Pap. Biol. Field Stn., State Univ. N.Y. Oneonta 6:1–276.

Buckley, D.E., and W.N. Harman. 1980. *Cipangopaludina chinensis* in the Adirondack Mountains of New York. Annu. Rep. Biol. Field Stn., State Univ. N.Y. Oneonta 13:55.

*Burch, J.B. 1975a. Freshwater sphaeracean clams (Mollusca: Pelecypoda) of North America. Rev. ed. Malacological Publ., Hamburg, Mich. 99 pp.

*——. 1975b. Freshwater unionacean clams (Mollusca: Pelecypoda) of North America. Rev. ed. Malacological Publ., Hamburg, Mich. 204 pp.

*——. 1982. Freshwater snails (Mollusca: Gastropoda) of North America. EPA-600/3-82-026. 294 pp.

Burch, J.B., and C.M. Patterson. 1976. Key to the genera of freshwater pelecypods of Michigan. Univ. Mich. Mus. Zool. Circ. 4:1–38.

Call, R.E. 1900. A descriptive illustrated catalog of the Mollusca of Indiana. Ind. Dep. Geol. Nat. Res. Annu. Rep. 24:335–535.

*Clarke, A.H. 1973. The freshwater mollusks of the Canadian Interior Basin. Malacologia 13:1–509.

*——. 1981a. The freshwater molluscs of Canada. Natl. Mus. Nat. Sci., Natl. Mus. Can., Ottawa. 446 pp.

*——. 1981b. The tribe Alasmidontini. Part I. *Pegias*, *Alasmidonta*, and *Arcidens*. Smithson. Contrib. Zool. 326:1–101.

——. 1982. Discussion of Thomas Freitag's presentation. *In* A.C. Miller (compiler). Report of freshwater mollusks workshop, p. 155. U.S. Army Engineers Waterway Experiment Station, Vicksburg, Miss.

——. 1985. The tribe Alasmidontini (Unionidae: Anodontinae). Part II. *Lasmigona* and *Simpsonais*. Smithson. Contrib. Zool. 399:1–75.

*Clarke, A.H., and C.O. Berg. 1959. The freshwater mussels of central New York. Mem. Cornell Univ. Agric. Exp. Stn. 367:1–79.

Clench, W.J. 1959. Mollusca. *In* W.T. Edmondson (ed.). Fresh-water biology, 2nd ed., pp. 1117–1160. John Wiley and Sons, New York.

——. 1962a. A catalog of the Viviparidae of North America with notes on the distribution of *Viviparus georgianus* Lea. Occas. Pap. Mollusks. Harv. Univ. Mus. Comp. Zool. 2:261–287.

——. 1962b. New records of the genus *Lioplax*. Occas. Pap. Mollusks. Harv. Univ. Mus. Comp. Zool. 2:288.

Clench, W.J., and S.L.H. Fuller. 1965. The genus *Viviparus* in North America. Occas. Pap. Mollusks. Harv. Univ. Mus. Comp. Zool. 2:385–412.

Clench, W.J., and R.J. Turner. 1955. The North American genus *Lioplax* in the family Viviparidae. Occas. Pap. Mollusks. Harv. Univ. Mus. Comp. Zool. 2:1–20.

Coker, R.E., A.F. Shira, H.W. Clark, and A.D. Howard. 1921. Natural history and propagation of freshwater mussels. Bull. U.S. Bur. Fish. 37:75–181.

Davis, G.M. 1982. Historical and ecological factors in the evolution, adaptive radiation, and biogeography of freshwater molluscs. Amer. Zool. 22:375–395.

——. 1984. Genetic relationships among some North American Unionidae (Bivalvia): sibling species, convergence, and cladistic relationships. Malacologia 25:629–648.

Davis, G.M., and S.L.H. Fuller. 1981. Genetic relationships among recent Unionacea of North America. Malacologia 20:217–253.

DeKay, C. 1844. Zoology of New York, or the New York fauna. Part V. Mollusca. Carroll and Cook, Albany, N.Y. 271 pp.

Foehrenbach, J., and D. Raihle. 1984. A further range extension of the Asiatic clam. N.Y. Fish Game J. 31:224–226.

Fuller, S.L.H. 1974. Clams and mussels (Mollusca: Bivalvia). *In* C.W. Hart and S.L.H. Fuller (eds.). Pollution ecology of freshwater invertebrates, pp. 215–273. Academic Press, N.Y.

Goodrich, C. 1932. The Mollusca of Michigan. Univ. Mich. Mus. Zool. Handbook Ser. 5:1–120.

Goodrich, C., and H. van der Schalie. 1939. Aquatic mollusks of the Upper Peninsula of Michigan. Misc. Publ. Univ. Mich. Mus. Zool. 43:1–45.

——. 1944. A revision of the Mollusca of Indiana. Amer. Midl. Nat. 32:257–326.

Harman, W.N. 1968a. Replacement of pleurocerids by *Bithynia* in polluted waters of central New York. Nautilus 81:77–83.

——. 1968b. Interspecific competition between *Bithynia* and pleurocerids. Nautilus 82:72–73.

——. 1970a. *Anodontoides ferussacianus* (Lea) in the Susquehanna River watershed in New York State. Nautilus 83:114–115.

——. 1970b. New distribution records and ecological notes on central New York Unionacea. Amer. Midl. Nat. 84:46–58.

——. 1973. The Mollusca of Canadarago Lake and a new record for *Lasmigona compressa* (Lea). Nautilus 87:114.

——. 1974. Snails (Mollusca: Gastropoda). *In* C.W. Hart and S.L.H. Fuller (eds.). Pollution ecology of freshwater invertebrates, pp. 275–312. Academic Press, N.Y.

——. 1982. Pictorial keys to the aquatic mollusks of the upper Susquehanna. Occas. Pap. Biol. Field Stn., State Univ. N.Y. Oneonta 9:1–13.

*Harman, W.N., and C.O. Berg. 1971. The freshwater snails of central New York with illustrated keys to the genera and species. Search: Cornell Univ. Agric. Exp. Stn., Entomol. 1(4):1–68.

Harman, W.N., and J.L. Forney. 1970. Fifty years of change in the molluscan fauna of Oneida Lake, New York. Limnol. Oceanogr. 15:454–460.

*Heard, W.H. 1982. Family Valvatidae. *In* J.B. Burch (ed.). Freshwater snails (Mollusca: Gastropoda) of North America, pp. 67–74. EPA-600/3-82-026.

Heard, W.H., and R.H. Guckert. 1970. A re-evaluation of the recent Unionacea of North America. Malacologia 10:333–355.

Hendelberg, J. 1960. The freshwater pearl mussel, *Margaritifera margaritifera* (L.). Rep. Inst. Freshwat. Res. Drottnigholm 41:149–171.

*Herrington, H.B. 1962. A revision of the Sphaeriidae of North America. Misc. Publ. Univ. Mich. Mus. Zool. 118:1–74.

Hillis, D.M., and J.C. Patton. 1982. Morphological and electrophoretic evidence for two species of *Corbicula* (Bivalvia: Corbiculidae) in North America. Amer. Midl. Nat. 108:74–80.

Hornbach, D.J., M.J. McLeod, and S.I. Guttman. 1980. On the validity of the genus *Musculium* (Bivalvia; Sphaeriidae): electrophoretic evidence. Can. J. Zool. 58:1703–1707.

Hubendick, B. 1951. Recent Lymnaeidae. Their variation, morphology, taxonomy, nomenclature, and distribution. K. Sven. Ventensk Handl. 3:1–225.

Inaba, A. 1969. Cytotaxonomic studies of lymnaeid snails. Malacologia 7:143–168.

Jacobson, M.K. 1974. How to study and collect shells: a symposium, 4th ed. American Malacological Union. 107 pp.

Jacobson, M.K., and W.K. Emerson. 1961. Shells of the New York City area. Argonaut Books, Larchmont, N.Y. 142 pp.

Johnson, C.W. 1915. Fauna of New England, 13. List of the Mollusca. Boston Soc. Nat. Hist. Occas. Pap. 7. 231 pp.

Johnson, R.I. 1946. *Anodonta implicata* Say. Occas. Pap. Mollusks. Harv. Univ. Mus. Comp. Zool. 1:109–116.

——. 1947. *Lampsilis cariosa* Say and *Lampsilis ochracea* Say. Occas. Pap. Mollusks. Harv. Univ. Mus. Comp. Zool. 1:145–156.

——. 1970. The systematics and zoogeography of the Unionidae of the southern Atlantic Slope region. Harv. Univ. Mus. Comp. Zool. Bull. 140:263–449.

——. 1978. Systematics and zoogeography of *Plagiola* (= *Dysnomia* = *Epioblasma*), an almost extinct genus of freshwater mussels from middle North America. Harv. Univ. Mus. Comp. Zool. Bull. 148:239–321.

——. 1980. Zoogeography of North American Unionacea north of the maximum Pleistocene glaciation. Harv. Univ. Mus. Comp. Zool. Bull. 149:77–189.

Jokinen, E.M. 1982. *Cipangopaludina chinensis* in North America, review and update. Nautilus 96:89–95.

*——. 1983. The freshwater snails of Connecticut. Bull. Conn. State Geol. Nat. Hist. Surv. 109:1–83.

——. 1987. Structure of freshwater snail communities: species-area relationships and incidence categories. Amer. Malacol. Bull. 5:9–19.

Jokinen, E.M., J. Guerrette, and R.W. Kortmann. 1982. The natural history of an ovoviviparous snail, *Viviparus georgianus*, in a soft-water eutrophic lake. Freshwat. Invert. Biol. 1:2–17.

Kat, P.W. 1983a. Genetic and morphological divergence among nominal species of North American *Anodonta* (Bivalvia: Unionidae). Malacologia 23:361–374.

——. 1983b. Sexual selection and simultaneous hermaphroditism among the Unionidae (Bivalvia: Mollusca). J. Zool. 201:395–416.

——. 1983c. Morphologic divergence, genetics, and speciation among *Lampsilis* (Bivalvia: Unionidae). J. Moll. Studies 49:133–145.

——. 1984. Parasitism and the Unionacea (Bivalvia). Biol. Rev. 59:189–207.

——. 1986. Hybridization in a unionid faunal suture zone. Malacologia 27:107–125.

LaMarche, A., P. Legendre, and A. Chodorowski. 1982. Facteurs responsables de la distributions des gastéropodes dulcicoles dans le fleuve Saint-Laurent. Hydrobiologia 89:61–76.

La Rocque, A. 1967. Pleistocene Mollusca of Ohio. Part 2. Ohio Div. Geol. Surv. Bull. 62:113–356.

——. 1968. Pleistocene Mollusca of Ohio. Part 3. Ohio Div. Geol. Surv. Bull. 62:357–553.

Letson, E.J. 1905. Check list of the Mollusca of New York. Bull. N.Y. State Mus. 88:1–112.

Mackie, G.L., D.S. White, and T.W. Zdeba. 1980. A guide to freshwater mollusks of the Laurentian Great Lakes with special emphasis on the genus *Pisidium*. EPA-600/3-80-068. 144 pp.

Marshall, W.B. 1895. Geographical distribution of New York Unionidae. Annu. Rep. N.Y. State Mus. 48:47–99.

Mathiak, H. 1979. A river survey of the unionid mussels of Wisconsin, 1973–1977. Sand Shell Press, Horicon, Wisc. 75 pp.

McMahon, R.F. 1982. The occurrence and spread of the introduced Asiatic freshwater clam, *Corbicula fluminea* (Muller) in North America: 1924–1982. Nautilus 96:134–141.

Merritt, R.W., and K.W. Cummins (eds.). 1984. An introduction to the aquatic insects of North America, 2nd ed. Kendall/Hunt Publ. Co., Dubuque, Iowa. 722 pp.

Ortmann, A.E. 1911. A monograph of the naiades of Pennsylvania. Parts I and II. Mem. Carnegie Mus. 4:279–374.

*——. 1919. A monograph of the naiades of Pennsylvania. Part III. Systematic account of the genera and species. Mem. Carnegie Mus. 8:1–384.

——. 1924. Distributional features of naiades in tributaries of Lake Erie. Amer. Midl. Nat. 9:101–117.

Parmalee, P.W. 1967. The freshwater mussels of Illinois. Ill. State Mus. Pop. Sci. Ser. 8:1–108.

Pennak, R.W. 1978. Freshwater invertebrates of the United States, 2nd ed. John Wiley and Sons, New York. 803 pp.

Robertson, I.C.S., and C.L. Blakeslee. 1948. The Mollusca of the Niagara Frontier region. Bull. Buffalo Soc. Nat. Sci. 19:1–191.

Russell-Hunter, W.D. 1978. Ecology of freshwater pulmonates. *In* V. Fretter and J. Peake (eds.). Pulmonates. Vol. 2: Systematics, evolution, and ecology, pp. 335–384. Academic Press, New York.

—— (ed.). 1983. The Mollusca. Vol. 6: Ecology. Academic Press, New York. 695 pp.

Sepkoski, J.J., and M.A. Rex. 1974. Distribution of fresh-water mussels: coastal rivers as biogeographic islands. Syst. Zool. 23:165–188.

Simpson, C.T. 1914. A descriptive catalog of the naiades or pearly mussels. Bryant Walker, Detroit, Mich. 1540 pp.

Smith, D.G. 1982. The zoogeography of the freshwater mussels of the Taconic and southern Green Mountain region of northeastern North America (Mollusca: Pelecypoda: Unionacea). Can. J. Zool. 60:261–267.

——. 1983. Notes on Mississippi River basin Mollusca presently occurring in the Hudson River system. Nautilus 97:128–131.

——. 1985. A study of the distribution of freshwater mussels (Mollusca: Pelecypoda: Unionoida) of the Lake Champlain drainage in northwestern New England. Amer. Midl. Nat. 114:19–29.

——. 1986. Keys to the freshwater macroinvertebrates of Massachusetts, no. 1. Mollusca Pelecypoda (clams, mussels). Mass. Div. Water Poll. Control, Westborough, Mass. 53 pp.

Smith, D.G., and W.P. Wall. 1984. The Margaritiferidae reinstated: a reply to Davis and Fuller (1981), "Genetic relationships among Recent Unionacea (Bivalvia) of North America." Occas. Pap. Mollusks. Harv. Univ. Mus. Comp. Zool. 4:321–330.

Stansbery, D.H. 1970. Eastern freshwater mollusks. I. The Mississippi and St. Lawrence River systems. American Malacological Union Symposium on Rare and Endangered Mollusks. Malacologia 10:9–22.

Strayer, D. 1987. Ecology and zoogeography of the freshwater mollusks of the Hudson River basin. Malacol. Rev. 20:1–68.

*Taylor, D.W., and E.H. Jokinen. 1984. A new species of freshwater snail *(Physa)* from seasonal habitats in Connecticut. Freshwat. Invert. Biol. 3:189–202.

Taylor, D.W., and N.F. Sohl. 1962. An outline of gastropod classification. Malacologia 1:7–32.

*Te, G.A. 1975. Michigan Physidae, with systematic notes on *Physella* and *Physodon*. Malacol. Rev. 8:7–30.

*——. 1978. The systematics of the family Physidae. Ph.D. dissertation. Univ. Michigan, Ann Arbor. 325 pp.

——. 1980. New classification system for the family Physidae. Arch. Molluskenkd. 110:179–184.

Thompson, F.G. 1968. The aquatic snails of the family Hydrobiidae of peninsular Florida. Univ. Fla. Press, Gainesville. 268 pp.

*——. 1984. North American freshwater snail genera of the hydrobiid subfamily Lithoglyphinae. Malacologia 25:109–142.

van der Schalie, H. 1938. The naiad fauna of the Huron River in southeastern Michigan. Misc. Publ. Univ. Mich. Mus. Zool. 40:1–83.

——. 1953. Nembutal as a relaxing agent for mollusks. Amer. Midl. Nat. 50:511–512.

——. 1970. Hermaphroditism among North American freshwater mussels. Malacologia 10:93–112.

Walter, H.J., and J.B. Burch. 1957. Key to the genera of freshwater gastropods occurring in Michigan. Univ. Mich. Mus. Zool. Circ. 3:1–8.

Way, C.M. 1988. An analysis of life histories in freshwater bivalves (Mollusca: Pisidiidae). Can. J. Zool. 66:1179–1190.

18 Aquatic Oligochaeta

David Strayer

Oligochaetes (phylum Annelida, class Oligochaeta) are common in most freshwater habitats, but they are often ignored by freshwater biologists because they are thought to be extraordinarily difficult to identify. The extensive taxonomic work done since 1960 by Brinkhurst and others, however, has enabled routine identification of most of our freshwater oligochaetes from simple whole mounts. Furthermore, a series of recent publications (Hiltunen and Klemm 1980, Stimpson et al. 1982, Klemm 1985, Brinkhurst 1986) has provided good, illustrated keys to the North American freshwater tubificids, naidids, and lumbriculids. Of the common freshwater oligochaetes, only the enchytraeids still require great taxonomic expertise and carefully prepared specimens to be identified.

Classification

Four families in the orders Tubificida and Lumbriculida are common in freshwater in northeastern North America: the Tubificidae, Naididae, Lumbriculidae, and Enchytraeidae. In addition, freshwater biologists sometimes encounter lumbricine oligochaetes (order Lumbricina; the familiar earthworms), haplotaxid oligochaetes (order Haplotaxida; rare inhabitants of groundwater), *Aeolosoma* (class Aphanoneura; small worms once classified with the oligochaetes), and *Manayunkia speciosa* (class Polychaeta) in waters of northeastern North America.

Biology: Habitat, Feeding, Respiration, Life Cycles

The tubificids probably are the best known of the freshwater oligochaetes. Tubificids are most commonly found in soft sediments rich in organic matter, and several

I thank Michael Smith, Ralph Brinkhurst, and Mark Wetzel for their careful reviews of an early draft of this chapter. They offered many useful suggestions for improving the keys and saved me from some embarrassing mistakes. Kurt Jirka was very helpful with editorial matters. This is a contribution to the program of the Institute of Ecosystem Studies of The New York Botanical Garden.

species characteristically live in sites that receive organic pollution. Like all aquatic oligochaetes, tubificids respire cutaneously, but a unique feature of this family is that some species can tolerate anoxic conditions. Most tubificids are deposit feeders, subsisting on organic detritus and its associated microflora. Because tubificids typically feed with their heads buried several centimeters below the sediment surface and their tails protruding above the sediment surface, their feeding activities can be important in mixing sediments, and consequently can affect the physical and chemical characteristics of the sediments. A few of the tubificids of northeastern North America reproduce predominantly by fragmentation (e.g., *Aulodrilus* spp. and *Tasserkidrilus harmani*), but most are sexually reproducing hermaphrodites. Tubificids enclose their eggs in cocoons and deposit them on the sediments.

The Naididae is an ecologically diverse family of worms common in both running and standing waters. Many naidids are sediment dwellers, like the tubificids, but other species are characteristically found among aquatic plants. The family Naididae includes detritivores (e.g., *Specaria josinae*), algivores (e.g., *Amphichaeta americana*), carnivores (e.g., *Chaetogaster diaphanus*), and even a parasite of snails (*Chaetogaster limnaei*). Sexual reproduction is rare in most species. Reproduction occurs predominantly by paratomy, where the posterior segments of a worm develop into a daughter that breaks free after development is complete. It is common to find worms consisting of two or more individuals that have not yet separated.

Only three species in the family Lumbriculidae occur in northeastern North America, and only two of them (*Stylodrilus heringianus* and *Lumbriculus variegatus*) are common and widespread. These large worms are found in standing and running waters, and are ecologically somewhat similar to tubificids. *Lumbriculus variegatus*, however, usually reproduces by fragmentation, and sexually mature individuals are rarely found.

Enchytraeids are common in marginal aquatic habitats—marshes, small streams, springs, and interstitial waters along the margins of streams—and they are found occasionally in the sediments of lakes and rivers as well. Because of taxonomic difficulties, very little work has been done on the ecology of freshwater enchytraeids in North America.

For more detailed information on aquatic oligochaete biology, see Brinkhurst and Jamieson 1971, Brinkhurst and Cook 1980, Bonomi and Erséus 1984, and Brinkhurst and Diaz 1987.

Collection and Preservation

The same methods used to collect insects and other macroinvertebrates generally are suitable for the collection of oligochaetes. However, samples should be sieved carefully; oligochaetes can be broken by vigorous sieving, and many oligochaetes, especially naidids, are too small to be retained on the relatively coarse (ca. 0.5-mm mesh) sieves often used for macrobenthos. Sieves as fine as 0.05-mm mesh may be required for work on naidids. Oligochaetes can be preserved in 5–10% formalin or

70% ethanol; ideally, the worms should be fixed in formalin for a day and then transferred to alcohol. Almost all oligochaetes should be mounted on microscope slides prior to identification. For critical taxonomic work, oligochaetes should be stained, dehydrated, and mounted in a resinous medium such as Canada balsam, Kleermount, or Permount. Most workers who deal with large numbers of oligochaetes in ecological studies, however, opt for the much simpler procedure of mounting worms directly from alcohol into a nonresinous medium such as polyvinyl lactophenol (Atomergic Chemicals Corp., 100 Fairchild Ave., Plainview, N.Y. 11803) or one of the CMC series of media (Master's Chemical Co., P.O. Box 2382, Des Plaines, Ill. 60018).

Identification

The following keys, based closely on those of Hiltunen and Klemm (1980), Stimpson et al. (1982), Klemm (1985), and Brinkhurst (1986), include all species of oligochaetes and their allies known from the freshwaters of New York State northeast to Newfoundland. Species not covered in these keys will likely be found in northeastern North America as it is explored more thoroughly. In addition, the brackish waters of this region contain many species of oligochaetes and polychaetes that are not treated here; Klemm (1985) provides a key to some of those polychaetes.

Members of the order Lumbricina (earthworms) and the family Enchytraeidae are not keyed to genus or species here. Reynolds (1977) can be used to identify the lumbricines of northeastern North America. Unfortunately, the taxonomy of the North American enchytraeids is not well advanced. Ambitious readers can get an introduction to the taxonomy of this family in the works of Nielsen and Christensen (1959), Tynen (1975), and Kasprzak (1981, 1984).

The user of these keys will need some background on oligochaete anatomy. The segments of an oligochaete are designated by Roman numerals, beginning with segment I at the anterior end of the worm. Sometimes it is difficult for a beginner to tell the front end of a worm from the back. The front end bears the prostomium (the anteriormost segment, which lacks chaetae and is often snout-shaped), the mouth, and the eyes (if present). Typically, each segment posterior to segment I bears four bundles of chaetae, two dorsal and two ventral. All four bundles will be visible on a slide-mounted oligochaete because the mounting media contain clearing agents that make the worm transparent. On a slide-mounted worm, the dorsal bundles often include two distinctly different kinds of chaetae (ventral bundles do not), and dorsal chaetae often are more elaborate than ventral chaetae; these characters enable distinction between the dorsal and ventral sides.

The keys rely largely on characters of the chaetae (= setae) and reproductive organs. The various kinds of chaetae are illustrated in Figs. 1–5. The dorsal chaetae that accompany hair chaetae often are referred to as crochets in tubificids and as needles in naidids. Important features of the reproductive system include the modified ventral chaetae known as genital chaetae and penis sheaths. Genital chaetae

Table 1. Some distinguishing characteristics of the common families of freshwater oligochaetes of northeastern North America

Family	Reproductive organs[a]	Genital chaetae[a]	Penis sheath[a]	Dorsal chaetae: hairs[c]	Dorsal chaetae: shape[b]	Dorsal chaetae: begin in segment	Dorsal chaetae: no. per bundle	Typical length (mm)	Eyes[c]
Enchytraeidae	V, XI, XII	−	−	−	s	II	2–10	1–25	−
Lumbriculidae	variable	−	−	−	b (s in 1 sp.)	II	2	5–50	−
Naididae	V, VI	V, VI	−	+/−	s,b,pe,pa,a	II–XX	usually > 2	0.5–10	+/−
Tubificidae	X, XI	X, XI	XI	+/−	s,b,pe,pa	II	often > 5	5–50	−

Note: Roman numerals indicate the segments in which a given structure is located.
[a]Present only in mature specimens, which are scarce in some collections.
[b]s, simple-pointed; b, bifurcate; pe, pectinate; pa, palmate; a, dorsal chaetae absent.
[c]+, present; −, absent.

replace the normal ventral chaetae in some mature tubificids in segments X (spermathecal chaetae) and XI (penial chaetae) and in some mature naidids on segments V (spermathecal chaetae) and VI (penial chaetae). Cuticular penis sheaths are found in segment XI of some mature tubificids. These features of the reproductive system are lacking in many species and are not developed in immature worms. When counting back to check specific segments for reproductive structures, remember that segment I does not bear chaetae.

Several areas of the key may present difficulties. First, although several of the families (e.g., Aeolosomatidae, Haplotaxidae) are instantly recognizable, differences among other families are more subtle, and it is not always possible to write simple rules for assigning worms to families. When starting to identify an oligochaete, note the form, number, and location of chaetae, the form and location of reproductive structures, the presence or absence of eyes, and use Table 1 as well as the key to families.

Second, identification of some species, especially in the Tubificidae, requires mature worms, which may be scarce or lacking in field collections. Although it is not possible to make a positive identification of immature worms of these species, it is sometimes possible to associate immatures with co-occurring adults.

Third, it is critical at several points in the naidid portion of the key to know on which segment the dorsal chaetae begin. Unfortunately, newly budded naidids, which occur often in field collections, may not yet have formed all of their anterior segments, so they will have dorsal chaetae anterior to where they are supposed to be. Thus, a young worm with dorsal chaetae on segment II may have to be keyed as having dorsal chaetae beginning on segment VI. The only way for a beginner to overcome this problem is to run the worm through both branches of the key. Sometimes, a newly budded worm can be recognized by its poorly developed mouth and prostomium.

Fourth, taxonomists are still uncertain about the systematics of several genera (*Limnodrilus*, *Nais*, *Pristinella*), and differences among named taxa are subtle and possibly unreliable. Here it may be necessary to settle for a certain generic identification rather than a dubious specific identification.

Finally, a worm that is difficult to identify may belong to a species not included in this key. In such a case, the keys of Klemm (1985) and Brinkhurst (1986) may be helpful. If the specimen still resists identification, consult the monographic treatment of Brinkhurst and Jamieson (1971) and its supplement (Brinkhurst and Wetzel 1984), which also provide an entry to the primary taxonomic literature.

Checklist of Aquatic Oligochaeta and Allies

Class Aphanoneura
Aeolosomatidae
Aeolosoma
Class Oligochaeta
Order Haplotaxida
Haplotaxidae
Haplotaxis gordioides
Order Lumbricina[a]
Order Lumbriculida
Lumbriculidae
Eclipidrilus lacustris
Lumbriculus variegatus
Stylodrilus heringianus
Order Tubificida
Enchytraeidae[a]
Naididae
Allonais pectinata
Amphichaeta americana
A. leydigi
Arcteonais lomondi
Bratislavia bilongata
B. unidentata
Chaetogaster diaphanus
C. diastrophus
C. cf. *krasnopolskiae*
C. limnaei
C. setosus
Dero digitata
D. flabelliger
D. furcata
D. lodeni
D. nivea
D. obtusa
D. vaga
Haemonais waldvogeli
Nais alpina
N. barbata
N. behningi
N. bretscheri
N. communis
N. elinguis
N. pardalis
N. pseudobtusa
N. simplex
N. variabilis
Ophidonais serpentina
Paranais frici
P. litoralis
Piguetiella blanci
P. michiganensis
Pristina aequiseta
P. breviseta
P. leidyi
P. plumaseta
P. synclites
Pristinella acuminata
P. jenkinae
P. osborni
P. sima
Ripistes parasita
Slavina appendiculata
Specaria josinae
Stephensoniana trivandrana
Stylaria lacustris
Uncinais uncinata
Vejdovskyella comata
V. intermedia
Tubificidae
Aulodrilus americanus
A. limnobius
A. pigueti
A. pluriseta
Bothrioneurum vejdovskyanum
Branchiura sowerbyi
Haber speciosus
Ilyodrilus templetoni
Isochaetides curvisetosus
I. freyi
Limnodrilus angustipenis
L. cervix
L. claparedianus
L. hoffmeisteri
L. maumeensis
L. profundicola
L. udekemianus
Phallodrilus hallae
Potamothrix bavaricus
P. bedoti
P. moldaviensis
P. vejdovskyi
Psammoryctides barbatus
P. californianus
Quistadrilus multisetosus
Rhyacodrilus coccineus
R. falciformis
R. montana
R. sodalis
R. cf. *subterraneus*
Spirosperma ferox
S. nikolskyi
Tasserkidrilus harmani
T. kessleri
T. superiorensis
Tubifex ignotus
T. nerthus
T. tubifex
Class Polychaeta
Order Sabellida
Sabellidae
Manayunkia speciosa

[a]Not keyed to genus.

Key to Species of Aquatic Oligochaeta and Allies

1a. Head with a distinctive branchial crown (Fig. 6); tube-dwelling . . . . Class POLYCHAETA, **Sabellidae**, *Manayunkia speciosa*
1b. Head without a branchial crown 2

2a (1b). Worm with hair chaetae in dorsal and ventral bundles; small and delicate, often with colored lipid globules . . . . Class APHANONEURA, **Aeolosomatidae**, *Aeolosoma*
2b. Worm with hair chaetae absent or restricted to dorsal bundles . . . . Class OLIGOCHAETA 3

3a (2b). Large worms (earthworms) with 2 simple-pointed, slightly sigmoid chaetae in each bundle (Fig. 7), all chaetae alike . . . . Order LUMBRICINA
The lumbricines are not treated further here. See Reynolds 1977 for identification.
3b. Worms relatively slender (mostly < 3 mm in diameter) and usually < 5 cm long; chaetae of various shapes and numbers 4

4a (3b). Worm very long and threadlike; ventral chaetae sickle-shaped and 1 per bundle (Fig. 8); dorsal chaetae small and 1 per bundle or absent; a rare species of underground waters . . . . **Haplotaxidae**, *Haplotaxis gordioides*
4b. Chaetae not as above, usually 2 or more per bundle 5

5a (4b). All chaetae simple-pointed; hair chaetae absent 6
5b. At least some chaetae bifurcate or pectinate; hair chaetae present or absent 7

6a (5a). Chaetae always 2 per bundle; spermathecae in or posterior to segment IX; rare . . . . **Lumbriculidae** (in part) 95
6b. Chaetae often more than 2 per bundle (Fig. 9); spermathecae in segment V; small to medium-sized worms especially common in small streams, wetlands, and marginal aquatic habitats . . . . **Enchytraeidae** (not keyed further)

7a (5b). Chaetae 2 per bundle, usually unequally bifurcate (Fig. 10); large worms (often too large to slide-mount) . . . . **Lumbriculidae** (in part) 95
7b. Chaetae usually more than 2 per bundle, of various shapes, often more than 1 shape of chaetae on a single worm; worms usually small enough to slide-mount 8

8a (7b). Small to medium-sized worms that *may* have eyes, that *may* be found budding, and that *may* bear genital chaetae in segments V and VI; dorsal chaetae *may* begin posterior to segment II, and the dorsal chaetae that accompany hair chaetae often differ strikingly from ventral chaetae; dorsal bundles often contain 1–2 hairs and 1–2 needles . . . . **Naididae** 47
8b. Medium-sized to large worms that never have eyes, that never bud, and that *may* bear specialized genital chaetae and/or penis sheaths in segments X and

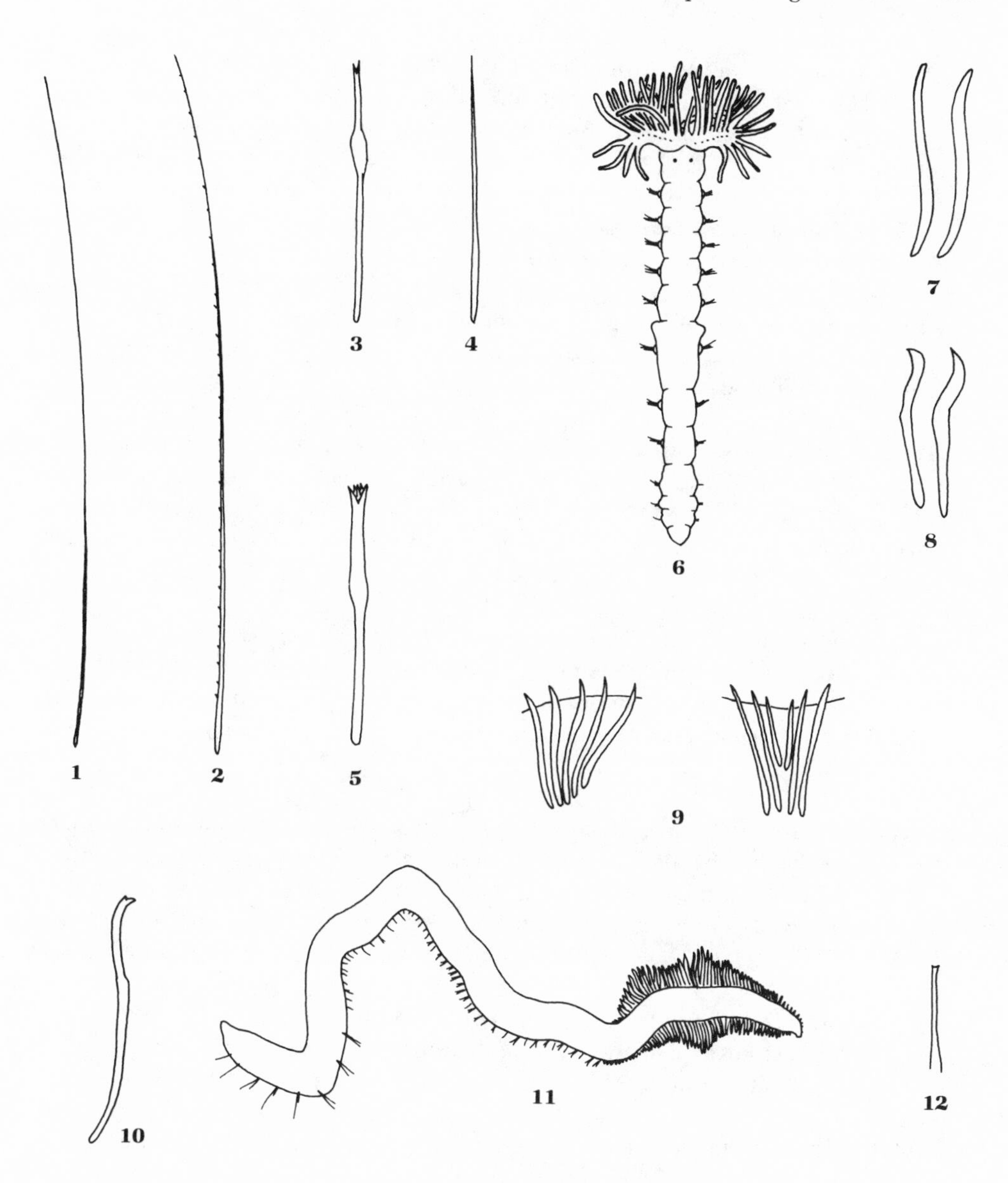

1. Hair chaeta. **2.** Serrate hair chaeta. **3.** Bifurcate chaeta. **4.** Simple-pointed chaeta. **5.** Pectinate chaeta. **6.** *Manayunkia speciosa* (Sabellidae), dv. **7.** Typical lumbricine bundle. **8.** Chaetae of *Haplotaxis gordioides* (Haplotaxidae). **9.** Typical enchytraeid bundles. **10.** Chaeta of *Lumbriculus variegatus* (Lumbriculidae). **11.** *Branchiura sowerbyi* (Tubificidae), dv. **12.** Chaeta of *Spirosperma nikolskyi* (Tubificidae). dv, dorsal view. (1–6 redrawn from Pennak 1978; 7 modified from Hickman et al. 1984; 8–10, 12 redrawn from Brinkhurst 1986; 11 modified from Brinkhurst and Jamieson 1971.)

XI (rarely on a few segments from VI to XII); dorsal chaetae begin on segment II, dorsal chaetae often broadly similar in form to ventral chetae; dorsal bundles often contain more than 2 hairs and more than 2 crotchets .... **Tubificidae** 9

9a (8b). **Tubificidae**: Hair chaetae present in dorsal bundles 10
9b. Hair chaetae absent from dorsal bundles 31

10a (9a). Large worms with conspicuous gills on posterior segments (Fig. 11) .... *Branchiura sowerbyi*
10b. Worms without posterior gills 11

11a (10b). Body wall with papillae and/or encrusted with foreign material; dorsal chaetae (which may be hard to see) pectinate; hair chaetae usually stout and saberlike 12
11b. Body wall without papillae and not encrusted with foreign material; dorsal chaetae bifurcate or pectinate 14

12a (11a). Dorsal chaetae short, thin, and without broadened tips (Fig. 12), although often hidden by encrusted foreign material; some ventral chaetae of segments II and III simple-pointed .... *Spirosperma nikolskyi*
12b. Dorsal chaetae pectinate (Figs. 13, 14); all ventral chaetae of segments II and III bifurcate 13

13a (12b). Body wall with rows of large papillae (these are especially conspicuous along the chaetal line in posterior segments); hair chaetae may be very numerous (up to 14 per bundle) .... *Quistadrilus multisetosus*
13b. Body wall densely covered with small papillae; less than 7 hairs per bundle .... *Spirosperma ferox*

14a (11b). Dorsal chaetae bifurcate or with a doubled upper tooth 15
14b. Dorsal chaetae palmate or pectinate (this character should be checked at 1,000X magnification) 17

15a (14a). Dorsal chaetae oar-shaped on segments posterior to about segment VII (Fig. 15) .... *Aulodrilus pigueti*
15b. Dorsal chaetae not oar-shaped 16

16a (15b). Upper tooth distinctly reduced in both dorsal and ventral chaetae (Fig. 16) .... *Aulodrilus pluriseta*
16b. Upper tooth nearly as large as lower in dorsal and ventral chaetae; hair chaetae characteristically bent (Fig. 17) .... *Potamothrix vejdovskyi*

17a (14b). Anterior dorsal chaetae palmate (Fig. 18); a European species reported in North America only from the St. Lawrence River .. *Psammoryctides barbatus*
17b. Anterior dorsal chaetae simply pectinate 18

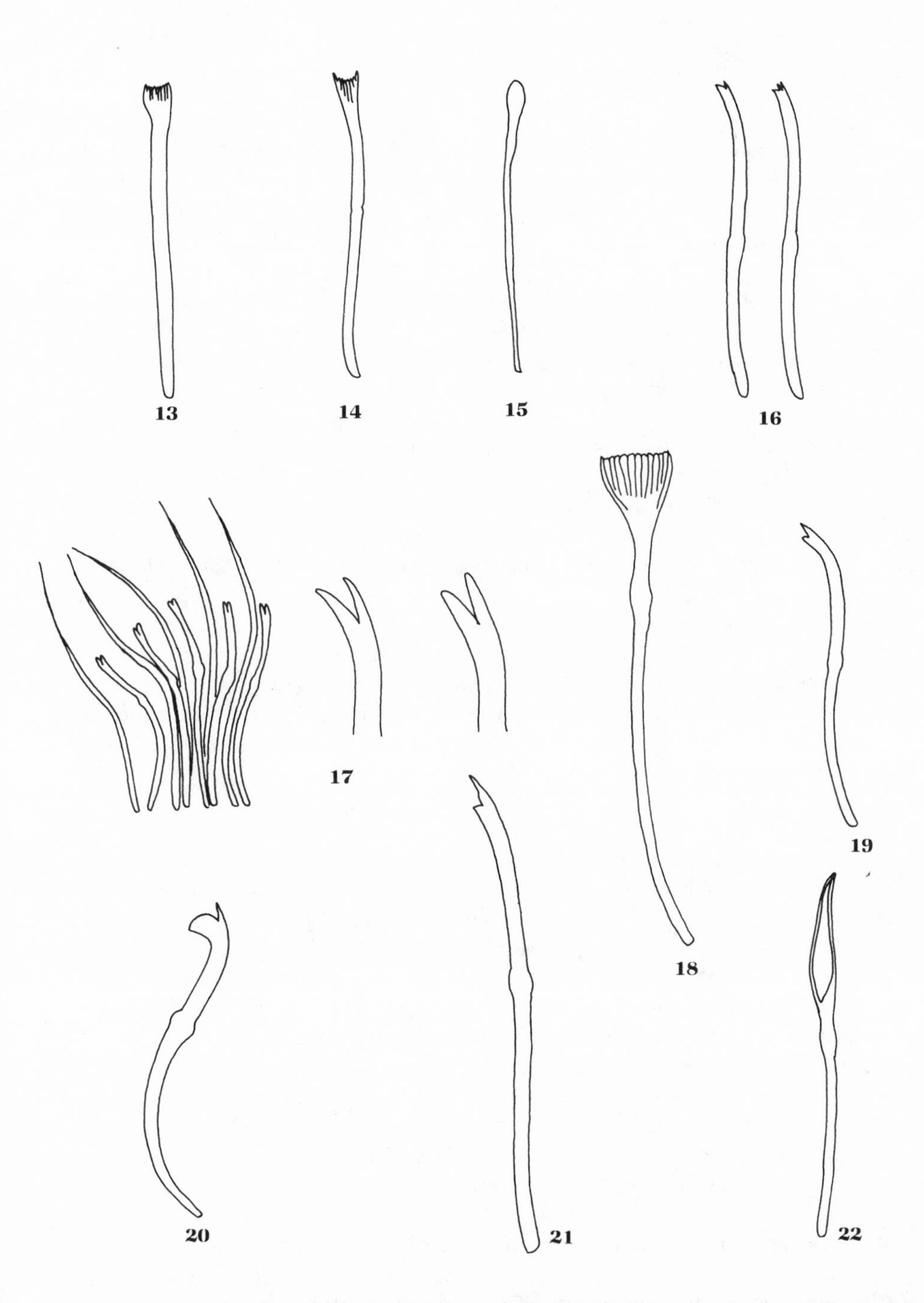

13. Dorsal chaeta of *Quistadrilus multisetosus* (Tubificidae). **14.** Dorsal chaeta of *Spirosperma ferox* (Tubificidae). **15.** Dorsal chaeta of *Aulodrilus pigueti* (Tubificidae). **16.** Chaetae of *Aulodrilus pluriseta* (Tubificidae). **17.** Bundle and tips of chaetae of *Potamothrix vejdovskyi* (Tubificidae). **18.** Chaeta of *Psammoryctides barbatus* (Tubificidae). **19.** Chaeta of *Rhyacodrilus montana* (Tubificidae). **20.** Chaeta of *Tasserkidrilus harmani* (Tubificidae). **21.** Chaeta of *Tubifex nerthus* (Tubificidae). **22.** Chaeta of *Potamothrix bedoti* (Tubificidae). (13–16, 17 [left], 18–21 redrawn from Stimpson et al. 1982; 17 [right and middle] redrawn from Brinkhurst 1986; 22 redrawn from Spencer 1978.)

18a (17b). Upper teeth of both anterior dorsal and anterior ventral chaetae much ($\geq 1\frac{1}{2}$ times) longer than lower (Fig. 19); hair chaetae of segment II often much longer than those of other segments .... *Rhyacodrilus montana*

Rhyacodrilus cf. *subterraneus*, recently found in hyporheic waters in New York (Strayer and Bannon-O'Donnell 1988), will key out here. It is distinguished from *R. montana* by its short (< 200 μm), sparse (0–2 per bundle) hair chaetae (Fig. 89).

18b. Teeth of anterior dorsal chaetae about equally long; hair chaetae of segment II not especially long 19

19a (18b). Posterior dorsal and ventral chaetae distinctively curved with large lower teeth (Fig. 20); rare .... *Tasserkidrilus harmani*

19b. Without such distinctive chaetae 20

20a (19b). Worm very long and slender; hair chaetae serrate and exceptionally long; rare .... *Tubifex ignotus*

20b. Not as above 21

21a (20b). Upper tooth of anterior ventral chaetae about twice as long as lower tooth (Fig. 21); reported from coastal waters in New Brunswick and Newfoundland .... *Tubifex nerthus*

21b. Upper tooth of anterior ventral chaetae only a little longer than lower tooth 22

22a (21b). Mature specimens are required to proceed. Scalpel-shaped chaetae (Fig. 22) present ventrally on any or all of segments VI–XII .... *Potamothrix bedoti*

22b. Specialized ventral chaetae absent or restricted to segments X and XI; not scalpel-shaped 23

23a (22b). With distinctive penial chaetae in segment XI; these chaetae are obviously different from the normal ventral chaetae (Figs. 23–25) 24

23b. Without penial chaetae 26

24a (23a). With spermathecal chaetae in segment X, as well as penial chaetae in segment XI (Fig. 23); rare .... *Haber speciosus*

24b. Spermathecal chaetae lacking 25

25a (24b). Dorsal pectinate chaetae with long, parallel outer teeth and fine intermediate teeth (Fig. 24) .... *Rhyacodrilus sodalis*

25b. Dorsal pectinate chaetae with divergent outer teeth of moderate length and distinct intermediate teeth (Fig. 25) .... *Rhyacodrilus coccineus*

26a (23b). With spermathecal chaetae (modified ventral chaetae obviously different from normal ventral chaetae) on segment X 27

26b. Without spermathecal chaetae 28

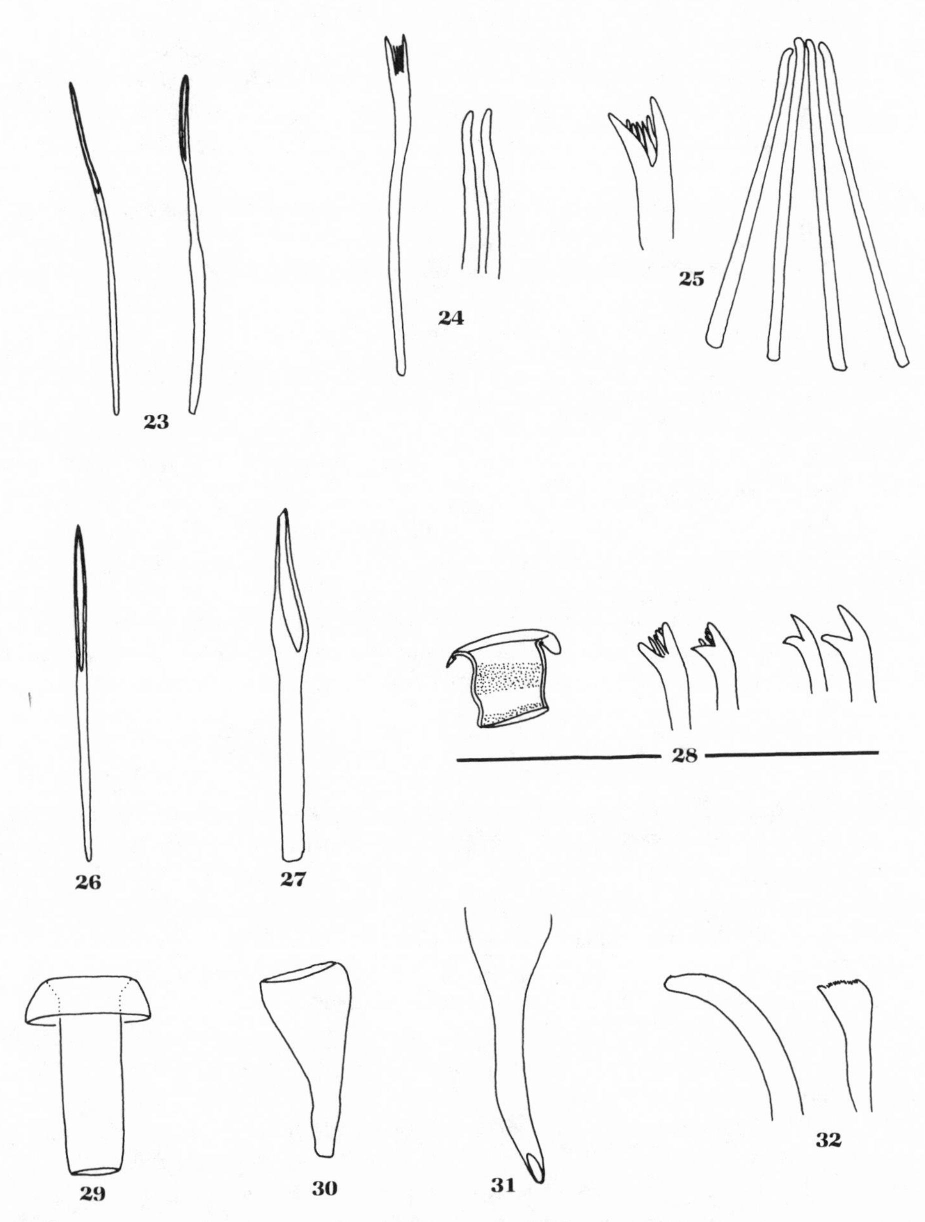

23. Penial chaetae of *Haber speciosus* (Tubificidae). **24.** Dorsal chaeta (left) and tips of penial chaetae of *Rhyacodrilus sodalis* (Tubificidae). **25.** Tip of dorsal chaeta (left) and penial chaetae of *Rhyacodrilus coccineus* (Tubificidae). **26.** Spermathecal chaeta of *Psammoryctides californianus* (Tubificidae). **27.** Spermathecal chaeta of *Potamothrix bavaricus* (Tubificidae). **28.** Penis sheath (left) tips of and chaetae of *Tubifex tubifex* (Tubificidae). **29.** Penis sheath of *Tasserkidrilus superiorensis* (Tubificidae). **30.** Penis sheath of *Ilyodrilus templetoni* (Tubificidae). **31.** Penis sheath of *Tasserkidrilus kessleri* (Tubificidae). **32.** Tips of chaetae of *Aulodrilus americanus* (Tubificidae). (23, 24 [left], 25 [right], 26, 29–31 redrawn from Stimpson et al. 1982; 24 [right], 25 [left], 28, 32 redrawn from Brinkhurst 1986; 27 redrawn from Spencer 1978.)

27a (26a). Spermathecal chaetae with parallel sides (Fig. 26) . . . . *Psammoryctides californianus*

27b. Spermathecal chaetae distinctly broadened in the middle (Fig. 27) . . . . *Potamothrix bavaricus*

28a (26b). Penis sheath (in segment XI) thin-walled (may be difficult to see in well-cleared specimens) and no longer than wide (Fig. 28) . . . . *Tubifex tubifex*

28b. Penis sheath distinctly longer than wide (Figs. 29–31) 29

29a (28b). Penis sheath cylindrical, flared at one end (Fig. 29) . . . . *Tasserkidrilus superiorensis*

29b. Penis sheath distinctly conical 30

30a (29b). Penis sheath relatively wide, evenly tapered from base to tip (Fig. 30); common and widespread . . . . *Ilyodrilus templetoni*

30b. Penis sheath relatively narrow, abruptly tapered near base (Fig. 31); typically in oligotrophic lakes . . . . *Tasserkidrilus kessleri*

31a (9b). Anterior chaetae simple-pointed or with a rudimentary upper tooth; dorsal chaetae posterior to segment VI palmate (Fig. 32) . . . . . *Aulodrilus americanus*

31b. Anterior chaetae not simple-pointed 32

32a (31b). Dorsal chaetae posterior to segment VI distinctively shaped: spatula-shaped in frontal view and compressed in lateral view (Fig. 33) . . . . *Aulodrilus limnobius*

32b. Dorsal chaetae posterior to segment VI not spatula-shaped and compressed 33

33a (32b). Upper tooth of ventral chaetae of segments II–IV large and oriented at right angles to the chaetal shaft (Fig. 34) . . . . *Limnodrilus udekemianus*

33b. Without such distinctive ventral chaetae in segments II–IV 34

34a (33b). Dorsal chaetae posterior to segment XX strongly curved and larger than the ventral chaetae of the same segment (Fig. 35) . . . . . . *Isochaetides curvisetosus*

34b. Without such distinctive dorsal chaetae posterior to segment XX 35

35a (34b). Prostomium with a dorsal sensory cavity, which appears as a depression (lateral view) or a light area (dorsal or ventral view) (Fig. 38); mature specimens with penial chaetae in segment XI (Fig. 36); 3–4 normal chaetae (Fig. 37) per bundle in segments II–IX and 2 per bundle posterior to segment IX . . . . *Bothrioneurum vejdovskyanum*

35b. Prostomium without a dorsal sensory cavity; chaetal characters various . . 36

36a (35b). (Mature specimens are required to proceed.) With spermathecal chaetae in segment X 37

36b. Without spermathecal chaetae in segment X 38

37a (36a). Spermathecal chaetae long and parallel-sided; with cuticular penis sheath (Fig. 39) . . . . *Isochaetides freyi*

37b. Spermathecal chaetae broad and tapering (Fig. 40); without penis sheath . . . . *Potamothrix moldaviensis*

38a (36b). With penial chaetae in segment XI (see also 35a) 39

38b. Without penial chaetae, but with conspicuous penis sheaths . . . . *Limnodrilus* (in part) 40

39a (38a). With 1 pair of sickle-shaped penial chaetae (Fig. 41); reported only from the Hudson River . . . . *Rhyacodrilus falciformis*

39b. With 3–6 pairs of hooked penial chaetae (Fig. 42); reported only from oligotrophic waters of the Great Lakes . . . . *Phallodrilus hallae*

40a (38b). Head of penis sheath more or less at a right angle to the shaft (Figs. 43–45) 41

40b. Head of penis sheath not at a right angle to the shaft (Figs. 46–48) 43

41a (40a). Penis sheath long (> 300 μm in fully developed specimens); base of sheath broadened, thin-walled, and often a little crumpled (Fig. 43) . . *L. angustipenis*

41b. Penis sheath short (Fig. 45) or cylindrical (Fig. 44) 42

42a (41b). Penis sheath long (> 300 μm in fully developed specimens) and cylindrical (Fig. 44) . . . . *L. hoffmeisteri*

42b. Penis sheath short (< 300 μm) (Fig. 45) . . . . *L. profundicola*

43a (40b). Head of penis sheath scalloped (Fig. 44) . . . . *L. hoffmeisteri* (variant)

43b. Head of penis sheath not scalloped 44

44a (43b). Wall of penis sheath markedly thickened (Figs. 46, 47) 45

44b. Wall of penis sheath not especially thick (Figs. 44, 48) 46

45a (44a). Head of penis sheath broad and more or less triangular; shaft of penis sheath often bent distally (Fig. 46) . . . . *L. maumeensis*

45b. Head of penis sheath a narrow triangle or broadly V-shaped; shaft more or less straight (Fig. 47) . . . . *L. cervix*

46a (44b). Penis sheath usually 300–600 μm long in fully developed specimens; head of penis sheath often with a hood or lip; distal end of penis sheath often a little flared (Fig. 44) . . . . *L. hoffmeisteri*

46b. Penis sheath 800–1,100 μm long in fully developed specimens; head of penis sheath roughly triangular and without hood or lip; distal end of penis sheath not flared (Fig. 48) . . . . *L. claparedianus*

47a (8a). **Naididae**: Worm entirely without dorsal chaetae (i.e., 2 bundles of chaetae per segment) . . . . *Chaetogaster* 48

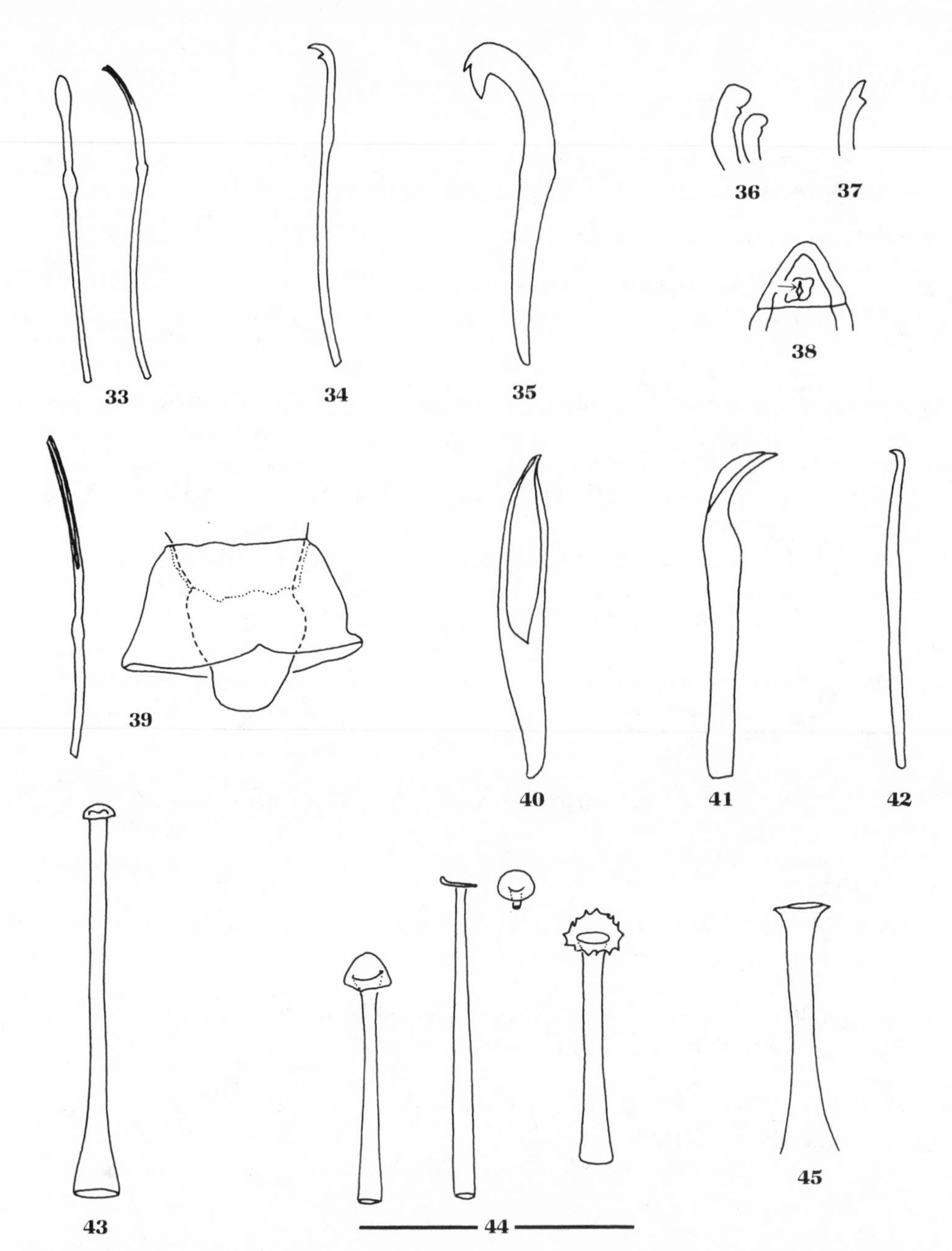

33. Chaetae of *Aulodrilus limnobius* (Tubificidae): frontal view (left), lateral view (right). **34.** Anterior ventral chaeta of *Limnodrilus udekemianus* (Tubificidae). **35.** Dorsal chaeta of *Isochaetides curvisetosus* (Tubificidae). **36.** Tips of penial chaetae of *Bothrioneurium vejdovskyi* (Tubificidae). **37.** Tip of somatic chaeta of *Bothrioneurium vejdovskyi* (Tubificidae). **38.** Prostomium of *Bothrioneurum vejdovskyanum* (Tubificidae), showing sensory cavity, dv. **39.** Penis sheath (right) and spermathecal chaeta of *Isochaetides freyi* (Tubificidae). **40.** Spermathecal chaeta of *Potamothrix moldaviensis* (Tubificidae). **41.** Penial chaeta of *Rhyacodrilus falciformis* (Tubificidae). **42.** Penial chaeta of *Phallodrilus hallae* (Tubificidae). **43.** Penis sheath of *Limnodrilus angustipenis* (Tubificidae). **44.** Penis sheaths of *Limnodrilus hoffmeisteri* (Tubificidae). **45.** Penis sheath of *Limnodrilus profundicola* (Tubificidae). dv, dorsal view. (33–35, 39–45 redrawn or modified from Stimpson et al. 1982; 36–38 redrawn from Brinkhurst and Jamieson 1971.)

47b. Worm with dorsal chaetae (i.e., 4 bundles of chaetae per segment) in at least some segments 51

48a (47a). Tip of chaetae simple-pointed (Fig. 49) or inconspicuously bifurcate .... *C. setosus*
48b. Tip of chaetae distinctly bifurcate at 400X (Figs. 50, 51) 49

49a (48b). Chaetal teeth strongly recurved (Fig. 50); animal commensal or parasitic on mollusks .... *C. limnaei*
49b. Chaetal teeth not recurved (Fig. 51); animal free-living 50

50a (49b). Chaetae of segment II ≥ 120 μm long .... *C. diaphanus*
50b. Chaetae of segment II < 120 μm long .... *C. diastrophus*
Chaetogaster cf. *krasnopolskiae*, found recently in hyporheic waters in New York (Strayer and Bannon-O'Donnell 1988), will key out here, but can be recognized by its distinctive chaetae, in which the upper tooth is reduced (Fig. 90).

51a (47b). Hair chaetae entirely lacking 52
51b. Hair chaetae present in some dorsal bundles 58

52a (51a). Dorsal chaetae begin in segment III .... *Amphichaeta* 53
52b. Dorsal chaetae begin in segments V or VI 54

53a (52a). Six chaetae per bundle in segment III and 3 per bundle in segment IV; upper tooth about twice as long as lower (Fig. 52) .... *A. americana*
53b. Five chaetae per bundle in segment III and 2 per bundle in segment IV; upper tooth a little longer than lower (Fig. 53) .... *A. leydigi*

54a (52b). Dorsal chaetae straight and blunt or indistinctly bifurcate (Fig. 54) .... *Ophidonais serpentina*
54b. Dorsal chaetae distinctly bifurcate (Fig. 55) and usually sigmoid 55

55a (54b). Dorsal chaetae begin in segment V; body often bears a thin layer of foreign material; usually from coastal fresh or brackish waters .... *Paranais* 56
55b. Dorsal chaetae usually begin in segment VI (rarely in V) 57

56a (55a). Usually with 1–2 (rarely 3) chaetae per bundle, except on segment II, which bears 2–4 chaetae per bundle; all chaetae with upper tooth much longer than lower (Fig. 55); chaetae of segment II ca. 100 μm long .... *P. frici* (= *Wapsa mobilis*)
56b. Usually with 3 chaetae per bundle, except on segment II, which bears 4–7 chaetae per bundle; chaetae posterior to segment II with teeth almost equal in length (Fig. 56); chaetae of segment II ca. 60 μm long .... *P. litoralis*

57a (55b). Chaetae of segment II ca. 60 μm long, with teeth of equal length (Fig. 57); worm never with eyes .... *Piguetiella michiganensis*

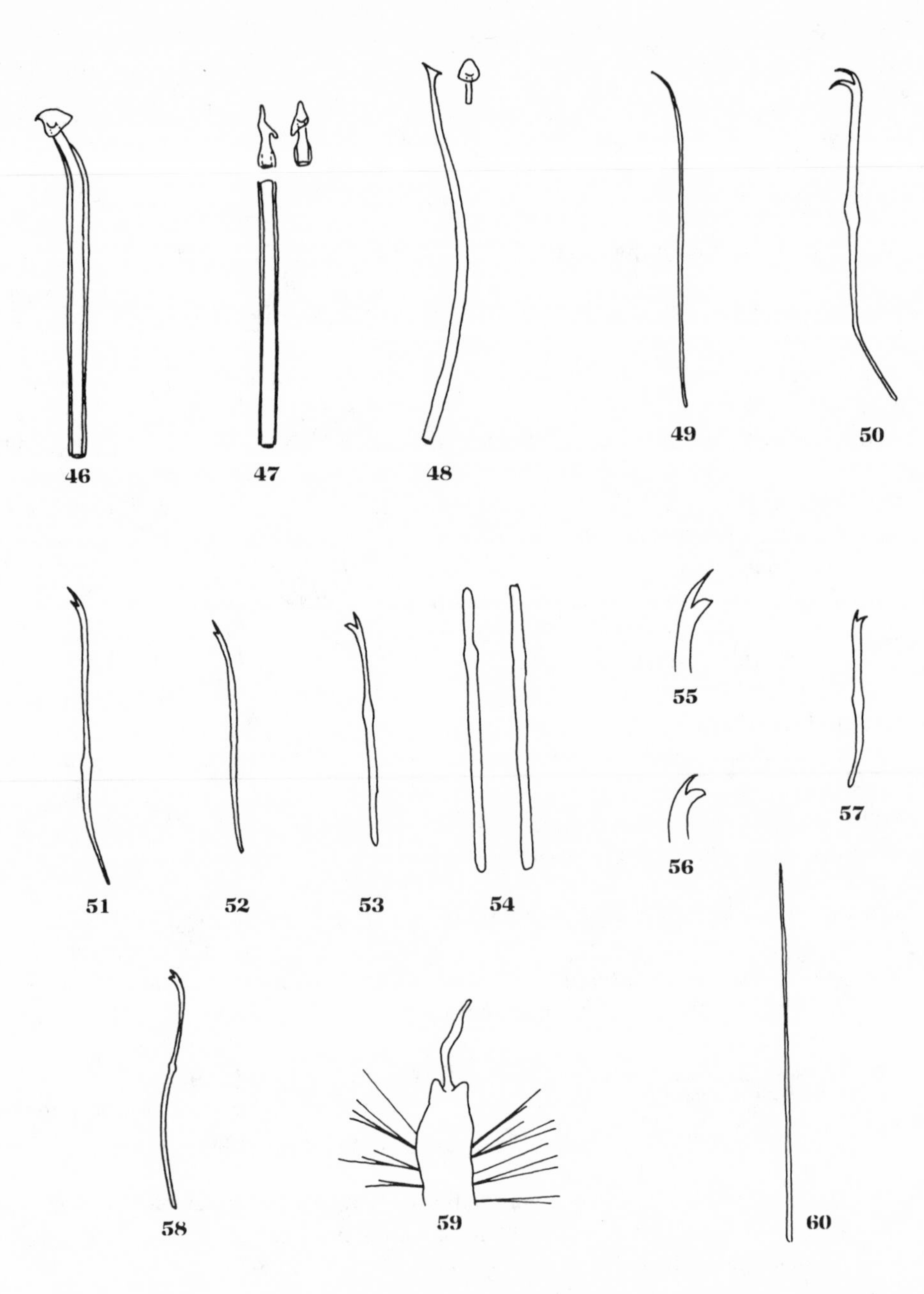

46. Penis sheath of *Limnodrilus maumeensis* (Tubificidae). **47.** Penis sheath of *Limnodrilus cervix* (Tubificidae). **48.** Penis sheath of *Limnodrilus claparedianus* (Tubificidae). **49.** Chaeta of *Chaetogaster setosus* (Naididae). **50.** Chaeta of *Chaetogaster limnaei* (Naididae). **51.** Chaeta of *Chaetogaster diastrophus* (Naididae). **52.** Chaeta of *Amphichaeta americana* (Naididae). **53.** Chaeta of *Amphichaeta leydigi* (Naididae). **54.** Dorsal chaetae of *Ophidonais serpentina* (Naididae). **55.** Tip of chaeta of *Paranais frici* (Naididae). **56.** Tip of chaeta of *Paranais litoralis* (Naididae). **57.** Chaeta from segment II of *Piguetiella michiganensis* (Naididae). **58.** Chaeta from segment II of *Uncinais uncinata* (Naididae). **59.** Anterior end of *Stylaria lacustris* (Naididae). **60.** Needle of *Pristina leidyi* (Naididae). (46–48 redrawn from Stimpson et al. 1982; 49, 50, 52–54, 58 redrawn from Hiltunen and Klemm 1980; 51, 55 redrawn from Brinkhurst and Jamieson 1971; 56, 57, 59 redrawn or modified from Brinkhurst 1986.)

57b. Upper tooth of chaetae of segment II much longer than lower (Fig. 58), chaetae of segment II ca. 100 µm long; worm may have eyes .... *Uncinais uncinata*

58a (51b). Worm with elongate proboscis (as in Fig. 59) 59
58b. Worm lacking proboscis 66

59a (58a). Segment II with dorsal chaetae .... *Pristina* 60
59b. Segment II without dorsal chaetae 64

60a (59a). Needles appear to be simple-pointed at 1,000X (Fig. 60) (actually, they bear a minute bifurcation) .... *P. leidyi*
60b. Needles distinctly bifurcate 61

61a (60b). Teeth of needles short (Fig. 61) .... *P. aequiseta* (including *P. foreli*)
61b. Teeth of needles long (Figs. 62, 63) 62

62a (61b). Hair chaetae serrate .... *P. plumaseta*
62b. Hair chaetae not serrate 63

63a (62b). Anterior ventral chaetae (e.g., those in segment II) with teeth unequal in length; needles with teeth equal in length (Fig. 62) .... *P. breviseta*
63b. Anterior ventral chaetae with teeth equal in length; teeth of needles slightly unequal in length (Fig. 63) .... *P. synclites*

64a (59b). With 1–3 hair chaetae per bundle; widespread and common (Fig. 59) .... *Stylaria lacustris*
64b. With more than 3 hair chaetae per bundle 65

65a (64b). Hair chaetae of segments VI–VIII long (more than 3 times as long as those of segment IX) .... *Ripistes parasita*
65b. Hair chaetae of segments VI–VIII not especially long .... *Arcteonais lomondi*

66a (58b). Dorsal chaetae absent anterior of segment X .... *Haemonais waldvogeli*
66b. Dorsal chaetae present anterior of segment X 67

67a (66b). Dorsal chaetae begin in segment II or III 68
67b. Dorsal chaetae begin in segments IV–VII 73

68a (67a). Body wall covered with foreign material; needles simple-pointed; an uncommon species chiefly of large rivers .... *Stephensoniana trivandrana*
68b. Body wall free from foreign material; needles simple-pointed or bifurcate 69

69a (68b). Dorsal chaetae begin in segment II; common .... *Pristinella* 70
69b. Dorsal chaetae begin in segment III; rare .... *Bratislavia* 97

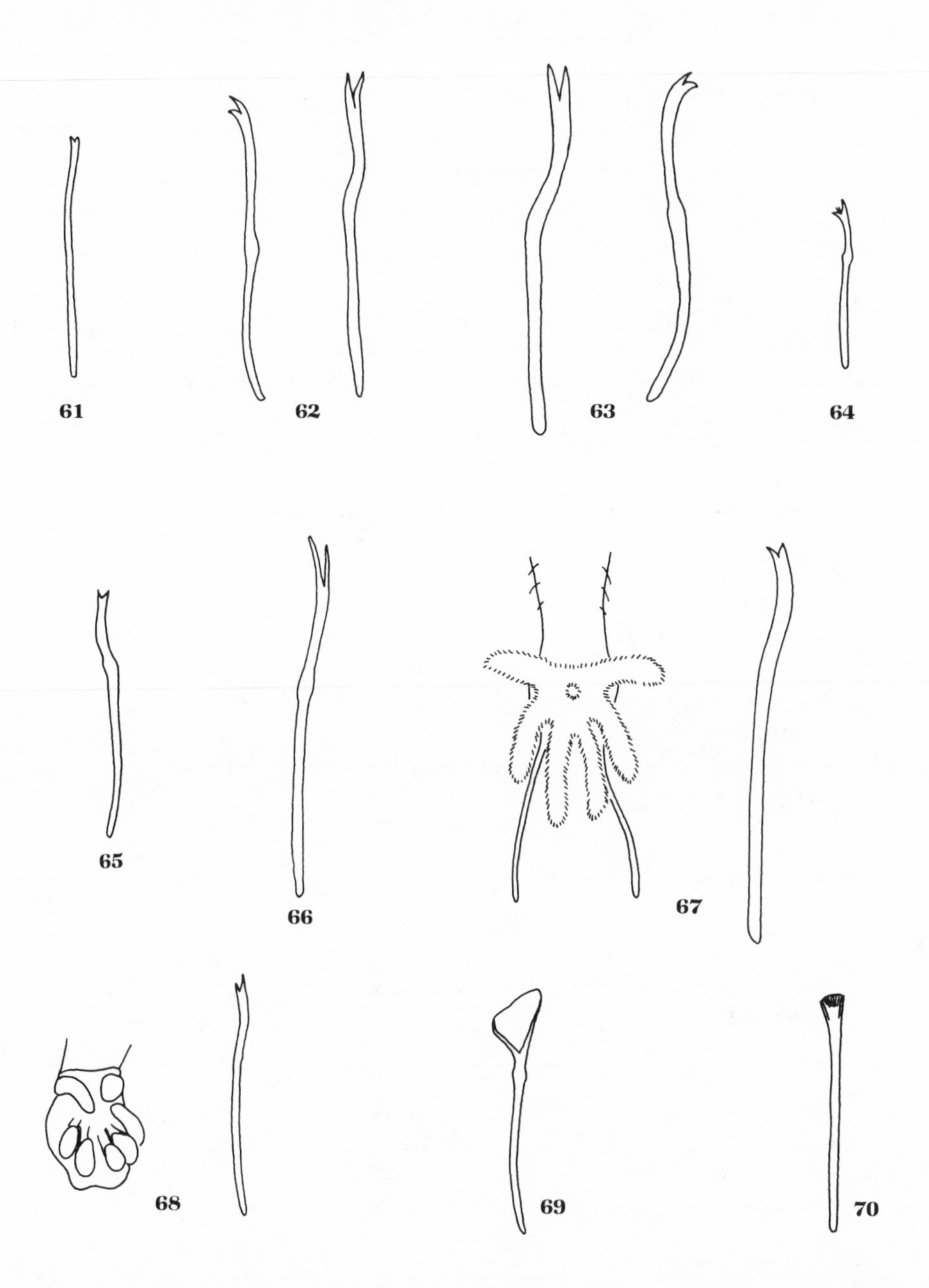

61. Needle of *Pristina aequiseta* (Naididae). **62.** Ventral chaeta (left) and needle of *Pristina breviseta* (Naididae). **63.** Needle (left) and ventral chaeta of *Pristina synclites* (Naididae). **64.** Needle of *Pristinella sima* (Naididae). **65.** Needle of *Pristinella osborni* (Naididae). **66.** Needle of *Pristinella jenkinae* (Naididae). **67.** Posterior end (left) and needle of *Dero furcata* (Naididae). **68.** Posterior end (left) and needle of *Dero digitata* (Naididae). **69.** Needle of *Dero flabelliger* (Naididae). **70.** Needle of *Dero vaga* (Naididae). (61, 64, 69, 70 redrawn from Hiltunen and Klemm 1980; 62, 63, 65, 67, 68 redrawn or modified from Brinkhurst 1986.)

70a (69a). Teeth of needles strongly divergent (Figs. 64, 65) 71
70b. Teeth of needles nearly parallel (Fig. 66) 72

71a (70a). Tip of needles with small intermediate teeth (Fig. 64) .... *P. sima*
71b. Tip of needles without small intermediate teeth (Fig. 65) .... *P. osborni*
P. sima and *P. osborni* may be ecophenotypes of a single species.

72a (70b). Hair chaetae serrate; rare .... *P. acuminata*
72b. Hair chaetae not serrate; common and widespread .... *P. jenkinae* (= *P. idrensis*) (Fig. 66)

73a (67b). Posterior end of body with finger-shaped lobes or palps (Figs. 67, 68) (may be hard to see on slide-mounts); dorsal chaetae begin on segments IV–VI; worm never with eyes .... *Dero* 74
73b. Posterior end of body without lobes or palps; dorsal chaetae begin on segments V–VII; worm with or without eyes 80

74a (73a). Posterior end with lobes and long palps (Fig. 67) 75
74b. Posterior end with only short lobes (Fig. 68) 78

75a (74a). Needles palmate (Figs. 69, 70); dorsal chaetae begin on segment VI 76
75b. Needles bifurcate or pectinate (Figs. 67, 68, 71, 72); dorsal chaetae begin in segment V 77

76a (75a). Needles very broad and asymmetrical (Fig. 69) .... *D. flabelliger*
76b. Needles moderately broad, more or less symmetrical (Fig. 70) .... *D. vaga*

77a (75b). Needles bifurcate (Fig. 67) .... *D. furcata*
77b. Needles pectinate (Fig. 71) .... *D. lodeni*

78a (74b). Teeth of needles unequal in length (Fig. 68); needles long (usually $> 70\ \mu m$); posterior end usually with 4 pairs of lobes .... *D. digitata*
78b. Teeth of needles equal in length (Fig. 72); needles short (usually $< 75\ \mu m$); posterior end usually with 3 pairs of lobes 79

79a (78b). Posterior lobes extend beyond the end of the anal segment; chaetae of moderate length (ventral chaetae of II–V usually $> 90\ \mu m$ long; needles usually $> 50\ \mu m$ long) .... *D. obtusa*
79b. Posterior lobes do not extend beyond the end of the anal segment; chaetae short (ventral chaetae of II–V usually $< 90\ \mu m$ long; needles usually $< 50\ \mu m$ long) .... *D. nivea*

80a (73b). Anterior segments with more than 4 hair chaetae (which are serrated) in each bundle .... *Vejdovskyella* 81
80b. Anterior segments with 0–3 hair chaetae in each bundle 82

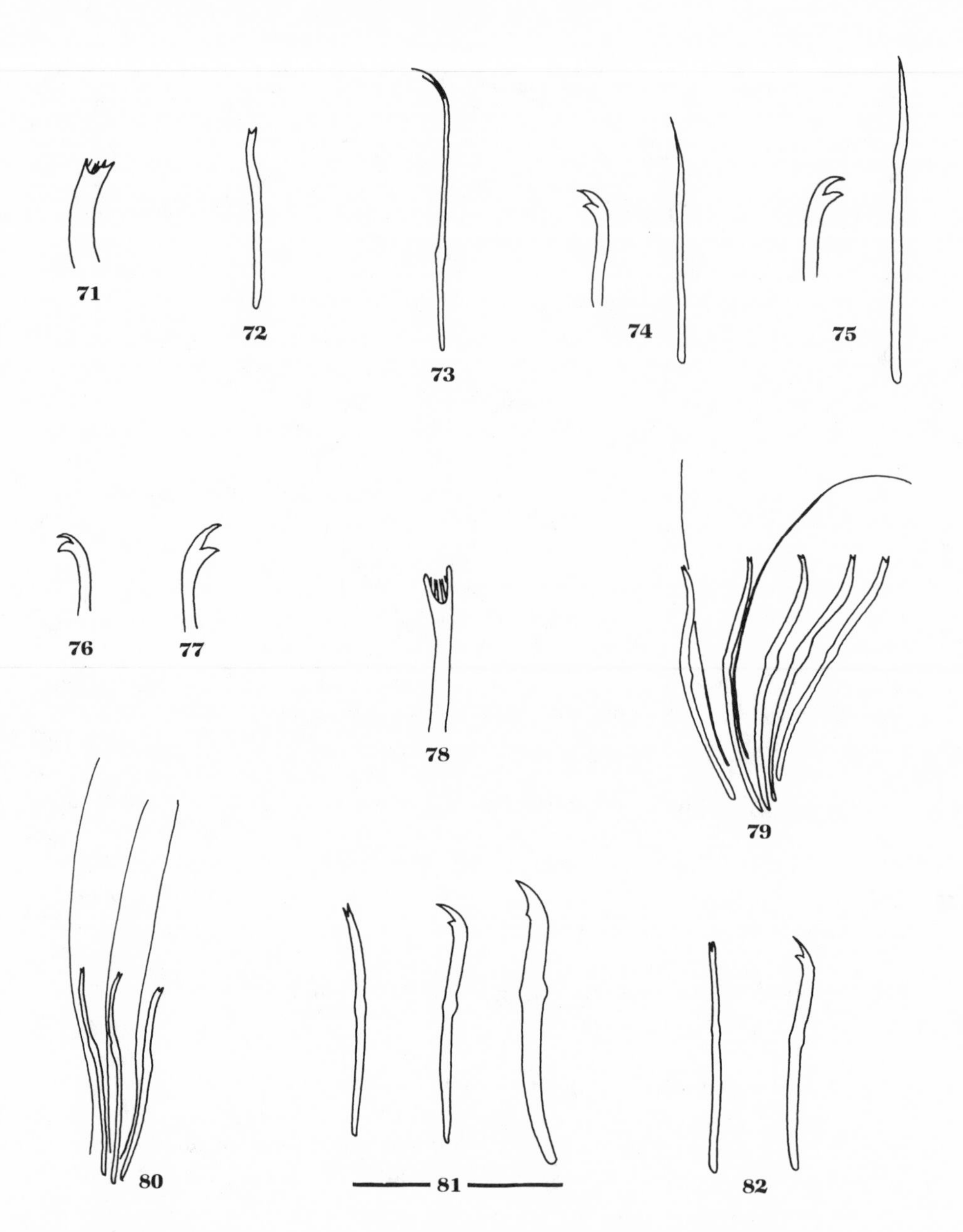

71. Tip of needle of *Dero lodeni* (Naididae). **72.** Needle of *Dero nivea* (Naididae). **73.** Anterior ventral chaeta of *Nais behningi* (Naididae). **74.** Tip of ventral chaeta (left) and needle of *Nais pseudobtusa* (Naididae). **75.** Tip of ventral chaeta (left) and needle of *Nais simplex* (Naididae). **76.** Tip of ventral chaeta of *Nais barbata* (Naididae). **77.** Tip of ventral chaeta of *Nais alpina* (Naididae). **78.** Tip of needle of *Allonais pectinata* (Naididae). **79.** Dorsal bundle of *Piguetiella blanci* (Naididae). **80.** Dorsal bundle of *Specaria josinae* (Naididae). **81.** Needle (leftmost) and ventral chaetae of *Nais bretscheri* (Naididae). **82.** Needle (left) and ventral chaeta of *Nais pardalis* (Naididae). (71, 72, 74–78 redrawn from Brinkhurst 1986; 73, 81, 82 redrawn from Hiltunen and Klemm 1980; 79, 80 redrawn from Strayer 1983.)

81a (80a). Ventral bundles posterior to segment V with only 1 chaeta; eyes absent . . . . *V. intermedia*

81b. Ventral bundles posterior to segment V with 2–3 chaetae; eyes usually present . . . . *V. comata*

82a (80b). Hair chaetae of segment VI remarkably long (ca. 2 times as long as hair chaetae of other segments); body wall usually covered with foreign material . . . . *Slavina appendiculata*

82b. Hair chaetae of segment VI not remarkably long 83

83a (82b). Needles simple-pointed . . . . *Nais* (in part) 84

83b. Needles bifurcate or pectinate 88

84a (83a). Teeth of ventral chaetae of segments II–V thin and elongate (Fig. 73) . . . . *N. behningi*

84b. Teeth of anterior ventral chaetae not especially long 85

85a (84b). Tips of needles drawn out into a hairlike tip (Fig. 74); needles usually ca. 90 μm long 86

85b. Tips of needles stout (Fig. 75); needles usually ca. 60 μm long 87

86a (85a). Ventral chaetae of segments VI and posterior with teeth unequal in length (Fig. 74); 1–3 hair chaetae and 1–3 needles per bundle . . . . . . *N. pseudobtusa*

86b. Ventral chaetae of segments VI and posterior with teeth equal in length (Fig. 76); up to 5 hair chaetae and 5 needles per bundle . . . . *N. barbata*

87a (85b). Ventral chaetae of segments VI and posterior with teeth unequal in length (Fig. 77) . . . . *N. alpina*

87b. Ventral chaetae of segments VI and posterior with teeth equal in length (Fig. 75) . . . . *N. simplex*

88a (83b). Needle teeth pectinate (Fig. 78) . . . . *Allonais pectinata*

88b. Needle teeth bifurcate 89

89a (88b). Needles sigmoid, usually more than 2 per bundle in anterior segments (Figs. 79, 80) 90

89b. Needles straight or a little bent (Figs. 81–85), usually 1–2 per bundle . . . . *Nais* (in part) 91

90a (89a). Hair chaetae short (less than 1.7 times as long as the needles of the same segment) (Fig. 79); eyes present or absent . . . . *Piguetiella blanci*

90b. Hair chaetae long (more than 1.7 times as long as the needles of the same segment) (Fig. 80); eyes absent . . . . *Specaria josinae*

91a (89b). Some anterior ventral bundles with enlarged chaetae present (Fig. 81) 92

91b. Anterior ventral bundles without enlarged chaetae 93

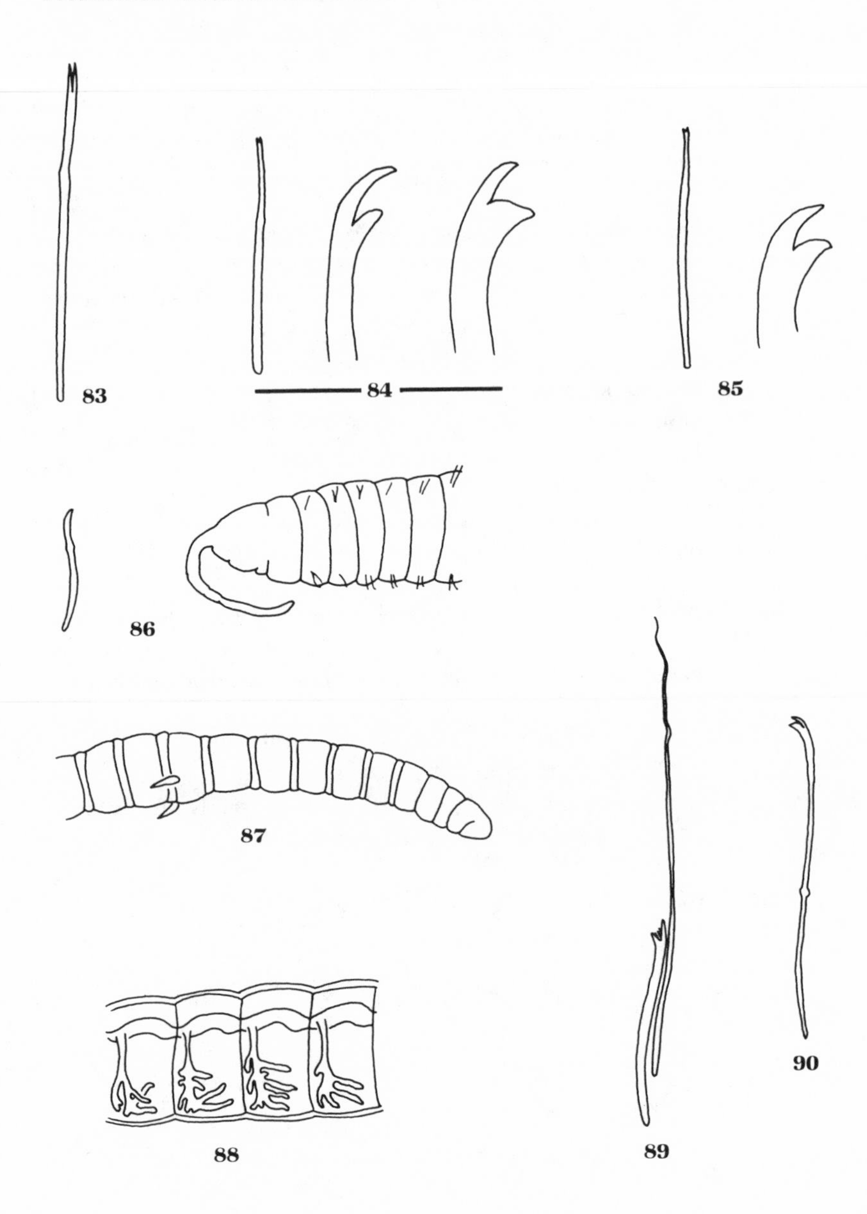

83. Needle of *Nais elinguis* (Naididae). **84.** Needle (left), tip of anterior ventral chaeta (center), and tip of posterior ventral chaeta (right) of *Nais variabilis* (Naididae). **85.** Needle (left) and tip of anterior ventral chaeta of *Nais communis* (Naididae). **86.** Anterior end (right) and chaeta of *Eclipidrilus lacustris* (Lumbriculidae), lv. **87.** Anterior end of *Stylodrilus heringianus* (Lumbriculidae), vlv. **88.** Posterior segments, showing branched blood vessels of *Lumbriculus variegatus* (Lumbriculidae), lv. **89.** *Rhyacodrilus* cf. *subterraneus* (Tubificidae), dorsal bundle of segment II. **90.** Chaeta of *Chaetogaster* cf. *krasnopolskiae* (Naididae). lv, lateral view; vlv, ventrolateral view. (83, 84 [left], 85 [left] redrawn from Hiltunen and Klemm 1980; 84 [middle, right], 85 [right], 86, 88 modified from Brinkhurst 1986; 87 redrawn from Brinkhurst and Jamieson 1971.)

92a (91a). Teeth of needles slightly divergent (Fig. 81); commonly, some enlarged ventral chaetae of segments VI and posterior bear a large tooth that is strongly bent . . . . *N. bretscheri*
92b. Teeth of needles parallel (Fig. 82); teeth of enlarged ventral chaetae of segments VI and posterior not strongly bent . . . . *N. pardalis*

93a (91b). Teeth of needles long and parallel (Fig. 83) . . . . *N. elinguis*
93b. Teeth of needles short (Figs. 84, 85) 94

94a (93b). Ventral chaetae of segments II–V with teeth strongly unequal in length and distinctly different from those of segments VI and posterior; teeth of needles small, obscure, and almost parallel (Fig. 84) . . . . *N. variabilis*
94b. Ventral chaetae of segments II–V with teeth more or less equal in length and similar to those of segments VI and posterior; teeth of needles small but distinct, and a little divergent (Fig. 85) . . . . *N. communis*
N. variabilis and *N. communis* are difficult to distinguish.

95a (6a, 7a). **Lumbriculidae**: Animal with a proboscis; chaetae simple pointed (Fig. 86) . . . . *Eclipidrilus lacustris*
95b. Animal without a proboscis (Fig. 87); chaetae bifurcate (Fig. 10) 96

96a (95b). Animal dark red in life, often greenish anteriorly; posterior blood vessels branched (Fig. 88); no permanently everted penes . . . *Lumbriculus variegatus*
96b. Animal pale in life; posterior blood vessels with short lateral pouches; mature specimens with permanently everted penes on segment X (Fig. 87) . . . . *Stylodrilus heringianus*

97a (69b). Needles simple-pointed . . . . *B. unidentata*
97b. Needles bifurcate . . . . *B. bilongata*

References on Aquatic Oligochaeta Systematics

(*Used in construction of key.)

Bonomi, G., and C. Erséus (eds.). 1984. Aquatic Oligochaeta. Hydrobiologia 115:1–240.

Brinkhurst, R.O. 1978. Freshwater Oligochaeta in Canada. Can. J. Zool. 56:2166–2175.

——. 1979. Addendum: freshwater Oligochaeta in Canada. Can. J. Zool. 57:1569.

——. 1981. A contribution to the taxonomy of the Tubificinae (Oligochaeta: Tubificidae). Proc. Biol. Soc. Wash. 94:1048–1067.

——. 1982. Evolution in the Annelida. Can. J. Zool. 60:1043–1059.

——. 1984. The position of the Haplotaxidae in the evolution of oligochaete annelids. Hydrobiologia 115:25–36.

——. 1985. The generic and subfamilial classification of the Naididae (Annelida: Oligochaeta). Proc. Biol. Soc. Wash. 98:470–475.

*——. 1986. Guide to the freshwater aquatic microdrile oligochaetes of North America. Can. Spec. Publ. Fish. Aquat. Sci. 84:1–259.

——. 1988. A taxonomic analysis of the Haplotaxidae (Annelida, Oligochaeta). Can. J. Zool. 66:2243–2252.

Brinkhurst, R.O., and D.G. Cook (eds.). 1980. Aquatic oligochaete biology. Plenum, New York. 529 pp.

Brinkhurst, R.O., and R.J. Diaz (eds.). 1987. Aquatic Oligochaeta. Hydrobiologia 155:1–323.

*Brinkhurst, R.O., and B.G. Jamieson. 1971. Aquatic Oligochaeta of the world. Oliver and Boyd, Edinburgh, England. 860 pp.

Brinkhurst, R.O., and R.D. Kathman. 1983. A contribution to the taxonomy of the Naididae (Oligochaeta) of North America. Can. J. Zool. 61:2307–2312.

Brinkhurst, R.O., and M.J. Wetzel. 1984. Aquatic Oligochaeta of the world: supplement. Can. Tech. Rep. Hydrogr. Ocean Sci. 44:1–101.

Bunke, D. 1967. Zur Morphologie und Systematik der Aeolosomatidae und Potamodrilidae nov. fam. (Oligochaeta). Zool. Jahrb. Abt. Syst. Okol. Geogr. 94:187–368.

Chekanovskaya, O.V. 1981. Aquatic Oligochaeta of the U.S.S.R. (English translation). Amerind Press, New Delhi. 513 pp.

Erséus, C. 1987. Phylogenetic analysis of the aquatic Oligochaeta under the principle of parsimony. Hydrobiologia 155:75–89.

Foster, N. 1972. Freshwater polychaetes (annelids) of North America. Biota Freshwat. Ecosyst. Ident. Man. 4:1–15.

*Hickman, C.P., Jr., L.S. Roberts, and F.M. Hickman. 1984. Integrated principles of zoology, 7th ed. Times Mirror/Mosby College Publ., St. Louis. 472 pp.

*Hiltunen, J.K., and D.J. Klemm. 1980. A guide to the Naididae (Annelida: Clitellata: Oligochaeta) of North America. EPA-600/4-80/031. 48 pp.

Holmquist, C. 1985. A revision of the genera *Tubifex* Lamarck, *Ilyodrilus* Eisen, and *Potamothrix* Vejdovsky and Mrazek (Oligochaeta, Tubificidae) with extensions to some connected genera. Zool. Jahrb. Abt. Syst. Okol. Geogr. 112:311–366.

Howmiller, R.P., and M.S. Loden. 1976. Identification of Wisconsin Tubificidae and Naididae. Trans. Wisc. Acad. Sci. Arts Lett. 64:185–197.

Kasprzak, K. 1981. Taxonomic problems of species determination of the Enchytraeidae (Oligochaeta), with key to genera. Biológia (Bratislava) 36:943–951.

——. 1984. Generic criteria in Enchytraeidae (Oligochaeta) family. Biológia (Bratislava) 39:163–172.

Kathman, R.D. 1985. Synonymy of *Pristinella jenkinae* (Oligochaeta: Naididae). Proc. Biol. Soc. Wash. 98:1022–1027.

*Klemm, D.J. (ed.). 1985. A guide to the freshwater Annelida (Polychaeta, naidid and tubificid Oligochaeta, and Hirudinea) of North America. Kendall/Hunt Publ. Co., Dubuque, Iowa. 198 pp.

Learner, M.A., G. Lochhead, and B.D. Hughes. 1978. A review of the biology of British Naididae, with emphasis on the lotic environment. Freshwat. Biol. 8:357–375.

Loden, M.S. 1981. Reproductive ecology of Naididae (Oligochaeta). Hydrobiologia 83:115–123.

Milligan, M.R. 1986. Separation of *Haber speciosus* (Hrabě) (Oligochaeta: Tubificidae) from its congeners, with a description of a new form from North America. Proc. Biol. Soc. Wash. 99:406–416.

Nemec, A.F.L., and R.O. Brinkhurst. 1987. A comparison of methodological approaches to the subfamilial classification of the Naididae (Oligochaeta). Can. J. Zool. 65:691–707.

Nielsen, C.O., and B. Christensen. 1959. The Enchytraeidae. Critical revision and taxonomy of European species. Nat. Jutland. 8–9:1–160.

*Pennak, R.W. 1978. Freshwater invertebrates of the United States, 2nd ed. John Wiley and Sons, New York. 803 pp.

*Spencer, D.R. 1978. The Oligochaeta of Cayuga Lake, New York, with a redescription of *Potamothrix bavaricus* and *P. bedoti*. Trans. Amer. Microscop. Soc. 97:139–147.

*Sperber, C. 1948. A taxonomical study of the Naididae. Zool. Bidr. Uppsala 28:1–296.

Steinlechner, R. 1987. Identification of immature tubificids (Oligochaeta) of Lake Constance and its influence on the evaluation of species distribution. Hydrobiologia 155:57–63.
Stephenson, J. 1930. The Oligochaeta. Clarendon Press, Oxford. 978 pp.
*Stimpson, K.S., D.J. Klemm, and J.K. Hiltunen. 1982. A guide to the freshwater Tubificidae (Annelida: Clitellata: Oligochaeta) of North America. EPA-600/3-82/033. 61 pp.
*Strayer, D. 1983. *Piguetiella blanci*, a naidid oligochaete new to North America, with notes on its relationship to *Piguetiella michiganensis* and *Specaria josinae*. Trans. Amer. Microscop. Soc. 102:349–354.
Strayer, D., and E. Bannon-O'Donnell. 1988. Aquatic microannelids (Oligochaeta and Aphanoneura) of underground waters of southeastern New York. Amer. Midl. Nat. 119:327–335.
Tynen, M.J. 1975. A checklist and bibliography of the North American Enchytraeidae (Annelida: Oligochaeta). Syllogeus 9:1–14.
Whitley, L.S. 1982. Aquatic Oligochaeta. *In* A.R. Brigham, W.U. Brigham, and A. Gnilka (eds.). Aquatic insects and oligochaetes of North and South Carolina, pp. 2.1–2.29. Midwest Aquatic Enterprises, Mahomet, Ill.

19 Hirudinea

Donald J. Klemm

Classification

The class Hirudinea (see note to Checklist of Freshwater Hirudinea) in the phylum Annelida (segmented worms) comprises the leeches, the most highly specialized of the major annelid groups (classes Hirudinea, Oligochaeta, and Polychaeta). Although leeches are thought to be closely related to oligochaetes, they are anatomically and behaviorally more advanced. Because of the close relationship of Hirudinea and Oligochaeta, various classification schemes of the higher taxa (above family group rank) have been proposed. Some authors place both classes in the superclass Clitellata, while others regard them as subclasses of the class Clitellata. Clark (1969) considers the Hirudinea a separate class, and Brinkhurst (1982) concluded from his study of annelid evolution that the leeches and oligochaetes should be kept as separate classes. McKey-Fender and Fender (1988) stated that the relationship of the leeches to other oligochaetes must be decided not on the basis of the pharynx, which is influenced by the mode of life, but by an exhaustive analysis of all features of leech morphology in their most primitive conditions. Most authors of invertebrate zoology textbooks and students of the leeches and oligochaetes regard them as separate classes.

Leeches are typically dorsoventrally flattened annelids with suckers at both ends and 34 body segments (designated I–XXXIV), which are externally divided into a number of annuli. Most species in North America are found in fresh and marine waters, but many terrestrial species occur in tropical regions. As predators, parasites of animals, vectors of parasites, and as food for semiaquatic and aquatic animals, leeches are important components of food webs. About 65 species are known to occur in North America (Davies 1973; Klemm 1982, 1985; Sawyer 1986b). In northeastern North America (i.e., New England, New York, New Jersey, Delaware, Maryland, Pennsylvania, and eastern Ohio and eastern Ontario to Newfoundland), there are about 42 species.

The classification scheme used here is modified from Moore 1959; Davies 1971; Klemm 1972a, 1976, 1977, 1982, 1985; Sawyer 1972, 1986b; and Pennak 1978. The class Hirudinea comprises two orders: Arhynchobdellida and Rhynchobdellida. The arhynchobdellids are divided into three families: Haemopidae, Hirudinidae, and Erpobdellidae. The haemopids and hirudinids have relatively large mouths, occupying the entire cavity of the oral sucker; a noneversible pharynx; muscular pharyngeal ridges; poorly developed jaws with vestigial or no teeth or well-developed jaws and teeth; five annuli on each segment, counted in the middle of the body; testisacs that are large and usually in 10 metameric pairs; and red blood. The haemopids are chiefly aquatic or amphibious, are good swimmers, and are considered blood-sucking or predaceous leeches. The hirudinids are aquatic, are also good swimmers, and are considered truly sanguivorous leeches. The erpobdellids have medium-sized mouths that occupy the entire cavity of the oral sucker; a muscular pharynx that is nonprotrusible; no jaws or teeth; five annuli on each segment, counted in the middle body region, and the annuli often are subdivided. The testisacs of the erpobdellids are very small and numerous in a grape-bunch arrangement, and the blood is red. The erpobdellids are strictly aquatic, are good swimmers, and prey on small invertebrates.

The rhynchobdellids are strictly aquatic leeches that have small, porelike mouths in the oral sucker, through which a muscular pharyngeal proboscis can be protruded. They have no jaws or teeth and have colorless blood. The families Glossiphoniidae and Piscicolidae belong to this order. The glossiphoniids have three annuli on each body segment. Many are ectoparasites on both invertebrates and vertebrates, and some forms are predaceous on invertebrates. The piscicolids usually have more than three annuli on each body segment and are parasites of many fishes and rarely of crustaceans.

Life History

Leeches are hermaphroditic but do not self-fertilize. In most species, one or both members of a copulating pair implant spermatophores on the ventral and dorsal epidermal surface of the other. Sperm leave the spermatophores and penetrate the recipient leech, fertilizing the eggs in the ovaries. In the Haemopidae and Hirudinidae, each member of the copulating pair deposits sperm from its penis into the vagina of the recipient. Some groups of leeches secrete ringlike structures that form cocoons around the developing eggs, as do terrestrial earthworms. Cocoons are fastened to the substrate or buried superficially. After a variable period, juvenile leeches emerge from the cocoons and become independent. The glossiphoniids carry their fertilized eggs in membranous sacs or in thin, transparent cocoons on the ventral surface of their bodies, or they deposit the cocoons containing the eggs on the substrate and lie on top of them. Hatched juvenile leeches remain attached to the parent's body, feeding on the mucus until they become rather large, and then leave the parent. The parasitic piscicolids and some species of glossiphoniids typically remain on the bodies of their fish hosts except during the breeding season; however, some species of piscicolids deposit their cocoons on their fish hosts (Pennak 1978) or use the carapace of crustaceans as a substrate for the deposition of their cocoons

(Sawyer 1986a). Copulation is common in the spring and summer, and sometimes into autumn. Some leeches have been reported to live for as many as 15 years; others can have two generations a year (Pennak 1978).

Habitat

Leeches are most common in warm, protected shallows where there is little disturbance from currents. Free-living leeches avoid light and generally hide and are active or inactive under stones or other inanimate objects, among aquatic plants, or in detritus. Some species are most active at night. Silted substrates are unsuitable for leeches because they cannot attach. Leeches are usually rare in calcium-poor waters (Pennak 1978). Some species can tolerate mild pollution. Broad ecological studies and reviews of the group in North America are given in Mann 1962, Klemm 1972b, and Sawyer 1974, 1986b. Freshwater leeches can be found in lakes, ponds, springs, streams, or marshes; others are considered amphibious, especially the genus *Haemopis*, because they commonly crawl around eating invertebrates on the shores of bodies of water and have been collected as far as 1.5 km from the water's edge.

Feeding

The occurrence and abundance of food organisms play an important role in the distribution of many leech species (Klemm 1972b, 1985; Sawyer 1974, 1986b). Generally, any factor that disrupts the distribution patterns of the host or prey directly affects the leeches. The order Rhynchobdellida (the leeches with a proboscis) contains species that feed on a variety of animals. Some species of the family Glossiphoniidae are carnivorous on snails, aquatic insects, or oligochaetes; others are temporary parasites on fish, amphibians, reptiles, and waterfowl and have on occasion been reported sucking human blood. The family Piscicolidae is exclusively parasitic, but only on fish or crustaceans. Glossiphoniids release digestive enzymes from their probosces to digest the epidermis and to prevent the blood of the host from clotting, permitting blood flow (Sawyer 1986b). The Erpobdellidae, order Arhynchobdellida, prey on small invertebrates. Species in the families Haemopidae and Hirudinidae, order Arhynchobdellida, are predatory or bloodsuckers. The sanguivorous forms have three sharp-toothed jaws in their large mouths that are used for making incisions in the host's epidermis. Two species of haemopids in North America, however, *Haemopis grandis* and *H. plumbea*, have vestigial teeth or no teeth and are predaceous only on other invertebrates. Species of the hirudinid genera *Macrobdella* and *Philobdella* are specialized for sucking the blood of vertebrates. *Philobdella gracilis* occurs in the southern United States, but has been reported from Michigan and Illinois. *Macrobdella* species occur throughout the United States and Canada. These species attach to the host (mammals and frogs) with their caudal suckers and seek a proper feeding spot with their anterior end. When engorged, they may have increased their body weight fivefold (Pennak 1978). After they are sated, the leeches leave their hosts, but the incision usually continues to bleed for some time because of the presence of hirudin, the salivary anticoagulant. The European species of Hirudinidae, *Hirudo medicinalis* (the medicinal leech), was used by doctors in the eighteenth and nineteenth centuries to stimulate bleeding, a presumed cure for human diseases. This species is imported today into the United States and Canada and is used for various medical purposes, including microsurgery.

Respiration

Freshwater leeches have no specialized respiratory structure and respire through a highly vascularized epidermis (except that some species of the genera *Piscicola* and *Cystobranchus* have locomotor pulsatile vesicles, filled with coelomic fluid, along the lateral margins of the body that pulsate rhythmically and also function as accessory respiratory organs). Behavioral regulation of oxygen intake may be one function of the body undulations commonly observed in leeches (Pennak 1978). Many marine forms have pulsatile vesicles or lateral gills.

Collection and Preservation

Leeches can be collected with dip nets, by examining objects in the water, or by collecting their hosts and examining them for leeches. To properly preserve live specimens for identification, one should first narcotize them. Pennak (1978), Richardson (1975), Madill (1983) and Klemm (1982, 1985) suggest several possible narcotizing agents. Specimens should be kept in the narcotic until all movement has stopped and they no longer respond to gentle probing. Some specimens may require up to several hours for narcotization. A simple alternative method is to place them in carbonated water and leave them there until they stop moving. After relaxation, the leeches should be placed in the bottom of a dish or pan for 24 hours in 5 or 10% formalin to fix the tissue. The leeches should then be washed and preserved in 70% ethyl alcohol or 10% buffered formalin.

Identification

The identification of leeches to species is difficult or impossible with improperly preserved specimens. The main external features used in the key are body shape and pigmentation, structure and size of the oral and caudal suckers, pedicle separating caudal sucker from the body, eyes, accessory eyes, and eyespots, pulsatile vesicles, papillae on the dorsal body surface, body annulation, copulatory gland pores, and annuli between gonopores. The internal characters used are the jaws and teeth, the velum, and the internal ridges of the pharynx.

More-Detailed Information

Sawyer 1972, 1986b; Klemm 1976, 1977, 1982, 1985.

Checklist of Hirudinea

Order Arhynchobdellida
- Haemopidae
  - *Haemopis grandis*
  - (= *Mollibdella grandis*)
  - *H. marmorata*
  - (= *Percymoorensis marmorata*)
  - *H. plumbea*
  - (= *Bdellarogatis plumbeus*)
  - *H. terrestris*[a]
  - (= *Percymoorensis lateralis*)
- Hirudinidae
  - *Macrobdella decora*
  - *M. sestertia*
- Erpobdellidae
  - *Erpobdella dubia*
  - (= *Dina dubia*)
  - *E. parva*
  - (= *D. parva*)
  - *E. punctata*
  - *Mooreobdella fervida*
  - *M. melanostoma*
  - *M. microstoma*
  - *M. tetragon*
  - *Nephelopsis obscura*

Order Rhynchobdellida
- Glossiphoniidae
  - *Actinobdella annectens*[b]
  - *A. inequiannulata*
  - *A. pediculata*
  - (= *Placobdella pediculata*)
  - *Alboglossiphonia heteroclita*
  - *Batracobdella paludosa*[c]
  - *B. phalera*
  - *Gloiobdella elongata*
  - (= *Helobdella elongata*)
  - *Glossiphonia complanata*
  - *Helobdella fusca*
  - *H. papillata*
  - *H. stagnalis*
  - *H. triserialis*
  - (= *H. lineata*)
  - *Marvinmeyeria lucida*
  - *Placobdella hollensis*
  - *P. montifera*
  - *P. nuchalis*
  - *P. ornata*
  - *P. papillifera*
  - *P. parasitica*
  - *P. picta*
  - (= *Batracobdella picta*)
  - *Theromyzon biannulatum*
  - *T. tessulatum*
- Piscicolidae
  - *Cystobranchus meyeri*
  - *C. verrilli*
  - *Myzobdella lugubris*
  - *Piscicola geometra*[d]
  - *P. milneri*
  - *P. punctata*
  - *Piscicolaria reducta*

Note: The class Hirudinoidea proposed by Soos (1965) and Stuart (1982) is suppressed here in accordance with the recommendation in the International Code of Zoological Nomenclature (Article 29A) and the illustration in Schmidt and Emerson 1970.

[a]Two mature examples of *Haemopis terrestris* were collected "in the woods near Jamestown, N.Y." in the autumn of 1983 (C. O. Berg, pers. comm., 1983), the first known record of this species from New York State. The specimens are in the aquatic invertebrate collection at Cornell University, Ithaca, N.Y.

[b]Described from a single specimen collected on a snapping turtle (*Chelydra serpentina*) from Lake Erie, Ontario; has not been reported since its original description (Moore 1906); may have been an aberrant form.

[c]Common in some localities of Eurasia. Pawlowski's (1948) report of this species from Pottle Lake near North Sydney, Nova Scotia. Identification is based on a single specimen that is no longer available for verification. A doubtful species in North America.

[d]Common but not abundant in Eurasia. Reported records of this species in North America are questionable; cannot be adequately distinguished from *Piscicola milneri*.

Key to Species of Hirudinea

1a. Mouth a small pore in the oral sucker through which a muscular proboscis can be protruded (Figs. 1–4) 2

1b. Mouth medium to large, occupying entire oral sucker, without a proboscis (Figs. 5–6) 30

2a (1a). Body flat and much wider than head region (Figs. 1, 7, 8), never cylindrical (Figs. 1, 7), except body subcylindrical for *Gloiobdella elongata* (Fig. 8); oral sucker ventral and more or less fused to body (Figs. 1–3); no pulsatile vesicles; eggs in membranous sacs on ventral side of adult, where young are brooded . . . . . **Glossiphoniidae** 3

2b. Body cylindrical and narrow but often divided into a narrow anterior region and a wider posterior region (Figs. 9–13); oral sucker expanded, distinctly separate from body (Figs. 4, 9–13); pulsatile vesicles along lateral margins present (Fig. 12) or absent (Fig. 13); no eggs or brooding young on ventral side of adult; almost invariably found as parasites on fish . . . . . **Piscicolidae** 24

3a (2a). **Glossiphoniidae**: With 1 pair of eyes (Figs. 14–18), but may have a series of paired accessory eyes (Fig. 18) 4

3b. With more than 1 pair of eyes (Figs. 19–23) 20

4a (3a). Oral sucker distinctly expanded to form a discoid head, set off from body by a narrow neck constriction (Fig. 15) 5

4b. Oral sucker not distinctly expanded or set off from body by a narrow neck, but more or less continuous with body (Fig. 14) 6

5a (4a). Dorsum with 3 prominent tuberculate keels or ridges (discernible when preserved) (Fig. 24); 9–16 mm long; free-living, parasitic on fish . . . . . *Placobdella montifera*

5b. Dorsum smooth, no keels or ridges (Fig. 25); 15–25 mm long; free-living, parasitic on fish . . . . . *P. nuchalis*

6a (4b). With a series of accessory eyes behind single pair of functional eyes in cephalic region (Fig. 18); 20–30 mm long; feeding habits unknown . . . . . *Placobdella hollensis*

6b. Without accessory eyes 7

7a (6b). Eyes distinctly separated by at least the diameter of 1 eye (Fig. 14) 8

7b. Eyes separated by less than the diameter of 1 eye (Fig. 16), or eyes touching or fused (Fig. 17) 13

8a (7a). With a brown chitinoid plate (scute) on the dorsal surface of segment VIII (Fig. 26); brown, pink, or green; 9–14 mm long; the most cosmopolitan of all North American leeches . . . . . *Helobdella stagnalis*

8b. Without a scute 9

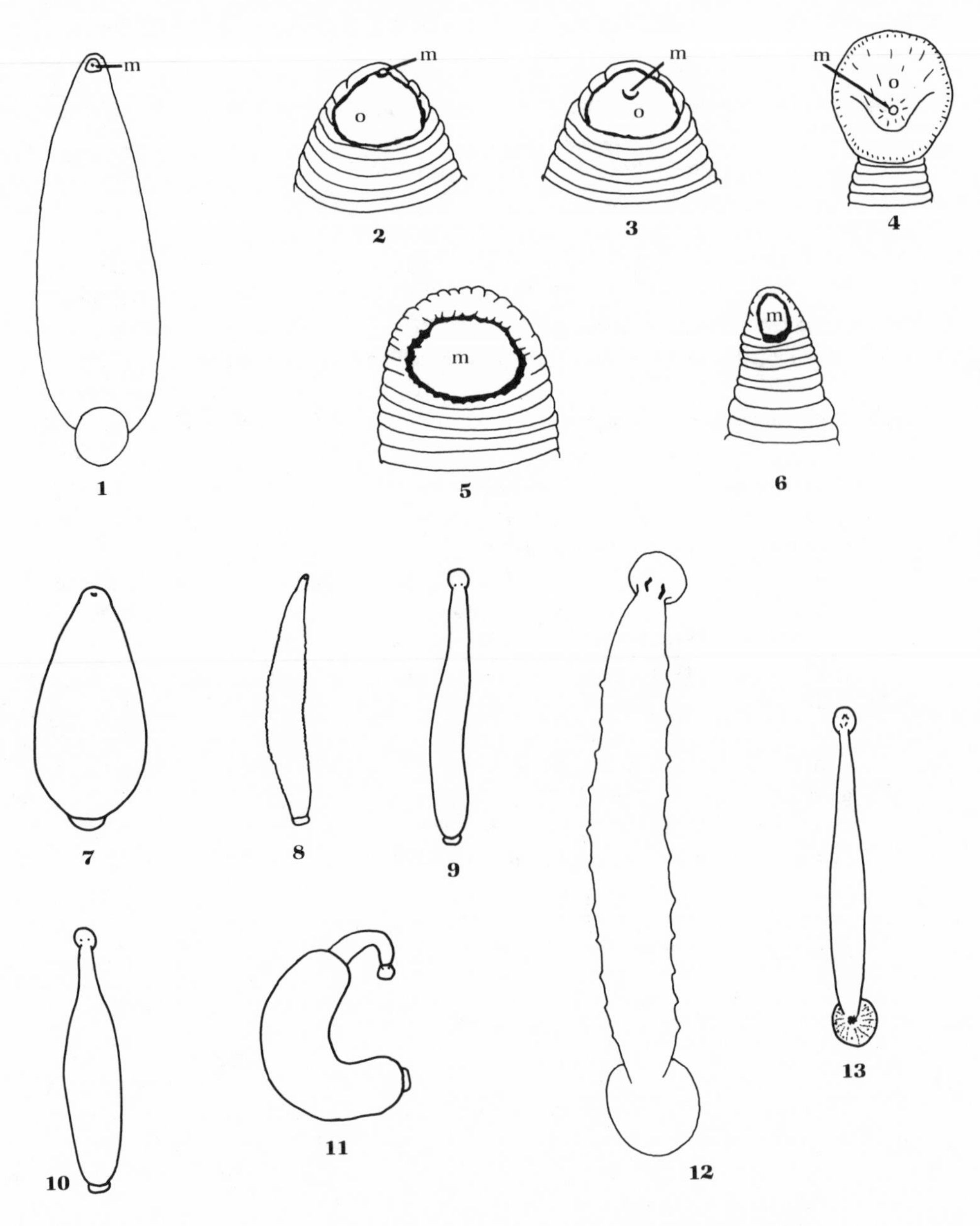

1. *Glossiphonia complanata* (Glossiphoniidae), vv: m, mouth. **2.** Oral sucker (o) with mouth pore (m) on rim, vv. **3.** Oral sucker (o) with mouth pore (m) in sucker, vv. **4.** Oral sucker (o) with mouth pore (m) near center, vv. **5.** Oral sucker with mouth pore (m) occupying entire sucker cavity, vv. **6.** Oral sucker with mouth pore (m) occupying entire sucker cavity, vv. **7.** Glossiphoniid, dv. **8.** *Gloiobdella elongata* (Glossiphoniidae), dv. **9.** *Myzobdella lugubris* (Piscicolidae), dv. **10.** *Myzobdella lugubris* (Piscicolidae), dv. **11.** *Myzobdella lugubris* (Piscicolidae), dv. **12.** *Piscicola punctata* (Piscicolidae) with pulsatile vesicles, dv. **13.** *Piscicola* sp. (Piscicolidae) without pulsatile vesicles, dv. dv, dorsal view; vv, ventral view. (1, 12 modified or redrawn from Pennak 1978; 2–11, 13 modified or redrawn from Klemm 1982.)

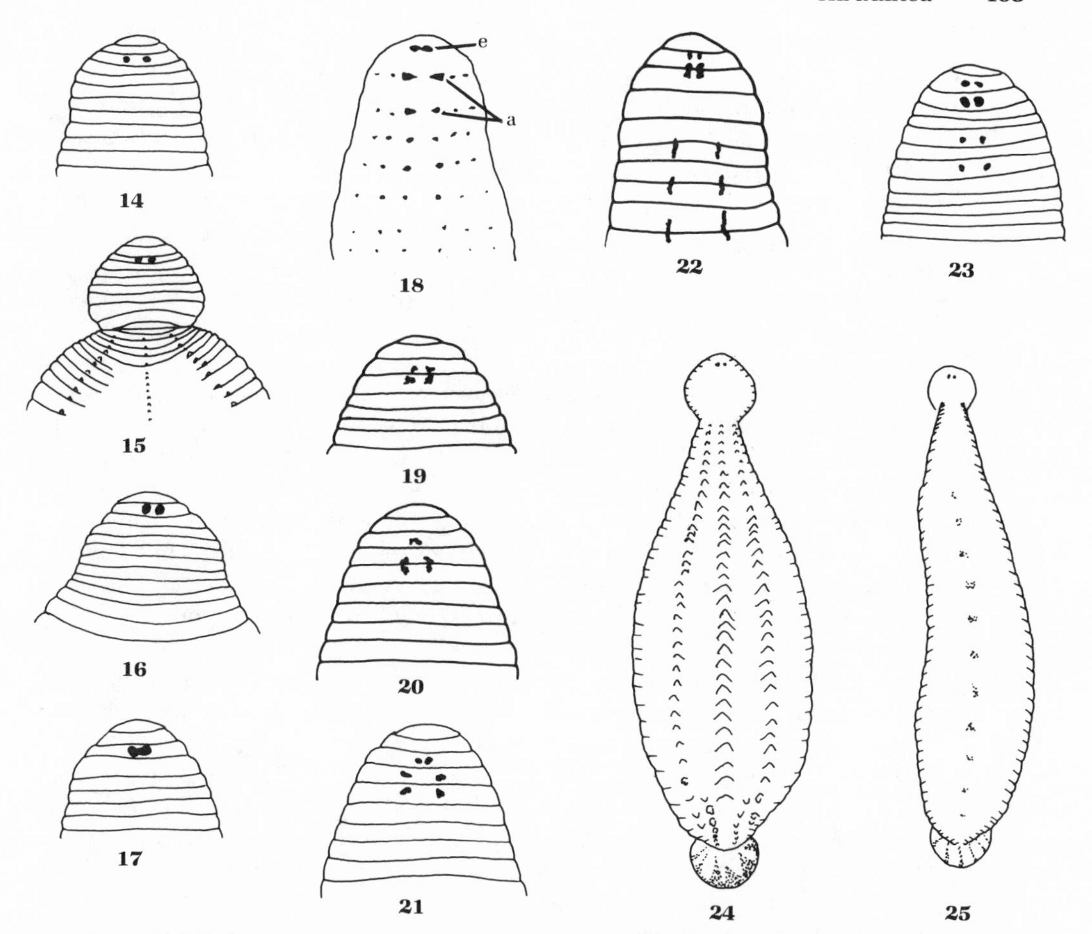

14. Glossiphoniid with eyes well separated, dv. **15.** Glossiphoniid with discoid head, dv. **16.** Glossiphoniid with eyes close together, dv. **17.** Glossiphoniid with eyes fused, dv. **18.** Head of *Placobdella hollensis* (Glossiphoniidae), dv: a, accessory eyes; e, eyes. **19.** Head of *Batracobdella paludosa* (Glossiphoniidae), dv. **20.** Head of *Alboglossiphonia heteroclita* (Glossiphoniidae), dv. **21.** Head of *Alboglossiphonia heteroclita* (Glossiphoniidae), dv. **22.** Head of *Glossiphonia complanata* (Glossiphoniidae), dv. **23.** Head of *Theromyzon* sp. (Glossiphoniidae), dv. **24.** *Placobdella montifera* (Glossiphoniidae), dv. **25.** *Placobdella nuchalis* (Glossiphoniidae), dv. dv, dorsal view. (14–25 modified or redrawn from Klemm 1982.)

9a (8b). Dorsal surface smooth 10

9b. Dorsal surface with 3–9 longitudinal series of papillae, or with scattered papillae 11

10a (9a). Body uniformly pigmented or pigmented with longitudinal lines or stripes; body flattened (Fig. 7) 11

10b. Body unpigmented, opaque white or gray, translucent, elongate, and subcylindrical (Fig. 8); body smoothly rounded with lateral margins almost parallel; dorsal surface without lines or stripes; 9–25 mm long .... *Gloiobdella elongata*

11a (10a). Dorsal surface heavily pigmented with uniform, minute, grayish blue or blackish chromatophores, with thin dark paramedial lines extending into neck region; 15–22 mm long . . . . *Marvinmeyeria lucida*

11b. Dorsal surface pigmented with longitudinal white stripes alternating with brown lines, stripes, or uniformly coffee brown chromatophores; 10–14 mm long . . . . *Helobdella fusca*

12a (9b). Dorsal surface with 3 or fewer incomplete series of papillae or with scattered papillae; papillae black-tipped or uniformly pale white; pigmentation variable; 10–29 mm long . . . . *H. triserialis*

12b. Dorsal surface with 5–9 longitudinal rows of papillae; papillae large, whitish, and rounded; lightly pigmented or unpigmented; 9–14 mm long . . . . *H. papillata*

13a (7b). Dorsum with conspicuous white genital and anal patches, 1 or more medial white patches, and a white ring (bar) in neck region (Fig. 27); 6–10 mm long; free-living or parasitic on fish . . . . *Batracobdella phalera*

13b. Dorsum without white patches and white ring (bar) in neck region 14

14a (13b). Caudal sucker separated from body on a distinct pedicel (Figs. 28, 29); rim of caudal sucker thick and bulbous 15

14b. Caudal sucker continuous with body, not separated from body on a distinct pedicel; rim of caudal sucker thin . . . . *Placobdella* (in part) 17

15a (14a). With a marginal circle of 30–60 digitate processes (retracted when preserved [Fig. 28]; position marked dorsally on outer rim by faint radiating ridges or whitish radiating bands) 16

15b. Without digitate processes on caudal sucker (Fig. 29); 20–35 mm long; parasite on freshwater drum (*Aplodinotus grunniens*) . . . . *Actinobdella pediculata*

16a (15a). Caudal sucker with about 30 digitate processes on rim; 7–22 mm long; free-living, parasitic on fish . . . . *Actinobdella inequiannulata*

16b. Caudal sucker with about 60 digitate processes on rim; 7–11 mm long (see Checklist, note b) . . . . *A. annectens*

17a (14b). Dorsum roughly papillate 18

17b. Dorsum not roughly papillate, either without papillae or with low smooth domes or suppressed papillae 19

18a (17a). Dorsal surface warty, entirely covered with numerous papillae, varying in size or randomly arranged, with or without a median interrupted band; dorsal surface a mixture of brown, green, and yellow, with ventral surface unstriped; up to 40 mm long; free-living or parasitic on turtles . . . . *P. ornata*

18b. Dorsal surface less warty, not entirely covered with papillae; larger papillae in 5–7 longitudinal rows; ventral surface with 2–8 bluish longitudinal stripes; 15–45 mm long; free-living or parasitic on turtles . . . . *P. papillifera*

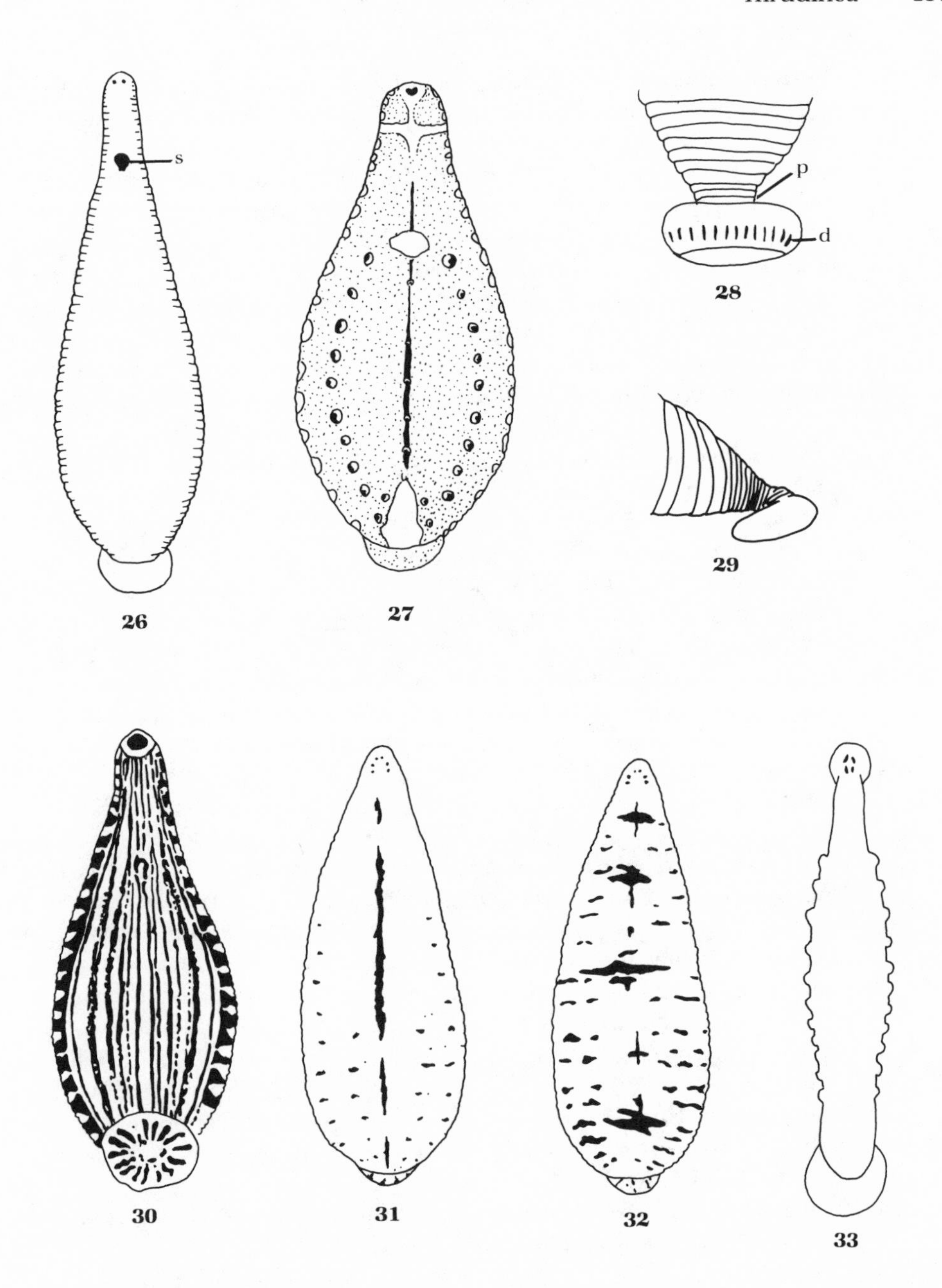

26. *Helobdella stagnalis* (Glossiphoniidae), dv: s, scute. **27.** *Batracobdella phalera* (Glossiphoniidae), dv. **28.** *Actinobdella inequiannulata* (Glossiphoniidae), dv: d, retracted digitate processes; p, pedicel. **29.** Caudal sucker of *Actinobdella pediculata* (Glossiphoniidae) without digitate processes, lv. **30.** *Placobdella parasitica* (Glossiphoniidae), vv. **31.** Pigmented form of *Alboglossiphonia heteroclita* (Glossiphoniidae), dv. **32.** Pigmented form of *Alboglossiphonia heteroclita* (Glossiphoniidae), dv. **33.** *Cystobranchus* (Piscicolidae), dv. dv, dorsal view; lv, lateral view; vv, ventral view. (26–33 modified or redrawn from Klemm 1982.)

19a (17b). Ventral surface with 8–12 bluish, greenish, or brownish longitudinal stripes (Fig. 30); dorsal pattern variable; 38–64 mm long; free-living or parasitic on turtles . . . . . *P. parasitica*

19b. Ventral surface not striped; dorsum dark greenish brown, finely variegated with orange, with a thin dark median line (can be absent in preserved individuals) and with 4 paramedial rows of yellowish or whitish metameric spots; 13–25 mm long; free-living or parasitic on amphibians . . . . . *P. picta*

20a (3b). Two pairs of eyes (Fig. 19); dorsum smooth; 7–20 mm long; 1 dubious record for North America . . . . . *Batracobdella paludosa*

20b. Three or 4 pairs of eyes (Figs. 20–23) 21

21a (20b). Three pairs of eyes (Figs. 20–22) 22

21b. Four pairs of eyes (Fig. 23) 23

22a (21a). Eyes arranged in a roughly triangular pattern, the 1st pair always closer together than the posterior pairs (Figs. 20, 21); body translucent with little pigmentation or with a dark median longitudinal stripe (Fig. 31), and sometimes with 4–7 middorsal, irregular, transverse bars (Fig. 32); 6–9 mm long; free-living or parasitic on mollusks and aquatic insects . . . . . *Alboglossiphonia heteroclita*

22b. Eyes in 2 nearly parallel rows near the median line (Fig. 22); body opaque and deeply pigmented; 14–25 mm long; free-living, predaceous on invertebrates . . . . . *Glossiphonia complanata*

23a (21b). Gonopores separated by 2 annuli; known distribution in central and eastern United States and Canada; free-living or parasitic on waterfowl; 20–26 mm long . . . . . *Theromyzon biannulatum*

23b. Gonopores separated by 4 annuli; known distribution in Europe, but has been reported only from Colorado in the United States and from British Columbia, Saskatchewan, Quebec, and Nova Scotia in Canada; free-living or parasitic on waterfowl; 15–30 mm long . . . . . *T. tessulatum*

24a (2b). **Piscicolidae**: Caudal sucker flattened, as wide or wider than the widest part of the body; with or without pulsatile vesicles on margin of trunk (Figs. 12, 13) 25

24b. Caudal sucker concave, weakly developed, and narrower than the widest part of the body (Figs. 9–11); pulsatile vesicles absent 29

25a (24a). Body not divided into anterior and posterior parts (Figs. 12, 13); pulsatile vesicles small, each covering 2 annuli (may be absent in preserved specimens) . . . . . *Piscicola* 26

25b. Body divided into anterior and posterior parts (Fig. 33); pulsatile vesicles large, each covering 4 annuli . . . . . *Cystobranchus* 28

26a (25a). Caudal sucker without eyespots; 14–16 mm long . . . . . *P. punctata*

26b. Caudal sucker with 8–14 eyespots 27

27a (26b). With 10–12 (usually 10) eyespots on caudal sucker; dark rays absent (Fig. 34); gonopores separated by 2 annuli; 16–24 mm long .... *P. milneri*

27b. With 12–14 eyespots on caudal sucker, separated by dark pigmented rays (Fig. 35); gonopores separated by 3 annuli; 20–30 mm long ...... *P. geometra*

28a (25b). With eyespots on caudal sucker and lateral margins of body; 4–7 mm long .... *C. meyeri*

28b. Without eyespots on caudal sucker and lateral margins; 10–30 mm long .... *C. verrilli*

29a (24b). Dorsum with a series of 6 brownish black longitudinal stripes, medial pair most conspicuous (Figs. 36, 37); 6–8 mm long .... *Piscicolaria reducta*

29b. Dorsum without longitudinal stripes (Figs. 38, 39); 9–30 mm long .... *Myzobdella lugubris*

30a (1b). Five pairs of eyes forming a regular arch (Fig. 40) .... **Haemopidae** and **Hirudinidae** 31

30b. Three or 4 pairs of eyes, never arranged in a regular arch (Figs. 41–43) .... **Erpobdellidae** 36

31a (30a). **Hirudinidae**: With 4 or 24 copulatory gland pores on ventral surface (Figs. 44, 45); ventral surface red or orange .... *Macrobdella* 32

31b. **Haemopidae**: Without copulatory gland pores on ventral surface; ventral surface not red or orange .... *Haemopis* 33

32a (31a). Four copulatory gland pores (2 rows of 2) on ventral surface (Fig. 44); 110–150 mm long .... *M. decora*

32b. Twenty-four copulatory gland pores (with 2 rows of 2 groups containing 6 gland pores each) (Fig. 45); 100–150 mm long; known only from Massachusetts (Smith 1981) .... *M. sestertia*

33a (31b). Dorsum uniformly black or slate grey, with median longitudinal black stripe and reddish to yellowish band along margins (Fig. 46); 150–200 mm long .... *H. terrestris*

33b. Dorsum with moderately to heavily blotched, spotted, or irregularly scattered black flecks, or else uniformly colored olive-green to slate grey; dorsum with no median black stripe and with or without reddish to yellowish bands along margins 34

34a (33b). With jaws and teeth (Fig. 47); 75–100 mm long .... *H. marmorata*

34b. Without jaws and teeth (Fig. 48) or, rarely, with vestigial teeth; 150–300 mm long 35

35a (34b). Margin of oral sucker thin, aperture elongate; lower surface of velum (Fig. 48)

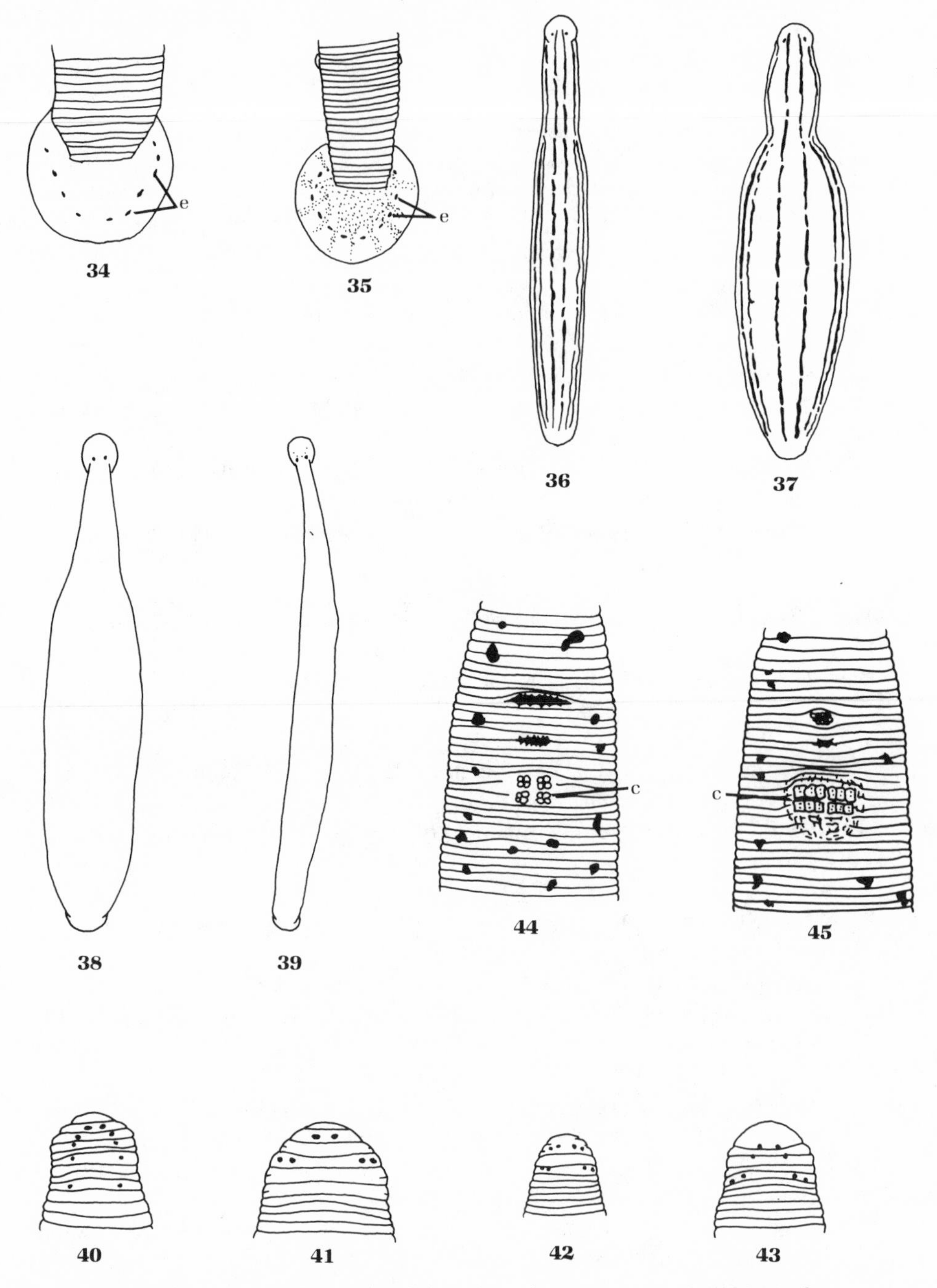

34. *Piscicola milneri* (Piscicolidae) caudal sucker, dv: e, eyespots. **35.** *Piscicola geometra* (Piscicolidae) caudal sucker, dv: e, eyespots and pigmented rays. **36.** *Piscicolaria reducta* (Piscicolidae), dv. **37.** *Piscicolaria reducta* (Piscicolidae), dv. **38.** *Myzobdella lugubris* (Piscicolidae), dv. **39.** *Myzobdella lugubris* (Piscicolidae), dv. **40.** Arrangement of eyes in Haemopidae and Hirudinidae, dv. **41.** Arrangement of eyes in Erpobdellidae, dv. **42.** Arrangement of eyes in erpobdellid, dv. **43.** Arrangement of eyes in erpobdellid, dv. **44.** *Macrobdella decora* (Hirudinidae), vv: c, copulatory gland pores. **45.** *Macrobdella sestertia* (Hirudinidae), vv: c, copulatory gland pores. dv, dorsal view; vv, ventral view. (34–45 modified or redrawn from Klemm 1982.)

smooth; pharynx with 12 folds (Fig. 49); common in northeastern United States and Canada; 150–225 mm long . . . . *H. grandis*

35b. Margin of oral sucker thick, rounded, aperture transverse; lower surface of velum (Fig. 48) closely and finely papillate; pharynx with 15 folds (Fig. 50); uncommon, known in United States from Great Lakes region, in Canada from Ontario and Quebec; 150–300 mm long . . . . *H. plumbea*

36a (30b). **Erpobdellidae**: Three pairs of eyes 37

36b. Four pairs of eyes 41

37a (36a). Dorsum with scattered black pigment concentrations, body whitish; gonopores in furrows, separated by 2 annuli; length to 55 mm . . . . *Mooreobdella melanostoma*

37b. Dorsum either lacking pigment or uniformly grey, almost black, or heavily barred, or with longitudinal rows of black pigment concentrations, but never with scattered black pigment concentrations 38

38a (37b). With 3–4½ annuli between gonopores (Figs. 51–53); gonopores in furrows or on rings 39

38b. With 2–2½ annuli between gonopores (Figs. 54, 55); gonopores in furrows or on rings 40

39a (38a). Gonopores separated by 3 annuli, usually in furrows (Fig. 51); 30–50 mm long . . . . *Mooreobdella microstoma*

39b. Gonopores separated by 4–4½ annuli, usually on rings (Figs. 52, 53); up to 40 mm long . . . . *M. tetragon*

40a (38b). Segments mostly composed of 5 equal annuli; dorsum with 2 or 4 longitudinal rows of black pigment concentrations, some individuals with black bars or almost black on dorsum (Figs. 56–60); up to 100 mm long . . . . *Erpobdella punctata*

40b. Segments mostly composed of 6 or 7 unequal annuli, with every 5th annulus wider and slightly subdivided ; dorsum a uniform smokey grey (Fig. 61), unpigmented, or with darker clouding in places, or with 2 dark longitudinal stripes always including a lighter median stripe (Fig. 62); 20–50 mm long . . . . *Mooreobdella fervida* (in part)

41a (36b). Gonopores separated by 2 annuli 42

41b. Gonopores separated by 2½–4 (usually 3½) annuli 43

42a (41a). Four pairs of eyes, with anterior 2 pairs and posterior 2 pairs arranged almost in parallel (Fig. 42); up to 100 mm long . . . . *Nephelopsis obscura*

42b. Sometimes 4 (usually 3) pairs of eyes, not arranged in parallel (Fig. 43); 20–50 mm long . . . . *Mooreobdella fervida* (in part)

43a (41b). Dorsum greenish with a variable dark brown or black middorsal stripe, ob-

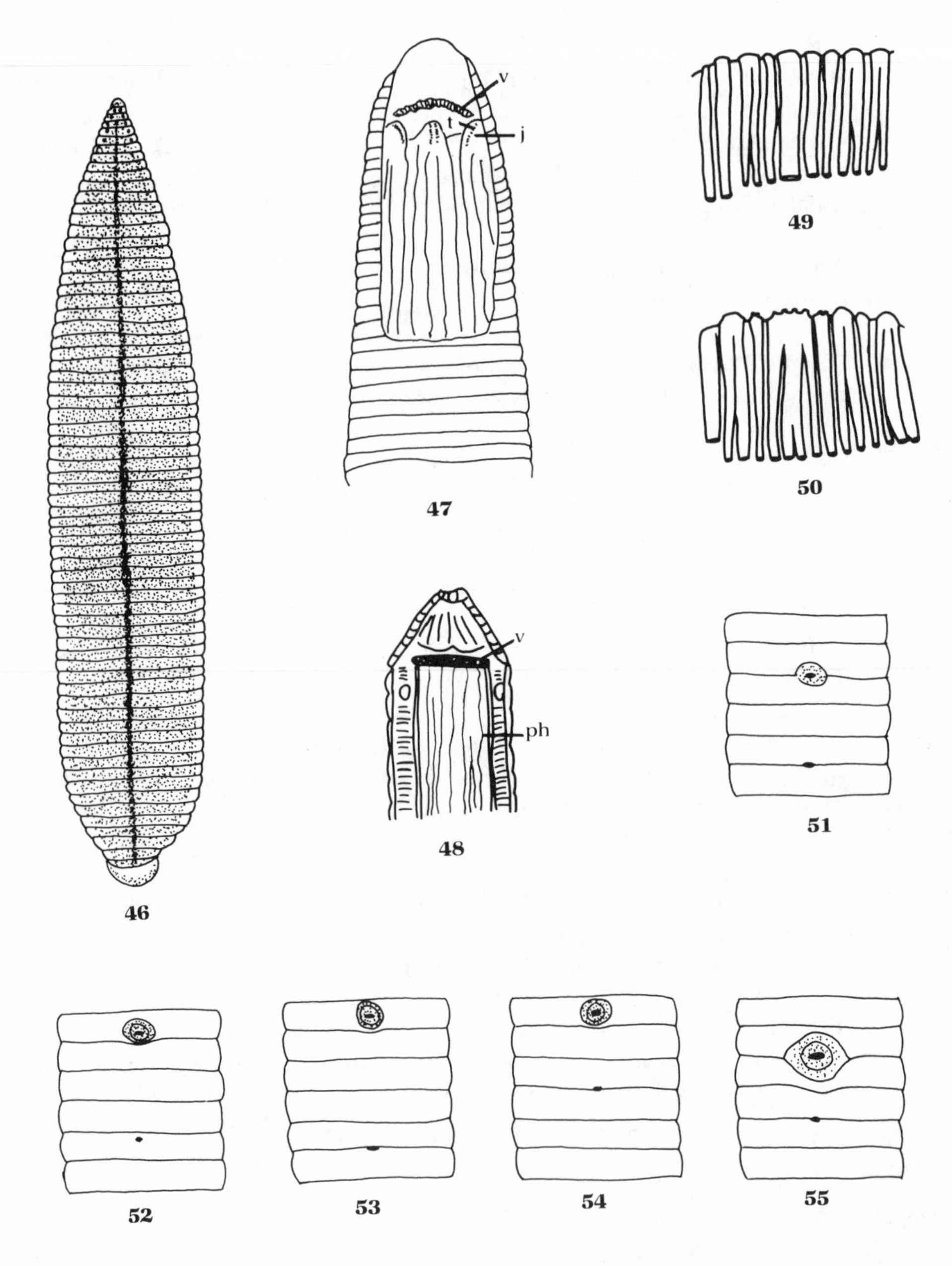

46. *Haemopis terrestris* (Haemopidae), dv. **47.** Dissected mouth of *Haemopis marmorata* (Haemopidae), vv: j, jaws; t, teeth; v, velum. **48.** Dissected mouth of *Haemopis* sp. (Haemopidae) without jaws and teeth, vv: ph, pharynx folds; v, velum. **49.** Pharynx folds of *Haemopis grandis* (Haemopidae), vv. **50.** Pharynx folds of *Haemopis plumbea* (Haemopidae), vv. **51.** Gonopores of *Mooreobdella microstoma* (Erpobdellidae), vv. **52.** Gonopores of *Mooreobdella tetragon* (Erpobdellidae), vv. **53.** Gonopores of *Mooreobdella tetragon* (Erpobdellidae), vv. **54.** Gonopores of *Erpobdella punctata* (Erpobdellidae), vv. **55.** Gonopores of *Erpobdella punctata* (Erpobdellidae), vv. dv, dorsal view; vv, ventral view. (46–55 modified or redrawn from Klemm 1982.)

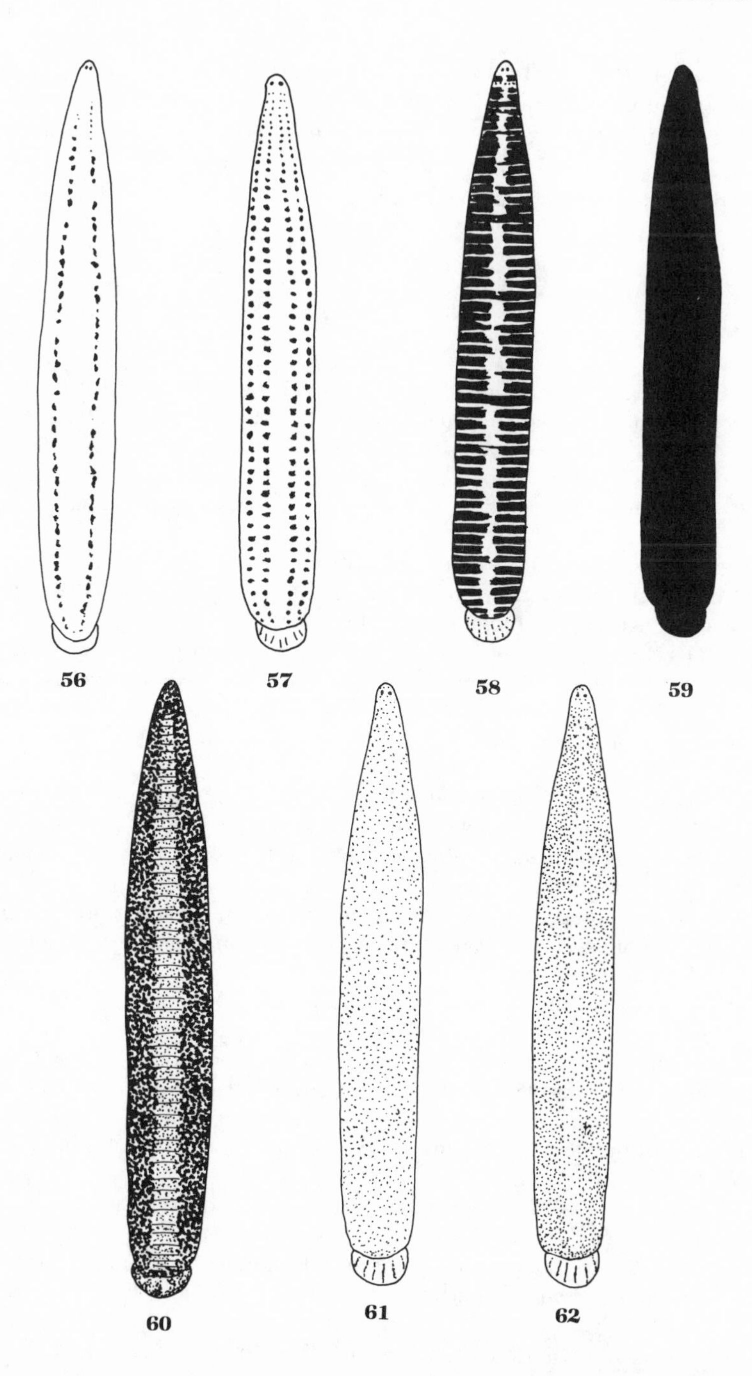

56. *Erpobdella punctata* (Erpobdellidae), dv. **57.** *Erpobdella punctata* (Erpobdellidae), dv. **58.** *Erpobdella punctata* (Erpobdellidae), dv. **59.** *Erpobdella punctata* (Erpobdellidae), dv. **60.** *Erpobdella punctata* (Erpobdellidae), dv. **61.** *Mooreobdella fervida* (Erpobdellidae), dv. **62.** *Mooreobdella fervida* (Erpobdellidae), dv. dv, dorsal view. (56–62 modified or redrawn from Klemm 1982.)

vious in anterior half, sometimes fading posteriorly; gonopores separated by 3½ (sometimes 4) annuli; 20–60 mm long .... *Erpobdella dubia*

43b. Dorsum unpigmented or uniformly smokey grey with variegated dark and light pigment, lacking a middorsal stripe; gonopores separated by 3½ (sometimes 2 or 2½) annuli; 25–30 mm long .... *E. parva*

References on Hirudinea Systematics

(*Used in construction of key.)

Brinkhurst, R.O. 1982. Evolution in the Annelida. Can. J. Zool. 60:1043–1059.

Clark, R.B. 1969. Systematics and phylogeny: Annelida, Echiura, Sipuncula. Chem. Zool. 4:1–68.

Davies, R.W. 1971. A key to the freshwater Hirudinoidea of Canada. J. Fish. Res. Bd. Can. 28:543–552.

——. 1973. The geographic distribution of freshwater Hirudinoidea in Canada. Can. J. Zool. 51:531–545.

Klemm, D.J. 1972a. Freshwater leeches (Annelida: Hirudinea) of North America. Ident. Man. no. 8, Biota of freshwater ecosystems. Water Poll. Contr. Res. Ser. 18050 ELD05/72, U.S. Environmental Protection Agency, Washington, D.C. 54 pp.

——. 1972b. The leeches (Annelida: Hirudinea) of Michigan. Mich. Acad. 4:405–444.

——. 1976. Leeches (Annelida: Hirudinea) found in North American mollusks. Malacol. Rev. 9:63–76.

——. 1977. A review of the leeches (Annelida: Hirudinea) in the Great Lakes Region. Mich. Acad. 9:397–418.

*——. 1982. Leeches (Annelida: Hirudinea) of North America. EPA-600/3-82-025. 195 pp.

——. 1985. Freshwater leeches (Annelida: Hirudinea). *In* D.J. Klemm (ed.). A guide to the freshwater Annelida (Polychaeta, Naidid and Tubificid Oligochaeta, and Hirudinea) of North America, pp. 70–173. Kendall/Hunt Publ. Co., Dubuque, Iowa.

Madill, J. 1983. The preparation of leech specimens: relaxation, the key to preservation. *In* D.J. Baber (ed.). Proceedings of 1981 workshop on care and maintenance of natural history collections. Syllogenus 44:37–41.

Mann, K.H. 1962. Leeches (Hirudinea) their structure, physiology, ecology and embryology. Pergamon, New York.

McKey-Fender, D., and W.M. Fender. 1988. *Phagodrilus* gen. nov. (Lumbriculidae): systematics and biology of a predaceous oligochaete from western North America. Can. J. Zool. 66:2304–2311.

Moore, J.P. 1906. Hirudinea and Oligochaeta collected in the Great Lakes Region. Bull. U.S. Bur. Fish. 25:153–172.

——. 1959. Hirudinea. *In* W.T. Edmondson (ed.). Fresh-water biology, 2nd ed., pp. 542–557. John Wiley and Sons, New York.

Pawlowski, L.K. 1948. Contribution a la connaissance des Sangsues (Hirudinea) de la Nouvelle-Ecosse, de Terre-Nueve et des iles francaises Saint-Pierre et Miquelon. Frag. Faun. (Warsaw) 5:317–353.

*Pennak, R.W. 1978. Freshwater invertebrates of the United States, 2nd ed. John Wiley and Sons, New York. 803 pp.

Richardson, L.R. 1975. A convenient technique for killing and preserving leeches for general study. J. Parasitol. 61:78.

Sawyer, R.T. 1972. North American freshwater leeches, exclusive of the Piscicolidae, with a key to all species. Ill. Biol. Monogr. 46:1–154.

——. 1974. Leeches (Annelida: Hirudinea). *In* C.W. Hart, Jr., and S.L.H. Fuller (eds.). Pollution ecology of freshwater invertebrates, pp. 81–142. Academic Press, New York.

——. 1986a. Leech biology and behavior. Vol. 1: Anatomy, physiology and behavior. Oxford Univ. Press, New York. 417 pp.

——. 1986b. Leech biology and behavior. Vol. 2: Feeding biology, ecology, and systematics, pp. 419–793. Oxford Univ. Press, New York.

Schmidt, K.P., and A.E. Emerson. 1970. Taxonomy. *In* Encyclopedia Britannica, vol. 21, pp. 728–730.

Smith, D.G. 1981. Selected freshwater invertebrate proposed for special concern status in Massachusetts (Molusca, Annelida, Arthropoda). Tech. Ser. Br. Dept. Environ. Qual. Engineering, Div. Water Pollution Control, Westborough, Mass. 26 pp.

Soos, A. 1965. Identification key to the leech (Hirudinoidea) genera of the world with a catalogue of the species: I. Family: Piscicolidae. Acta Zool. Acad. Sci. Hung. 11:417–463.

Stuart, J. 1982. Hirudinoidea. *In* S.P. Parker (ed.). Synopsis and classification of living organisms, vol. 2, pp. 43–50. McGraw-Hill, New York.

Glossary

abdomen The most posterior body region, where the appendages do not form legs; in malacostracans, the last six body segments plus the telson.

accessory flagellum The smaller flagellum in a biramous antenna.

acetabulum (acetabula) Papilla.

acumen The median tooth or point on the rostrum of a crayfish.

acuminate Tapering to a long point.

acute Forming an angle of less than 90°.

aestivation A state of dormancy or torpor during summer or periods of drought.

air straps Retractile respiratory appendages at posterior end of some hemipterans.

ametabolous Without metamorphosis.

amphipneustic With anterior and posterior spiracles functional.

anal tubules Fleshy appendages of the anal segment located near the base of the posterior prolegs of larval chironomids.

anautogenous In adult biting dipterans, requiring a blood meal to acquire enough protein to produce eggs.

angulate Forming an angle; not rounded.

annulate Ringed; surrounded by a ring of a different color; formed in ringlike segments.

annulus A ring or division.

annulus ventralis In female crayfish, a median structure on the posterior part of the ventral surface; contains the sperm in the interval between mating and egg laying; spermatheca.

antenna (antennae) The first pair of appendages in insects, and first two pairs in crustaceans; adapted for feeling and smelling objects in front of the animal.

antennal blade A setalike structure usually arising from the apex of the basal segment of the antenna.

antennule See **first antenna**.

anterior In the direction of the head; the opposite of *posterior*.

anterolateral Toward the anterior and side.

Ap Maximum diameter of the aperture on a gastropod shell.

aperture The large opening in a snail shell through which the animal protrudes.

apex The tip.

apical Toward the tip of a structure or appendage. Apical whorls are those at the very tip of the shell of a gastropod.

apical corona In nymphal plecopterans, a ring of setae encircling the distal end of a cercal segment.

apneustic Lacking spiracles entirely.
apotome A smaller sclerite appearing separated from the main sclerite. In larval trichopterans, the gula.
appendage-b A two- or three-segmented appendage situated at apex of basal maxillary palp segment.
apterous Without wings.
AR (antennal ratio) Length of the basal antennal segment divided by the combined length of the remaining segments.
arcuate Arched; bowlike.
areola The median area of the thoracic part of the carapace of a crayfish separated from the gill covers by the branchiostegal grooves.
atrium A cavity.
attenuate Tapering gradually to a point.
basad The adverbial form of basal.
basal At or pertaining to the base or point of attachment to or nearest the main body.
beak The hump on the dorsal margin of a bivalve shell; sometimes called the *umbo*.
beak sculpture Tiny raised lines on the tips of the beaks of unionid shells. The pattern of beak sculpture is an important taxonomic character.
beard Fringed with hair or long setae.
bicuspid Having two points.
bifid Cleft, or divided into two parts; two-pronged.
bifurcate Divided partly, or forked into two.
bilamellate Divided into two lamellae or plates.
biramous Having two axes; Y-shaped, as in appendages of crustaceans.
body whorl The last, and largest, whorl of a snail shell.
bothriotricha Very fine sensory hairs arising from cuplike pits in the integument of collembolans.
brachypterous With short or abbreviated wings.
branchiostegal groove In crayfish, a groove separating a gill cover from the areola.
bristle A stiff hair, usually short and blunt.
buccal mass In snails, a mass of tissue in the head consisting of the radula and associated structures.
bulbous Bulblike; swollen.
calceolus (calceoli) Linear or paddle-shaped sensory structures found on the antennae in many subterranean amphipods.
capitulum In water mites, a complex sclerite to which the palps attach and which encloses the chelicerae.
carapace A chitinous or bony shield covering the whole or portions of the head, thorax, and abdomen in crustaceans and insects.
cardinal beard Group of setae located near base of ventromental plates in larval chironomids.
cardinal teeth The hinge teeth located near the beaks of sphaeriid shells.
carina (carinae) An elevated ridge or keel.
carinate Ridged or keeled.
carpus The fifth segment of a crustacean leg, equivalent to the tibia of other arthropods.
caudal Toward or in the direction of the tail end of the body along the median line.
cephalothorax In mysidaceans and decapods, the part of the body covered by the carapace; the head and thorax combined.
cercus (cerci) A paired appendage of the last abdominal segment.
cervical Ventral neck region.

cervical groove In crayfish, a groove separating the head region from the thoracic region of the carapace.
cervical spine In crayfish, a projecting spine or tooth located just posterior to the cervical groove.
chaeta (chaetae) Hairlike structures or appendages. In oligochaetes, the term *hair chaetae* refers to long, thin filaments.
chela (chelae) The fourth thoracic appendage of a crayfish, which is greatly enlarged and adapted for grasping.
chelate With a movable claw; pincerlike; adapted for grasping by having a finger of the propodus opposed to the dactyl, as in the chela of a crayfish.
chelicera (chelicerae) In water mites, the most anterior paired appendages, used to pierce, tear, or anchor prey.
chitinoid Composed of chitin.
chloride epithelia Structures involved in ion absorption for osmoregulation in larval trichopterans.
cilia Fringes; series of moderate or thin hair arranged in tufts or single lines.
ciliate Fringed with a row of parallel hairs or cilia.
claval suture Suture between clavus and corium.
clavus Part of hemelytron next to scutellum.
cleft Split; partly divided longitudinally.
clypeus The part of the head between the frons and labrum.
colinear Lying along the same line.
collophore tube Ventral tube in collembolans used for respiration, osmoregulation, and adhesion.
color rays Bands of color, usually green, extending from the beaks of a unionid shell toward the shell margin.
compound eye An eye with many separate units or facets, each with its own lenslike structure.
compressed Having a low width/height ratio; flattened in such a way that the body is deep and narrow, as in a flea; the opposite of *depressed*.
concentric A series of circles or ellipses having a common center.
copulatory stylet Modified first pleopod of male crayfish.
cordate Heart-shaped.
corium Middle division of an elytron.
corona A crown shape.
corrugated Containing many small folds or ridges.
costal A riblike structure in the wing; wing vein.
coxa (coxae) The basal segment of an arthropod leg. In isopods, fused to the body.
coxal gill In an amphipod, a flaplike gill attached to the inner surface of a coxa; in the female it is lateral to the oostegite.
creeping welt A slightly raised, often darkened structure on dipteran larvae.
crenulate With small scallops, evenly rounded and rather deeply curved.
crepuscular Active at dawn and dusk.
crochet Curved spine or hook, as on the prolegs of lepidopteran and dipteran larvae.
cupule The cup-shaped segment at the base of the club on some antennae.
cuticle The material forming the exoskeleton of an insect.
dactyl The most distal segment of a crustacean leg, the equivalent of the pretarsus of other arthropods.
deflexed Bent downward.
demibranchs The gills of a unionid, in which developing embryos are brooded.

dens The middle segment of the furcula or springtail of collembolans.
dentate Toothed or with toothlike projections.
denticle A small tooth.
denticulate Set with little teeth or notches.
dentiform Toothlike.
depressed Flattened in such a way that the body is low and wide; having a low height/width ratio; the opposite of *compressed*.
dextral To the right of the median line, or, for snails, coiled to the right.
diapause A period during which growth or development is suspended.
digitate Formed or shaped like a finger.
digitiform See *digitate*.
discoid Disk-shaped.
distal Near or toward the free end of any appendage; the part farthest from the body.
divergent Spreading out from a common base.
dorsad The adverbial form of dorsal.
dorsal Toward the upper surface when the body is in normal walking position; the opposite of *ventral*.
dorsocentralia Dorsomedial series of paired plates in water mites.
dorsolateral Toward the dorsum and side.
dorsomental teeth Teeth occurring as anterior part of mentum in some larval tanypodines.
dorsum The top of back; tergum.
dun See **subimago**.
ecdysial line Suture that splits during molting.
effaced Obliterated; rubbed out.
elevated Having a high height/width ratio.
elytron (elytra) Hardened shell-like mesothoracic wing of adult coleopterans.
emarginate Having the margin notched or indented.
embolium A narrow piece of the corium along the costal margin.
entire With an even, unbroken margin.
epicranial suture A Y-shaped suture on the dorsal surface of the head.
epipleural fold The portion of the elytra bent down along the lateral margin.
epipleuron (epipleura) Part of the elytra (underside) bordering the inner edge of the epipleural fold.
epiproct A median terminal abdominal appendage above the anus.
episternum The anterior sclerite of a thoracic pleuron.
eurytherm An organism that tolerates a wide range of temperatures.
exarate Having appendages that are free, not appressed to the body.
exopod In crustaceans, the outer ramus or branch of a pleopod.
exuvia (exuviae) The cast skin of an arthropod.
facet A single unit in a compound eye.
femur (femora) The third segment of an arthropod leg, connected to the trochanter.
fibrilliform In the form of many threads.
filiform Threadlike; slender and of equal diameter.
first antenna Primitively, the most anterior appendage of a crustacean, though in some amphipods it actually arises slightly posterior to the second antenna. The first antenna usually has two flagella, in contrast to the second antenna, which has only one. Sometimes called the *antennule*.
fixed finger In crustaceans, the projection from the propodus of a chelate appendage, adapted to oppose the dactyl.
flagellum (flagella) A whiplike process, often associated with antennae. In crustaceans, the

outer part of an antenna, usually divided into many small segments. In larval chironomids, antennal segments distal to the basal segment.

flattened Whorls on a snail shell with straight sides; compare with *inflated* and *shouldered*.

fluted Channeled or grooved.

form I Male crayfish in breeding condition, characterized by having pleopods hardened at the tips.

form II Male crayfish in nonbreeding condition.

fossa A cavity or sinus in the surface of the seminal receptacle.

fossorial Formed for or with the habit of digging or burrowing.

frons Front of the head between the arms of epicranial sutures.

frontal horn Shelflike forward-projecting portion of head between the antennae of certain nymphal odonates.

frontal sutures The arms of the epicranial suture.

frontoclypeus The fused frontal and clypeus sclerites or near these parts of the head, forehead.

furcal pits Pits on the thoracic sterna, especially in plecopterans, where they may be connected by a Y ridge.

furcula The forked springtail of collembolans; composed of the manubrium, dens, and mucro.

galea The outer lobe of the maxilla.

gastric caeca Outpocketings in the gut wall.

gena (genae) The part of the head on each side below the eyes, extending to the gular suture.

geniculate Abruptly bent at an angle; like a knee joint.

genital field The gonopore, genital acetabula, and associated flaps or plates of water mites.

genu In water mites, a segment in the leg between the femur and the tibia.

gill Any structure especially adapted for the exchange of dissolved gases between an animal and a surrounding liquid. In isopods, a transverse flap on the posterior face of a pleopod. In amphipods, a longitudinal flap on the inner face of a coxa or a longitudinal or tubular flap near the midline on the ventral surface of the thorax. In decapods, a featherlike structure hidden by the gill cover, and arising either from a coxa or from the lateral surface of the thorax.

gill cover In crayfish, the part of the carapace that protects the gills; located posterior to the cervical groove and lateral to the branchiostegal groove.

glabrous Smooth, without hairs.

glandular Consisting of, bearing, or resembling a gland.

glandularium In water mites, special paired glands each of which has an associated seta.

globose Approaching the shape of a sphere.

globular See **globose**.

glochidium (glochidia) A larval unionacean.

glossa (glossae) One of the two median terminal lobes of the labium.

gnathopod In crustaceans, a thoracic appendage that is subchelate and adapted for grasping. In asellids, the appendage of the first free thoracic segment. In amphipods, the appendage of the first or second free thoracic segment.

gnathosoma In water mites, the anterior-most section of the body; includes the palps, chelicerae, and capitulum; primarily used during feeding.

gnathosomal bay In water mites, a U-shaped or V-shaped gap between the first pair of coxal sclerites; the capitulum is typically located there.

gonopore A genital opening, also called genital pore.

gravid In bivalves, pertaining to females brooding embryos or glochidia within the demibranchs.

growth lines In gastropods, raised lines running across the whorls; not to be confused with spiral striae.

gula The sclerite forming the central part of the venter of the head between the genae.

H Height.

hair In naidid oligochaetes, the longer of the two kinds of chaetae that may be present in dorsal bundles (compare *needle*); hairs are long, slender, and simple-pointed.

head The most anterior part of the body, bearing the eyes, antennae, mouth, and jaws. The term is imprecise for crustaceans, where the boundary between head and thorax has shifted as varying numbers of thoracic appendages have been incorporated into the mouthparts. In this book, the head of a crayfish is defined as the region anterior to the first free thoracic segment, thus including the narrow ringlike segment that bears the maxillipeds.

hemelytron (hemelytra) The front wing of adult hemipterans.

hemimetabolous Having simple metamorphosis such as that in odonates, ephemeropterans, and plecopterans.

hepatic spine In crayfish, spines located on the carapace between the orbit and the cervical groove.

hermaphroditic Having both sexes in one individual.

hinge teeth Any of various ridges or thickenings on the dorsal side of the interior of a bivalve shell that aid in holding the two halves of the shell together; includes lateral teeth, pseudocardinal teeth, cardinal teeth, and interdental teeth.

holometabolous With complete metamorphosis (egg, larva, pupa, adult).

hydrofuge Water-repellent.

hypopharynx A structure on the upper and inner part of the labium.

hyporheic zone In streams, the deep interstitial zone.

hypostomium (hypostoma) A sclerotized, anteriorly toothed plate on the mentum or on the ligula.

incised Notched or deeply cut into.

inflated In a snail shell, *inflated* means swollen, as if the whorls had been pumped full of air. In a bivalve, *inflated* means laterally expanded, as opposed to *compressed*.

infuscate Smokey gray-brown, with a blackish tinge.

instar The insect between successive molts.

integument Skin; outer body wall.

interdental tooth A hinge tooth, present in some unionids, that lies between the lateral teeth and the pseudocardinal teeth. Shells with interdental teeth look as if they have three pseudocardinal teeth in the left valve.

interocular space The space between the eyes.

intrasegmental Within one segment.

keel An elevated ridge; a carina.

L Length.

labium Lower lip; mouthpart lying behind the maxillae.

labral Of or pertaining to the labrum.

labral fans Mouthparts of black fly larvae modified for filtering suspended food particles from flowing water.

labral lamellae Small plates near base of SI setae.

labroclypeus The fused sclerites of the labrum and clypeus or the area near these parts of the head.

labrum Upper lip; located below the clypeus on the anterior side of the head.

lacinia (laciniae) The inner bladelike segment of the maxilla that bears brushes of hairs or spines.

lamella (lamellae) A thin plate or leaflike process.

lamellar Of or pertaining to a lamella.
laminate Composed of or covered with thin plates.
lanceolate Lance- or spear-shaped; oblong and tapering to the end.
lateral Toward the side of the body; the opposite of *mesal*.
lateral teeth Hinge teeth that flank or lie posterior to the beak of a bivalve shell.
lauterborn organ Small sensory organ in larval dipterans usually situated at apex of second antennal segment, often ovate and sometimes stalked.
left valve (of a unionid) Hold the shell with the hinge up and the short end of the shell forward. The left valve is on your left. In almost all species of unionids that have lateral teeth, the left valve has two lateral teeth.
leg (of crustaceans) As used in this book, a pereiopod that is adapted for walking. This definition excludes the gnathopods of amphipods and isopods and the chelae of crayfish, which are counted as legs by some authors.
lentic Of or pertaining to standing water.
ligula The mesal, apical segment of the labium.
linear Straight; elongate; in the form of a straight line.
lotic Of or pertaining to flowing water.
macropterous With long wings.
malleated Having small flattened areas, as if having been hit with a hammer.
mandibles The most anterior of paired mouthparts that are adapted for feeding. In malacostracans the third pair of appendages.
mantle In mollusks, a thin layer of tissue that secretes the shell and lines the shell's interior.
manubrium The proximal segment of the furcula (springtail) in collembolans.
marginal spine One of a pair of spines located on either side of the base of the acumen of a crayfish rostrum.
marsupium A pouch or cavity in which young develop.
maxilla (maxillae) One of the fourth or fifth pairs of appendages in malacostracans, adapted as supplementary jaws. They are softer and more complex than the mandibles.
maxillary palp Segmented appendage of maxilla in the mouth.
maxilliped In crustaceans, one of the appendages posterior to the maxillae that is adapted as a supplementary jaw. In decapods, there are three pairs of maxillipeds, constituting the sixth, seventh, and eighth appendages. In other malacostracans, there is only one pair of maxillipeds, forming the sixth pair of appendages.
median Lying in the midline of the body.
mentum The distal segment of the labium bearing the movable parts and attached to the submentum.
mesal Pertaining to the middle; toward the middle; the opposite of *lateral*.
mesal process In crayfish, a terminal process located mesally on the male's first pleopod.
meso- Prefix, pertaining to the middle (e.g., the mesothorax is the middle thoracic segment).
meta- Prefix, pertaining to the last or third in a series (e.g., the metathorax is the last thoracic segment).
metapneustic Having functional spiracles only on the terminal segment.
metasternal "wings" A wing-shaped pattern on the metasternum of some adult dytiscids.
mouth hook Vertically oriented mandible-like structure in dipteran larvae.
movable hook The hook on the lateral edge of the palpal lobe in odonate nymphs.
mucro The distal segment of the furcula or springtail of collembolans.
multispiral Marked by a spiral with many turns.
multivoltine Having more than one flight period per year.
muscle scar A dark or light ovoid mark that contrasts with the background.
nacre The pearly white or colored material covering the inside of a unionid shell.

naked Without hairs or setae.
nauplius (nauplii) The free-swimming first larval stage of certain crustaceans.
needle In naidid oligochaetes, the longer of the two kinds of chaetae that may be present in dorsal bundles (compare *hair*); needles may be simple-pointed or bifurcate, pectinate or palmate.
nodal furrow A shallow groove in the lateral edge of the hemelytron made by a vestigial crossvein in the nodus.
nodus In adult odonates, a notch marking the position of a prominent crossvein near the middle of the anterior margin of the front wing.
notum The dorsal surface of any segment.
nuclear whorl The first whorl on the apex of a snail shell.
oblique At or from an angle.
obovate Inversely egg-shaped.
obsolete Mostly or entirely absent; indistinct; not fully developed.
obtuse Forming an angle exceeding 90° but less than 180°.
occipital Of or pertaining to the occiput.
occiput The back part of the head.
ocellus (ocelli) A small visual organ with only one lens. In many arthropods, ocelli occur in addition to compound eyes.
oligopneustic With few spiracles functional.
oostegite A projecting plate on the inner side of the coxa of a mature female malacostracan. Each pair of oostegites overlaps in the midline, forming a chamber in which the developing eggs or young are carried.
operculate Having an operculum.
operculate gill Covering other gills, modified for protection.
operculum (opercula) A lid or cover.
orbit In insects, the region along the inner border of the eye; in crustaceans, the cavity that contains the eyestalk.
ovate, ovoid Somewhat oval in shape.
ovigerous Egg-carrying.
ovoviviparous Having eggs that complete development in the body of the female and hatch before or upon oviposition.
pala The much-dilated anterior tarsal joint in adult corixids.
palm In crayfish, the broadened portion of the propodus.
palmate Like the palm of the hand, with fingerlike processes (palps).
palp A segmented process of the maxilla or labium.
palpal lobes The paired terminal segments of the odonate labium, adapted for grapsing prey.
PAO Post antennal organ of collembolans, located near the base of the antenna.
papilla (papillae) A soft projection.
papilliform Shaped like papillae.
paraglossa (paraglossae) The lateral terminal lobe of the labium.
paraprocts A pair of terminal abdominal appendages lying below the cerci and anus.
parthenogenetic Reproduces by development of an unfertilized gamete.
paucispiral Marked by a spiral with few turns.
paurometabolous In hemipterans, undergoing gradual, incomplete metamorphosis.
PE Pecten epipharyngis; sclerotized scales located ventral to the labral margin.
pecten Comb.
pectinate Comblike.
pedicel (pedicle) A slender stalk or process that serves for support or attachment; the second segment of an arthropod antenna.

peduncle A pedicel.
penial Of or pertaining to the penis.
penial sheath In gastropods, a structure of the male reproductive tract between the preputium and the vas deferens.
penis sheath The thickened cuticle that covers the penis of some species of tubificid oligochaetes; the penis sheath, if present, occurs in segment XI.
penultimate Next to last.
pereiopod In crustaceans, a thoracic appendage that has not been converted into a maxilliped. Legs, chelae, and gnathopods are all pereiopods.
periostracum The colored, proteinaceous layer that covers the outside of mollusk shells.
peripneustic With many spiracles functional; polypneustic.
peritreme Long, thin sclerite associated with stigmal opening of some aquatic mites.
petiole Slender stalk.
pharyngeal Related to or located in the throat region.
pharynx Throat.
phoretic Symbiotic in being transported on the body of the host.
pilose Hairy; covered with close hairs.
pilosity Hairiness.
pinnate Having branches or parts arranged on each side of a central axis as a feather.
plastron Fine hydrofuge hairs that support gas gills of adults of some species of Hemiptera and Coleoptera.
pleopod In crustaceans, an abdominal appendage that has not been converted into a uropod; three pairs in amphipods, five pairs in the other orders.
pleurite A sclerite of the pleural area (side) of an insect.
pleuron The lateral sclerotized wall of a segment, composed of one or more pleurites.
plumose Featherlike (said of a hair that has lateral branches).
polypneustic With many spiracles functional; peripneustic.
posterior In the direction away from the head; the opposite of *anterior*.
posterior ridge A more-or-less distinct ridge that runs from the beak to the posterior-ventral margin of a unionid shell. It separates the main face of the shell from the posterior slope.
posterior slope The area of a unionid shell posterior of and dorsal to the posterior ridge.
posterolateral Toward the posterior and the side.
postmentum Specialized submentum of nymphal odonates.
postoccipital margin Posterior sclerotized margin of head capsule.
postocular space Space between the back of the eyes and the occipital opening.
preanal On or of the segment anterior to that segment where the anus is located.
preapical Located before the tip or apex.
prehensile Fitted or adapted for grasping, holding, or seizing.
premandible Movable appendage attached to ventral surface of labrum.
prementum Specialized mentum of nymphal odonates.
preputium In gastropods, a structure of the male reproductive tract, connected to the penial sheath.
pretarsus Terminal limb segment in arthropods.
primary hairs Those hairs found on setiferous tubercles, definite in number and position.
pro- Prefix, pertaining to the forwardmost or first in a series (e.g., the prothorax is the first thoracic segment).
proboscis (probosces) A tubular process of the head.
procercus (procerci) Preanal tubercle usually bearing several apical setae.
process An elongation of the surface, or a margin, or an appendage; any prominent part of the body not otherwise definable.

produced Drawn out, elongated, projecting.
proleg Any process or appendage that serves for support, locomotion, or attachment; the fleshy, unjointed thoracic or abdominal appendages of larval trichopterans, lepidopterans, and dipterans; may be sclerotized.
propneustic With anterior spiracles functional.
propodus The sixth segment of a crustacean leg (the second to last segment), corresponding to the tarsus of another arthropod. In isopods, appearing to be the fifth segment because the coxa is incorporated into the body.
prosternal horn An unsclerotized process on the prosternum of some larval trichopterans.
prostomium In annelids, the portion of the head situated anteriorly to the mouth.
prothoracic lobe In adult hemipterans especially, a ventrolateral projection of the prothorax.
prothorax First thoracic segment.
protuberant Rising or produced above the surface or general level.
proximal Away from the tip of a structure of appendage; the opposite of *distal*.
proximate Close to.
pruinose As if frosted or covered with a fine dust.
pseudobasal Appearing to be basal.
pseudocardinal teeth The anterior set of hinge teeth of a unionid shell.
pubescence Short, fine, erect hair or down.
pulsatile vesicle Locomotive lateral vesicles of some leeches.
punctate Set with impressed points or punctures.
puparium In some dipterans, the last larval exuvia that hardens to enclose and protect the pupa.
pustule A large projecting lump.
quadrate Square, or nearly so.
radula The tough, tonguelike strip of tissue that bears the teeth of snails.
ramus (rami) A branch of a forked appendage.
raptorial Used for catching prey.
rastrate Covered with longitudinal scratches.
recurved Bent backward.
reniform Shaped like a kidney or a bean; shaped like an oval, but with one side slightly concave.
reticulate Covered with a network of lines.
retractile Capable of being drawn in or back.
rheocene A region or area of flowing water.
right valve (of a unionid) See **left valve**.
ring organ Small circular depression on basal antennal segment.
riparian Living on the bank of a lake, pond, or stream.
rostrum A beak; a snoutlike projection of the head bearing the mouthparts. In decapods, a projection from the anterior margin of the carapace that projects forward between the eyes.
SI setae The most anterior pair of dorsal labral setae.
SII setae Dorsal labral setae immediately posterior and lateral to SI setae.
SIII setae Labral setae posterior and medial to SII setae.
scalloped With the edge marked with rounded hollows, without intervening angles.
sclerite Any piece of the insect body wall bounded by sutures.
sclerotized Hardened and usually darkened.
scute Dorsal chitinoid plate in some leeches.
scutellum In adult coleopterans and hemipterans, the triangular sclerite between the bases of the elytra or hemelytra.

second antenna The second appendage on a crustacean. In contrast to the *first antenna*, which has two flagella, the second antenna has only one. The other flagellum, if present, is in the form of an unsegmented scale.
secondary hairs Scattered hairs without a constant or fixed position.
semichelate In crustaceans, intermediate in form between chelate and subchelate, with the propodus having a small rounded fixed finger that opposes the tip of the dactyl.
seminal receptacle A sperm storage structure; spermatheca.
semivoltine Requiring two years for development.
septum (septa) A dividing wall or membrane.
serrate Sawlike; with notched edges.
sessile Attached by the base; without stem or stalk.
seta (setae) Slender, hairlike appendage; hair.
seta interna Group of mandibular setae on inner dorsal side; mandibular brush.
seta subdentalis Mandibular seta on inner margin ventral to mandibular teeth.
setiferous Bearing setae.
setose Furnished or covered with setae or stiff hairs.
shouldered Whorls on a gastropod shell that are not evenly rounded.
sigmoid S-shaped.
sinistral To the left of the median line, or, for snails, coiled to the left. See **dextral**.
sinuate Curving in and then out, as in the hind margin of the basis of the fifth leg in *Gammarus fasciatus*.
spermatheca (spermathecae) In oligochaetes, a sac for sperm storage in the female reproductive tract; seminal receptacle.
spine A multicellular, thornlike process or outgrowth of the cuticle not separated from it by a joint.
spinose (spinous) With many spines.
spinule A very small spine.
spiracle The external opening to a tracheal system.
spiracular plate Structure surrounding the terminal spiracles, especially in tipulids.
spiral striae Fine lines running parallel to the sutures on a snail shell.
spire The entire snail shell, excepting the body whorl.
spur A spinelike appendage of the cuticle, connected to the body wall by a joint.
stenotherm Organism that tolerates a narrow range of temperatures.
sternellum A small ventral sclerite of the thorax.
sternite The ventral sclerotized part of a segment.
sternum (sterna) The entire ventral division of any segment.
stigmal opening In water mites, the spiracle.
stipes The second segment of the maxilla, the segment to which movable parts are attached.
stria (striae) A fine, longitudinally impressed line.
striate Marked with striae.
strigil A dark, roughened structure on the dorsolateral portion of the abdomen of adult corixids.
stylet A needle-shaped structure.
sub- Prefix, more-or-less.
subchelate Adapted for grasping by having the dactyl fold back against the edge of the expanded propodus, as in the gnathopods of amphipods and asellids.
subimago In mayflies, the first of two winged instars emerging from the water.
submental gill Lateral gill situated immediately posterior to the head.
submentum The basal segment of the labium.
suborbital Below the eye.

subulate Long and thin, tapered from base to a fine point.
sulcate Marked with a sulcus.
sulcus A furrow or groove; or a flattening or slight depression on the face of a unionid shell.
supraanal seta Dorsal seta situated anterior to anal tubules.
suture A seam or impressed line indicating the division of the distinct parts of the body wall or, on a snail shell, separating the whorls.
tarsal formula A three-digit sequence indicating the number of tarsal segments on the anterior, middle, and posterior legs, respectively (claws are not counted as segments).
tarsomeres Segments of the tarsi.
tarsus The fifth (last) segment of an arthropod leg, connected to the tibia (may be composed of several tarsomeres).
telson The most posterior segment of a crustacean. It is dorsal and posterior to the anus, and never bears appendages (not visible in isopods).
tenaculum A catchlike structure on the venter of the third abdominal segment of collembolans; holds the furcula in place.
tenent hairs Sensory hairs found near the apex of the tibiotarsus on the legs of collembolans.
tentacles The pair of "feelers" at the anterior end of a gastropod.
tentorium (tentoria) Support structure within the head.
tergite The dorsal sclerotized part of a segment.
tergum (terga) The upper or dorsal surface of any body segment of an insect.
thoracic Of the thorax.
thorax The part of the body between the head and the abdomen.
tibia The fourth segment of an arthropod leg, connected to the femur.
tibiotarsus In collembolans, the fused tibia and tarsus.
tomentum A form of pubescence composed of matted, woolly hair.
trachea An air-filled tube running into the body from the exoskeleton and forming a means of exchanging gases between the tissues of the animal and the air of the surrounding environment.
tracheation A network of respiratory tubules.
translucent Semitransparent; can be seen through, but not clearly.
transverse At right angles to the long axis of the body.
tricuspid Having three points.
trifid Three-pronged; divided into three parts.
tritosternum A forked projection on the sternum of some aquatic mites.
trochanter The second segment of an arthropod leg, connected to the coxa.
trochantin A small, forward-projecting sclerite on the lateral side of the coxa.
truncate With a shortened end; often squarish.
tubercle A small bump or pimplelike structure.
tubule A small, elongate tubelike structure.
tunica Membranous structure located on the dorsal side of the unguis of collembolans.
tusk Modified mandible of some nymphal ephemeropterans.
umbilicus The small opening present in some snail shells behind the base of the aperture.
umbo The hump on the dorsal margin of a bivalve shell; also called the *beak*.
uncate Refering to the palp of a water mite in which the tibia is laterally expanded so that the tarsus can fold against it.
unguiculus (unguiculi) The inferior opposing claw of the unguis of collembolans.
unguis (ungues) The superior claw on the leg of a collembolan.
uniramous Having one axis. See **biramous**.
univoltine Having one flight period per year.

uropod In amphipods, an appendage of any of the last three abdominal segments; in other orders of Crustacea, an appendage of the last segment. It is usually adapted to be used in association with flexion and extension of the abdomen, often combined with the telson to form a tail fan. In isopods, it is primarily tactile in function.
uronites Last three body segments in crustaceans.
valves The terminal abdominal processes in nymphal anisopterans.
vas deferens The principal duct conveying sperm from the testis to the outside of the body.
velum Internal transverse structure anterior to the pharyngial folds in leeches.
venter Sternum.
ventral Toward the lower surface when the body is in normal walking position; the opposite of *dorsal*.
ventral groove In male asellids, a groove on the ventral surface of the inner ramus of the second pleopod (when the pleopod is in normal position with its tip to the rear; if the pleopod is directed ventrally, then the ventral groove lies on the anterior face).
ventromental plate A separate lateral plate situated ventral to the mentum.
vertex Top of the head.
vesicle Bump or papilla.
vestigial Small or degenerate.
vestiture A surface covering of scales, hairs, or similar structures.
W Width.
whorl One turn of the spiral of a snail shell.
wingpad The developing wing of an immature insect.
Y ridge Mesosternal pattern in nymphal plecopterans that connects the furcal pits.

Index

Page numbers in italics refer to pages with illustrations.